Soil Improving Cropping Systems for Sustainable and Profitable Farming in Europe

Soil Improving Cropping Systems for Sustainable and Profitable Farming in Europe

Editors

Guido Wyseure
Julián Cuevas González
Jean Poesen

MDPI • Basel • Beijing • Wuhan • Barcelona • Belgrade • Manchester • Tokyo • Cluj • Tianjin

Editors

Guido Wyseure
KU Leuven
Leuven
Belgium

Julián Cuevas González
University of Almería
Almería
Spain

Jean Poesen
KU Leuven
Leuven, Belgium
UMCS
Lublin, Poland

Editorial Office
MDPI
St. Alban-Anlage 66
4052 Basel, Switzerland

This is a reprint of articles from the Special Issue published online in the open access journal *Land* (ISSN 2073-445X) (available at: https://www.mdpi.com/si/land/soil_improving_cropping_systems).

For citation purposes, cite each article independently as indicated on the article page online and as indicated below:

LastName, A.A.; LastName, B.B.; LastName, C.C. Article Title. *Journal Name* **Year**, *Volume Number*, Page Range.

ISBN 978-3-0365-6253-7 (Hbk)
ISBN 978-3-0365-6254-4 (PDF)

Contents

land MDPI

Article

Soil-Improving Cropping Systems for Sustainable and Profitable Farming in Europe

Rudi Hessel [1,*], Guido Wyseure [2], Ioanna S. Panagea [2], Abdallah Alaoui [3], Mark S. Reed [4,5], Hedwig van Delden [6], Melanie Muro [7], Jane Mills [8], Oene Oenema [1], Francisco Areal [4], Erik van den Elsen [1], Simone Verzandvoort [1], Falentijn Assinck [1], Annemie Elsen [9], Jerzy Lipiec [10], Aristeidis Koutroulis [11], Lilian O'Sullivan [12], Martin A. Bolinder [13], Luuk Fleskens [14], Ellen Kandeler [15], Luca Montanarella [16], Marius Heinen [1], Zoltan Toth [17], Moritz Hallama [15], Julián Cuevas [18], Jantiene E. M. Baartman [14], Ilaria Piccoli [19], Tommy Dalgaard [20], Jannes Stolte [21], Jasmine E. Black [8] and Charlotte-Anne Chivers [8]

1 Wageningen Environmental Research, Droevendaalsesteeg 3, 6708 PB Wageningen, The Netherlands; oene.oenema@wur.nl (O.O.); erik.vandenelsen@wur.nl (E.v.d.E.); simone.verzandvoort@wur.nl (S.V.); falentijn.assinck@wur.nl (F.A.); marius.heinen@wur.nl (M.H.)
2 Department of Earth and Environmental Sciences, KU Leuven, 3001 Leuven, Belgium; guido.wyseure@kuleuven.be (G.W.); ioanna.panagea@kuleuven.be (I.S.P.)
3 Institute of Geography, University of Bern, Hallerstrasse 12, 3012 Bern, Switzerland; abdallah.alaoui@giub.unibe.ch
4 Centre for Rural Economy, School of Natural and Environmental Sciences, Newcastle University, Newcastle upon Tyne NE1 7RU, UK; mark.reed@sruc.ac.uk (M.S.R.); francisco.areal-borrego@newcastle.ac.uk (F.A.)
5 Thriving Natural Capital Challenge Centre, Scotland's Rural College (SRUC), King's Buildings, W Mains Rd, Edinburgh EH9 3JG, UK
6 Research Institute for Knowledge Systems (RIKS), Hertogsingel 11A, 6211 NC Maastricht, The Netherlands; hvdelden@riks.nl
7 Milieu Consulting SRL, Chau. de Charleroi 112, 1060 Brussels, Belgium; melanie.muro@web.de
8 Countryside and Community Research Institute, University of Gloucestershire, Cheltenham GL50 4AZ, UK; jmills@glos.ac.uk (J.M.); jblack2@glos.ac.uk (J.E.B.); cchivers@glos.ac.uk (C.-A.C.)
9 Bodemkundige Dienst van België, Willem de Croylaan 48, 3001 Heverlee, Belgium; aelsen@bdb.be
10 Institute of Agrophysics, Polish Academy of Sciences, Doświadczalna 4, 20-290 Lublin, Poland; j.lipiec@ipan.lublin.pl
11 School of Chemical and Environmental Engineering, Technical University of Crete, Akrotiri Campus, Chania 731 00, Crete, Greece; akoutroulis@isc.tuc.gr
12 Teagasc, Environment, Crops, Environment & Land Use Programme, Johnstown Castle, Co., Y35 TC97 Wexford, Ireland; lilian.osullivan@teagasc.ie
13 Department of Ecology, Swedish University of Agricultural Sciences (SLU), P.O. Box 7044, 75007 Uppsala, Sweden; martin.bolinder@slu.se
14 Soil Physics and Land Management Group, Wageningen University, Droevendaalsesteeg 3, 6708 PB Wageningen, The Netherlands; luuk.fleskens@wur.nl (L.F.); jantiene.baartman@wur.nl (J.E.M.B.)
15 Institute of Soil Science and Land Evaluation, Soil Biology Department, University of Hohenheim, Emil Wolff Str. 27, 70599 Stuttgart, Germany; ellen.kandeler@uni-hohenheim.de (E.K.); hallama@uni-hohenheim.de (M.H.)
16 Joint Research Centre, European Commission, Via E. Fermi, 2749, 21027 Ispra, Italy; luca.montanarella@ec.europa.eu
17 Georgikon Campus Keszthely, Institute of Agronomy, Hungarian University of Agriculture and Life Sciences, 2100 Gödöllő, Hungary; toth.zoltan@uni-mate.hu
18 Agronomy Department, CeiA3, University of Almería, 04120 Almería, Spain; jcuevas@ual.es
19 Department of Agronomy, Food, Natural Resources, Animals and Environment, University of Padova, Viale dell'Università 16, 35020 Legnaro, Italy; ilaria.piccoli@unipd.it
20 Department of Agroecology, Aarhus University, Blichers Alle 20, 8830 Tjele, Denmark; tommy.dalgaard@agro.au.dk
21 Norwegian Institute of Bioeconomy Research (NIBIO), Oluf Thesens vei 43, 1433 Ås, Norway; jannes.stolte@nibio.no
* Correspondence: rudi.hessel@wur.nl; Tel.: +31-3174-86-530

Citation: Hessel, R.; Wyseure, G.; Panagea, I.S.; Alaoui, A.; Reed, M.S.; van Delden, H.; Muro, M.; Mills, J.; Oenema, O.; Areal, F.; et al. Soil-Improving Cropping Systems for Sustainable and Profitable Farming in Europe. *Land* **2022**, *11*, 780. https://doi.org/10.3390/land11060780

Academic Editor: Richard Cruse

Received: 14 April 2022
Accepted: 23 May 2022
Published: 25 May 2022

Abstract: Soils form the basis for agricultural production and other ecosystem services, and soil management should aim at improving their quality and resilience. Within the SoilCare project, the concept of soil-improving cropping systems (SICS) was developed as a holistic approach to facilitate the adoption of soil management that is sustainable and profitable. SICS selected with stakeholders were monitored and evaluated for environmental, sociocultural, and economic effects to determine profitability and sustainability. Monitoring results were upscaled to European level using modelling and Europe-wide data, and a mapping tool was developed to assist in selection of appropriate SICS across Europe. Furthermore, biophysical, sociocultural, economic, and policy reasons for (non)adoption were studied. Results at the plot/farm scale showed a small positive impact of SICS on environment and soil, no effect on sustainability, and small negative impacts on economic and sociocultural dimensions. Modelling showed that different SICS had different impacts across Europe—indicating the importance of understanding local dynamics in Europe-wide assessments. Work on adoption of SICS confirmed the role economic considerations play in the uptake of SICS, but also highlighted social factors such as trust. The project's results underlined the need for policies that support and enable a transition to more sustainable agricultural practices in a coherent way.

Keywords: soil quality; sustainable soil management; adoption; crop management; environmental dimension; sociocultural dimension; economic dimension

1. Introduction

Crop production in Europe faces the challenge to remain profitable while at the same time achieving environmental sustainability. Average wheat yields in several European. countries are less than what is locally attainable [1–4], possibly because of suboptimal management and/or impairment caused by poor soil quality (defined as 'the capacity of a soil to function within ecosystem and land-use boundaries to sustain biological productivity, maintain environmental quality, and promote plant and animal health', following [5]). In addition, agricultural land faces a number of other threats that may lead to physical, chemical, and biological degradation of the soil [6–9]. These include erosion, compaction, salinization [10], soil pollution, loss of organic matter [11], and loss of soil biodiversity [12]. For example, the use of heavy machinery can lead to soil compaction and impaired root growth [13]; increased soil cultivation and climate change can lead to soil organic matter decline [14]; and narrow rotations may cause biodiversity decline and increased incidence of soil-borne diseases [15]. These forms of soil degradation are often neglected by land managers because of low awareness, low visibility during initial stages of degradation, and a lack of appropriate tools, benchmark values, and policies. As a result, production levels in some cropping systems are maintained by high input (e.g., nutrients and pesticides) and technology (e.g., machinery and breeding), which may mask losses in long-term productivity due to reduced soil quality [16,17]. Such increased use of agricultural inputs may reduce long-term farm profitability because of their costs while also negatively affecting the environment because of unsustainable use of energy and resources in producing inputs [18] and as a consequence of their application (e.g., [19–21]). Soil improvement is necessary to break the negative spiral of degradation, increased inputs, increased costs, and damage to soil and the environment [22]. Maintaining or improving soil quality is crucial for crop production [23] and can especially contribute to remediating forms of soil degradation that are initially hardly visible, such as gradual loss of soil biodiversity and soil organic matter.

Soils are at the intersection of a broad range of land use and environmental challenges. They are critical for economic and environmental well-being, because they form the basis for agricultural production, support high-quality food output [24], and provide a range of other ecosystem services. For example, good-quality soils are more resilient to weather extremes [25] and provide better buffering and cycling of nutrients [26], water purification and regulation, and resilience to pests [27] and climate variability/change [28]. Other ecosystem services provided by soils [29] include provision of biodiversity [30,31] and carbon sequestration, cycling, and regulation [32,33]. Thus, to ensure that sufficient healthy food for expanding human populations can be grown within planetary boundaries [34], soil management should aim at improving the quality and resilience of land and soil [35].

Attention on soil quality is increasing (e.g., [5,7,36–43]). In Europe, various projects (see, e.g., CORDIS | European Commission (europa.eu), domain 'Food and Natural Resources') have worked on soil threats, prevention of soil degradation, sustainable land management, agricultural management practices, soil functions, and soil quality. There is also increasing recognition of the fact that crop production should be enhanced without compromising the environment [44,45]. More than ever, the important role that soil plays in sustaining life on the planet is being recognized, with high-level objectives at the E.U. scale (e.g., [46]) and the UN Sustainable Development Goals (SDGs) being reliant in large part on sustainable land and soil management [47].

More sustainable farming systems (defined as 'Farming systems that use land resources, including soils, water, and plants, for the production of crops, while simultaneously ensuring the long-term productive potential of these resources and the maintenance of their environmental functions', following the definition of sustainable land management given by WOCAT (www.wocat.net/en/slm (accessed on 13 April 2022)) ' and practices, such as organic farming, conservation agriculture, and precision farming have taken a foothold in Europe [48,49]. For example, Bioland, an association for organic farmers in Germany and Austria, already had more than 5800 members in 2014 [50] and 8500 in 2021 (see https://www.bioland.de/fileadmin/user_upload/Verband/Entwicklung_Betriebe_und_Flaeche_01.svg (accessed on 13 April 2022)). However, these farming systems were not adopted to their full potential and were in some cases even abandoned [51]. Reasons behind this may be the possible negative effect of conservation agriculture on crop yield [39]; the complexity of conservation agriculture, which is management and knowledge intensive [52]; problems with weed and residue management [51]; or the increased occurrence of pests and diseases. There are also cultural and political barriers to the adoption of more sustainable agricultural practices [53]. Barriers to adoption often involve issues around land tenure, access to credit and inputs [7], and other socioeconomic factors, and the lack of knowledge, credible scientific evidence, and good-quality technical advice has also been highlighted [54].

This paper proposed and operationalized a multidisciplinary, multi-actor approach to identifying soil-improving cropping systems (SICS) that are both sustainable and profitable, and hence are more likely to achieve mainstream adoption in agriculture. The focus is on two main aspects, namely evaluation of SICS based on field experiments and modelling and adoption of SICS. To do this, we:

- Present the concept of SICS, as developed in the H2020 SoilCare project (2016–2021) https://www.soilcare-project.eu/ (accessed on 13 April 2022);
- Review literature on factors influencing farmer adoption of SICS;
- Propose a methodological framework for identifying and evaluating SICS that have a high likelihood of adoption;
- Present findings from the application of this framework in 16 study sites across Europe and from its upscaling to E.U. scale.

The paper starts by describing the concepts and methodology used for evaluating SICS and studying their adoption (Section 2) and then proceeds by presenting and discussing key findings from SoilCare (Sections 3 and 4). For a literature review that summarizes the main findings of published meta-analyses on SICS, the reader is referred to [55].

2. Concepts and Methodology

2.1. Conceptualization of SICS

The term 'cropping system' refers to the crop type, crop rotation, and agronomic management techniques used on a particular field over a period of years [56]. Choices made for these factors can influence the profitability and sustainability of crop production [57–59]. We considered these systems soil-improving if they resulted in a durable increased ability of the soil to maintain its functions, including food and biomass production, buffering and filtering capacity, and provision of other ecosystem services.

The basic concept adopted in the SoilCare project was that profitability and sustainability of crop production in Europe should be integrated and enhanced. Both are influenced by choices made in farm management, which are in turn influenced by external drivers and factors (Figure 1). External drivers and factors include E.U. policies and international agreements, supply chain and market effects (suppliers, industry, processing, retail, and consumers), macroeconomic conditions, society (public opinion), and pedoclimatic conditions. These external drivers and factors are dynamic and change because of socioeconomic developments, geopolitics, and climate change. As the focus of SoilCare was on arable cropping systems, grazing systems, multisystem farms, and other on-farm activities were not considered.

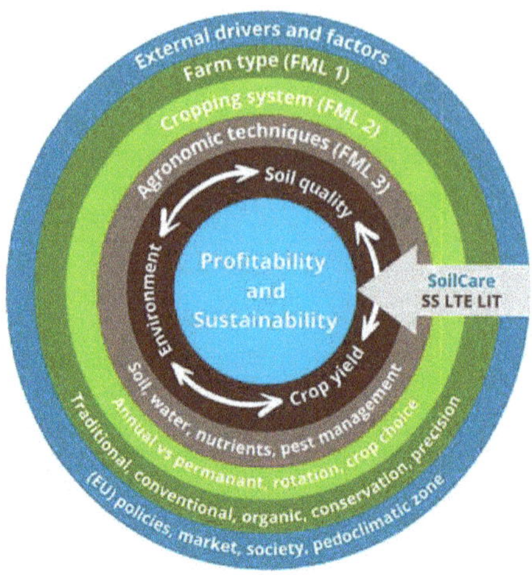

Figure 1. Methodological framework for assessing sustainability and adoptability of soil-improving cropping systems, showing the influence of farm management levels (FML 1–3) on soil quality, environment, crop yield, profitability, and sustainability. LIT refers to literature and other published data, LTE to long-term experiments, and SS to work in the study sites.

At the highest farm management level (FML1, see Figure 1) a choice is made among different types of farming; cropping systems are decided on at FML2, while choices regarding agronomic techniques that are used for management of soil, water, nutrients, and pests are made at FML3. Which farm type is chosen depends on external factors but also on the farm's ownership, resources and social context, such as the education, age, and preferences of the farmer (e.g., [60]). Choices made at this level also influence FML2 and FML3. For example, a choice for organic farming made at FML1 implies crop rotation at FML2 and biological pest management at FML3.

Choices made at all three FMLs have impacts on soil quality, on the environment, and on yield (thus farm economy) (Figure 1). These also influence each other. For example, the occurrence of a soil threat such as erosion influences soil quality as well as crop yield [61]. Crop yield can also influence soil quality, for example, through nutrient mining, rooting effects, and below-ground biomass. When impacts on soil quality and environment are positive, and the balance between production costs and revenues is also positive, the dual targets of farm profitability and environmental sustainability are reached.

The use of SICS improves soil quality and environmental benefits and has positive impacts on the farm economy (Figure 2). Some benefits result directly from the application of proper agronomic techniques; for example, avoiding overapplication of nutrients reduces greenhouse gasses (GHGs) and pollution (soil degradation). Other benefits of

SICS are indirect, as they result from improved soil quality brought about by application of the SICS. For example, improved soil quality improves infiltration and hydrological properties, increases rooting depths and resilience to climate change impacts, and stimulates soil biodiversity [11]. Finally, SICS also have above-ground impacts on vegetation and landscape (e.g., through the use of hedges, buffer strips, trees, terraces, ditches). Such impacts may also contribute to the conservation of biodiversity and wildlife, which may in turn positively influence soil quality.

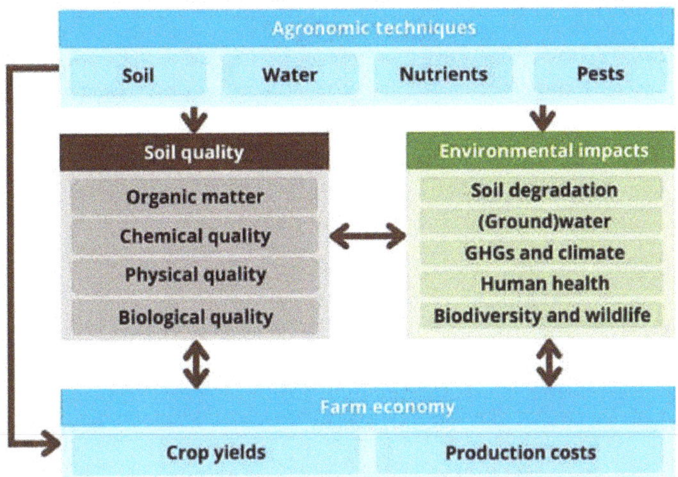

Figure 2. Impacts of agronomic techniques for managing soil, water, nutrients, and pests. One-sided arrows indicate impact, while two-sided arrows indicate that factors influence each other. Note that agronomic techniques are part of cropping systems and correspond to FML3 in Figure 1.

Profitability is a key factor influencing the adoption of SICS [62–66] that is partly influenced by the choice of cropping system and its management and partly by factors that farmers (in Europe) cannot typically control, such as global markets and policies [53]. A key aspect of profitability is production costs, as farmers have more control over this aspect than over the prices they get for their products. Different cropping systems require different types and levels of inputs (e.g., [67]) with different costs. In addition, the choice of cropping system influences the price of the product, which is often higher for organic than for conventional farming.

Conventional farming may become increasingly costly because of rising costs for external inputs and/or for mitigation/restoration measures against soil degradation. In addition, prices of external inputs fluctuate. For example, refinery curtailments due to the COVID-19 pandemic have limited supplies of raw materials, raising input costs by increasing the price of fertilizers for farmers [68]. Price fluctuations of agricultural products are expected to persist and continue to challenge the ability of consumers, producers and authorities to cope with the consequences [69]. In this context of rapid change and long-term challenges, farm profitability is at risk. In line with the Europe 2020 Strategy [70] on achieving smart, sustainable, and inclusive growth, boosting profitability is not only about reducing production costs, or increasing productivity, but also about more sustainable agriculture and the transformation of the food market to green, high-quality products. Smarter and greener agriculture also has the potential to contribute to a more circular bioeconomy and increase the value of agricultural products and the willingness of consumers to buy European agricultural products both inside and outside of the European Union [71,72].

SICS have the potential to reduce costs in the long run by reducing the need for external, costly inputs such as fertilizers and pesticides, reducing energy use for operating machinery, and/or reducing labour input [73–75]. While some SICS may lead to reduced productivity, they may make more efficient used of inputs and thus be more profitable. Costs associated with current unsustainable land use and management are estimated to be in excess of EUR 50 billion per year in the European Union [46]. In the long term, adoption of SICS should help reverse the current trajectory, and when soil quality has improved, efficiency is expected to increase further as a consequence of the reduced need for external inputs and possibly higher production. Additional long-term benefits lie in the reduction of expenditures due to reduced land degradation, GHG emissions, and risk to damages from natural disasters such as storms, droughts, or floods [25].

Various factors influence where SICS are most needed and best suitable and thereby determine the balance between the benefits and drawbacks of SICS and the ways in which these drawbacks can be minimized. These factors include the pedoclimatic zone (zones that are relatively homogeneous concerning climate and soil; see, e.g., [76]), the type of problem that constrains soil quality and crop production, biophysical conditions, and socioeconomic and political conditions. These different conditions require the use of different SICS and determine the applicability, profitability, and environmental impacts of the SICS across Europe. Hence, an assessment of SICS should incorporate environmental, economic, social, and policy aspects while also taking into account future trends in land use and climate change.

2.2. Methods Used for Evaluation of SICS

The first step in evaluating selected SICS was an in-depth analysis of the benefits and drawbacks of SICS as reported in literature and other published sources [55,77]. This was followed by investigating data from existing long-term experiments (LTEs). Next, we conducted field experiments and stakeholder research in 16 study sites located in different parts of Europe (Table 1, Figure 3), covering different pedoclimatic, socioeconomic, and policy conditions. Literature and other published data were mainly used to assess external drivers and factors (Figure 1). This was supplemented by stakeholder consultation at the E.U. level and modelling. Data from LTEs were mainly used to investigate SICS that show effects only in the long term. The focus of field experiments and stakeholder research in the study sites was primarily on FML3, since soil, water, nutrient and pest management can be adapted in the course of the year and these choices generally have more immediate effects than choices made at FML1 and FML2.

Table 1. Overview of SoilCare study sites. Types of crops listed here represent the study site region, not the sites where monitoring was conducted.

Study Site	Types of Crop	Pedoclimatic Zone [1]	Problems That Caused Reduced Soil Quality or Crop Yield or Increased Cost
1. Flanders, Belgium	Winter wheat, sugar beet, potato, vegetables, forage crops, orchards	Atlantic Central, soil depends on site	N and P leaching, erosion, compaction, SOC [2]
2. Viken, Norway	Cereals	Nemoral/Boreal, marine clay soils	Erosion, nutrient loss, pests, disease, SOC, compaction
3. Keszthely, Hungary	Cereals, maize	Pannonian, sandy loam, Eutric Cambisol	Soil compaction, humus degradation, nitrate leaching, acidity, weeds
4. Frauenfeld, Switzerland	Grass, cereals, maize, rape, potato, sugar beet, vegetables	Continental/Alpine South, Fluvisol	Soil structure, subsoil compaction, pounding risk
5. Viborg, Denmark	Winter cereals (wheat, 25%), forage crops	Atlantic North, sandy–loamy soils	SOC, compaction, erosion, nutrient losses (N and P)
6. Loddington, United Kingdom	Cereals, oilseeds, pulses, grass/clover leys	Atlantic Central/North, clay soils	Compaction, SOC
7. Tachenhausen, Germany	Maize, wheat, barley, oilseed rape, soya	Atlantic Central, karst, silty loam	Soil structure, compaction, reduced infiltration

Table 1. *Cont.*

Study Site	Types of Crop	Pedoclimatic Zone [1]	Problems That Caused Reduced Soil Quality or Crop Yield or Increased Cost
8. Draganesti Vlasca, Romania	Cereals, sunflower	Panonnian, Phaeozem	Soil compaction
9. Legnaro, Italy	Maize, wheat, sugar beet, soybean, alfalfa	Mediterranean North, Cambisol	SOC, compaction, climate variations
10. Szaniawy, Poland	Barley, rye, wheat, oats, potatoes, maize, grassland.	Continental, Sandy, loamy soils	Water deficit, SOC, acidity, compaction, weeds.
11. Caldeirão, Portugal	Cereals (maize and rice), vineyards	Lusitanean, silty–clayey soils	Water availability
12. Chania, Crete, Greece	Olive, citrus vineyards	Mediterranean South, Calcisol	Erosion, compaction, water availability
13. Orup, Sweden	Winter wheat, spring barley, spring oilseed rape, peas	Nemoral, sandy loams	Compaction
14. Prague-Ruzyně, Czech Republic	Barley, rye, wheat, oats, potatoes, maize, grassland	Continental, Luvisol	Erosion, compaction, SOC, acidification, reduced water retention
15. Almeria, Spain	Olive, stone fruit crops	Mediterranean South, Regosol, Leptosol	Erosion, salinization, water shortage
16. Brittany, France	Wheat, maize, grassland	Lusitanian/Atlantic Central, Cambisol	Compaction, weeds

[1] climatic zones based on the Environmental Stratification of Europe (version 8) [76]; [2] SOC = soil organic carbon decline.

Figure 3. The 16 SoilCare study sites. Details on each study site can be found in Table 1.

Within the study sites, different SICS were selected, tested in field, and evaluated in collaboration with stakeholders. Evaluation of SICS was conducted by applying the same assessment methodology at each study site. This general methodology was based on a shared database [78], a common monitoring plan, a unified statistical analysis (according to the experimental design of each experiment) and sustainability assessment. In the field experiments, SICS were compared with a control (usually a standard conventional practice) [79], and SICS were monitored for 2–4 years. Data from the field trials were assessed using a decision tree in terms of soil quality (physical, biological, and chemical); environmental, economic, and sociocultural dimensions; and sustainability, resulting in a score between −1 and 1 for each dimension [80]. For the three dimensions, the following methods were used for scoring:

- Environmental (including soil quality): Monitoring results compared SICS and control for several chemical, physical, and biological soil properties such as infiltration, aggregate stability, bulk density, mineral nitrogen, soil organic carbon (SOC), pH, earthworm density, crop yield, yield quality, crop cover, pests, root diseases, and weed diseases (see [79,80]). For each parameter, it was determined whether there was a statistically significant difference between SICS and control using mixed-effects models adjusted to the different experimental designs. For each experiment, the status of the soil was also evaluated as 'good' or 'bad' using threshold values based on expert opinion. A score of 1 was assigned if the SICS resulted in improvement, 0 if there was no change, and −1 if there was a deterioration. The overall environmental score (Table 2) was then obtained by averaging the scores for the individual parameters.
- Economic: The impact score compared costs and benefits for SICS and control (see [80]), where costs were calculated as the sum of investment costs, maintenance costs, and production costs. Equipment costs were not included. The analysis was conducted at the field/farm level and did not consider (monetization of) off-site effects of SICS.
- Sociocultural: Sociocultural impact was based on workload, perceived risk, and farmer reputation. Workload and farmer reputation were scored between −1 and 1, where negative values indicated a deterioration for the SICS compared with the control or usual practice. Perceived risk was scored between −1 and 0, where 0 meant no risks were perceived to be associated with the SICS. However, we did not assess whether a SICS reduced risks compared to the control, and therefore, no positive values were possible. This was a shortcoming of the assessment methodology and led to a 'negative bias' in assessing the sociocultural dimension of SICS.

Detailed results of the evaluation of environmental, economic, and sociocultural dimensions were presented in [79]. For SICS for which data on all three dimensions were available, we calculated the impact on sustainability as the average of the impact on the three dimensions [80].

Finally, the study site results were upscaled to the European level using a storyline, simulation, and policy support process [81–83]. This process combined participation and modelling to better understand the impacts of SICS across Europe and to provide policy support to facilitate the uptake of SICS under different contexts and conditions. As part of the approach, an integrated assessment model (IAM) consisting of spatial, socioeconomic, and environmental simulation models (i.e., the AGMEMOD [84], METRONAMICA [85], PESERA [86], dyna-QUEFTS [87], and MITERRA [88] models) was developed [81]. The IAM was used to simulate possible effects of four scenarios that captured diverse pathways for European agriculture until 2050 (Figure 4). These scenarios differed with regard to challenges to voluntary instruments and mandatory instruments. We used a combination of qualitative and quantitative techniques in a multi-actor approach to develop these scenarios in order to assess how agricultural practices could contribute to sustainable and profitable European agriculture and, finally, to discuss what is needed to enable adoption and implementation of these practices. In addition, for a range of 27 SICS, Europe-wide maps and modelling were combined with expert judgement from study site partners and their stakeholders to provide a SICS potential index based on the applicability, relevance, and impact of each SICS [82]. An interactive web-based tool was developed to help land users and decision makers select suitable SICS throughout Europe (imt.soilcare-project.eu; accessed on 13 April 2022) [83]. This tool allows users to compare different SICS with regard to various aspects, including IAM results and the SICS potential index.

Table 2. Results of SICS analysis based on the developed assessment methodology [80]. Values were scored on a range from −1 to 1 for those experiments where data on all three dimensions were available (see [79]). Details on experiments can be found in [79]. Impact on sustainability was the average of environmental impact, economic impact, and sociocultural impact. Negative impacts are indicated by red and positive ones by green. More details are provided in Table S1.

Country	SICS Treatment	Environmental Impact	Economic Impact	Sociocultural Impact	Impact on Sustainability
Belgium	Wood chips	0.00	−0.93	−0.33	−0.38
Norway	Spring-sown cover crop/root mix	0.00	0.03	−0.26	−0.07
Hungary	N (maize 210, winter wheat 150, winter barley 120 kg/ha) + farmyard manure	0.34	−0.12	−0.13	0.06
Hungary	N fertilization (as above) + straw/stalk	0.37	0.38	0.60	0.44
Hungary	Minimum tillage + N (maize 180, winter wheat 160 kg/ha)	0.00	0.04	0.20	0.07
Switzerland	Controlled Uptake Long-Term Ammonium Nutrition (CULTAN) method	−0.10	−0.60	0.20	−0.16
Switzerland	Green manure, no pesticide	−0.15	−0.01	0.10	−0.03
Germany	Glyphosate + cover crops	0.00	−0.03	0.07	0.01
Romania	Rotation + mouldboard ploughing	0.24	0.31	−0.20	0.13
Italy	No-till, radish cover crop	0.00	0.07	0.00	0.02
Portugal	Conventional maize, Urban Sludge amendment	0.35	0.15	−0.56	0.02
Portugal	Maize with legume winter cover crop	0.11	0.03	−0.26	−0.03
Greece	Conversion from orange to avocado	0.03	0.76	0.00	0.24
Spain	Deficit irrigation with minimum tillage and pruning chips or temporal cover crops	0.30	−0.90	−0.03	−0.16
France	Early wheat sowing (Aug)	−0.08	−0.89	−0.20	−0.36
France	Sowing on the row of maize–buckwheat	−0.07	−0.33	0.10	−0.10
	average	0.08	−0.13	−0.05	−0.01
	median	0.00	0.01	−0.03	0.01
	# positive (>0.1)	6	4	3	3
	# negative (<−0.1)	1	6	7	4
	# no change (−0.1 to 0.1)	9	6	6	9

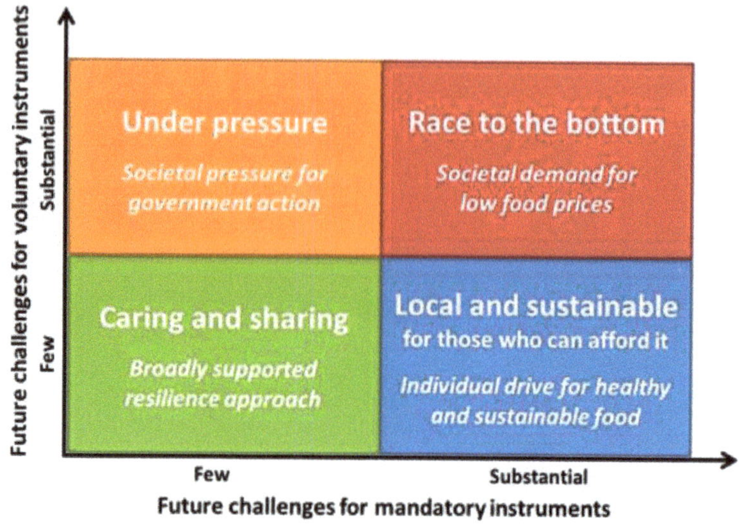

Figure 4. Overview of scenario framing linked with scenario titles and motivating factors [82].

2.3. Concepts and Methodology Used to Study Adoption of SICS

In the last decade, there have been numerous policy initiatives at the European level that, directly or indirectly, promoted the adoption of beneficial agricultural practices [89,90]. Most recently, the European Green Deal (COM/2019/640 final. https://eur-lex.europa.eu/legal-content/EN/TXT/?qid=1576150542719&uri=COM%3A2019%3A640%3AFIN (accessed on 13 April 2022) and the new Soil Strategy (COM/2021/699 final. https://eur-lex.europa.eu/legal-content/EN/TXT/?uri=CELEX%3A52021DC0699 (accessed on 13 April 2022)) set out the roadmap for making the European Union's economy more sustainable and identified several key actions that will be crucial in advancing land and soil protection in Europe. With this shift to more sustainable practices comes increasing pressure on farmers to change how they operate and adopt new techniques and practices. However, innovations associated with potential benefits to soil quality have not yet been adopted to their full potential and have, in some cases, even been abandoned, raising the question of why support for and adoption of these practices by European farmers is still weak.

Adoption of new or modified agricultural practices by farmers is a complex process that is governed not only by physical effectiveness and economics of agricultural practices but by a range of other factors, including individual, social, cultural, and policy-related factors [91]. These include internal factors, such as the farmer's own views on farming, the influence of peers and advisers, their perceived difficulties in implementing practices, and sociodemographic characteristics, and external factors, such as pedoclimatic conditions, markets, and policies [91,92]. Economics is an important factor and is often considered to be the main driver for adoption. However, overlooking some of the other factors may be one of the main reasons why seemingly advantageous measures have not been adopted widely by farming communities (e.g., [93,94]). Factors influencing the adoption of sustainable farming practices in Europe range from the land managers' access to information, training, and technical advice [95], to the performance of a particular practice in terms of yield increase or reduction in production costs or work time [96,97], to aspects rooted in the social and cultural context or in the personality of the individual land user. Social factors include the underlying motives (e.g., social or personal rewards) and attitude towards risks [98]; personality traits such as openness to new experience or resistance to change; what land users perceive others expect from them; and land users' perceptions of the relative benefits, costs, and risks associated with a particular practice [97,99]. In addition, farming practices, e.g., conservation measures, must be compatible with the values of landowners [97], cultural constructions of 'good farming' [100,101], and farmers' sense of professional identity and aesthetic preferences [102]. Finally, social factors such as trust and acceptability also influence adoption [59]. The dynamics of trust (across space, time, social groups, and culture) can explain how innovations are adopted through social learning and collaborative learning processes. The speed and spatial scale at which trust can develop likely depends on the extent to which it is possible to find or develop shared values, converge towards compatible epistemologies, and find common interests that can transcend sociocultural, political, and economic differences. It should be noted that engagement processes work differently and can lead to different outcomes when they operate over different spatial and temporal scales [103] so that engagement processes should be adapted to local conditions.

To understand all the factors that influence adoption and take them into account, a multidisciplinary integrated approach is needed, including, e.g., soil science (physics, chemistry, and biology), agronomy, hydrology, ecology, climatology, economics, and social sciences. In addition, a variety of stakeholders should be involved, as multiple stakeholders influence the ways in which crops are produced. This makes adoption site-specific, as every area has its own unique combination of biophysical, sociocultural, economic, and policy factors, as well as its own set of stakeholders. Thus, adoption research necessitates the involvement of scientists and practitioners from multiple disciplines, as well as active involvement of stakeholders. For SoilCare, this contextual nature of sociocultural and political drivers meant, on the one hand, that a robust assessment of adoption factors could be performed only at the study-site scale, so the broader suitability of SICS across Europe was considered primarily based on biophysical and environmental characteristics. On the other hand, the adoption work could still offer insights into more general trends with respect to the typical factors that can influence the adoption of particular SICS.

The SoilCare research on the adoption of SICS focussed on understanding the reasons why SICS are being adopted or not adopted and how farmers can be encouraged through appropriate incentives to adopt suitable SICS. The methods applied addressed four types of factors affecting adoption:

- Biophysical factors, which followed from the evaluation of monitoring results [79] as well as from literature reviews [55,77]. This included the effects that SICS had on soil quality but also on crop yield. Results of the evaluation of monitoring of SICS were presented to stakeholders and were discussed with them;
- Economic factors, which followed from a cost–benefit analysis of SICS implemented for monitoring [79] in combination with macroeconomic modelling using the AGMEMOD model [84]. Results of the economic analysis of SICS performed at the plot/farm scale were presented to, and discussed with, stakeholders;
- Social factors, which were studied in a selection of study sites via work with farmers and agricultural stakeholders in the United Kingdom and Norway to understand their perceptions of causes of and potential solutions to soil degradation and how they perceived SICS in relation to alternative approaches to increasing the sustainability of cropping systems in Europe [104]. An assessment of the role of the farming press and social media in decisions to adopt SICS and other sustainable agricultural practices was based on content analysis of media and interviews with U.K. farmers and agricultural advisers [105,106]. A wider analysis of social factors influencing adoption decisions, including an in-depth analysis of the role of social capital and trust, was based on literature review [91] and interviews with farmers and agricultural advisers in the United Kingdom and Hungary [107];
- Policy factors, which were studied through analyses of soil-related agricultural and environmental policies at both the E.U. and study site levels, through workshops and interviews.

Adoption should be considered not only with regard to a range of factors but also at different scales, from the farm scale to the European scale, because operations and actors in the agricultural value chain stretch out over these scales in the supply, purchase, processing, and distribution of agricultural products. Furthermore, socioeconomic developments, such as changing public awareness of the importance of sustainable production and the consequences this has for the prices consumers and companies are willing to pay for sustainably produced food, have an influence on adoption.

The storyline, simulation, and policy support approach presented in Section 2.2 was used to assess the adoption potential of SICS at the European scale. By developing different scenarios or pathways for European agriculture using a combination of sociocultural, technological, economic, environmental, and political factors and drivers of change, the impact of (policy) actions on enhancing adoption of SICS was assessed under various current and future conditions to arrive to options that would be robust across scenarios or target specific factors/barriers and enablers within scenarios.

3. Key findings

3.1. Main Effects of SICS

Table 2 provides an overview of monitoring results from 11 countries, derived from [79], which contained details on the experiments. Overall, these results showed a small positive impact of SICS (when compared with the control) on environment (including soil quality), no effect on sustainability, and a small negative impact on economics and the sociocultural dimension. Some treatments showed both high and low values of impact scores on the dimensions of the sustainability assessment, which illustrated trade-offs in the performance of a SICS. Some treatments yielded only zero or negative impacts (e.g., early wheat sowing, FR), and other treatments gave positive impact scores in all dimensions (e.g., N fertilization with straw/stalk, HU).

3.1.1. Environmental Dimension

In general, the SoilCare field experiments were too short to show clear statistically significant effects on productivity (yield or relative yield), SOC, structure stability (water stable aggregates), infiltration rate (hydraulic conductivity), biological activity (earthworm counting), or soil bulk density. Hydraulic conductivity and bulk density have large spatial and temporal variability in the field, which made it difficult to detect significant differences without dramatically increasing the number of measurements. The study site in Poland illustrated this spatial variability well [108]. Overall, SICS showed a small but positive effect on soil properties and the environmental dimension (Table 2); 6 out of 16 experiments showed a positive impact of SICS, 1 experiment showed a negative impact, and 9 experiments showed no change. Although not significant from a statistical point of view, slight improvements were found for most of the experiments. In addition, stakeholders and scientists in many cases could visually detect and evaluate positive effects of SICS, in properties such as soil structure or infiltration, or negative effects, such as weed infestation.

In addition, the SoilCare monitoring results provided the following insights based on the evaluation of the environmental dimension for all SICS [79].

Tillage: For most experiments, reduced tillage and noninversion tillage had a positive effect on soil characteristics and did not lead to lower yields. The noninversion tillage in a Belgian experiment presented better physical characteristics (hydraulic conductivity and aggregate stability). The minimized tillage in a Hungarian LTE [109] also improved the aggregate stability and SOC content when compared with conventional ploughing and increased the plant available water content [110]. A Czech experiment [111] showed that zero tillage was difficult for heavy soils and root crops but significantly improved the topsoil SOC, bulk density and aggregate stability when compared with conventional ploughing. However, the increase in SOC did not affect the plant available water content [110]. Pest and weed control was a challenge in the Belgian experiments under strip tillage and significantly impacted plant growth and crop yield. Weed control was also a major issue in several no-tillage systems; this resulted in increasing use of herbicides.

Soil compaction: Subsoiling is a means to alleviating compaction [112] by breaking up the compaction of deeper soil layers. In a Romanian experiment, subsoiling was suggested to a depth of 60 cm every 3 to 4 years to improve the aggregate stability and hydraulic conductivity and reduce the soil bulk density while maintaining a good crop yield. A Swedish experiment on a naturally compacted soil found that mechanical subsoiling, with or without incorporation of organic materials, had a positive impact on root growth and rooting depth. In a U.K. experiment, different physical and biological methods for compaction alleviation were explored. Ploughing was the most effective method for opening up the soil structure and alleviating topsoil compaction, but no effect on crop yield was observed in the two years of study [113]. The results of an Italian experiment that used different crops and tillage methods to reduce soil compaction indicated a higher risk of crop failure and difficulties with weed control (requiring herbicides) under no-tillage systems. Nevertheless, reduced-tillage systems had the potential to increase farm environmental and agronomic sustainability according to the relative sustainability index, which was based on 11 physical chemical and biological properties [114].

Fertilizers and amendments: An LTE in Hungary [115] showed significant positive effects on yield and soil structure (water stable aggregates and bulk density) when incorporating crop residues into the soil or when applying farmyard manure. The SOC content and plant available water content were not significantly increased [110] despite the positive effects on yield and soil structure. A Belgian experiment compared adding woodchips, compost, and pig manure with a control (no additions). The C/N ratio of the amendments helped to explain the availability of nutrients for crops. In a Portuguese experiment, urban sludge from wastewater treatment plants increased SOC and soil nutrient contents and earthworm population without affecting the heavy metal concentration in the soil in the short term. In a Danish experiment [116], the use of manure helped to reduce the crop yield gap between organic cultivation treatments and conventional control treatment with mineral fertilizers and to reduce soil bulk density. A study in Italy [117] examined the effects of SICS with different crop residue management and concluded that crop residues reduced the need for fertilizers. The Controlled Uptake Long-Term Ammonium Nutrition (CULTAN) method in Switzerland reduced the risk of nitrate leaching.

Data from LTEs in Belgium, Denmark, the United Kingdom, and Hungary indicated that soil management influenced soil biota, which in turn influenced soil quality [118]. The fungal communities were found to be very variable across sites located in different soil types and climatic regions, and only fertilization showed a consistent effect on arbuscular mycorrhizal fungi and plant pathogenic fungi, whereas the responses to tillage, cover crops, and organic amendments were site, soil, and crop-species specific. A study in Poland [119] examined the effects of adding spent mushroom substrate and chicken manure to soils on soil fungal community composition and mycobiome diversity. Both increased the abundance of fungi and reduced the relative abundance of several potential crop pathogens. These results provided a novel insight into the fungal communities associated with organic additives, which should be beneficial in the task of managing the soil mycobiome as well as crop protection and productivity. Both additives were also found to result in increased SOC [120].

Cover crops: Over the last decade, the increased use of cover crops between growing seasons has motivated the inclusion of this practice in the field experiments of many study sites. The benefits of cover crops are generally well accepted, and recent research has indicated that they can also enhance the availability of soil P and have positive effects on the soil microbial community [121–123] and earthworm abundance [116]. Positive effects were also illustrated by experiments in the study sites in Norway, Portugal, Denmark, France, Italy, and Germany [79]. However, because of global warming, which was visible in the results of the meteorological analyses for these study sites, the lack of freezing during recent winters meant that cover crops survived the winter. In that case, either herbicides or mechanical measures were required to kill them in spring. This is an important issue for further investigation. In the German experiment, the possible negative effect of glyphosate on soil quality was investigated by using different soil microbiological methods. An increase was found in ß-glucosidase activity (C-cycling enzyme) as a stress response of soil microorganisms after a period of seven days of application (unpublished data). Since no significant changes in microbial community composition occurred after the application of glyphosate in the field experiment, these effects were considered minor. Nevertheless, transport of glyphosate by preferential flow into deeper zones of soils might hinder the fast decay of this compound by bacterial glyphosate degraders [124]. Banning herbicides would require high-precision shallow tillage/mechanical weeding before seeding of the crops so as not to destroy the benefits of cover crops on soils again. Furthermore, mechanical weeding might mean more fuel use and GHG emissions.

In Greece and Spain, the tested cropping systems were vineyards, stone fruit, and olive orchards. In Crete (Greece), erosion reduction was the major challenge. Crete had historically high rainfall in October 2017 and some other heavy rainfall events afterwards. It was concluded that cover crops in vineyards and minimum tillage in olive orchards could reduce the erosion rates during extreme rainfall events and increase the earthworm density. The conversion of the traditional orange orchards to avocado cultivation resulted in a statistically significant reduction in erosion and increased SOC content and hydraulic conductivity [125]. Almería (southeast Spain), as the driest and hottest place in Europe, focused on water savings by deficit irrigation and erosion reduction with different soil cover or cultivation methods. The application of different combinations of irrigation led to water savings of up to 15%, but topsoil management did not cause significant differences in yield, fruit quality, or soil quality apart from an unexplained increase in the electrical conductivity when cover crops were used. [79].

3.1.2. Economic Impact (Profitability)

Table 2 indicates that the economic impact was positive for 4 out of 16 experiments, while it was negative for 6 and did not show change for the remaining 6. The average impact was −0.13, but the median impact was 0.01. Closer inspection of detailed data on costs and benefits (available for 15 SICS in Table S2) reveals that:

- For nine SICS, costs were higher than for the control; for five, they were lower; and for one, there was no change (defined as values between −25 and +25 EUR per ha). Hence, our hypothesis that SICS would reduce costs because of the lesser need for external inputs was not confirmed.
- For seven SICS, the benefits are higher than for the control; for two, there was no change compared with control; and for six, the benefits were lower.
- For seven SICS, the benefits minus the costs were higher than for the control; for seven, they were lower; and for one, there was no change.
- For 13 out of the 15 SICS for which detailed data were available, profitability was above 0.

This indicates that, at the field/farm level, short-term profitability was generally positive for the SICS (13 out of 15), but in half of the cases, it was lower for the SICS than for the control.

3.1.3. Sociocultural Impact

Table 2 indicates that for 3 out of 16 SICS, the sociocultural impact was positive; for 7, it was negative; and for 6, there was no change. The average impact was −0.04, and the median impact was −0.02. Analysis of data from 16 SICS showed (Table S3):

- Workload: Five SICS scored positive (required less work), six SICS scored negative (required more work), and for four SICS, there was no change.
- Perceived risk: 12 SICS were perceived to imply risks, and 3 were perceived to be riskless.
- Farmer reputation: Eight SICS scored positive (farmer implementing the SICS had a better reputation than farmer who did not), one scored negative (farmer had a worse reputation; the SICS in this case was the application of sewage sludge), and six registered no change.

This indicates that application of SICS had a positive impact on farmer reputation, as land users applying SICS were usually considered to be innovative. Workload did not show a clear trend, as for some SICS it was higher, while for others, it was lower. Many SICS are perceived to be associated with potential risks, most importantly the risk of crop failure and/or other economic risks (such as, e.g., high investment costs). The respondents often related the risk of crop failure to specific weather conditions such as prolonged dry spells or heavy rainfalls.

3.1.4. Main Results Upscaling SICS

Upscaling results included the potential for applying SICS across Europe as well as an assessment of the impact of SICS application under future uncertainty using the four developed scenarios (Figure 4). Figure 5 shows the SICS Potential Index for cover crops (for 2018) as an example result of the first type of upscaling activity. The figure shows that differences in climate, soil, and land use conditions resulted in differences in the applicability, relevance, and impact (on SOM, erosion, and yield) of cover crop use and hence the potential to apply them across Europe. Regarding the second type of upscaling activity, the results of the IAM indicated that over time (until 2050), in the different scenarios, different changes were expected in consumption, production and net exports, yield, gross margin, SOC, and erosion. This was due to, amongst other factors, growth in population, changes in diets, trade flows, climate change, technological changes, and changes in agricultural practices (i.e., through application of SICS). While some drivers were expected to result in impacts in the same direction in all scenarios (e.g., population growth was likely to lead to more consumption), other drivers could impact in very different ways. This was caused by regional differences such as, e.g., climate change impacting on yield levels and gross margins based on country-specific crop prices and location-specific biophysical conditions.

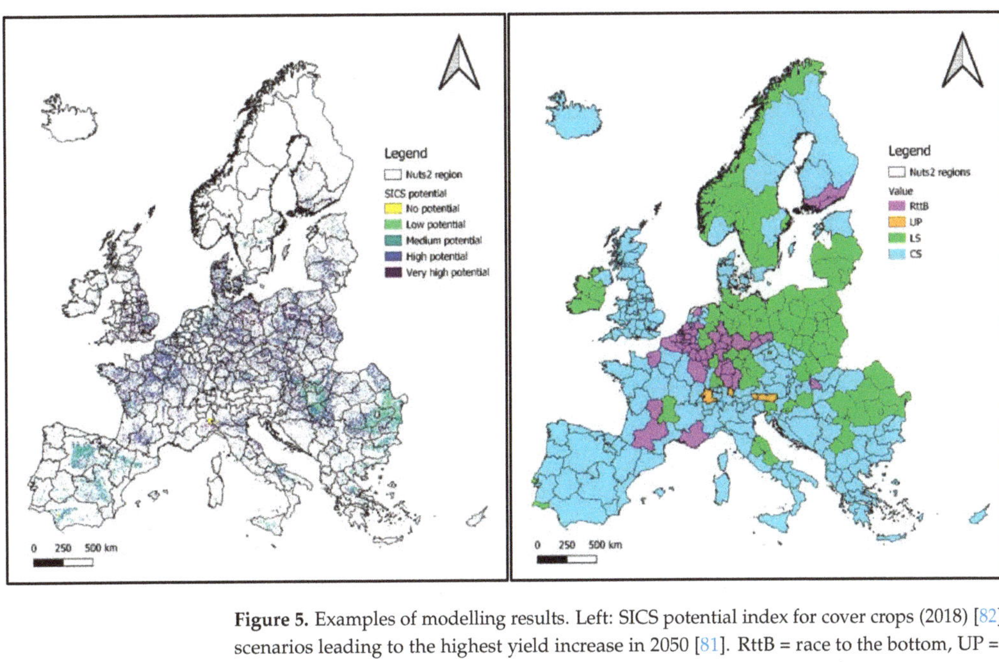

Figure 5. Examples of modelling results. Left: SICS potential index for cover crops (2018) [82]; right: scenarios leading to the highest yield increase in 2050 [81]. RttB = race to the bottom, UP = under pressure, LS = local and sustainable, CS = caring and sharing (see Figure 4).

As expected because of its formulation, the Caring and Sharing (CS) scenario, which assumed wide application of SICS (Figure 4), was likely to provide the best environmental impacts (i.e., increased, or stable SOC content and reduced erosion rates), and the Race to the Bottom scenario, assuming limited application, was likely to provide the worst.

An important finding, however, is that although the CS scenario in most regions led to highest yield impacts (Figure 5), the gross margin of SICS uptake under this scenario was negative in many NUTS-2 regions [81]. The most important factor contributing to this was the high implementation costs assumed when combinations of SICS were implemented. Despite sustainability being high on the agenda in the CS scenario, (financial) policy support would therefore likely be needed to enhance uptake of SICS. Alternatively, value added through additional products and services and valuation of environmental co-benefits could be a pathway to widespread SICS adoption.

The cost–benefit analysis showed a mixed spatial pattern of scenarios that had the highest gross margin across Europe. The reason for this was that the combination of drivers played out differently in different parts of Europe, indicating the complexity of the issue and the importance of understanding local dynamics. Using these scenarios for policy support also illustrated the importance of tailored/context-specific policy design/development, as selected options were often expected to have different performance under different scenarios.

3.2. Adoption of SICS

As illustrated in Table 3, there is a wide range of issues affecting adoption of sustainable soil management. Following this, country-specific issues stem from the fundamental E.U.-level factors listed below:

- **Sociocultural Factors:** A lack of awareness of soil in society and its framing as a resource to be exploited for humankind and economy engenders a disconnect between publics and impacts of agricultural production on soil. Further, mechanization creates distance between farmers and their fields and soil, making it difficult for farmers to see ecosystem changes. Some SoilCare stakeholders stressed ethical convictions favouring ecological approaches to farming as an important force for change with respect to these issues.
- **Economic Factors:** The financially difficult transition period from conventional to organic or more sustainable soil management practices can prove too risky for many farmers to undertake, as yields can reduce during this period. Farmers therefore need funding to support them through this. Further, financial incentives from policy and public demand can motivate a change in practice. Global trade systems favouring monocultures also inhibit change, as power is accumulated in the retailers rather than the producers.
- **Institutional/Policy Factors:** Change via regulation was thought by SoilCare stakeholders to be both positive and negative. Possible inadvertent effects can be avoided by closely working with farmers. Currently, advisory services are seen as a tool for safeguarding business as usual and do not reflect scientific evidence for sustainable soil management. Regular training is needed for both farmers and advisers. Publics education and accessibility of sustainably produced food also needs prioritizing.

Table 3. Adoption factors in SoilCare study sites.

Sociocultural Factors	Economic Factors
Society's awareness and valuing of soil—Consumers need to better understand the impacts production methods have on soil for more informed purchasing decisions and increase willingness to pay prices reflecting costs of sustainable production *New generation of farmers open to change*—Habit made many farmers reluctant to change practices. However, there were also pioneers who want to try out new practices *Social factors*—Results reiterated the value of social learning from different peers and networks and the dynamics of trust and social acceptability it can engender [107]. Influencers and champions have a critical role to play in lending legitimacy to important sources of information	*High investment and/or implementation costs*—Change in practices involves high costs for, e.g., organic fertilizer, equipping machinery with the right tools, and purchase of new crops as well as additional seeds on top of main crop for cover crops *Holistic approaches and cobenefits to soil*—UK: changes in arable rotations due to weed and disease control have been mainstreamed and have coincidentally benefited the soil *Market pressures/demands*—BE: policy encourages farmers to plant cover crops and rotate crops, but because of the high demand, too many potatoes were grown; in addition, crop residues and organic materials have been used for biofuels and other bioproducts instead of being returned to the soil
Institutional/policy factors	**Knowledge and education**
Adverse effects of policy design—Policies were perceived to dictate practices that needed to be adopted regardless of feasibility/practicability, sometimes resulting in adverse behaviour, e.g., converting existing grassland to avoid the 'permanent grassland' status *Lack of coherence between legislation/conflicting objectives*—UK: targets and subsidies for increasing woodland areas for growing biofuel crops fail to specify that land must be suitable for these purposes; BE: because of the fragmentation in public services and departments, farmers often receive contradictory advice (Nitrates Directive versus CAP)	*Insufficient resources*—Advisory services need more resources for experimental and demonstration farms. Advice providers were often reliant on project funding, which has continuity problems *Adviser expertise and quality*—ES: quality of advice was heterogeneous, and advice was given on ad hoc basis; BE: physical and biological soil management was often neglected because of a focus on nutrients and fertilizers/manures; NO: quality of advice from NLR (independent membership organisation) is good; these people know a lot about soil and try to incorporate advice to enhance soil and environmental conditions when they can

4. Discussion and Conclusions

4.1. Evaluation of SICS

SoilCare provided scientific evidence on the potential of SICS at 16 study sites and Europe-wide. Although monitoring in study sites did not provide conclusive results in all cases, it did show positive effects on most soil properties as well as a small positive impact on the environmental dimension. This was in line with the main results reported by meta-analyses such as those reviewed in [55]. No significant changes were observed for sustainability or for the economic dimension at the farm level. Nevertheless, most SICS were found to be profitable, since benefits were often higher than costs. However, in a small majority of cases, the profitability of the SICS was lower than for the control. The sociocultural dimension was slightly negative on average, mainly because SICS were perceived to be risky by farmers. The respondents often related the risk of crop failure to specific weather conditions, such as prolonged dry spells or heavy rainfalls. Indeed, it is known that some SICS are more sensitive to yearly variations than conventional practices, such as, for example, organic farming (e.g., [126–128]). On the other hand, weather conditions would in most cases also challenge the performance of the controls, but the risks associated with these practices were not assessed in our study. As described in Section 3.2, risks can also be higher during the transition period from conventional to more sustainable practices, although our economic data overall showed similar revenues for SICS and control. A final reason why SICS are perceived to be risky may have to do with uncertainty and risk aversion on the part of farmers, as switching from normal practices to

SICS means a switch from familiar ways to something new. A repeated questionnaire after a few years of implementation of SICS might help to investigate whether risk perception of SICS changes over time.

It should be noted that our results were obtained at the plot/farm level and based on only 3 (max. 4 for some study sites) years of monitoring. This has several implications:

- Not all SICS may have reached their full potential within such a short period, and long-term monitoring is needed. In LTEs, several similar SICS proved to increase sustainability and crop yield when managed to optimize soil fertility [129]. Thus, LTEs provide useful information but cannot be used to directly compare with the exact SICS that were tested in SoilCare, as these SICS were selected within the project through interaction with stakeholders to cover specific local needs and preferences.
- Furthermore, specific conditions during the years of monitoring had an impact on the outcomes. For example, in 2018, droughts occurred at several study sites. Moreover, all the years had sometimes record-breaking high summer temperatures and less cold weather during the winter. Longer-term monitoring is needed to obtain reliable data on the effects of SICS.
- The economic analysis was conducted based on short-term SICS application, whereas the slow accrual of soil fertility enhancement and soil conservation effects are expected to lead to increasing yield impacts in the long term [130,131]. The short timeframe also carried, e.g., the risk that initial investments for implementation of SICS were given too much weight (though in our study we could not include equipment costs, which could be significant for some SICS) or the risk that workload was overestimated since farmers need time to find the most efficient ways for managing SICS. Furthermore, economic analysis should be based on the full rotation, which takes several years [132,133].
- Economic analysis should not be restricted to farm economics but should also consider other ecosystem services, both on-site (e.g., nutrient cycling, weed suppression, [134]) and off-site (e.g., sedimentation, [135]), to be able to assess societal costs and benefits of the application of SICS. Preference-based rather than cost-based valuation methods should be used to better capture this diverse set of impacts and offer credible policy support [136].
- As monitoring was conducted at the plot/farm scale, it did not study diversification. However, diversification could contribute to more sustainable agricultural production through, e.g., the reallocation of some farming resources/material, such as lands, equipment, and labour, to other fields; other social or natural services, including changes in productive goals; and switching to nonfarming activities at both spatial and temporal scales [137]. In addition, diversification may alter soil chemical, physical, and/or biological properties, supporting large and sustainable production [138].
- Analysis of the social dimension was, by necessity, based on the views of farmers, and these might change over time as the farmers become more familiar with SICS. In addition, there may have been a bias in farmers participating in SoilCare experiments, as for the most part only farmers open to innovation took part in this work.

In addition, the assessment methodology for SICS that was applied may need further development and refinement. Both the assessment methodology and its application relied on expert opinion, not only with regard to the weights assigned to different parameters and to the environmental, economic, and sociocultural dimensions but with regard to the underlying concepts. For example, the economic dimension did not give very positive results for the SICS, which was at least partly due to the fact that more importance was attached to the relative difference between SICS and a control than to the difference between benefits and costs. As a result, SICS with a positive benefit/cost ratio scored negatively on the economic dimension because the control had a more positive benefit/cost ratio. This may actually reflect reality, as this meant that farmers would earn less by applying SICS, but the point here is to illustrate that assumptions made in the assessment methodology did have an impact on the outcome. Such assumptions are open to discussion and can be subject to revision as more data become available.

Furthermore, the outcome of the assessment was, of course, influenced by the input. Although this may seem trivial, it is not, as the input by necessity has to be a combination of different types of data (quantitative as well as qualitative) originating from different sources (including scientific experiments but also stakeholder perceptions), sometimes with gaps or limitations.

For all of these reasons, the results of the evaluation should not be seen as a final result, but rather as an indication that forms a starting point for discussion with stakeholders (from farmers to scientists and policy makers).

4.2. Adoption of SICS

SoilCare also delivered knowledge on how to promote the adoption of SICS to individual farmers, European institutions, member state authorities, and agricultural advisory services. The analyses carried out in SoilCare delivered increased insight into biophysical, economic, social, and political barriers to adoption, several of which corresponded to barriers already identified in [52] for conservation agriculture. SoilCare also provided solutions that could help to overcome such barriers. The results confirmed the crucial role of social factors such as trust in adoption and underlined the need for policies that support and enable a transition to more sustainable agricultural practices in a coherent way.

Historically, soil has been an overlooked component in studies on ecosystem service and policy decision making [139]. At a policy level, the removal of the proposed Soil Framework Directive (COM (2006) 232 final) in 2014 highlighted a need and an opportunity to think about soils differently [140]. The SoilCare project represents a short timeline when set against its objectives; however, it is also noteworthy that the role of soils transitioned to being at the heart of high-level ambitious European policies such as the European Green Deal and the CAP Farm-to-Fork and Biodiversity Strategies during the project lifetime. This was complemented by a focus on soil research and innovation in the European Joint Programme on soil and a mission in the area of soil health and food. E.U. policies to target soil and environmental objectives have been criticized for their lack of nuance to account for localized conditions in the past. In this regard, the SoilCare project has framed a methodology for SICS that reflects the key dimensions that must be considered in governance for local but also wider-scale dynamics. Although more work is required, the lessons learned, particularly in relation to those SICS that exhibited promise, should be further explored and leveraged under the new opportunities that now exist within the policy, research, and innovation space. Table 4 provides an overview of policy recommendations resulting from SoilCare.

Table 4. Policy recommendations resulting from SoilCare, after [141].

Recommendation I: Define long-term ambitions and targets

- Develop horizontal, holistic, long-term strategies for sustainable agriculture
- Raise and clearly define the level of ambitions in existing policies
- Define binding soil targets and promote sustainable practices through either dedicated soil policies or mainstreaming of soil objectives in existing and new environmental/sectoral policy instruments

Recommendation II: Increase coherence and exploit synergies between policies more effectively
There are many different pieces of legislation that can work better together if coherence and integration between them is improved. In addition, stakeholders noted that some SICS might not align with existing policy objectives. At the E.U. and country levels, policy conflicts and synergies need to be carefully analysed and aligned to avoid discouraging a transition to sustainable farming.

Table 4. *Cont.*

Recommendation III: Design targeted economic instruments that facilitate a transition to sustainable practices and reward environmental benefits delivered The CAP should strive to be less prescriptive and avoid one-size-fits-all approaches, instead providing farmers with a general direction clearly defined by targets and empowering them to take steps towards these targets. There is a need to consider the different conditions in which farmers operate (e.g., differences in tenure), and measures need to be flexible enough to allow for regional differences. Priority should be given to farming techniques that are also means of food production and are both profitable and sustainable.
Recommendation IV: Strengthen existing and establish new opportunities for learning and knowledge exchange for farmers Strengthen capacity of Farm Advisory Services: These are valuable sources of information for farmers, but their independence and neutrality should be ensured. Advisers need to learn about new practices, their practical application and costs, and benefits to support farmers. Ref. [142] gave suggestions for achieving more effective advisory services. Inform farmers about new developments and insights: Dissemination of knowledge, awareness raising, and education are important components of policy interventions, and they should be used in parallel with economic and legislative instruments [143].
Recommendation V: Strengthen monitoring and enforcement At the E.U. level, there is a need to establish a clear, robust, and reliable monitoring and enforcement system for the CAP. At the country level, stronger monitoring and enforcement systems require the training of farm inspectors, who, like farmers, need to understand regulatory requirements and their practical implementation.

4.3. Sustainability and Profitability

Results obtained at the farm level indicated a small decrease in profitability and a small positive effect on the environmental dimension (Table 2). As discussed above, however, there is a need to consider larger temporal and spatial scales. This was done in the modelling approach, which was used to upscale results from the different study sites and integrate these results with factors operating at the European scale, such as policy development, macroeconomy, societal developments, and climate change. Several scenarios of possible developments with a time horizon of 2050 were simulated. Simulations showed that scenarios in which sustainability was given priority resulted in better soil quality and better environmental conditions. However, while SICS would be profitable to society in the long term, they may not always be profitable to farmers in the short term. As short-term benefit over conventional practice is a key point for farmers [63], and as modelling suggested that SICS outperformed control treatments in the longer term, some form of compensation and support to farmers would be required to stimulate adoption of SICS, for example, in the form of bridge payments.

4.4. Conclusions

The need for sustainable soil management is evident from the literature. Soils are critical for economic and environmental well-being because they provide a range of ecosystem services and form the basis for agricultural production. They are at the intersection of a broad range of agricultural and land use challenges. Soil management should aim at improving the quality and resilience of land and soil. Within the SoilCare project, the concept of soil-improving cropping systems (SICS) was developed and applied. SICS can play an important role in the transition towards more sustainable agricultural production that can also be profitable. In practice, the effectiveness of SICS is difficult to demonstrate within the lifespan of a single project, as results vary from year to year because of different conditions, such as different weather and price fluctuations of inputs and crops. Furthermore, many SICS are expected to reach their full potential only after a long time. SoilCare paved the way for further research on SICS by developing an assessment methodology for SICS, a database for SICS data, and a modelling approach for upscaling and scenario evaluation. In addition, SoilCare contributed to the understanding of adoption factors and provided a first assessment of a range of SICS. Whilst our work on adoption confirmed the role economic considerations play in the uptake of SICS, it also highlighted the influence of social factors, such as trust, and of knowledge. This underlines the need for policies that support and enable a transition to more sustainable agricultural practices in a coherent way.

Supplementary Materials: The following are available online at https://www.mdpi.com/article/10.3390/land11060780/s1, Table S1: Results of environmental dimension, Table S2: Results of economic dimension, Table S3: Results of sociocultural dimension.

Author Contributions: Conceptualization: R.H., O.O., M.S.R., A.A., G.W., H.v.D., M.M., J.M., E.v.d.E., S.V. and F.A. (Falentijn Assinck); methodology: R.H., M.S.R., A.A., G.W., H.v.D., M.M., J.M., I.S.P., E.v.d.E., S.V., F.A. (Francisco Areal) and J.S.; software: I.S.P. and H.v.D.; validation: H.v.D., A.A., A.E., J.L., A.K., M.A.B., E.K., Z.T., M.H. (Moritz Hallama) and J.S.; investigation: A.A., A.E., J.L., A.K., M.A.B., E.K., Z.T., M.H. (Moritz Hallama), J.C., I.P., T.D. and J.S.; data curation: S.V., E.v.d.E., G.W., I.S.P. and H.v.D.; writing—original draft preparation: R.H., O.O., M.S.R., A.A., G.W., H.v.D., M.M., J.M. and J.E.B.; writing—review and editing: F.A. (Francisco Areal), M.H. (Marius Heinen), I.S.P., A.E., M.A.B., E.K., M.H. (Moritz Hallama), J.C., L.O., L.F., L.M., J.E.M.B. and J.S.; visualization: R.H., H.v.D., J.M. and C.-A.C. All authors have read and agreed to the published version of the manuscript.

Funding: This research received funding from the European Union's Horizon 2020 research and innovation programme under grant agreement No 677407 (SoilCare project). R.H., O.O., E.v.d.E., S.V., F.A. (Falentijn Assinck) and M.H. (Marius Heinen) also received funding from the Dutch Ministry of LNV via Kennis Basis programma 34, project KB-34-008-005. The contribution of the University of Hohenheim was partly financially supported by the German Research Foundation (DFG) under the Collaborative Research Centre 1253 CAMPOS (DFG grant SFB 1253/1 2017).

Institutional Review Board Statement: Not applicable in this study.

Informed Consent Statement: Ethical standards and guidelines have been applied to the collection, processing and storage of data about persons, in accordance with the agreed project ethical statement. All subjects gave their informed consent for inclusion before they participated in the study.

Data Availability Statement: The data presented in this study are available in the supplementary material.

Acknowledgments: We thank the stakeholders in the SoilCare study sites. Without their collaboration, this research would not have been possible.

Conflicts of Interest: The authors declare no conflict of interest.

References

1. Van Ittersum, M.K.; Cassman, K.G.; Grassini, P.; Wolf, J.; Tittonell, P.; Hochman, Z. Yield gap analysis with local to global relevance—A review. *Field Crops Res.* **2013**, *143*, 4–17. [CrossRef]
2. Boogaard, H.; Wolf, J.; Supit, I.; Niemeyer, S.; van Ittersum, M. A regional implementation of WOFOST for calculating yield gaps of autumn-sown wheat across the European Union. *Field Crops Res.* **2013**, *143*, 130–142. [CrossRef]
3. Wiesmeier, M.; Hübner, R.; Kögel-Knabner, I. Stagnating crop yields: An overlooked risk for the carbon balance of agricultural soils? *Sci. Total Environ.* **2015**, *636*, 1045–1051. [CrossRef] [PubMed]
4. Schils, R.; Olesen, J.E.; Kersebaum, K.-C.; Rijk, B.; Oberforster, M.; Kalyada, V.; Khitrykau, M.; Gobin, A.; Kirchev, H.; Manolova, V.; et al. Cereal yield gaps across Europe. *Eur. J. Agron.* **2018**, *101*, 109–120. [CrossRef]
5. Bünemann, E.; Bongiorno, G.; Bai, Z.; Creamer, R.E.; de Deyn, G.; de Goede, R.; Fleskens, L.; Geissen, V.; Kuyper, T.W.; Mäder, P.; et al. Soil quality—A critical review. *Soil Biol. Biochem.* **2018**, *120*, 105–125. [CrossRef]
6. Attard, E.; Recous, S.; Chabbi, A.; De Berranger, C.; Guillaumaud, N.; Labreuche, J.; Philippot, L.; Schmid, B.; Le Roux, X. Soil environmental conditions rather than denitrifier abundance and diversity drive potential denitrification after changes in land uses. *Glob. Chang. Biol.* **2011**, *17*, 1975–1989. [CrossRef]
7. Cassman, K.G. Ecological intensification of cereal production systems: Yield potential, soil quality, and precision agriculture. *Proc. Natl. Acad. Sci. USA* **1999**, *96*, 5952–5959. [CrossRef]
8. Gasso, V.; Sørensen, C.A.G.; Oudshoorn, F.W.; Green, O. Controlled traffic farming: A review of the environmental impacts. *Eur. J. Agron.* **2013**, *48*, 66–73. [CrossRef]
9. Sapkota, T.B.; Mazzoncini, M.; Barberi, P.; Antichi, D.; Silvestri, N. Fifteen years of no till increase soil organic matter, microbial biomass and arthropod diversity in cover crop-based arable cropping systems. *Agron. Sustain. Dev.* **2012**, *32*, 853–863. [CrossRef]
10. Cuevas, J.; Daliakopoulos, I.N.; del Moral, F.; Hueso, J.J.; Tsanis, I.K. A Review of Soil-Improving Cropping Systems for Soil Salinization. *Agronomy* **2019**, *9*, 295. [CrossRef]

11. Bolinder, M.A.; Crotty, F.; Elsen, A.; Frac, M.; Kismányoky, T.; Lipiec, J.; Tits, M.; Tóth, Z.; Kätterer, T. The effect of crop residues, cover crops, manures and nitrogen fertilization on soil organic carbon changes in agroecosystems: A synthesis of reviews. *Mitig. Adapt. Strat. Glob. Chang.* **2020**, *25*, 929–952. [CrossRef]
12. Crotty, F.; Hannula, E.; Hallama, M.; Kandeler, E. Can soil improving cropping systems reduce the loss of soil biodiversity within agricultural soils? In *Sustainable Soil Management as a Key to Preserving Soil Biodiversity and Stopping Its Degradation*; Reyes-Sánchez, L.B., Horn, R., Costantini, E.A.C., Eds.; International Union of Soil Sciences (IUSS): Vienna, Austria, 2022; pp. 187–220.
13. Nosalewicz, A.; Lipiec, J. The effect of compacted soil layers on vertical root distribution and water uptake by wheat. *Plant Soil* **2014**, *375*, 229–240. [CrossRef]
14. Mehra, P.; Baker, J.; Sojka, R.E.; Bolan, N.; Desbiolles, J.; Kirkham, M.B.; Ross, C.; Gupta, R. A review of tillage practices and their potential to impact the soil carbon dynamics. *Adv. Agron.* **2018**, *150*, 185–230.
15. Schneider, M.K.; Lüscher, G.; Jeanneret, P.; Arndorfer, M.; Ammari, Y.; Bailey, D.; Balázs, K.; Báldi, A.; Choisis, J.P.; Dennis, P.; et al. Gains to species diversity in organically farmed fields are not propagated at the farm level. *Nat. Commun.* **2014**, *5*, 41–51. [CrossRef]
16. Reeves, D.W. The role of soil organic matter in maintaining soil quality in continuous cropping systems. *Soil Tillage Res.* **1997**, *43*, 131–167. [CrossRef]
17. Panagos, P.; Barcelo, S.; Bouraoui, F.; Bosco, C.; Dewitte, O.; Gardi, C.; Erhard, M.; Hervas De Diego, F.; Hiederer, R.; Jeffery, S.; et al. *The State of Soil in Europe: A Contribution of the JRC to the European Environment Agency's Environment State and Outlook Report—SOER 2010*; EUR 25186 EN; JRC68418; Jones, A., Ed.; Publications Office of the European Union: Luxembourg, 2012.
18. Rockström, J.; Steffen, W.; Noone, K.; Persson, Å.; Chapin, F.S., III; Lambin, E.; Lenton, T.M.; Scheffer, M.; Folke, C.; Schellnhuber, H.; et al. Planetary boundaries: Exploring the safe operating space for humanity. *Ecol. Soc.* **2009**, *14*, 32. [CrossRef]
19. Tang, F.H.; Lenzen, M.; McBratney, A.; Maggi, F. Risk of pesticide pollution at the global scale. *Nat. Geosci.* **2021**, *14*, 206–210. [CrossRef]
20. Kanter, D.R.; Chodos, O.; Nordland, O.; Rutigliano, M.; Winiwarter, W. Gaps and opportunities in nitrogen pollution policies around the world. *Nat. Sustain.* **2020**, *3*, 956–963. [CrossRef]
21. Zhang, X.; Davidson, E.A.; Mauzerall, D.L.; Searchinger, T.D.; Dumas, P.; Shen, Y. Managing nitrogen for sustainable development. *Nature* **2015**, *528*, 51–59. [CrossRef]
22. Sørensen, C.G.; Halberg, N.; Oudshoorn, F.W.; Petersen, B.M.; Dalgaard, R. Energy inputs and GHG emissions of tillage systems. *Biosyst. Eng.* **2014**, *120*, 2–14. [CrossRef]
23. Mendes, I.C.; Sousa, D.M.G.; Dantas, O.D.; Lopes, A.A.C.; Junior, F.B.R.; Oliveira, M.I.; Chaer, G.M. Soil quality and grain yield: A win–win combination in clayey tropical oxisols. *Geoderma* **2021**, *388*, 114880. [CrossRef]
24. De Vries, F.T.; Thébault, E.; Liiri, M.; Birkhofer, K.; Tsiafouli, M.A.; Bjørnlund, L.; Bracht Jørgensen, H.; Brady, M.V.; Christensen, S.; de Ruiter, P.C.; et al. Soil food web properties explain ecosystem services across European land use systems. *Proc. Natl. Acad. Sci. USA* **2013**, *110*, 14296–14301. [CrossRef] [PubMed]
25. Webb, N.P.; Marshall, N.A.; Stringer, L.C.; Reed, M.S.; Chappell, A.; Herrick, J.E. Land degradation and climate change: Building climate resilience in agriculture. *Front. Ecol. Environ.* **2017**, *15*, 450–459. [CrossRef]
26. Lori, M.; Symanczik, S.; Mäder, P.; Efosa, N.; Jaenicke, S.; Buegger, F.; Tresch, S.; Goessmann, A.; Gattinger, A. Distinct nitrogen provisioning from organic amendments in soil as influenced by farming system and water regime. *Front. Environ. Sci.* **2018**, *6*, 40. [CrossRef]
27. Bongiorno, G.; Postma, J.; Bünemann, E.K.; Brussaard, L.; de Goede, R.G.M.; Mäder, P.; Tamm, L.; Thuerig, B. Soil suppressiveness to Pythium ultimum in ten European long-term field experiments and its relation with soil parameters. *Soil Biol. Biochem.* **2019**, *133*, 174–187. [CrossRef]
28. Maynard, J.J.; Levi, M.R. Hyper-temporal remote sensing for digital soil mapping: Characterizing soil-vegetation response to climatic variability. *Geoderma* **2017**, *285*, 94–109. [CrossRef]
29. O'Sullivan, L.; Wall, D.; Creamer, R.; Bampa, F.; Schulte, R.P.O. Functional Land Management: Bridging the Think-Do-Gap using a multi-stakeholder science policy interface. *Ambio* **2018**, *47*, 216–230. [CrossRef]
30. Rillig, M.C.; Ryo, M.; Lehmann, A.; Aguilar-Trigueros, C.A.; Buchert, S.; Wulf, A.; Iwasaki, A.; Roy, J.; Yang, G. The role of multiple global change factors in driving soil functions and microbial biodiversity. *Science* **2019**, *366*, 886–890. [CrossRef]
31. Oehlmann, Y.; Lange, M.; Leimer, S.; Roscher, C.; Aburto, F.; Alt, F.; Dassen, S.; De Deyn, G.; Eisenhauer, N.; Gleixner, G.; et al. Above- and Belowground Biodiversity Jointly Tighten the P Cycle. *Nat. Commun.* **2021**, *12*, 4431. [CrossRef]
32. Bossio, D.A.; Cook-Patton, S.C.; Ellis, P.W.; Fargione, J.; Sanderman, J.; Smith, P.; Wood, S.; Zomer, R.J.; von Unger, M.; Emmer, I.M.; et al. The role of soil carbon in natural climate solutions. *Nat. Sustain.* **2020**, *3*, 391–398. [CrossRef]
33. Griscom, B.W.; Adams, J.; Ellis, P.W.; Houghton, R.A.; Lomax, G.; Miteva, D.A.; Schlesinger, W.H.; Schoch, D.; Siikamäki, J.V.; Smith, P.; et al. Natural climate solutions. *Proc. Natl. Acad. Sci. USA* **2017**, *114*, 11645–11650. [CrossRef] [PubMed]
34. Willett, W.; Rockström, J.; Loken, B.; Springmann, M.; Lang, T.; Vermeulen, S.; Garnett, T.; Tilman, D.; DeClerck, F.; Wood, A.; et al. Food in the Anthropocene: The EAT-Lancet Commission on healthy diets from sustainable food systems. *Lancet* **2019**, *393*, 447–492. [CrossRef]
35. Thomas, R.; Reed, M.; Clifton, K.; Appadurai, N.; Mills, A.; Zucca, C.; Kodsi, E.; Sircely, J.; Haddad, F.; Hagen, C.; et al. A framework for scaling sustainable land management options. *Land Degrad. Dev.* **2018**, *29*, 3272–3284. [CrossRef]

36. Karlen, D.L.; Mausbach, M.J.; Doran, J.W.; Cline, R.G.; Harris, R.F.; Schuman, G.E. Soil Quality: A concept, definition and framework for evaluation (a guest editorial). *Soil Sci. Soc. Am. J.* **1997**, *61*, 4–10. [CrossRef]
37. Teasdale, J.R.; Coffman, C.B.; Mangum, R.W. Potential long-term benefits of no-tillage and organic cropping systems for grain production and soil improvement. *Agron. J.* **2007**, *99*, 1297–1305. [CrossRef]
38. Seufert, V.; Ramankutty, N.; Foley, J.A. Comparing the yields of organic and conventional agriculture. *Nature* **2012**, *485*, 229–232. [CrossRef]
39. Pittelkow, C.M.; Liang, X.; Linquist, B.A.; van Groenigen, K.J.; Lee, J.; Lundy, M.E.; van Gestel, N.; Six, J.; Venterea, R.T.; van Kessel, C. Productivity limits and potentials of the principles of conservation agriculture. *Nature* **2014**, *517*, 365–368. [CrossRef]
40. Hatfield, J.L.; Sauer, T.; Kruse, R. Soil: The forgotten piece of the water, food, energy nexus. *Adv. Agron.* **2017**, *143*, 1–46.
41. Vogel, H.J.; Bartke, S.; Daedlow, K.; Helming, K.; Kögel-Knabner, I.; Lang, B.; Rabot, E.; Russell, D.; Stößel, B.; Weller, U.; et al. A systemic approach for modelling soil functions. *Soil* **2018**, *4*, 83–92. [CrossRef]
42. Jones, A.; Fernandes-Ugalde, O.; Scarpa, S.; Eiselt, B. *LUCAS 2022 ISSG Planning Document*; Publications Office of the European Union: Luxembourg, 2021.
43. Veerman, C.; Pinto Correia, T.; Bastioli, C.; Biró, B.; Bouma, J.; Cienciala, E.; Emmett, B.; Frison, E.; Grand, A.; Hristov, L.; et al. *Caring for Soil Is Caring for Life: Ensure 75% of Soils Are Healthy by 2030 for Food, People, Nature and Climate: Report of the Mission Board for Soil Health and Food*; Publications Office: Luxembourg, 2020. [CrossRef]
44. Tittonell, P. Ecological intensification of agriculture—Sustainable by nature. *Curr. Opin. Environ. Sustain.* **2014**, *8*, 53–61. [CrossRef]
45. Mottet, A.; Bicksler, A.; Lucantoni, D.; De Rosa, F.; Scherf, B.; Scopel, E.; López-Ridaura, S.; Gemmil-Herren, B.; Bezner Kerr, R.; Sourisseau, J.-M.; et al. Assessing transitions to sustainable agricultural and food systems: A Tool for Agroecology Performance Evaluation (TAPE). *Front. Sustain. Food Syst.* **2020**, *4*, 252. [CrossRef]
46. EU. *Caring for Soils Is Caring for Life, Report of the Mission Board For Soil Health and Food*; European Commission: Luxembourg, 2020.
47. Bouma, J.; Montanarella, L.; Evanylo, G. The challenge for the soil science community to contribute to the implementation of the UN Sustainable Development Goals. *Soil Use Manag.* **2019**, *35*, 538–546. [CrossRef]
48. Kassam, A.; Friedrich, T.; Derpsch, R.; Kienzle, J. Overview of the Worldwide Spread of Conservation Agriculture. *Field Actions Sci. Rep.* **2015**, *8*. Available online: http://factsreports.revues.org/3966 (accessed on 13 April 2022).
49. Anken, T.; Weisskopf, P.; Zihlmann, U.; Forrer, H.; Jansa, J.; Perhacova, K. Long-term tillage system effects under moist cool conditions in Switzerland. *Soil Tillage Res.* **2004**, *78*, 171–183. [CrossRef]
50. Bioland. Sieben Prinzipien für die Landwirtshaft der Zukunft, pp. 22. 2014. Available online: http://bioland.de/ueber-uns/sieben-prinzipien.html (accessed on 22 December 2014).
51. Lahmar, R. Adoption of conservation agriculture in Europe. Lessons of the KASSA project. *Land Use Policy* **2010**, *27*, 4–10. [CrossRef]
52. Kassam, A.; Friedrich, T.; Shaxson, F.; Bartz, H.; Mello, I.; Kienzle, J.; Pretty, J. The spread of Conservation Agriculture: Policy and institutional support for adoption and uptake. *Field Actions Sci. Rep.* **2014**, *7*. Available online: http://factsreports.revues.org/3720 (accessed on 13 April 2022).
53. Stoate, C.; Báldi, A.; Beja, P.; Boatman, N.D.; Herzon, I.; van Doorn, A.; de Snoo, G.R.; Rakosy, L.; Ramwell, C. Ecological impacts of early 21st century agricultural change in Europe—A review. *J. Environ. Manag.* **2009**, *91*, 22–46. [CrossRef]
54. Ingram, J.; Mills, J.; Frelih-Larsen, A.; Davis, M.; Merante, P.; Ringrose, S.; Molnar, A.; Sánchez, B.; Bahadur Ghaley, B.; Karaczun, Z. Managing Soil Organic Carbon: A Farm Perspective. *Eurochoices* **2013**, *13*, 12–19. [CrossRef]
55. Rietra, R.P.J.J.; Heinen, M.; Oenema, O. A Review of Crop Husbandry and Soil Management Practices Using Meta-Analysis Studies: Towards Soil-Improving Cropping Systems. *Land* **2022**, *11*, 255. [CrossRef]
56. Nafziger, E. Cropping Systems. Ch 5 in Illinois Agronomy Handbook. 2012, pp. 49–63. Available online: http://extension.cropsci.illinois.edu/handbook/ (accessed on 23 December 2014).
57. Deike, S.; Pallutt, B.; Melander, B.; Strassemeyer, J.; Christen, O. Long-term productivity and environmental effects of arable farming as affected by crop rotation, soil tillage intensity and strategy of pesticide use: A case-study of two long-term field experiments in Germany and Denmark. *Eur. J. Agron.* **2008**, *29*, 191–199. [CrossRef]
58. De Vita, P.; Di Paolo, E.; Fecondo, G.; Di Fonzo, N.; Pisante, M. No-tillage and conventional tillage effects on durum wheat yield, grain quality and soil moisture content in southern Italy. *Soil Tillage Res.* **2007**, *92*, 69–78. [CrossRef]
59. Jensen, H.G.; Jacobsen, L.B.; Pedersen, S.M.; Tavella, E. Socioeconomic impact of widespread adoption of precision farming and controlled traffic systems in Denmark. *Precis. Agric.* **2012**, *13*, 661–677. [CrossRef]
60. Oldfield, E.E.; Bradford, M.A.; Wood, S.A. Global meta-analysis of the relationship between soil organic matter and crop yields. *Soil* **2019**, *5*, 15–32. [CrossRef]
61. Panagos, P.; Standardi, G.; Borrelli, P.; Lugato, E.; Montanarella, L.; Bosello, F. Cost of agricultural productivity loss due to soil erosion in the European Union: From direct cost evaluation approaches to the use of macroeconomic models. *Land Degrad. Dev.* **2018**, *29*, 471–484. [CrossRef]
62. Alonge, A.J.; Martin, R.A. Assessment of the adoption of sustainable agriculture practices: Implications for agricultural education. *J. Agric. Educ.* **1995**, *36*, 34–42. [CrossRef]
63. Rodriguez, J.M.; Molnar, J.J.; Fazio, R.A.; Sydnor, E.; Lowe, M.J. Barriers to adoption of sustainable agriculture practices: Change agent perspectives. *Renew. Agric. Food Syst.* **2009**, *24*, 60–71. [CrossRef]

64. Piñeiro, V.; Arias, J.; Dürr, J.; Elverdin, P.; Ibáñez, A.M.; Kinengyere, A.; Morales Opazo, C.; Owoo, N.; Page, J.R.; Prager, S.D.; et al. A scoping review on incentives for adoption of sustainable agricultural practices and their outcomes. *Nat. Sustain.* **2020**, *3*, 809–820. [CrossRef]
65. Mensah, A.; Asiamah, M.; Wongnaa, C.A.; Adams, F.; Etuah, S.; Gaveh, E.; Appiah, P. Adoption impact of maize seed technology on farm profitability: Evidence from Ghana. *J. Agribus. Dev. Emerg. Econ.* **2021**, *11*, 578–598. [CrossRef]
66. Jat, H.S.; Datta, A.; Choudhary, M.; Sharma, P.C.; Jat, M.L. Conservation Agriculture: Factors and drivers of adoption and scalable innovative practices in Indo-Gangetic plains of India–a review. *Int. J. Agric. Sustain.* **2021**, *19*, 40–55. [CrossRef]
67. Lechenet, M.; Bretagnolle, V.; Bockstaller, C.; Boissinot, F.; Petit, M.S.; Petit, S.; Munier-Jolain, N.M. Reconciling pesticide reduction with economic and environmental sustainability in arable farming. *PLoS ONE* **2014**, *9*, e97922. [CrossRef]
68. Baffes, J.; Koh, W.C. Soaring Fertilizer Prices Add to Inflationary Pressures and Food Security Concerns. Available online: https://blogs.worldbank.org/opendata/soaring-fertilizer-prices-add-inflationary-pressures-and-food-security-concerns (accessed on 13 April 2022).
69. FAO. Price volatility from a global perspective. In *Technical Background Document for the High-Level Event on: "Food Price Volatility and the Role of Speculation"*; FAO Headquarters: Rome, Italy, 6 July 2012.
70. EC. Europe 2020 Strategy. 2021. Available online: https://ec.europa.eu/environment/green-growth/index_en.htm (accessed on 13 April 2022).
71. Katt, F.; Meixner, O. A systematic review of drivers influencing consumer willingness to pay for organic food. *Trends Food Sci. Technol.* **2020**, *100*, 374–388. [CrossRef]
72. Meyerding, S.G.; Trajer, N.; Lehberger, M. What is local food? The case of consumer preferences for local food labeling of tomatoes in Germany. *J. Clean. Prod.* **2019**, *207*, 30–43. [CrossRef]
73. De Leijster, V.; Verburg, R.W.; Santos, M.J.; Wassen, M.J.; Martínez-Mena, M.; De Vente, J.; Verweij, P.A. Almond farm profitability under agroecological management in south-eastern Spain: Accounting for externalities and opportunity costs. *Agric. Syst.* **2020**, *183*, 102878. [CrossRef]
74. Schütte, R.; Plaas, E.; Gómez, J.A.; Guzmán, G. Profitability of erosion control with cover crops in European vineyards under consideration of environmental costs. *Environ. Dev.* **2020**, *35*, 100521. [CrossRef]
75. Stuart, A.M.; Pame, A.R.P.; Vithoonjit, D.; Viriyangkura, L.; Pithuncharurnlap, J.; Meesang, N.; Suksiri, P.; Singleton, G.R.; Lampayan, R.M. The application of best management practices increases the profitability and sustainability of rice farming in the central plains of Thailand. *Field Crops Res.* **2018**, *220*, 78–87. [CrossRef]
76. Metzger, M.J. *The Environmental Stratification of Europe, [Dataset]*; University of Edinburgh: Edinburgh, UK, 2018. [CrossRef]
77. Oenema, O.; Heinen, M.; Rietra, R.; Hessel, R. A Review of Soil-Improving Cropping Systems (Full Report). SoilCare Scientific Report 07, Deliverable D2.1, SoilCare Project, Wageningen Environmental Research, The Netherlands. 2017. Available online: https://ec.europa.eu/research/participants/documents/downloadPublic?documentIds=080166e5b4c22812&appId=PPGMS (accessed on 13 April 2022).
78. Panagea, I.; Dangol, A.; Olijslagers, M.; Wyseure, G. *SoilCare Database 3: Schema (Empty Database) and Report 34 (D5.1): Database with Monitoring data (1.1)*; Zenodo: Geneva, Switzerland, 2021. [CrossRef]
79. Panagea, I.; Wyseure, G.; Hessel, R. Report on Monitoring Results and Analysis. SoilCare Report 35, p. 616, SoilCare Project, Wageningen Environmental Research, The Netherlands. 2021. Available online: https://research.wur.nl/en/publications/report-on-monitoring-results-and-analysis-d53 (accessed on 13 April 2022).
80. Alaoui, A.; Hallama, M.; Bär, R.; Panagea, I.; Bachmann, F.; Pekrun, C.; Fleskens, L.; Kandeler, E.; Hessel, R. A New Framework to Assess Sustainability of Soil Improving Cropping Systems in Europe. *Land* **2022**, *11*, 729. [CrossRef]
81. Van Delden, H.; Fleskens, L.; Muro, M.; Tugran, T.; Vanhout, R.; Baartman, J.; Nunes, J.P.; Vanermen, I.; Salputra, G.; Verzandvoort, S.; et al. Report on the Potential for Applying Soil-Improving CS across Europe SoilCare Report 43, p. 224. 2021. Available online: https://soilcare-project.eu/downloads/public-documents/soilcare-reports-and-deliverables/433-report-43-d6-2-report-on-the-potential-for-applying-sics-across-europe-riks-full/file (accessed on 13 April 2022).
82. Van Delden, H.; Fleskens, F.; Vanhout, R.; Nunes, J.P.; Baartman, J.; Lesschen, J.P.; Verzandvoort, S.; Hessel, R.; All Study Site Partners. Report on the Integration and Synthesis of Study Site Results and Their Potential for Upscaling. SoilCare Report 42, p. 197. 2021. Available online: https://library.wur.nl/WebQuery/wurpubs/fulltext/570318 (accessed on 13 April 2022).
83. Van Delden, H.; Vanhout, R.; Fleskens, L.; Nunes, J.P.; Baartman, J.; Verzandvoort, S.; Hessel, R.; All Study Site Partners. Interactive Mapping Tool for the Application of Soil Improving Cropping Systems across Europe. SoilCare Report 44, p. 16. 2021. Available online: https://www.soilcare-project.eu/downloads/public-documents/soilcare-reports-and-deliverables/434-report-44-d6-3-interactive-mapping-tool-for-the-application-of-sics-across-europe-riks-full/file (accessed on 13 April 2022).
84. Salamon, P.; Banse, M.; Donnellan, T.; Hass, M.; Jongeneel, R.; Laquai, V.; van Leeuwen, M.; Reziti, I.; Salputra, G.; Zirngibl, M.-E. *AGMEMOD Outlook for Agricultural and Food Markets in EU Member States 2018–2030*; Thünen Working Paper 114; Johann Heinrich von Thünen-Institut: Braunschweig, Germany, 2019. [CrossRef]
85. Van Delden, H.; Hurkens, J. A generic Integrated Spatial Decision Support System for urban and regional planning. In Proceedings of the 19th International Congress on Modelling and Simulation, Perth, Australia, 12–16 December 2011.
86. Kirkby, M.J.; Irvine, B.J.; Jones, R.J.A.; Govers, G.; PESERA Team. The PESERA coarse scale erosion model for Europe. I.—Model rationale and implementation. *Eur. J. Soil Sci.* **2008**, *59*, 1293–1306. [CrossRef]

87. Fleskens, L.; Baartman, J.; Van Delden, H.; Vanhout, R. *Madagascar: Land Use Planning for Enhanced Resilience of Landscapes (LAUREL)*; Final Report National LANDSIM-P; World Bank Project; Wageningen University: Wageningen, The Netherlands, 2020; p. 150.
88. Velthof, G.L.; Oudendag, D.; Witzke, H.P.; Asman, W.A.H.; Klimont, Z.; Oenema, O. Integrated assessment of nitrogen emissions from agriculture in EU-27 using MITERRA-EUROPE. *J. Environ. Qual.* **2009**, *38*, 402–417. [CrossRef]
89. Glaesner, N.; Helming, K.; de Vries, W. Do current European policies prevent soil threats and support soil functions? *Sustainability* **2014**, *6*, 9538–9563. [CrossRef]
90. Vrebos, D.; Bampa, F.; Creamer, R.E.; Gardi, C.; Ghaley, B.B.; Jones, A.; Rutgers, M.; Sandén, T.; Staes, J.; Meire, P. The Impact of Policy Instruments on Soil Multifunctionality in the European Union. *Sustainability* **2017**, *9*, 407. [CrossRef]
91. Rust, N.; Ptak, E.N.; Graversgaard, M.; Iversen, S.; Reed, M.S.; de Vries, J.; Ingram, J.; Mills, J.; Neumann, R.; Kjeldsen, C.; et al. Social capital factors affecting uptake of Soil-Improving management practices. A review. *Emerald Open Res. Sustain. Food Syst.* **2020**, *2*, 8. [CrossRef]
92. Leeuwis, C. *Communication for Rural Innovation: Rethinking Agricultural Extension with Contributions from Anne van den Ban*; Blackwell Science Limited: Oxford, UK, 2004; ISBN 0-632-05249-X.
93. Duesberg, S.; Dhubháin, A.N.; O'Connor, D. Assessing policy tools for encouraging farm afforestation in Ireland. *Land Use Policy* **2014**, *38*, 194–203. [CrossRef]
94. Greiner, R.; Gregg, D. Farmers' intrinsic motivations, barriers to the adoption of conservation practices and effectiveness of policy instruments: Empirical evidence from northern Australia. *Land Use Policy* **2011**, *28*, 257–265. [CrossRef]
95. Bavorová, M.; Unay-Gailhard, I.; Ponkina, E.V.; Pilařová, T. How sources of agriculture information shape the adoption of reduced tillage practices? *J. Rural Stud.* **2020**, *79*, 88–101. [CrossRef]
96. Fantappiè, M.; Lorenzetti, R.; De Meo, I.; Costantini, E.A.C. How to improve the adoption of soil conservation practices? Suggestions from farmers' perception in western Sicily. *J. Rural Stud.* **2020**, *73*, 186–202. [CrossRef]
97. Sattler, C.; Nagel, U.J. Factors affecting farmers' acceptance of conservation measures—A case study from north-eastern Germany. *Land Use Policy* **2010**, *27*, 70–77. [CrossRef]
98. Trujillo-Barrera, A.; Pennings, J.M.E.; Hofenk, D. Understanding producers' motives for adopting sustainable practices: The role of expected rewards, risk perception and risk tolerance. *Eur. Rev. Agric. Econ.* **2016**, *43*, 359–382. [CrossRef]
99. Dessart, F.J.; Barreiro-Hurlé, J.; van Bavel, R. Behavioural factors affecting the adoption of sustainable farming practices: A policy-oriented review. *Eur. Rev. Agric. Econ.* **2019**, *46*, 417–471. [CrossRef]
100. Sutherland, L.-A.; Darnhofer, I. Of organic farmers and 'good farmers': Changing habitus in rural England. *J. Rural Stud.* **2012**, *28*, 232–240. [CrossRef]
101. Schneider, F.; Ledermann, T.; Fry, P.; Rist, S. Soil conservation in Swiss agriculture—approaching abstract and symbolic meanings in farmers' life-worlds. *Land Use Policy* **2010**, *27*, 332–339. [CrossRef]
102. Schneider, F.; Rist, S. The significance of aesthetics for the adoption of no-tillage farming. *Agrar. Schweiz* **2012**, *3*, 216–223.
103. Reed, M.S.; Vella, S.; Challies, E.; de Vente, J.; Frewer, L.; Hohenwallner-Ries, D.; Huber, T.; Neumann, R.K.; Oughton, E.A.; Sidoli del Ceno, J.; et al. A theory of participation: What makes stakeholder and public engagement in environmental management work? *Restor. Ecol.* **2010**, *27*, 332–339. [CrossRef]
104. Rust, N.; Lunder, O.E.; Iversen, S.; Vella, S.; Oughton, E.A.; Breland, T.A.; Glass, J.H.; Maynard, C.M.; McMorran, R.; Reed, M.S. Perceived Causes and Solutions to Soil Degradation in the UK and Norway. *Land* **2022**, *11*, 131. [CrossRef]
105. Mills, J.; Reed, M.; Skaalsveen, K.; Ingram, J. The use of Twitter for knowledge exchange on sustainable soil management. *Soil Use Manag.* **2019**, *35*, 195–203. [CrossRef]
106. Rust, N.; Iversen, S.; Vella, S.; Hansda, R.; Reed, M.; Areal, F. Social Factors Influencing Adoption. SoilCare Report 12, 116p. 2021. Available online: https://www.soilcare-project.eu/downloads/public-documents/soilcare-reports-and-deliverables/130-report-12-d3-3-section-problems-causing-and-solutions-to-declining-soil-quality-in-the-uk-niki-rust-nu/file (accessed on 13 April 2022).
107. Rust, N.A.; Stankovics, P.; Jarvis, R.M.; Morris-Trainor, Z.; de Vries, J.R.; Ingram, J.; Mills, J.; Glikman, J.A.; Parkinson, J.; Toth, Z.; et al. Have farmers had enough of experts? *Environ. Manag.* **2021**, *69*, 31–44. [CrossRef]
108. Usowicz, B.; Lipiec, J. Spatial variability of saturated hydraulic conductivity and its links with other soil properties at the regional scale. *Sci. Rep.* **2021**, *11*, 8293. [CrossRef]
109. Hoffmann, S.; Kismányoky, T. Soil fertility in a long-term fertilizer trial with different tillage systems. *Arch. Agron. Soil Sci.* **2001**, *46*, 239–250. [CrossRef]
110. Panagea, I.S.; Berti, A.; Čermak, P.; Diels, J.; Elsen, A.; Kusá, H.; Piccoli, I.; Poesen, J.; Stoate, C.; Tits, M.; et al. Soil Water Retention as Affected by Management Induced Changes of Soil Organic Carbon: Analysis of Long-Term Experiments in Europe. *Land* **2021**, *10*, 1362. [CrossRef]
111. Mühlbachová, G.; Kusá, H.; Růžek, P. Soil characteristics and crop yields under different tillage techniques. *Plant Soil Environ.* **2015**, *61*, 566–572. [CrossRef]
112. Piccoli, I.; Seehusen, T.; Bussell, J.; Vizitu, O.; Calciu, I.; Berti, A.; Börjesson, G.; Kirchmann, H.; Kätterer, T.; Sartori, F.; et al. Opportunities for mitigating soil compaction in Europe—Case studies from the SoilCare project using soil improving cropping systems. *Land* **2022**, *11*, 223. [CrossRef]

113. Bussell, J.; Crotty, F.; Stoate, C. Comparison of Compaction Alleviation Methods on Soil Health and Greenhouse Gas Emissions. *Land* **2021**, *10*, 1397. [CrossRef]
114. Sartori, F.; Piccoli, I.; Polese, R.; Berti, A. A Multivariate Approach to Evaluate Reduced Tillage Systems and Cover Crop Sustainability. *Land* **2022**, *11*, 55. [CrossRef]
115. Kismányoky, T.; Tóth, Z. Effect of mineral and organic fertilization on soil organic carbon content as well as on grain production of cereals in the IOSDV (ILTE) long-term field experiment, Keszthely, Hungary. *Arch. Agron. Soil Sci.* **2013**, *59*, 1121–1131. [CrossRef]
116. De Notaris, C.; Jensen, J.L.; Olesen, J.E.; Stumpf da Silva, T.; Rasmussen, J.; Panagea, I.; Rubæk, G.H. Long-term soil quality effects of soil and crop management in organic and conventional arable cropping systems. *Geoderma* **2021**, *403*, 115383. [CrossRef]
117. Piccoli, I.; Sartori, F.; Polese, R.; Berti, A. Crop yield after 5 decades of contrasting residue management. *Nutr. Cycl. Agroecosystems* **2020**, *117*, 231–241. [CrossRef]
118. Hannula, S.E.; Di Lonardo, D.P.; Christensen, B.T.; Crotty, F.V.; Elsen, A.; Erp, P.J.; Hansen, E.M.; Rubæk, G.H.; Tits, M.; Toth, Z.; et al. Inconsistent effects of agricultural practices on soil fungal communities across twelve European long-term experiments. *Eur. J. Soil Sci.* **2021**, *72*, 1902–1923. [CrossRef]
119. Frąc, M.; Pertile, G.; Panek, J.; Gryta, A.; Oszust, K.; Lipiec, J.; Usowicz, B. Mycobiome Composition and Diversity under the Long-Term Application of Spent Mushroom Substrate and Chicken Manure. *Agronomy* **2021**, *11*, 410. [CrossRef]
120. Lipiec, J.; Usowicz, B.; Kłopotek, J.; Turski, M.; Frąc, M. Effects of Application of Recycled Chicken Manure and Spent Mushroom Substrate on Organic Matter, Acidity, and Hydraulic Properties of Sandy Soils. *Materials* **2021**, *14*, 4036. [CrossRef]
121. Hallama, M.; Pekrun, C.; Lambers, H.; Kandeler, E. Hidden miners—The roles of cover crops and soil microorganisms in phosphorus cycling through agroecosystems. *Plant Soil* **2019**, *434*, 7–45. [CrossRef]
122. Hallama, M.; Pekrun, C.; Pilz, S.; Jarosch, K.A.; Frąc, M.; Uksa, M.; Marhan, S.; Kandeler, E. Interactions between cover crops and soil microorganisms increase phosphorus availability in conservation agriculture. *Plant Soil* **2021**, *463*, 307–328. [CrossRef]
123. Christensen, J.T.; Hansen, E.M.; Kandeler, E.; Hallama, M.; Christensen, B.; Rubæk, G.H. Effect of soil P status on barley growth, P uptake and soil microbial properties after incorporation of cover crop shoot and root residues. *J. Plant Nutr. Soil Sci.* **2021**, *184*, 657–667. [CrossRef]
124. Wirsching, J.; Wimmer, B.; Ditterich, F.; Schlögl, J.; Martin-Laurent, F.; Huhn, C.; Haderlein, S.; Kandeler, E.; Poll, C. ^{13}C assimilation as well as functional gene abundance and expression elucidate the biodegradation of glyphosate in a field experiment. *Environ. Pollut.* **2022**, *under revision*.
125. Tsanis, I.K.; Seiradakis, K.D.; Sarchani, S.; Panagea, I.S.; Alexakis, D.D.; Koutroulis, A.G. The Impact of Soil-Improving Cropping Practices on Erosion Rates: A Stakeholder-Oriented Field Experiment Assessment. *Land* **2021**, *10*, 964. [CrossRef]
126. Knapp, S.; van der Heijden, M.G. A global meta-analysis of yield stability in organic and conservation agriculture. *Nat. Commun.* **2018**, *9*, 3632. [CrossRef] [PubMed]
127. Raseduzzaman, M.D.; Jensen, E.S. Does intercropping enhance yield stability in arable crop production? A meta-analysis. *Eur. J. Agron.* **2017**, *91*, 25–33. [CrossRef]
128. Smith, O.M.; Cohen, A.L.; Rieser, C.J.; Davis, A.G.; Taylor, J.M.; Adesanya, A.W.; Jones, M.S.; Meier, A.R.; Renagold, J.P.; Orpet, R.J.; et al. Organic farming provides reliable environmental benefits but increases variability in crop yields: A global meta-analysis. *Front. Sustain. Food Syst.* **2019**, *3*, 82. [CrossRef]
129. Johnston, A.E.; Poulton, P.R. The importance of long-term experiments in agriculture: Their management to ensure continued crop production and soil fertility; the Rothamsted experience. *Eur. J. Soil Sci.* **2018**, *69*, 113–125. [CrossRef]
130. Rubio, V.; Diaz-Rossello, R.; Quincke, J.A.; van Es, H.M. Quantifying soil organic carbon's critical role in cereal productivity losses under annualized crop rotations. *Agric. Ecosyst. Environ.* **2021**, *321*, 107607. [CrossRef]
131. Thierfelder, C.; Mhlanga, B. Short-term yield gains or long-term sustainability?—A synthesis of Conservation Agriculture long-term experiments in Southern Africa. *Agric. Ecosyst. Environ.* **2022**, *326*, 107812. [CrossRef]
132. Behnke, G.D.; Zuber, S.M.; Pittelkow, C.M.; Nafziger, E.D.; Villamil, M.B. Long-term crop rotation and tillage effects on soil greenhouse gas emissions and crop production in Illinois, USA. *Agric. Ecosyst. Environ.* **2018**, *261*, 62–70. [CrossRef]
133. Huynh, H.T.; Hufnagel, J.; Wurbs, A.; Bellingrath-Kimura, S.D. Influences of soil tillage, irrigation and crop rotation on maize biomass yield in a 9-year field study in Müncheberg, Germany. *Field Crops Res.* **2019**, *241*, 107565. [CrossRef]
134. Norris, C.E.; Congreves, K.A. Alternative management practices improve soil health indices in intensive vegetable cropping systems: A review. *Front. Environ. Sci.* **2018**, *6*, 50. [CrossRef]
135. Borrelli, P.; Van Oost, K.; Meusburger, K.; Alewell, C.; Lugato, E.; Panagos, P. A step towards a holistic assessment of soil degradation in Europe: Coupling on-site erosion with sediment transfer and carbon fluxes. *Environ. Res.* **2018**, *161*, 291–298. [CrossRef]
136. Bartkowski, B.; Bartke, S.; Helming, K.; Paul, C.; Techen, A.K.; Hansjürgens, B. Potential of the economic valuation of soil-based ecosystem services to inform sustainable soil management and policy. *PeerJ* **2020**, *8*, e8749. [CrossRef]
137. Tamburini, G.; Bommarco, R.; Wanger, T.C.; Kremen, C.; van der Heijden, M.G.A.; Liebman, M.; Hallin, S. Agricultural diversification promotes multiple ecosystem services without compromising yield. *Sci. Adv.* **2020**, *6*, eaba1715. [CrossRef]
138. Francaviglia, R.; Álvaro-Fuentes, J.; Di Bene, C.; Gai, L.; Regina, K.; Turtola, E. Diversification and Management Practices in Selected European Regions. A Data Analysis of Arable Crops Production. *Agronomy* **2020**, *10*, 297. [CrossRef]
139. Hewitt, A.; Dominati, E.E.; Webb, T.T.; Cuthill, T. Soil natural capital quantification by the stock adequacy method. *Geoderma* **2015**, *241–242*, 107–114. [CrossRef]

140. Bouma, J.; Montanarella, L. Facing policy challenges with inter- and transdisciplinary soil research focused on the UN Sustainable Development Goals. *Soil* **2016**, *2*, 135–145. Available online: https://www.soil-journal.net/2/135/2016/soil-2-135-2016.pdf (accessed on 13 April 2022). [CrossRef]
141. McNeill, A.; Muro, M.; Tugran, T.; Lukacova, Z. Report on the Selection of Good Policy Alternatives at EU and Study Site Level. SoilCare Report 13, p. 144. 2021. Available online: https://www.soilcare-project.eu/downloads/public-documents/soilcare-reports-and-deliverables/186-report-13-d7-2-milieu-full-v2/file (accessed on 13 April 2022).
142. Ingram, J.; Mills, J. Are advisory services "fit for purpose" to support sustainable soil management? An assessment of advice in Europe. *Soil Use Manag.* **2019**, *35*, 21–31. [CrossRef]
143. Aznar-Sánchez, J.A.; Velasco-Muñoz, J.F.; López-Felices, B.; del Moral-Torres, F. Barriers and Facilitators for Adopting Sustainable Soil Management Practices in Mediterranean Olive Groves. *Agronomy* **2020**, *10*, 506. [CrossRef]

land

Article

The Effects of Soil Improving Cropping Systems (SICS) on Soil Erosion and Soil Organic Carbon Stocks across Europe: A Simulation Study

Jantiene E. M. Baartman [1,*], Joao Pedro Nunes [1], Hedwig van Delden [2], Roel Vanhout [2] and Luuk Fleskens [1]

[1] Soil Physics and Land Management Group, Wageningen University, P.O. Box 47, 6700 AA Wageningen, The Netherlands; joao.carvalhonunes@wur.nl (J.P.N.); luuk.fleskens@wur.nl (L.F.)
[2] Research Institute for Knowledge Systems, P.O. Box 463, 6200 AL Maastricht, The Netherlands; hvdelden@riks.nl (H.v.D.); rvanhout@riks.nl (R.V.)
* Correspondence: jantiene.baartman@wur.nl

Abstract: Healthy soils are fundamental for sustainable agriculture. Soil Improving Cropping Systems (SICS) aim to make land use and food production more sustainable. To evaluate the effect of SICS at EU scale, a modelling approach was taken. This study simulated the effects of SICS on two principal indicators of soil health (Soil Organic Carbon stocks) and land degradation (soil erosion) across Europe using the spatially explicit PESERA model. Four scenarios with varying levels and combinations of cover crops, mulching, soil compaction alleviation and minimum tillage were implemented and simulated until 2050. Results showed that while in the scenario without SICS, erosion slightly increased on average across Europe, it significantly decreased in the scenario with the highest level of SICS applied, especially in the cropping areas in the central European Loess Belt. Regarding SOC stocks, the simulations show a substantial decrease for the scenario without SICS and a slight overall decrease for the medium level scenario and the scenario with a mix of high, medium and no SICS. The scenario with a high level of SICS implementation showed an overall increase in SOC stocks across Europe. Potential future improvements include incorporating dynamic land use, climate change and an optimal spatial allocation of SICS.

Keywords: large-scale modelling; Europe; soil health; SOC stocks; soil erosion; scenarios; sustainable soil management

Citation: Baartman, J.E.M.; Nunes, J.P.; van Delden, H.; Vanhout, R.; Fleskens, L. The Effects of Soil Improving Cropping Systems (SICS) on Soil Erosion and Soil Organic Carbon Stocks across Europe: A Simulation Study. *Land* **2022**, *11*, 943. https://doi.org/10.3390/land11060943

Academic Editors: Guido Wyseure, Julián Cuevas González and Jean Poesen

Received: 28 April 2022
Accepted: 14 June 2022
Published: 19 June 2022

1. Introduction

A well-functioning, healthy soil is fundamental for sustainable agriculture. Soil quality and soil health are increasingly considered important topics on the political and public agenda (e.g., [1,2]), and are also getting attention in the scientific community (e.g., [3,4]). This is reflected in, among others, the Sustainable Development Goals (SDGs; https://sdgs.un.org/goals (accessed on 12 April 2022)), where soil together with land use and management play an important role in SDG 1 (no poverty), 2 (zero hunger), 12 (responsible consumption and production), 13 (climate action) and especially SDG 15 (Life on Land) [2,5]. Moreover, in the current Farm to Fork Strategy (F2F, [6]), as part of the European Green Deal [7], sustainable food production is an important goal; the F2F aims at neutral or positive environmental impact, mitigating climate change, reversing the loss of biodiversity and ensuring food security. Land use and land management play a key role in achieving these policy aims and reversing the current trend of land degradation [8]. For example, the F2F strategy targets to 'bring back at least 10% of agricultural areas under high-diversity landscape features (with buffer strips, rotational or non-rotational fallow land, hedges, non-productive trees, terrace walls and ponds)' and 'have 25% of the EU's agricultural land as organic farming by 2030' [9]. These strategies are also developed as the costs of unsustainable land management are estimated to exceed €50 billion per year [10].

The measures mentioned in the F2F strategy are only a few of the very many existing land management options to improve soil health and reverse or prevent land degradation, ranging from farm and field to village and watershed or community scales (e.g., [11,12] and https://qcat.wocat.net/en/wocat/ (accessed on 12 April 2022)). Among those many options, some measures are common in annual and perennial agriculture across Europe. For example, maintaining a (winter) cover crop is widely applied [13–15]. No-tillage or minimum tillage has been estimated to be applied on 25% of the agricultural land in the EU [16]. Mulching is applied to reduce splash erosion and increase soil moisture [17,18]. Crop residue management [19] and/or maintaining a minimum soil cover is also widely applied [12,17]. Grass strips are applied at field borders [20] to reduce runoff and catch sediments [18,21] and as a means to reduce leaching of nutrients [21,22] and/or pesticides [23]. Rodrigues et al. [19] for example show that reduced tillage and soil protective measures can play an important role in soil carbon sequestration across the EU. Maetens et al. [18] investigated the effect of various soil and water conservation measures on runoff and soil loss across Europe.

These practices affect the farming and cropping systems, aiming to make land use and food production more sustainable. As defined in Hessel et al. [24], cropping system refers to crop type, crop rotation and the agronomic management techniques used. Soil improving cropping systems (SICS) can be defined as cropping systems that result in a durable increased ability of the soil to fulfil its functions, including food and biomass production, buffering and filtering capacity and provision of other ecosystem services [24]. However, the uptake and choice of SICS will vary due to external factors, such as EU policies, market effects, society and pedo-climatic conditions. In addition, these factors are dynamic in time as they are affected by e.g., climate change, geo-politics, consumer purchase power and preferences, technological advances and other developments [25,26]. Hence, when assessing the effects of SICS on improving soil health and combatting land degradation at continental scale it is important to consider divergent trends in these factors that affect the uptake of SICS (e.g., [26,27]).

Soil health and land degradation are both broad terms [2] that include many aspects. Soil health has been defined as the continued capacity of soil to function as a vital living ecosystem that sustains plants, animals and humans [3]; encompasses biological productivity, soil life and biodiversity; enhances its role in water quality and regulation and mitigates climate change. Similarly, land degradation entails many different processes, such as salinisation, nutrient depletion, dehydration, erosion by water or wind, compaction, soil pollution, loss of soil organic matter and soil biodiversity etc., [28–31]. In this study, we focused on one principal indicator for each aspect: Soil Organic Carbon (SOC) as a principal indicator for soil health [32,33] and soil erosion (by water) as an indicator and widely occurring process of land degradation [34,35]. Moreover, Kutter et al. [12] in their review on policy measures for agricultural soil conservation in the EU, found that most measures focused on erosion by water, followed by decline in organic matter.

Upscaling the assessment of the impact of measures from e.g., field or farm level to country or wider (e.g., EU) scale is challenging as measuring at this scale is infeasible [34]. Modelling is a common approach and can also include simulation of scenarios of e.g., climate change effects and policy adoption [36,37]. At EU wide scale, soil erosion was estimated by Panagos et al. [34], based on the RUSLE approach. EU wide SOC estimates include e.g., [38–40]. The RUSLE-based erosion estimates by Panagos et al. [34] also include the effect of mitigation options such as conservation tillage, plant residues and winter crop cover [16] and contour farming, stone walls and grass margins [41]. Modelling estimates of climate change and land use change effects on SOC are abundant, e.g., [42–45], and various studies quantified the effects of agricultural practices on carbon sequestration [46–48]. Lugato et al. [49] included straw incorporation, reduced tillage, their combination, ley cropping systems and cover crops into their spatially explicit modelling scenarios.

The SoilCare project (https://www.soilcare-project.eu/ (accessed on 12 April 2022) [24]) aimed to identify and evaluate promising soil improving cropping systems and agronomic techniques that increase the profitability and sustainability of agriculture across Europe. In addition to field trials [50,51], the project used a modelling approach to upscale the effects of SICS to EU scale, in a spatially explicit way. To ensure that sufficient healthy food for expanding human populations can be grown within planetary boundaries [52], soil management should aim at improving the health and resilience of land and soil [8]. In this study we evaluated how soil improving cropping systems (SICS) impact land degradation (specifically erosion) and soil health (specifically SOC stocks) across Europe, through the application of the PESERA model. For this purpose, we improved and further developed the PESERA model both in terms of input data improvements and in parameterisation and calibration of SICS and a range of crops, in four climate zones. Moreover, to be able to assess the impacts of SICS, existing land management options have been adapted in the model. Four scenarios, developed within the SoilCare project, were simulated until 2050, with varying application of (combinations of) SICS in each scenario.

2. Methods

2.1. PESERA Model Description

The Pan-European Soil Erosion Risk Assessment (PESERA) model simulates biophysical processes including above-ground biomass production, soil erosion risk, soil water deficit and soil humus content, using a monthly time-step. The model was originally developed by Kirkby et al., [53] and has been applied in various agro-ecological zones e.g., [54–57]. A brief technical description is given here, based on Kirkby et al. [53], where all details can be found. PESERA is a process-based and spatially distributed model which combines the effect of topography, climate and soil properties. A schematic model structure is provided in Figure 1. The model has three conceptual stages: (i) A storage threshold model to convert daily rainfall to daily total overland flow runoff; (ii) a power law to estimate sediment transport from runoff and slope gradient. The model interprets sediment transported to the base of a hillslope as average erosion loss. No flow or sediment routing over multiple cells is included; and (iii) integration of daily rates over the frequency distribution of daily rainfalls to estimate monthly erosion rates.

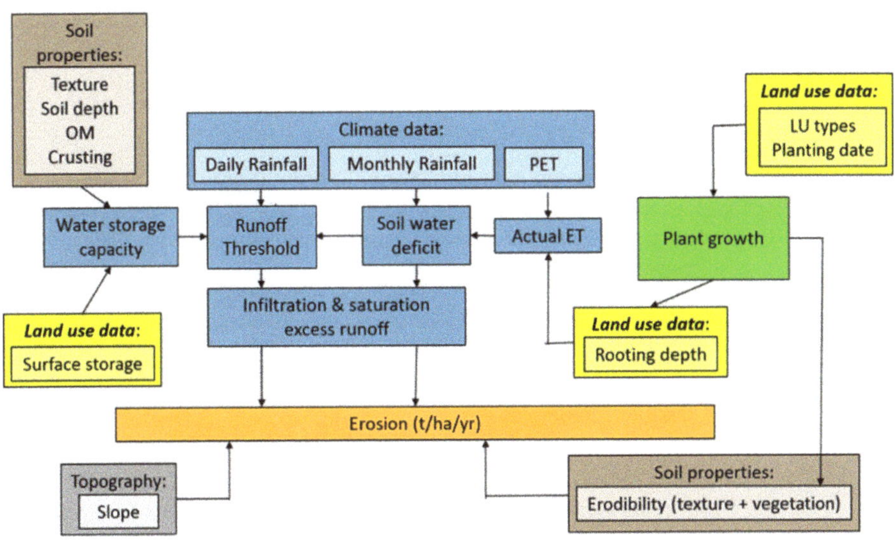

Figure 1. Schematic overview of processes in the PESERA model.

In the first step, a simple storage or bucket model is used to convert daily rainfall into daily runoff, which is estimated as the rainfall minus the threshold storage. The threshold storage depends dynamically on soil properties, vegetation cover and soil moisture status, varying over the year. The most important soil factors that determine the threshold storage beneath the vegetation-covered fraction of the surface are texture, depth (if shallow) and organic matter. Where the surface is not protected by vegetation, the susceptibility of the soil to crusting and the duration of crusting conditions generally determine a lower threshold. The final threshold is a weighted average from vegetated and bare fractions of the surface. Corrections are made for the soil water deficit, which may reduce the threshold where the soil is close to saturation. Transpiration is used to drive a generic plant growth model for biomass, constrained as necessary by land use decisions, primarily on a monthly time step. Leaf fall also drives a simple model to estimate soil organic matter.

Precipitation is divided into daily storm events, expressed as a frequency distribution. The distribution of daily rainfall totals is fitted to a Gamma distribution for each month. The rainfall distribution, reflected by the coefficient of variation of rainfall per rain day is given for each month of the simulation period and may be adapted for (future) climate change scenarios. Daily precipitation drives infiltration, excess overland flow and soil erosion, and monthly precipitation, driving saturation levels in the soil. Infiltration excess overland flow is estimated from storm rainfall and soil moisture. Sediment transport is then estimated using a power law approach driven by erodibility, gradient and runoff discharge. Soil erodibility is derived from soil classification data, primarily texture (see Section 2.2.7). Local relief is defined as the standard deviation of elevation within a defined radius around each point (Section 2.2.1 and Figure 2). Accumulated runoff is derived from a biophysical model that combines the frequency of daily storm sizes with an assessment of runoff thresholds based on seasonal water deficit and vegetation growth. Estimates of sediment transport are based on infiltration excess overland flow discharge. In the PESERA model, sediment transport is interpreted as the mean sediment yield delivered to stream channels and includes no downstream routing within the channel network.

The role of vegetation and soil organic matter can modify the infiltration rates through changes in soil structure and/or the development over time of surface or near-surface crusting. Three models are coupled to provide the dynamics of these responses: (i) A vertical hydrological balance, which partitions precipitation between evapotranspiration, overland flow, subsurface flow and changes in soil moisture; (ii) a vegetation growth model, which budgets living biomass and organic matter subject to the constraints of land use and cultivation choices; and (iii) a soil model, which estimates the required hydrological variables from moisture, vegetation and seasonal rainfall history.

The PESERA model works with two phases: an equilibrium phase and a simulation phase. The equilibrium phase model is run first: it calculates long-term average values, using long-term input data on e.g., climate. The equilibrium phase model is calibrated using long-term average data (see Section 2.3). Then, these long-term output maps are used to initiate the simulation phase model. This model uses monthly climate data to run future scenarios (see Section 2.4).

PESERA outputs consist of monthly maps of: vegetation biomass (ton/ha), erosion (risk) (ton/ha/y) and soil organic matter content (ton/ha) for each simulation year. Within the SoilCare project the following improvements were made in the PESERA model: additional crop types (sugar beets, rice, fodder versus consumption maize) have been parameterised and calibrated for Europe; all crops were parameterised and calibrated for four main climate zones across Europe and biomass and SOM were calibrated for each land use/crop type; irrigation has been added as an option in the model; erodibility information for the Northern countries (Norway, Sweden and Iceland) has been updated to solve issues with existing Europe-wide data (see Section 2.2) and soil management options (i.e., SICS) have been defined, parameterised and calibrated (see Section 2.3).

2.2. Input Data

The required model input data and their sources are summarised in Table 1. All input maps have a spatial resolution of 500 m and projection ETRS 1989 LAEA (Lambert

Azimuthal Equal Area). The area modelled is the EU-28, i.e., the current 27 EU countries plus UK. Basic details and the most important maps are given here; a full description and all input maps are given in Supplementary Material S1.

Table 1. Overview of PESERA input requirements.

Category	Variable	Number of Maps	Data Source
Topography	Local relief—st. dev. of elevation	1	ESDAC database (RECARE project)
Climate	*Equilibrium phase model (long-term current climate)*		
	Mean monthly temperature	12	Based on E-OBS version 21.0e, at 0.1° spatial resolution and daily scale. [58]. 1981–2010
	Mean monthly temperature range	12	
	Mean monthly rainfall	12	
	Mean monthly rainfall per rain day	12	
	Coefficient of variation of mean monthly rainfall per rain day	12	
	Mean monthly PET	12	Calculated from monthly Tmean and Trange following [59].
	Simulation phase model (climate scenarios)		
	Mean monthly temperature	12 * n_years	E-OBS version 21.0e, at 0.1° spatial resolution and daily scale. [58]. 2018–2050; RCP4.5 MPI-ES-LR + CCLM4-8-17 Data: JRC EU High Resolution and Precipitation dataset: https://data.jrc.ec.europa.eu/dataset/jrc-liscoast-10011 (accessed on 18 December 2020) [60]
	Mean monthly temperature range	12 * n_years	
	Monthly rainfall	12 * n_years	
	Maximum daily rainfall	12 * n_years	
Soil properties	Erodibility class (sensitivity to erosion)	1	Classified RUSLE K-factor map by Panagos et al. [61] https://esdac.jrc.ec.europa.eu/content/soil-erodibility-k-factor-high-resolution-dataset-europe (accessed on 1 July 2021)
	Crusting class (sensitivity to soil surface crusting)	1	Pedotransfer functions based on soil type and texture (ESDB)
	Scale depth (proxy for infiltration)	1	Based on Texture classes (ESDB)
	Soil water available to plants (0–300 mm)	1	Pedotransfer functions based on Available Water Content, Texture, Soil packing density and restriction of soil to bedrock; ESDB and SWAT-HWSD [62] for Iceland and Cyprus
	Soil water available to plants (300–1000 mm)	1	
	Effective soil water storage capacity	1	
Land use & crop data	Land use map	1	From Metronamica application, processed data from Eurostat and Corine Land Cover
	Crop map	1	
	Planting month (for crops only)	1	Grouped per climate region (see Table 2)
	Initial ground cover (%)	12	Following PESERA project manual estimations; adapted where needed

Table 1. *Cont.*

Category	Variable	Number of Maps	Data Source
	Initial surface storage (mm)	1	Following PESERA project manual
	Surface storage reduction (%)	1	Following PESERA project manual: 50% for crops, 0% for other land uses
	Rooting depth	1	Combined approach following PESERA project manual, FAO data http://www.fao.org/land-water/databases-and-software/crop-information/maize/en/ (accessed on 15 October 2020) and SWAT database.

2.2.1. Topography

One of the main variables in the model is local relief (Figure 2). It is estimated from the digital elevation model (DEM) as the standard deviation of elevation with a circle of 1.5 km (5 cell radius) diameter around each cell.

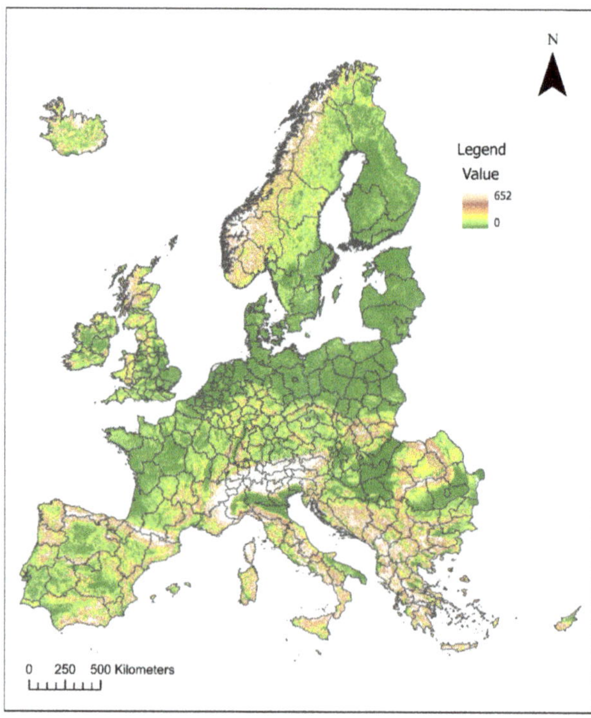

Figure 2. Local relief (standard deviation of elevation in a 1500 m radius) for Europe.

2.2.2. Climate

Climate input data differs slightly between the equilibrium and simulation phase models. For the equilibrium phase model, E-OBS version 21.0e data, at 0.1° spatial resolution and daily scale was used. Daily data for the ensemble mean of mean temperature, minimum temperature, maximum temperature and rainfall were collected for 1981–2010, representing the reference period used to bias-correct climate scenarios. The monthly parameters shown in Table 1 were calculated from these values, after being interpolated to a 500 m resolution. The data source is Cornes et al. [58]. Maps of the equilibrium climate input data are presented in Supplementary Material S1.

To minimise bias, climate scenarios at high resolution (0.1°), and already bias corrected with present-day climate (E-OBS) were used in the simulation phase model. The considered emission scenario was RCP4.5 (closer to the average of all emission scenarios). The selected GCM-RCM combination was MPI-ES-LR + CCLM4-8-17. This means that we used the MPI-ES-LR GCM, which has a median sensitivity to climate change [63] combined with the CCLM RCM, which appears to have less bias for temperature and rainfall in several European regions [64]. We used data from the JRC EU High Resolution and Precipitation dataset, which is already bias-corrected using E-OBS [60].

2.2.3. Land Use and Crop Data

The land use and crop map (Figure 3) was made within the SoilCare project, based on Corine Land Cover 2018 (CLC2018) (https://land.copernicus.eu/pan-european/corine-land-cover (accessed on 15 September 2021)) crop data from Eurostat (https://ec.europa.eu/eurostat (accessed on 15 September 2021)), and infrastructure (e.g., roads), zoning (e.g., protected natural areas, urban expansion plans) and crop suitability maps from Metronamica. Details on how these data were used to derive the SoilCare land use and crop map are given in [65].

2.2.4. Crop Calendars: Planting Month, WUE and Cover

As crop calendars for the same crop may differ per climate region, we created four major agro-climatic regions in Europe, for which crop calendars were constructed for each crop. We did not use existing maps for cropping calendars, as they are either too coarse [66], not crop-specific [67], or represent related variables which are difficult to translate into planting month [68,69]. We decided instead to aggregate areas per climate region. The existing Köppen-Geiger system determines 19 different climate types in Europe [70]. These were aggregated into the six most representative classes, each occupying at least 5% of the SoilCare study area, and together occupying 92% of the total area; the remainder were assigned to the closest climate class. It should be noted that the division between climate regions is not sharp, and there are often climatic gradients. The six classes were then transformed into four classes with two further aggregations: (1) For cropping purposes, the dry climate regions are similar to the Mediterranean climate regions, so they were reclassified as the latter; and (2) polar climate is important in a large part of mountain regions, but agriculture is not practiced there, so for the model they were reclassified as subarctic climate. Figure 4 shows the climate zones as used in the modelling; they are similar to the environmental stratification of Europe proposed by Metzger et al. [71].

Finally, we aggregated existing crop calendar information for different countries in Europe for the four climate zones using the following datasets according to the dominant climate in the country, in decreasing order of preference:

(a) JRC crop calendars for winter wheat, grain maize and rice: https://agri4cast.jrc.ec.europa.eu/DataPortal/Index.aspx?o=sd (accessed on 15 December 2021)
(b) USDA crop calendars for Europe: https://ipad.fas.usda.gov/rssiws/al/crop_calendar/europe.aspx (accessed on 15 December 2021) and https://ipad.fas.usda.gov/countrysummary/Default.aspx?id=E4 (accessed on 15 December 2021)
(c) Boons-Prins et al. [72] with crop calendars for many crops in Europe: https://edepot.wur.nl/308997 (accessed on 15 December 2021)

When extended (>1 month) planting and harvesting dates were given, the latest planting and earliest harvesting date were chosen. The aggregation of calendars gave consistent planting and harvest dates for each region, with the Mediterranean region showing differences from the three other regions, either in earlier planting dates or shorter growing seasons. Cropping calendars were discussed with local partners from the SoilCare project and adapted according to their experience.

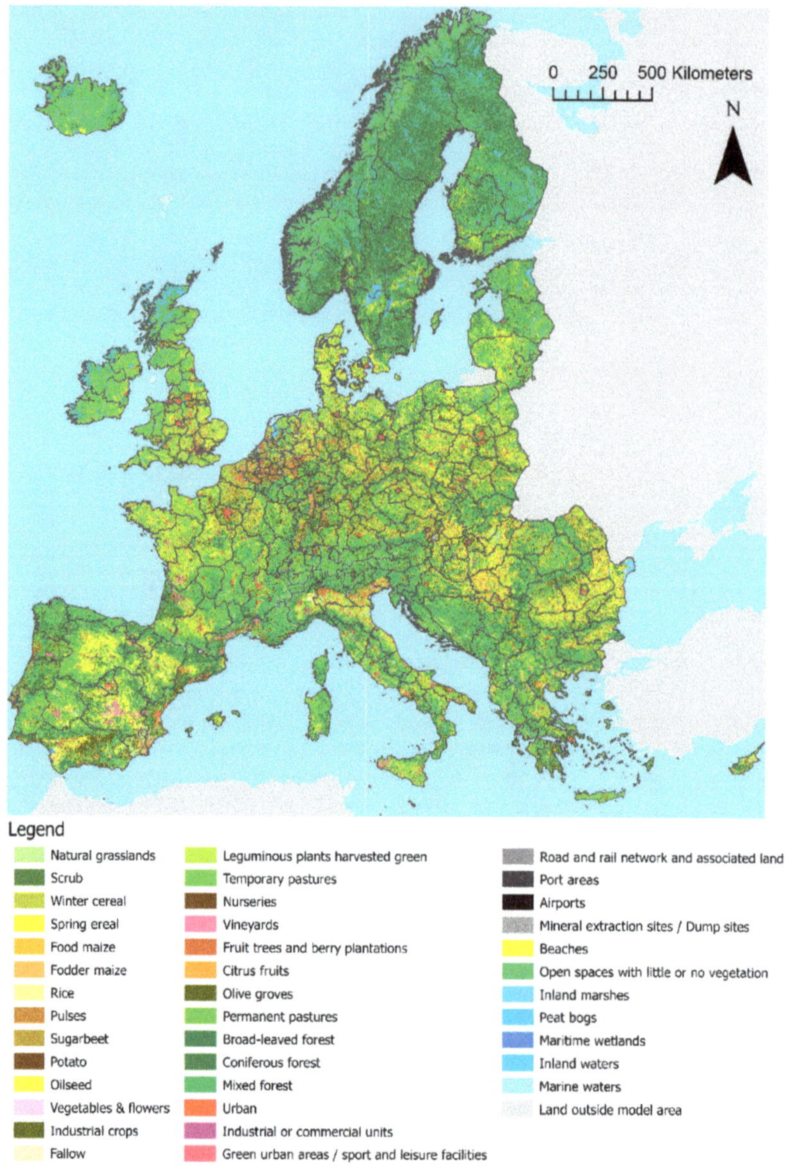

Figure 3. SoilCare land use and crop map (year: 2018). A GIS compatible version of this map is available in the Supplementary Materials.

Monthly ground cover (%) for each crop was derived mostly from the PESERA project manual, with some exceptions or additions:

- Sugarbeet: estimated and adapted from potato
- Oilseed: estimates based on pictures in Corlouer et al. [73] and comparison with winter wheat
- Rice: taken from FAO http://www.fao.org/docrep/S2022E/s2022e07.htm (accessed on 15 December 2021)

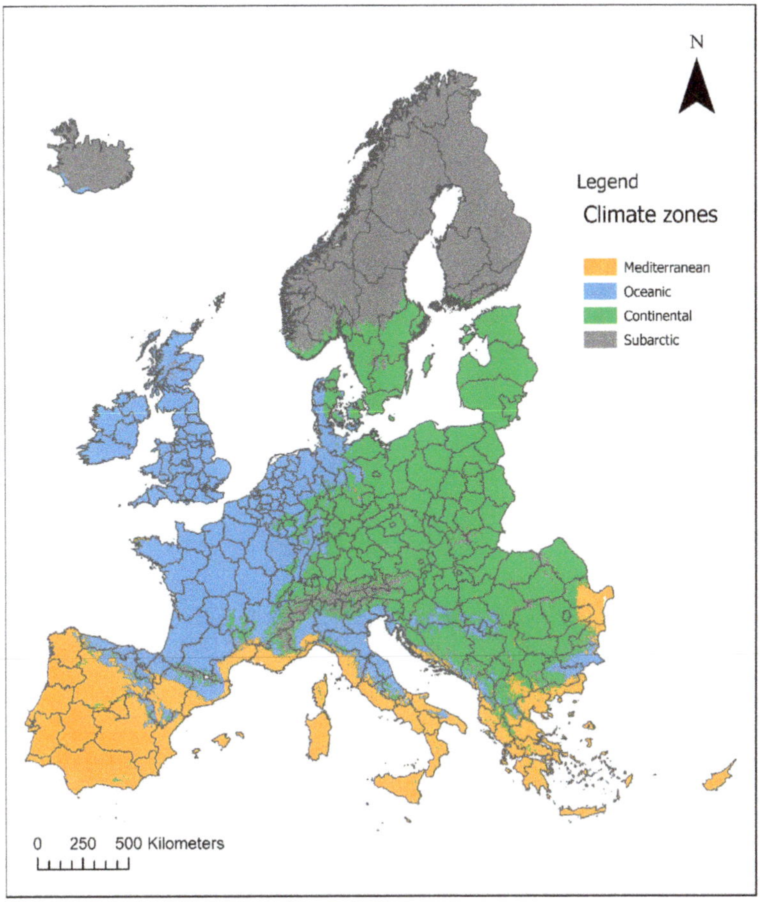

Figure 4. Climate zones as used in SoilCare to vary crop calendars by agroclimatic zone.

These cover calendars were then adjusted to the crop calendars. In most cases, the cover calendars fit inside the planting and harvest dates. When they did not fit, they were adjusted to keep the same shape as the PESERA growth curves but fitting a shorter or longer interval as needed. When the crop calendars indicated planting or harvesting seasons longer than one month, the cover values of these seasons were extended by repeating the first or last month value (respectively). Table 2 shows the crop calendars per agroclimatic zone and crop, with the cover indicated as value. Monthly canopy cover for permanent crops were based on the PESERA project estimations [74] for Europe and are given in Table S1.

Table 2. Crop calendar and ground cover values (%) per crop and agroclimatic zone. Dark green cells indicate the start of the growing season (planting month), orange cells indicate the last month of the growing season.

		Y1												Y2											
Crop	**Agroclimatic Zone**	**1**	**2**	**3**	**4**	**5**	**6**	**7**	**8**	**9**	**10**	**11**	**12**	**1**	**2**	**3**	**4**	**5**	**6**	**7**	**8**	**9**	**10**	**11**	**12**
Spring cereal	Continental					10	50	90	95	40															
	Mediterranean				10	50	90	95	40																
	Oceanic				10	50	90	95	40																
	Subartic					10	50	90	95	40															

Table 2. *Cont.*

		Y1												Y2											
Crop	**Agroclimatic Zone**	1	2	3	4	5	6	7	8	9	10	11	12	1	2	3	4	5	6	7	8	9	10	11	12
Winter cereal	Continental										5	5	25	50	75	90	95	95	85	30					
	Mediterranean											5	35	60	80	90	95	90	30						
	Oceanic										5	5	25	50	75	90	95	95	85	30					
	Subartic										5	5	10	15	25	50	75	90	95	90	30				
Maize	Continental				20	50	75	95	95	40	10														
	Mediterranean				20	60	95	95	40																
	Oceanic				20	50	75	95	95	40	10														
	Subartic																								
Pulses	Continental			20	65	95	70																		
	Mediterranean												20	65	95	70									
	Oceanic			20	65	95	70																		
	Subartic																								
Sugarbeet	Continental				10	50	70	90	95	85	50														
	Mediterranean			10	35	60	75	90	95	85	50														
	Oceanic				10	50	70	90	95	85	50														
	Subartic																								
Potato	Continental				10	70	95	95	85	35	10														
	Mediterranean				10	70	95	95	85	35															
	Oceanic				10	70	95	95	85	35	10														
	Subartic																								
Oilseed	Continental									10	50	80	90	90	90	95	95	85	50						
	Mediterranean											5	35	60	80	90	95	90	30						
	Oceanic									10	50	80	90	90	90	95	95	85	50	50					
	Subartic										10	50	80	90	90	90	95	90	50						
Veg & Flowers (sunflowers)	Continental					10	75	95	30																
	Mediterranean				10	65	80	95	30																
	Oceanic					10	75	95	30																
	Subartic																								
Rice	Continental					10	40	65	85	90	60														
	Mediterranean					10	40	65	85	90	60														
	Oceanic					10	40	65	85	90	60														
	Subartic																								
Forage	Continental			10	65	70	70	75	80	70	50														
	Mediterranean											10	10	70	70	80	50								
	Oceanic			10	65	70	70	75	80	70	50														
	Subartic																								

Water use efficiency values were calculated for different crops based on the following sources:

- For spring wheat, winter wheat, potato, sugarbeet, sunflower/tomato, bean (pulses): FAO http://www.fao.org/land-water/databases-and-software/crop-information/maize/en/ (accessed on 15 December 2021);
- For consumption maize (sweet maize) and fodder maize (grain maize): FAO http://www.fao.org/3/S2022E/s2022e07.htm (accessed on 15 December 2021);
- For oilseed (winter oilseed rape):
    - Length of the growing stages: (Marjanović-Jeromela et al., 2019)
    - Kc values: (Corlouer et al., 2019) (Figure 2 in their suppl. Material) [73]
- For rice: FAO paddy rice: http://www.fao.org/3/S2022E/s2022e07.htm (accessed on 15 December 2021)
- For forage: taken from PESERA manual [74].

WUE calendars, with crop- and growth stage specific WUE values, were also based on planting and harvest dates, and used the same method as that for cover calendars, including stretching or shortening curves to match planting and harvesting dates (Table S2).

2.2.5. Rooting Depth and Surface Storage

Rooting depth was estimated based on three sources: the PESERA project manual [74]; estimates from FAO: http://www.fao.org/land-water/databases-and-software/crop-information/maize/en/ (accessed on 15 October 2020). These estimates start at 30 cm root depth going to 100 cm at the end of the growing season. As PESERA estimates were lower, a conservative estimate was taken and cross-checked with the third source; the SWAT database, which also estimates slightly deeper (maximum) rooting depths. For initial surface storage (either 0, 5 or 10 mm) and reduction of surface storage (either 0 or 50%), the

PESERA project manual was followed. Values of rooting depth, initial surface storage and reduction of surface storage used in this study are given in Table S3.

2.2.6. Soil Properties

Soil property data are used to calculate storage capacity and therefore the runoff threshold and affect plant growth through soil water availability. Six layers of soil data are required: (1) Erodibility, which is the sensitivity of the soil for erosion; (2) crusting, which is the sensitivity of the soil to surface crusting and affects the infiltration; (3) scale depth, which is a proxy for infiltration; (4) the effective soil water storage capacity; and soil water available to plants for depths 0–300 mm (5) and 300–1000 mm (6) respectively.

2.2.7. Erodibility

The erodibility map has five classes. We used the RUSLE erodibility K-factor, as prepared by Panagos et al., [61], with stoniness effects incorporated, grouped into five classes (Table S4). As indicated earlier (Section 2.3), based on discussions with local partners, the erodibility map for Norway, Sweden and Iceland was adapted. Details of the method used can be found in Supplementary Material S1. Figure 5 shows the final erodibility map as used in this study.

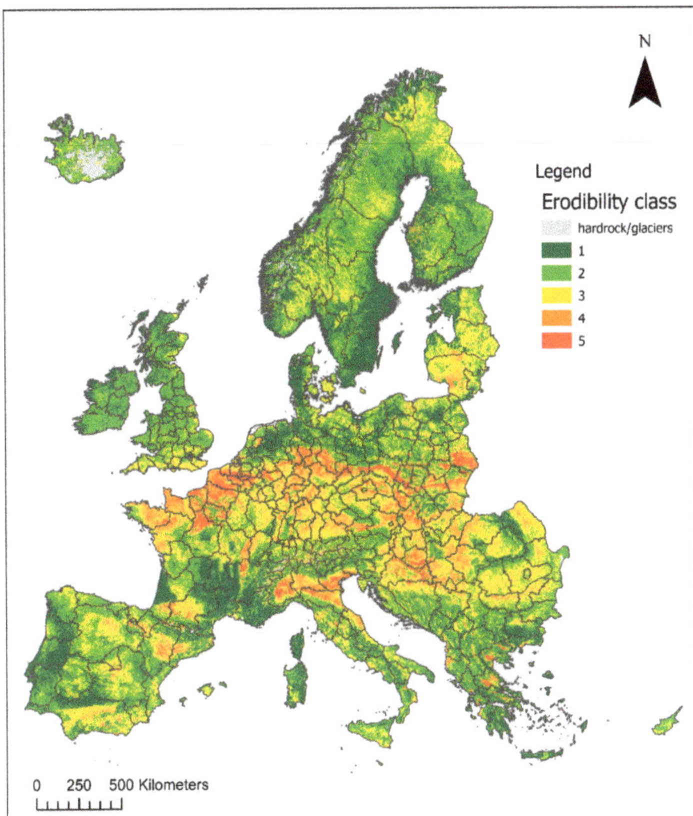

Figure 5. Erodibility map as used in SoilCare. Note that bare rock and glacier areas (according to CLC2018) were excluded (grey colours).

2.2.8. Crusting and Scale Depth Maps

The soil sensitivity to crusting index map was created using pedotransfer functions on texture, parent material and physical–chemical soil properties (Figure S1). The scale depth input map (Figure S2) was derived from soil texture classes (Table S5). Texture data were derived from the ESDB database.

2.2.9. Soil Water Availability and Storage Maps

Soil water available to plants (both 0–30 and 30–100 cm) and effective soil water storage capacity maps were derived based on the instructions from the PESERA project [74] and using ESDB data. Available Water Content for topsoil and subsoil (AWC_top and AWC_sub) maps of ESDB were used as a starting point. Additional soil property data used in the pedotransfer functions include texture, packing density and restriction of soil depth by bedrock.

The effective soil water storage capacity was then calculated from the soil water available to plants in the top- and subsoil following the PESERA project instructions [74]. Estimations for Iceland and Cyprus, that are not included in the ESDB maps, were derived using the SWAT data in combination with the FAO Harmonized World Soil Database (HWSD), available at https://doi.pangaea.de/10.1594/PANGAEA.901309 (accessed on 15 August 2020). All three maps are shown in Supplementary Material S1 (Figures S3–S5).

2.3. *Model Calibration and Evaluation*

During the equilibrium phase, long-term average output of the model was calibrated for erosion estimates and soil organic matter. As it was not feasible to calibrate the model for all countries, calibration was carried out for four countries in various climate zones across Europe (Belgium Spain, Slovakia and Norway), and the Greek island of Crete. Tuning parameters for calibration were: (1) The biomass conversion factor used in the model to calculate gross primary production—affecting ground cover and thereby erosion; and (2) the decomposition factor used in the model to calculate soil organic matter from plant residues. Both parameters are specific for each crop and land use, but generic for all regions. For soil organic matter calibration, the LUCAS topsoil soil organic carbon point data was used: https://esdac.jrc.ec.europa.eu/projects/lucas (accessed on 1 December 2020), which was aggregated to crops and land covers per climate zone (Table S6). In addition, to cross-validate and make use of the knowledge of the SoilCare local partners, both the spatial patterns and numerical (aggregated) results were shared with selected countries across Europe and their feedback was used for further fine-tuning. Preliminary results were sent to partners in Belgium, Germany, Greece, Spain, Italy, Norway, Poland and Romania. Based on their feedback:

- The crop calendars were adapted for some crops and regions (Table 2 and Table S2).
- The erodibility map for Norway was adapted because it had too high erodibility in the central mountain areas where soils are very shallow and granite bedrock is very often at the surface; hardly any erosion occurs in these areas. The existing K-factor map from JRC was adapted for certain land uses (following Corine Land Cover 2018), as detailed in Section 2.2

The model output at EU scale was evaluated by comparing the ranges and spatial patterns of the equilibrium phase PESERA erosion and SOC output maps to existing maps reported in the literature (for SOC e.g., [38,75–77]; for erosion e.g., [34,78,79]; see Supplementary Material S2.

2.4. *Parameterisation of SICS*

The PESERA model was used to investigate four SICS [80], each representing a different category: soil improving crops, soil amendments, soil cultivation and compaction alleviation. Respectively they were:

- Cover crops: these are non-harvested crops grown to protect the structural aspects of soil fertility and reduce erosion [13,81]. They can be applied in combination with annual crops, planted in the fallow period; or between the rows of permanent crops. They can also be incorporated into the soil as green manure.
- Mulching: application of various types of dead plant material on the soil surface, such as straw mulch, pruning residues or wood chips [17]. They are used to cover the soil to protect it against erosion, reduce evaporation from bare soil, increase local soil temperature and add organic material to the soil. It can be applied between the harvest and sowing of annual crops, or between rows of perennial crops.
- Minimum tillage: minimise soil disturbance by using less frequent or less intensive tillage operations, benefiting soil structure and preventing further compaction [82]. It can, especially when combined with soil cover by plant residues, reduce water and wind erosion and evaporation, leading to higher soil moisture before the growing season. It can also mitigate declines in soil carbon compared with conventional tillage.
- Compaction alleviation: reduced use of heavy machinery, preventing soil compaction and therefore improving soil water holding capacity and rooting depth [51,83].

These SICS were simulated individually, and in two combination scenarios, combining compaction reduction and minimum tillage with either cover crops or mulching (assuming that cover crops and mulching cannot be combined). The combination measures assumed that no additive effects would occur for each parameter, taking instead the most intensive effect of each individual measure on each parameter. The implementation of each measure in PESERA is described, in general terms, in Table 3. The model implementation of these measures was tested on a synthetic dataset, representative of climatic and crop conditions in the Oceanic climate regions of Europe. The differences between the application of the measure over the control conditions were compared with results taken from a survey of meta-analyses published in indexed journals, on soil erosion and soil organic matter; a detailed list of references is presented in Supplementary Material S3.

Table 3. Implementation of soil improving measures in PESERA.

Parameter	Cover Crops (CC)	Mulching (M)	Compaction Reduction (CR)	Minimum Tillage (MT)	Cover Crops + CR & MT	Mulching + CR & MT
General description	Annual crops: cover crop in fallow period Permanent crops: cover crop in interrows	0.2 kg/m^2 mulching added each year	Decrease in use of heavy machinery	Tillage depth reduced by 40% (except root crops); 40% stubble cover left	Cover crops, compaction reduction and minimum tillage	Mulching, compaction reduction and minimum tillage
Soil surface						
Erodibility	=	=	=	−1 class	−1 class	−1 class
Cover	80% of bare soil	80% of bare soil	=	40% of bare soil	80% of bare soil	80% of bare soil
Roughness	+5 mm	+10 mm	=	+5 mm	+5 mm	+10 mm
Hydrological properties						
Water storage capacity *	+25%	+30%	+10%	=	+25%	+30%
Soil evaporation	=	−40%	=	=	=	−40%
Root depth	=	=	+10%	=	+10%	+10%
Vegetation						
Water use (wue)	Permanent crops: +0.1	=	=	=	Permanent crops: +0.1	=
Active period	Annual crops: cover crop in fallow period (0.6 kg/m^2)	=	=	=	Annual crops: cover crop in fallow period (0.6 kg/m^2)	=
Soil Organic Matter						
SOM breakdown rate	=	=	=	Decreased in tillage month	Decreased in tillage month	Decreased in tillage month
SOM added to soil	0.06 kg/m^2 at tillage	0.01 kg/m^2 each month (except tillage and harvest)	=	=	0.06 kg/m^2 at tillage	0.01 kg/m^2 each month (except tillage and harvest)

2.5. Scenario Description

Socio-economic scenarios were developed in the SoilCare project in multiple workshops and feedback rounds, including all relevant stakeholders, with the aim to explore plausible agricultural pathways for Europe and assessing their sustainability and profitability impacts. It is beyond the scope of the current study to describe this in detail. The scope of the scenarios and full descriptions can be found in [65]. Here, the scenarios are very briefly described, with emphasis on how the SICS were included in each scenario:

- Race to the Bottom (RttB): existing agricultural practices are continued and increasing amounts of inputs are used. The focus is on optimising outputs and quick financial gains, with low attention for improvements in soil quality. This scenario entails low sustainability farming everywhere. No SICS are applied.
- Under Pressure (UP): a set of rules and regulations to ensure sustainable production and support for farmers is created, but only the large-scale farmers can comply with these rules. In this scenario, medium sustainability farming occurs everywhere. All farmers apply 1 SICS: either mulching, cover crops, minimum tillage or compaction alleviation (25% each).
- Caring & Sharing (CS): climate-resilient agriculture is prioritised and a widespread awareness and support for investment in sustainable practices exists. This scenario entails high sustainability farming everywhere. All farmers apply a combination of SICS: minimum tillage, compaction reduction, and either cover crops (50%) or mulching (50%).
- Local & Sustainable (LS): locally sourced, sustainably produced food is highly valued, but not everyone is able or willing to afford this, leading to pockets of self-sufficient communities, but also mainstream conventional farms. In this scenario a mix of low, medium and high sustainability farming areas exist. One-third of farmers apply low sustainability, one-third medium sustainability and one-third high sustainability, as described in the previous scenarios.

The actual location of which SICS were applied where within the scenarios on the map was randomly distributed within the arable land and perennial crops (i.e., olive groves, vineyards and fruit trees).

These four scenarios were run for the period 2020–2050 and erosion and SOC simulated maps were analysed for the year 2050, and compared to the baseline situation in 2020 with no SICS applied. Note that for erosion calculations, the climate (especially rainfall) of a specific year can affect results (e.g., a large rainfall event in a specific region may lead to high erosion estimates for that year and location, but this does not happen in other years). Therefore, to evaluate erosion output estimates, the average of 2020–2025 was used to represent 2020 and the average of 2045–2050 was taken to represent erosion in 2050.

3. Results

3.1. Model Calibration Results

3.1.1. Baseline Long-Term Erosion

Figure 6 shows the calibrated model output for erosion (t/ha/y). These are the equilibrium phase simulation results, based on average long-term climate input data (see Table 1). Overall average erosion across the whole of Europe was simulated at 2.54 t/ha/y, with erosion in arable land estimated at 4.3 t/ha/y on average across Europe. The highest erosion rates were simulated in sugar beet and potato crops and lowest in spring cereals. For the permanent crops, olive groves showed high erosion rates, followed by fruit trees, with mixed and coniferous forest having the lowest erosion rates. This aligns well with estimates by Panagos et al. [34] of 2.46 t/ha/y for erosion prone land covers and 2.22 t/ha/y for all land covers. In line with expectations, the general spatial pattern shows relatively high erosion values in the zone from Northern France and Belgium, across Germany and Poland, known as the Loess Belt with soils susceptible to erosion. Moreover, the mountain areas (Alps, Norway, Apennines, Pyrenees) are visible as areas with high erosion. A third zone of relatively high erosion is visible in the south of Spain and Italy, where low cover and erodible soils are present. The overall pattern across Europe compares well with estimates using RUSLE2015 [34] (see Figure S16), who also estimate relatively high erosion in the mountain areas (although Norway and Switzerland are not included in their calculations), in southern Spain and Italy and Northern UK. The RUSLE erosion map predicts less erosion in the Loess Belt than the PESERA estimates. Borrelli et al. [79] predict similar areas of relatively high erosion in southern Spain, Italy, across the Loess Belt, but less erosion in the mountain areas and Northern UK (Figure S18). Cerdan et al.'s [78] estimate of more erosion in the Loess Belt is comparable to the PESERA map. However, in the Cerdan et al. [78] map (Figure S17), more areas with relatively high erosion are visible, e.g., in Eastern Europe.

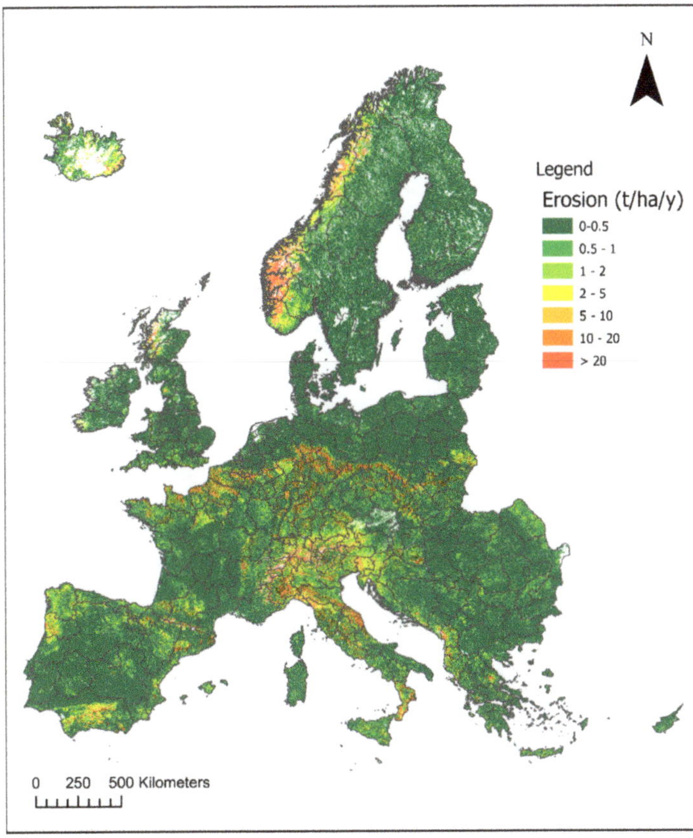

Figure 6. Simulated long-term average erosion rates across Europe using the PESERA model.

SoilCare partners' feedback on PESERA simulated erosion maps included for Spain that it seemed relatively low, compared to the national soil erosion map [84]. However, areas in the south show relatively high erosion in both maps. Belgium partners indicated that the relatively high erosion values for row crops like potato and sugar beet seemed valid, but that simulated erosion values for maize and vegetables, which have a wide spacing, were too low compared to their experience, especially when compared to simulated higher erosion in cereals. For Poland and Germany, simulated patterns of erosion were found to be plausible and matching e.g., the German national erosion map [85] with higher erosion in central Germany and very low to no erosion in the northern half of the country.

3.1.2. Baseline Long-Term SOC Stocks

Figure 7 shows the PESERA simulated maps of SOC stocks for Europe based on long-term average climate conditions (equilibrium phase model output). Overall estimates amount to 50 Gt, which is in line with estimates by Aagaard Kristensen et al. [77] (60 Gt), but somewhat higher than estimates of Yigini and Panagos [38] (38 Gt). The Nordic countries (Sweden, Finland) as well as the higher altitude areas clearly show higher SOC stocks (except where soil depth is shallow), while lower SOC stocks are simulated in for example inland Spain, parts of Italy, France and Eastern Europe. This coincides with the patterns of other SOC estimates (see Figures S12–S15). However, the SOC stock map based on the soil profile analytical database for Europe (SPADE; [77]; Figure S15), shows a slightly different pattern with lower SOC stock estimates for Sweden and parts of Finland, where our estimates show high SOC stocks. However, the intermediate stocks are similarly simulated to occur in the wet north-western Iberian Peninsula, the Massif Central in France and relatively low SOC stocks in the Norwegian mountain areas. Highest SOC stocks were simulated for forests, followed by grassland and shrubs. This matches estimates by other studies [38,39,77], although our estimates for grassland (11 Gt) are somewhat higher than those by Yigini and Panagos [38] (6.7 Gt). SOC stocks for fruit trees, olive groves and vineyard were estimated at around 60 t/ha on average across Europe, while the average SOC stocks for arable land across Europe was estimated at 43 t/ha or 5 Gt, which is lower than e.g., Lugato et al. [39] and Yigini and Panagos' [38] estimates of 17.6 and 12.8 Gt respectively.

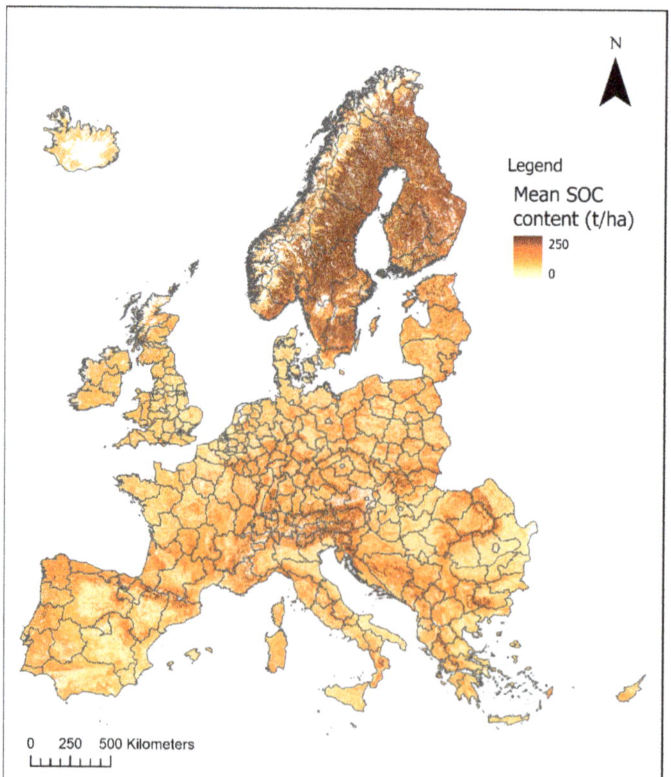

Figure 7. Simulated long-term average SOC stocks across Europe using the PESERA model.

Calibration results for SOC stocks for the three countries and Crete for which the model was calibrated are shown in Table 4. Overall, results were close to the observed data, derived from the LUCAS database. However, for some crops or land uses it was difficult to simulate good values across climate zones. For example, while SOC stocks for potato cultivation were well estimated for Spain, values were too high for Belgium and Slovakia, but too low for Crete. For maize, values were overestimated for Belgium and Slovakia, but underestimated for Spain.

Table 4. Calibrated (PESERA baseline long-term results) versus observed (LUCAS database) results for SOC (t/ha) for three countries and the Greek island of Crete in different climate zones. Note: X = crop does not occur. NA = not available.

	BELGIUM			SPAIN			SLOVAKIA			CRETE		
Land Use/Crop	**LUCAS**	**PESERA**	**Ratio (%)**	**LUCAS**	**PESERA**	**Ratio (%)**	**LUCAS**	**PESERA**	**Ratio (%)**	**LUCAS**	**PESERA**	**Ratio (%)**
spring cereal	50.8	46.7	91.9	36.6	35.5	97.0	49.6	46.5	93.6	X	X	X
winter cereal	50.8	51.1	100.5	36.6	48.9	133.8	49.6	50.7	102.2	36.6	47.5	129.8
consumption maize	46.7	52.2	111.6	51.4	41.0	79.8	47.3	57.7	121.9	X	X	X
fodder maize	46.7	57.6	123.2	51.4	40.6	79.1	47.3	65.8	139.1	51.4	12.5	24.3
pulses	46.0	51.4	111.8	44.2	43.3	97.9	55.4	52.6	94.9	44.2	41.5	93.9
sugarbeet	47.0	56.8	120.8	40.0	36.2	90.4	44.2	54.4	123.0	X	X	X
potato	47.0	61.9	131.8	40.0	40.8	101.8	44.2	71.3	161.4	40.0	15.2	37.9
oilseed	51.1	50.6	99.0	21.2	47.1	221.7	51.3	49.4	96.2	X	X	X
veg&flowers	41.1	39.9	97.0	29.9	29.2	97.5	49.8	43.3	86.9	29.9	23.8	79.5
forage	62.2	59.8	96.0	38.2	41.7	109.2	40.6	64.8	159.7	38.2	21.1	55.2
fallow	(NA)	21.3	(NA)	(NA)	12.1	(NA)	(NA)	25.9	(NA)	(NA)	8.1	(NA)
rice	X	X	X	44.4	39.5	88.9	X	X	X	X	X	X
vineyards	X	X	X	35.3	46.4	131.6	47.9	57.9	121.0	35.3	38.9	110.4
fruit trees	90.1	58.1	64.5	49.8	42.1	84.6	57.8	61.1	105.6	49.8	37.5	75.3
olives	X	X	X	46.6	42.7	91.5	X	X	X	46.6	38.5	82.5
pasture	149.1	105.7	70.9	72.8	103.0	141.4	103.2	109.0	105.6	72.8	95.8	131.5
broadleaf forest	174.1	147.6	84.8	111.2	119.4	107.5	134.3	149.4	111.3	111.2	84.4	75.9
coniferous forest	267.8	180.8	67.5	135.0	138.5	102.6	176.4	205.0	116.2	135.0	130.5	96.7
mixed forest	216.6	189.2	87.4	151.0	147.3	97.5	204.0	207.3	101.6	151.0	140.3	92.9
scrub	276.0	107.5	38.9	99.6	108.6	109.1	126.9	109.3	86.2	99.6	107.2	107.7

SoilCare partners' feedback on the PESERA calibrated SOC results indicated that they were in line with national estimates or maps (e.g., Belgium, Poland, Norway, Germany). For example, the German partners provided a German national map with organic matter [86], on which the spatial patterns were similar as those simulated by PESERA.

3.1.3. Calibration of the SICS

Figure 8 shows the simulated changes by the model for the individual SICS, compared with expected values from the literature. It should be noted that expected impacts on soil erosion were only found for cover crops, while the expected impacts on SOC were found for every SICS except compaction reduction; and that specific information for root crops and vegetables was less available than for cereals and permanent crops.

As can be seen, the simulated measures broadly followed what was expected from the literature in terms of erosion reduction and increase in SOC. When analysing per crop type, results for permanent crops tend not to be very good: no changes are simulated to erosion, because the baseline values were zero when using the test dataset; and changes to SOC are very limited. This indicates that the model is better adapted to simulate SOC changes for cereals than permanent crops. There is insufficient data to analyse model performance for root crops and vegetables.

In terms of impacts, PESERA simulates a small effect of compaction reduction when compared to other measures. For soil erosion control, mulching seems to have a limited effect in comparison to cover crops and minimum tillage; this results from the simulated wetter soil conditions when applying mulch, which increase biomass growth (and, indirectly, SOC) by limiting water stress, but also create the right conditions for more frequent runoff generation, counteracting beneficial soil protection effects. For SOC, mulching has a slightly larger benefit than cover crops or minimum tillage.

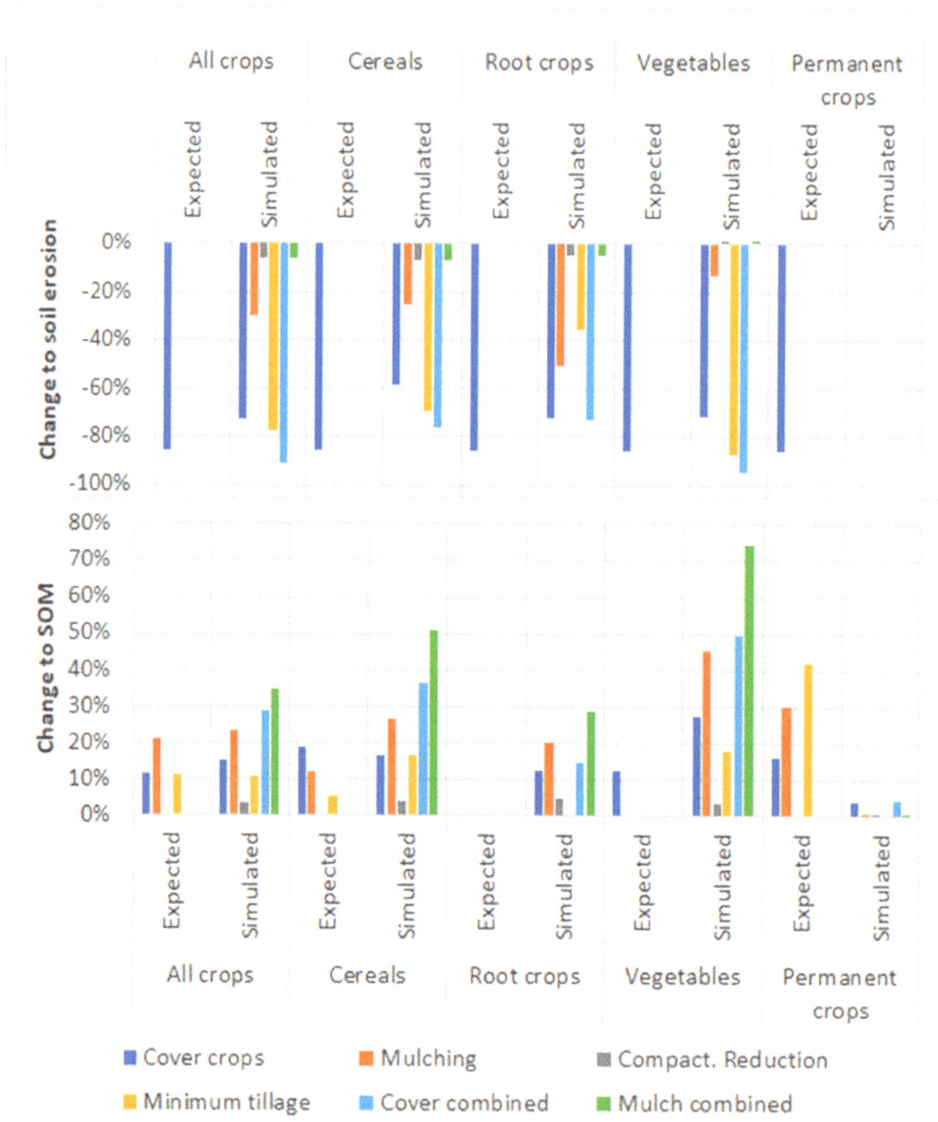

Figure 8. Impact of different SICS (expected; based on literature review, see Table S7, and simulated with PESERA on a test dataset) on different types of crops, for erosion (**top**) and soil organic matter (**bottom**).

As for the combination SICS, both tend to lead to higher increases of SOC when compared with the individual components. For soil erosion, the combinations involving cover crops led to larger reductions than the individual components. However, the combined mulch approach had a very limited impact on soil erosion, despite the erosion decrease expected when applying the individual components. As both individual measures increase soil moisture, the wetter soil conditions and increased runoff counteract the soil protection effects of the measures. In short, results suggest that the combined cover crop approach appears to have a better balance between SOC increase and erosion control, while the combined mulching approach has larger increases of SOC at the expense of the effects on erosion control.

3.2. Simulated Results for SICS Scenarios

Figure 9 shows the simulated difference in annual erosion between year 2050 and current (2020) for the four scenarios. Note that differences in erosion are affected by differences in climate (e.g., wet months in certain years) as well as by the application of SICS. Some areas show a consistent slight increase in erosion; these are mainly the steep mountain areas (Alps, Pyrenees) and areas that receive a lot of rainfall (e.g., Norwegian coastal zone), where SICS are not applied, as they are covered by e.g., pasture or shrubland. However, for example in the central European Loess Belt, southern Spain and eastern Europe, erosion was simulated to increase in the RttB scenario, while it decreases in the CS scenario due to application of SICS. Overall, across Europe and taking all land uses into account, erosion was simulated to increase slightly for the RttB scenario (+1.3% compared to 2020), while a decrease was simulated for the UP, LS and CS scenarios (75, 79 and 59% respectively, compared to the 2020 situation). When taking only the arable and orchard (fruit trees, olive groves and vineyards) areas into account, where SICS are applied, simulated erosion decreased to 43, 49 and 6.6% (compared to 2020) for the UP, LS and CS scenarios respectively, which is an average decrease of about 1.5 t/ha/y in both the UP and LS scenarios, and 2.6 t/ha/y in the CS scenario. So, especially for the CS scenario, a large decrease in erosion was simulated, which is in line with the large reductions that were parameterised for e.g., cover crops (Figure 8). Simulated erosion maps for RttB 2020 and the four scenarios for 2050 are given in Supplementary Material S4.

Figure 10 shows the simulated changes in SOC content for 2050, relative to the 2020 situation, for each of the four scenarios. All maps show both areas of decrease of SOC as well as areas of SOC increase. However, comparing between the scenarios, the results clearly show a more severe decrease in SOC for the RttB scenario, followed by UP, LS and CS scenarios. The average simulated SOC change across Europe, taking only the arable areas into account, was −23% for RttB, −4.5% for UP, −1.5% for LS and +22% for the CS scenario. This can also be seen in the maps (Figure 10): the CS scenario shows most increases in SOC content. For example, the arable areas in north-central Europe and north-central Spain that in the RttB show a strong decrease in SOC, turned into an increase in SOC in the CS scenario. This reflects the simulated SICS, where in CS all farmers apply a combination of minimum tillage, compaction reduction and either cover crops or mulching. In the UP scenario, all farmers apply only one type of SICS, while in the LS scenario, the application of SICS is mixed. Overall, it seems that, in terms of SOC content, the LS scenario leads to better results than the UP scenario, although local differences are likely greater in LS.

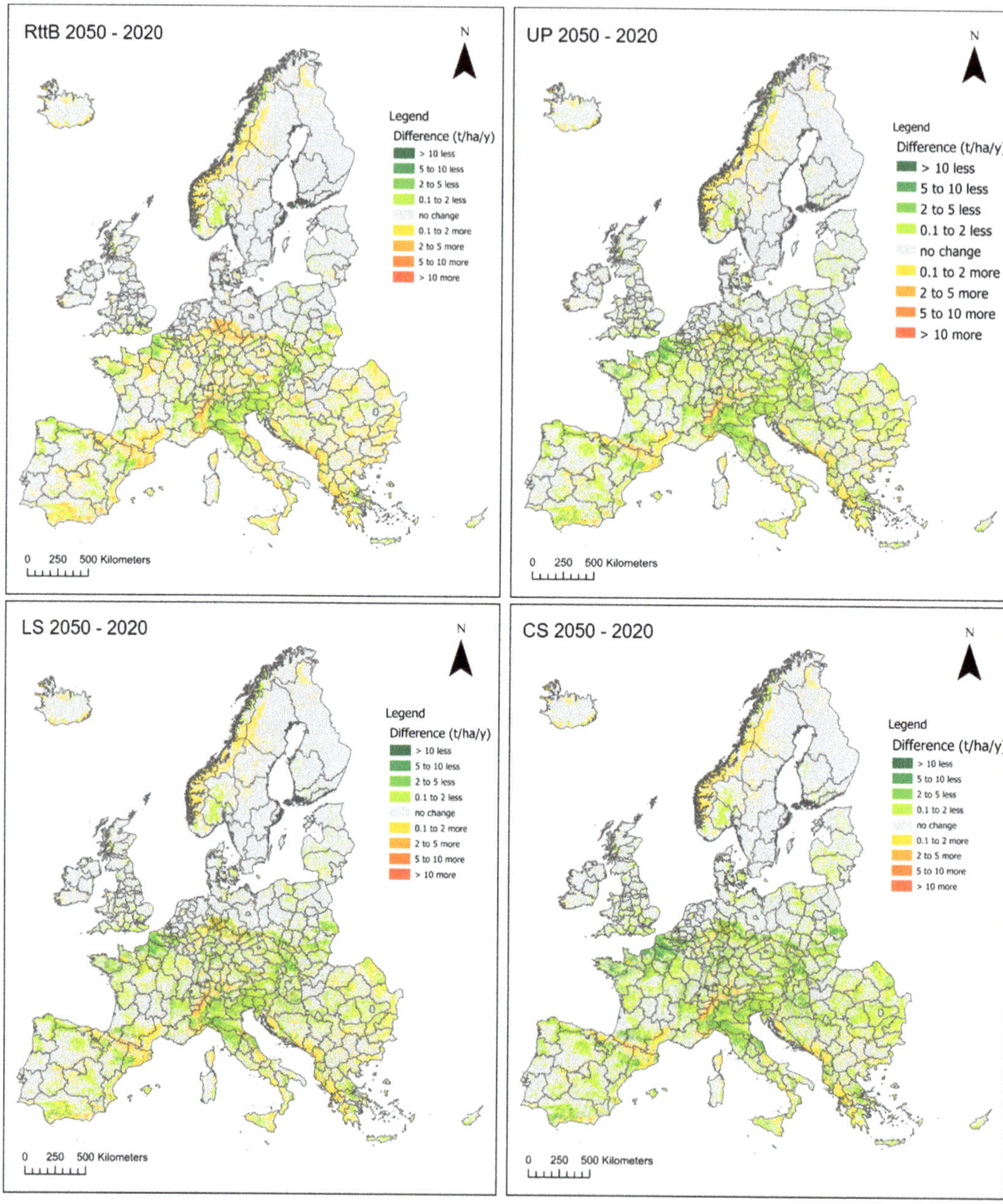

Figure 9. Simulated difference in erosion (t/ha/y) between 2020 and 2050 for each scenario.

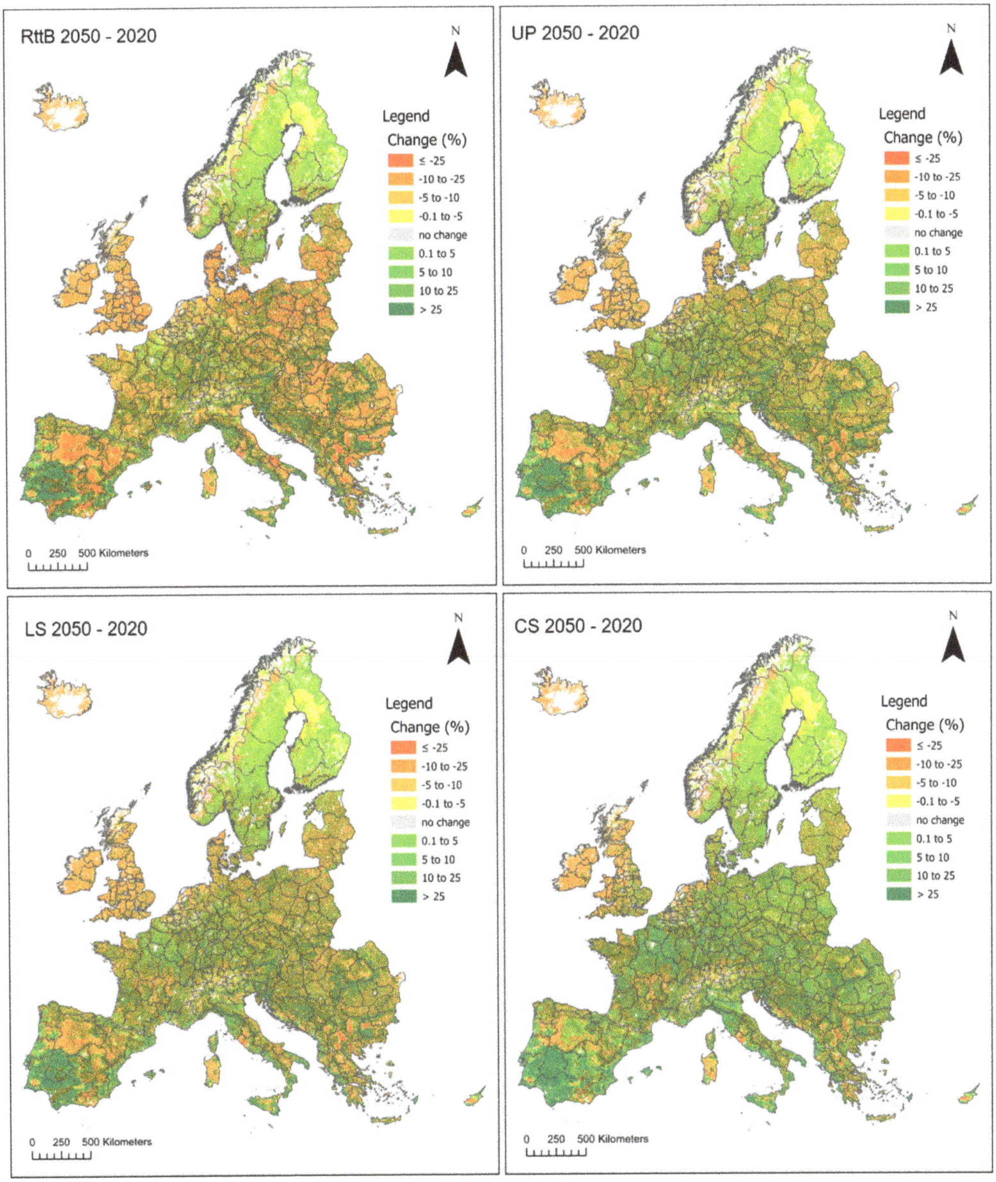

Figure 10. Simulated change in SOC stock (%) between 2020 and 2050 for each scenario.

4. Discussion

Using the PESERA model, we simulated the effects of SICS on erosion and SOC stock changes across Europe, based on scenarios in which either no SICS were applied (RttB scenario; low sustainability level), or where a medium (UP scenario), high (CS scenario) or a mix of these three levels of sustainability was assumed (LS scenario). Comparing the effects of the simulated scenarios clearly shows that the application of a high level of sustainability, where all farmers apply a combination of SICS: minimum tillage, compaction reduction, and either cover crops (50%) or mulching (50%), results in highest and most widespread erosion reduction (Figure 9) and a shift from a continuous reduction in SOC stocks to an increase in SOC stocks (Figure 10). In general, our results imply that erosion can be quite well prevented by application of SICS (Figure 9), but less pronounced effects are simulated for SOC change. This is in line with often reported findings of relatively quick effects of measures on runoff and erosion while changing SOC is a slower process [87].

The scenarios we simulated contain a mix of measures, so direct comparison with other estimates is difficult. Panagos et al. [16,41] calculated the effect of the C (cover management) and P (conservation practices) factors. Panagos et al. [16] found that conservation tillage reduced the C-factor (and thus, indirectly erosion, if all other factors remain the same) by 17%, application of crop residues reduced the C-factor by 1.2% and cover crops by 1.3%. Note that these numbers are affected by the (relatively small) area where these practices were found to be applied and that large differences between countries were found [16]. Panagos et al. [41] estimated the P-factor (conservation practices), when including contouring, stone walls and grass margins, to be 0.9702 across Europe, meaning that erosion would be reduced by 3% (if all other factors remain the same). This factor has a wide spatial variation, being lower (i.e., more effective erosion protection) in for example Portugal, Spain and Belgium, due to a high number of stone walls and grass margins [41]. These findings are somewhat different than our results, as they do not include mulching. In our results, cover crops were estimated to reduce erosion significantly (also due to the calibration, see Figure 8). In a review study based on a large database of plot-scale erosion and runoff, including the effects of SWCTs (Soil and Water Conservation Techniques), Maetens et al. [18] found that overall, application of SWCTs reduced the exceedance probability for a soil loss tolerance of 5 t/ha/y and 12 t/ha/y by 14 and 12% respectively. The individual measures ranked in the order (more to less effective) of geotextiles, buffer strips, mulching, contour bunds, cover crops, conservation tillage and strip cropping [18]. They concluded that crop and vegetation management (mulching, cover crops) and mechanical measures (terraces, contour bunds) are more effective than soil management techniques (reduced tillage). While our study did not include mechanical measures in the scenarios, our results are in line with this as the CS scenario, where cover crops and mulching is always applied (in combination with minimum tillage and compaction reduction) is clearly more effective in reducing erosion than the UP scenario, where only half of the farmers applies mulching or cover crops. However, it should be noted that, for soil erosion, even a low intervention scenario (UP) can decrease erosion below 1 t/ha/y (Figure S20), which can be considered as a threshold for sustainability [88]. There is some variability between climate regions, and within them, between regions with different topography and soil types. Nevertheless, these results indicate that the UP scenario might be good enough for most agricultural crops in Europe; and that special attention, and stronger intervention measures, could focus on remaining crop types (pulses, root crops, etc.,) and on areas with higher erosion rates. The results from this work could be used as a first approach to define priority areas for different levels of intervention across Europe.

Similar as for erosion, also a direct comparison with other studies regarding the simulated changes in SOC stocks for our scenarios is difficult. Lugato et al. [49] simulated the effect of six management practices scenarios on possible carbon sequestration, including spatially explicit maps across Europe. They found that, besides conversion of arable land to grassland which showed the highest SOC sequestration rates, ley cropping systems and cover crops results in higher SOC sequestration than straw incorporation and reduced tillage, which is in line with our results. Aertsen et al. [47] investigated the effect of agroforestry, hedges along field boundaries, cover crops and no/low tillage on carbon sequestration for the EU27, concluding that agroforestry has the highest potential, and no spatial maps of Europe were presented. Bellassen et al. [48] did not include no-tillage practices, as they only redistribute SOC instead of sequestering it. They also state that cover crops have a substantial potential for carbon sequestration, but that the large-scale potential of other practices such as hedges and crop residues is probably limited. Lessmann et al. [89] combined global meta-analytical results with spatially explicit data on current management practices and potential areas for implementation of measures at a global scale and found that organic matter inputs led to highest mean SOC changes, followed by crop residue incorporation, reduced tillage and increased crop diversity [89].

While in general terms the simulated values and spatial patterns are in line with other studies, local experiments and observations might deviate. This is a difficulty in any upscaling to large (continental) scales. Factors that play a role include assumptions in the model (e.g., biomass and humus conversion factors are crop-specific, but not adaptable per region), lack of (spatially explicit) input data (for example the difficulty of deriving a reliable erodibility map) and scarcity of (observed; i.e., non-modelled) calibration and validation data across Europe [79], but see [18,77]. Therefore, the absolute values should be taken with caution, but a qualitative and comparative analysis over time and across Europe can be insightful.

In this study, we focussed on erosion, as one of the most important processes of land degradation [35] and SOC changes, as one of the most widely used and important indicators of soil health [32,33]. While these are important indicators, many other indicators and processes play a role in a healthy functioning soil [2,5]. However, simulation of all these functions together is almost impossible, especially at a large (e.g., EU) scale. A few studies are beginning to attempt this. For example, the Soil Navigator decision support system was developed to assess and optimise various soil functions [90] on farm scale, incorporating soil management strategies. This was applied to monitor multiple soil functions at 94 sites across 13 European countries [91]. Vrebos et al. [92] analysed and mapped four soil functions on agricultural lands across the EU. Borelli et al. [79] evaluated soil degradation in Europe, including both erosion and soil carbon fluxes using the WaTEM/SEDEM modelling approach, but did not include the effect of soil and water conservation measures.

Potential additions to the modelling approach that we simulated in this study, would be to include additional indicators, such as biomass growth and effects on yields. While this is possible in the current PESERA model, preliminary results showed some difficulties. However, coupling of PESERA with more sophisticated biomass and yield models such as QUEFTS [93] is feasible and ongoing. This would also allow to evaluate the effects of (changes in) nutrient supply to the crops within the SICS. Another improvement within the PESERA model is to enable the parameterisation/calibration in the SOC calculations in the model (e.g., the decomposition rates) to be both spatially and crop specific (they are at the moment only crop specific), for example by including a spatial map of annual decomposition rates in Europe [46]. In addition to this, land use change as well as climate change can be included in the modelling framework. Finally, while in this study we only evaluated the effects of SICS on environmental indictors, in a really comprehensive analysis and modelling framework, also socio-economic factors and indicators should be taken into account, such as economic profitability and adoption of measures. In the SoilCare project, an important finding was that although the CS scenario leads to highest impacts, the gross margin of SICS uptake under this scenario is negative in many NUTS-2 regions [24,65]. Moreover, note that the spatial allocation of SICS application (e.g., where which SICS was applied within the scenarios) was randomly allocated. Interestingly, overall, results of the UP scenario (medium sustainability level with one SICS applied in all arable lands) were close to those of the LS scenario (a mix of low (no measures), medium (one measure) and high (multiple measures) sustainability levels). However, the spatial variability in LS will be higher, meaning that areas with high erosion and low increase (or decrease) of SOC will be offset by other areas with high erosion reduction and increase in SOC stocks. To avoid this and reach land degradation neutrality (LDN, [94]), careful planning is required and in terms of the scenarios simulated here, regarding the allocation of measures there is room for improvement in the scenarios, for example to base the allocation of certain SICS in areas that need them most and/or are most suitable [95].

5. Conclusions

In this study we simulated the effects of Soil Improving Cropping Systems (SICS) on SOC stocks and erosion on EU scale using the PESERA model. Four scenarios with varying levels and combinations of SICS were simulated for the time period 2020–2050. We can conclude that, for both SOC stocks, as an indicator for soil health, and erosion, as indicator for land degradation, the scenario with the highest level of SICS, i.e., application of minimum tillage and compaction alleviation in combination with either mulch or cover crops, clearly decreases erosion levels substantially across Europe as well as turning a decreasing trend of SOC stocks (when no SICS are applied) into an increase in SOC stocks, on average across Europe. Scenarios with medium level of SICS application as well as a scenario that implemented a mix of no SICS, medium level and high level SICS throughout Europe showed a decrease in erosion, while SOC stocks remained at the current level.

Future improvements for this modelling study would include to add climate change and dynamic land use. Furthermore, SICS were now randomly allocated in the arable lands; further scenarios including a more targeted spatial allocation of the levels of SICS would be interesting to conduct.

Supplementary Materials: The following are available online at https://www.mdpi.com/article/10.3390/land11060943/s1. Supplementary material S1: Input data maps for PESERA (Figures S1–S11; Tables S1–S5). S2: EU scale maps of SOC stocks (Figures S12–S15) and erosion (Figures S16–S18) used for model evaluation and SOC calibration data (Table S6). S3: Literature used to compile effect of measures (Table S7); and S4: Simulated erosion maps for the baseline (2020; Figure S19), RttB, UP LS and CS 2050 (Figure S20).

Author Contributions: Conceptualisation, J.E.M.B., J.P.N., L.F. and H.v.D.; methodology, J.E.M.B., J.P.N., L.F., H.v.D. and R.V.; software, J.E.M.B., J.P.N., H.v.D. and R.V.; validation, J.E.M.B. and J.P.N.; formal analysis, J.E.M.B. and J.P.N.; investigation, J.E.M.B., J.P.N., L.F. and H.v.D.; resources, J.E.M.B., J.P.N., L.F. and H.v.D.; data curation, J.E.M.B. and J.P.N.; writing—original draft preparation, J.E.M.B. and J.P.N.; writing—review and editing, J.E.M.B., J.P.N., L.F. and H.v.D.; visualisation, J.E.M.B. and J.P.N.; project administration, J.E.M.B., J.P.N., L.F. and H.v.D.; funding acquisition, L.F. and H.v.D. All authors have read and agreed to the published version of the manuscript.

Funding: This research has received funding from the European Union's Horizon 2020 research and innovation programme under grant agreement No 677407 (SoilCare project).

Institutional Review Board Statement: Not applicable in this study.

Informed Consent Statement: Not applicable in this study.

Data Availability Statement: The data presented in this study are available in the Figures in the paper and in the Supplementary Material S1–S4.

Acknowledgments: We acknowledge the E-OBS dataset from the EU-FP6 project UERRA (https://www.uerra.eu (accessed on 18 December 2020)) and the Copernicus Climate Change Service, and the data providers in the ECA&D project (https://www.ecad.eu (accessed on 18 December 2020)). We also acknowledge the feedback on preliminary model results by SoilCare partners, especially J. Stolte and R.J. Barneveld (NIBIO, Norway), M Tits and A. Elsen (Bodemkundige Dienst van Belgie, Belgium), P. Mayer-Gruner, E. Kandeler (University of Hohemheim, Germany), J. Cuevas, V. Pinillos (University of Almeria, Spain), M. Frac, J. Lipiec, B. Usowic (Institute of Agrophysics of the Polish Academy of Sciences, Poland); A. Berti, I. Piccoli (University of Padova, Italy); I. Calciu, O. Vizitiu (National Research and Development Institute for Soil Science, Agrochemistry and Environmental Protection, Romania), D. Alexakis, I. Tsanis (Technical University of Crete, Greece).

Conflicts of Interest: The authors declare no conflict of interest.

References

1. Rust, N.A.; Jarvis, R.M.; Reed, M.S.; Cooper, J. Framing of sustainable agricultural practices by the farming press and its effect on adoption. *Agric. Human Values* **2021**, *38*, 753–765. [CrossRef]
2. Bonfante, A.; Basile, A.; Bouma, J. Targeting the soil quality and soil health concepts when aiming for the United Nations Sustainable Development Goals and the EU Green Deal. *SOIL* **2020**, *6*, 453–466. [CrossRef]

3. Lehmann, J.; Bossio, D.A.; Kögel-Knabner, I.; Rillig, M.C. The concept and future prospects of soil health. *Nat. Rev. Earth Environ.* **2020**, *1*, 544–553. [CrossRef]
4. Yoder, D.C.; Jagadamma, S.; Singh, S.; Nouri, A.; Xu, S.; Saha, D.; Schaeffer, S.M.; Adotey, N.; Walker, F.R.; Lee, J.; et al. Soil health: Meaning, measurement, and value through a critical zone lens. *J. Soil Water Conserv.* **2022**, *77*, 88–99. [CrossRef]
5. Bouma, J.; Montanarella, L.; Evanylo, G. The challenge for the soil science community to contribute to the implementation of the UN Sustainable Development Goals. *Soil Use Manag.* **2019**, *35*, 538–546. [CrossRef]
6. European Commission. *Farm to Fork Strategy—For a Fair, Healthy and Environmentally-Friendly Food System*; European Commission: Brussels, Belgium, 2020.
7. COM/2019/640. Final The European Green Deal. Available online: https://eur-lex.europa.eu/legal-content/EN/TXT/?qid=1576150542719&uri=COM%3A2019%3A640%3AFIN (accessed on 12 April 2022).
8. Thomas, R.; Reed, M.; Clifton, K.; Appadurai, N.; Mills, A.; Zucca, C.; Kodsi, E.; Sircely, J.; Haddad, F.; Hagen, C.; et al. A framework for scaling sustainable land management options. *L. Degrad. Dev.* **2018**, *29*, 3272–3284. [CrossRef]
9. Europarc Federation EU Farm to Fork. Available online: https://www.europarc.org/european-policy/farm-to-fork-protectes-areas/ (accessed on 12 April 2022).
10. Veerman, C.; Pinto Correia, T.; Bastioli, C.; Biro, B.; Bouma, J.; Cienciala, E.; Emmett, B.; Frison, E.; Grand, A.; Hristov, L.; et al. *Caring for Soil Is Caring for Life: Ensure 75% of Soils Are Healthy by 2030 for Food, People, nature and Climate: Report of the Mission Board for Soil Health and Food*; European Commission: Brussels, Belgium, 2020. [CrossRef]
11. World Overview of Conservation Approaches and Technologies (WOCAT). *Where the Land Is Greener—Case Studies and Analysis of Soil and Water Conservation Initiatives Worldwide*; Liniger, H., Critchley, W., Eds.; World Overview of Conservation Approaches and Technologies: Bern, Switserland, 2007.
12. Kutter, T.; Louwagie, G.; Schuler, J.; Zander, P.; Helming, K.; Hecker, J.-M. Policy measures for agricultural soil conservation in the European Union and its member states: Policy review and classification. *L. Degrad. Dev.* **2011**, *22*, 18–31. [CrossRef]
13. Lavergne, S.; Vanasse, A.; Thivierge, M.-N.; Halde, C. Using fall-seeded cover crop mixtures to enhance agroecosystem services: A review. *Agrosystems, Geosci. Environ.* **2021**, *4*, e20161. [CrossRef]
14. Osipitan, O.A.; Dille, J.A.; Assefa, Y.; Knezevic, S.Z. Cover Crop for Early Season Weed Suppression in Crops: Systematic Review and Meta-Analysis. *Agron. J.* **2018**, *110*, 2211–2221. [CrossRef]
15. Kathage, J.; Smit, B.; Janssens, B.; Haagsma, W.; Adrados, J.L. How much is policy driving the adoption of cover crops? Evidence from four EU regions. *Land Use Policy* **2022**, *116*, 106016. [CrossRef]
16. Panagos, P.; Borrelli, P.; Meusburger, K.; Alewell, C.; Lugato, E.; Montanarella, L. Estimating the soil erosion cover-management factor at the European scale. *Land Use Policy* **2015**, *48*, 38–50. [CrossRef]
17. Qin, W.; Hu, C.; Oenema, O. Soil mulching significantly enhances yields and water and nitrogen use efficiencies of maize and wheat: A meta-analysis. *Sci. Rep.* **2015**, *5*, 16210. [CrossRef] [PubMed]
18. Maetens, W.; Poesen, J.; Vanmaercke, M. How effective are soil conservation techniques in reducing plot runoff and soil loss in Europe and the Mediterranean? *Earth Sci. Rev.* **2012**, *115*, 21–36. [CrossRef]
19. Rodrigues, L.; Hardy, B.; Huyghebeart, B.; Fohrafellner, J.; Fornara, D.; Barančíková, G.; Bárcena, T.G.; De Boever, M.; Di Bene, C.; Feizienė, D.; et al. Achievable agricultural soil carbon sequestration across Europe from country-specific estimates. *Glob. Chang. Biol.* **2021**, *27*, 6363–6380. [CrossRef]
20. Marshall, E.J.P.; Moonen, A.C. Field margins in northern Europe: Their functions and interactions with agriculture. *Agric. Ecosyst. Environ.* **2002**, *89*, 5–21. [CrossRef]
21. Van Vooren, L.; Reubens, B.; Broekx, S.; De Frenne, P.; Nelissen, V.; Pardon, P.; Kris, V. Ecosystem service delivery of agri-environment measures: A synthesis for hedgerows and grass strips on arable land. *Agric. Ecosyst. Environ.* **2017**, *244*, 32–51. [CrossRef]
22. Dorioz, J.M.; Wang, D.; Poulenard, J.; Trévisan, D. The effect of grass buffer strips on phosphorus dynamics—A critical review and synthesis as a basis for application in agricultural landscapes in France. *Agric. Ecosyst. Environ.* **2006**, *117*, 4–21. [CrossRef]
23. Felsot, A.S.; Mitchell, J.K.; Kenimer, A.L. Assessment of Management Practices for Reducing Pesticide Runoff from Sloping Cropland in Illinois. *J. Environ. Qual.* **1990**, *19*, 539–545. [CrossRef]
24. Hessel, R.; Wyseure, G.; Panagea, I.S.; Alaoui, A.; Reed, M.S.; van Delden, H.; Muro, M.; Mills, J.; Oenema, O.; Areal, F.; et al. Soil Improving cropping systems for sustainable and profitable farming in Europe. *Land* **2022**, *11*, 780. [CrossRef]
25. Berhe, A.A. Chapter 3—Drivers of Soil Change. In *Global Change and Forest Soils*; Busse, M., Giardina, C.P., Morris, D.M., Page-Dumroese, D.S., Eds.; Elsevier: Amsterdam, The Netherlands, 2019; Volume 36, pp. 27–42. ISBN 0166-2481. [CrossRef]
26. Lahmar, R. Adoption of conservation agriculture in Europe: Lessons of the KASSA project. *Land Use Policy* **2010**, *27*, 4–10. [CrossRef]
27. Hijbeek, R.; Pronk, A.A.; van Ittersum, M.K.; Verhagen, A.; Ruysschaert, G.; Bijttebier, J.; Zavattaro, L.; Bechini, L.; Schlatter, N.; ten Berge, H.F.M. Use of organic inputs by arable farmers in six agro-ecological zones across Europe: Drivers and barriers. *Agric. Ecosyst. Environ.* **2019**, *275*, 42–53. [CrossRef]
28. Montanarella, L. Trends in Land Degradation in Europe. In *Climate and Land Degradation*; Sivakumar, M.V.K., Ndiang'ui, N., Eds.; Springer: Berlin/Heidelberg, Germany, 2007; pp. 83–104, ISBN 978-3-540-72438-4. [CrossRef]

29. Prăvălie, R.; Patriche, C.; Borrelli, P.; Panagos, P.; Roșca, B.; Dumitrașcu, M.; Nita, I.-A.; Săvulescu, I.; Birsan, M.-V.; Bandoc, G. Arable lands under the pressure of multiple land degradation processes. A global perspective. *Environ. Res.* **2021**, *194*, 110697. [CrossRef] [PubMed]
30. Prăvălie, R. Exploring the multiple land degradation pathways across the planet. *Earth Sci. Rev.* **2021**, *220*, 103689. [CrossRef]
31. Smiraglia, D.; Ceccarelli, T.; Bajocco, S.; Salvati, L.; Perini, L. Linking trajectories of land change, land degradation processes and ecosystem services. *Environ. Res.* **2016**, *147*, 590–600. [CrossRef] [PubMed]
32. Bolinder, M.A.; Crotty, F.; Elsen, A.; Frac, M.; Kismányoky, T.; Lipiec, J.; Tits, M.; Tóth, Z.; Kätterer, T. The effect of crop residues, cover crops, manures and nitrogen fertilization on soil organic carbon changes in agroecosystems: A synthesis of reviews. *Mitig. Adapt. Strateg. Glob. Chang.* **2020**, *25*, 929–952. [CrossRef]
33. Evenson, G.; Osterholz, W.R.; Shedekar, V.S.; King, K.; Mehan, S.; Kalcic, M. Representing soil health practice effects on soil properties and nutrient loss in a watershed-scale hydrologic model. *J. Environ. Qual.* **2022**, 1–12. [CrossRef]
34. Panagos, P.; Borrelli, P.; Poesen, J.; Ballabio, C.; Lugato, E.; Meusburger, K.; Montanarella, L.; Alewell, C. The new assessment of soil loss by water erosion in Europe. *Environ. Sci. Policy* **2015**, *54*, 438–447. [CrossRef]
35. FAO. *Soil Erosion: The Greatest Challenge to Sustainable Land Management*; FAO: Rome, Italy, 2019.
36. Van Delden, H.; Luja, P.; Engelen, G. Integration of multi-scale dynamic spatial models of socio-economic and physical processes for river basin management. *Environ. Model. Softw.* **2007**, *22*, 223–238. [CrossRef]
37. Van Delden, H.; Stuczynski, T.; Ciaian, P.; Paracchini, M.; Hurkens, J.; Lopatka, A.; Shi, Y.; Calvo, S.; Van Vliet, J.; Vanhout, R.; et al. Integrated Assessment of Agricultural Policies with Dynamic Land Use Change Modelling. *Ecol. Modell.* **2010**, *221*, 2153–2166. [CrossRef]
38. Yigini, Y.; Panagos, P. Assessment of soil organic carbon stocks under future climate and land cover changes in Europe. *Sci. Total Environ.* **2016**, *557–558*, 838–850. [CrossRef]
39. Lugato, E.; Panagos, P.; Bampa, F.; Jones, A.; Montanarella, L. A new baseline of organic carbon stock in European agricultural soils using a modelling approach. *Glob. Chang. Biol.* **2014**, *20*, 313–326. [CrossRef] [PubMed]
40. Morais, T.G.; Domingos, T.; Teixeira, R.F.M. A spatially explicit life cycle assessment midpoint indicator for soil quality in the European Union using soil organic carbon. *Int. J. Life Cycle Assess.* **2016**, *21*, 1076–1091. [CrossRef]
41. Panagos, P.; Borrelli, P.; Meusburger, K.; van der Zanden, E.H.; Poesen, J.; Alewell, C. Modelling the effect of support practices (P-factor) on the reduction of soil erosion by water at European scale. *Environ. Sci. Policy* **2015**, *51*, 23–34. [CrossRef]
42. Schulp, C.J.E.; Nabuurs, G.-J.; Verburg, P.H. Future carbon sequestration in Europe—Effects of land use change. *Agric. Ecosyst. Environ.* **2008**, *127*, 251–264. [CrossRef]
43. Zaehle, S.; Bondeau, A.; Carter, T.R.; Cramer, W.; Erhard, M.; Prentice, I.C.; Reginster, I.; Rounsevell, M.D.A.; Sitch, S.; Smith, B.; et al. Projected Changes in Terrestrial Carbon Storage in Europe under Climate and Land-use Change, 1990–2100. *Ecosystems* **2007**, *10*, 380–401. [CrossRef]
44. Smith, J.O.; Smith, P.; Wattenbach, M.; Zaehle, S.; Hiederer, R.; Jones, R.J.A.; Montanarella, L.; Rounsevell, M.D.A.; Reginster, I.; Ewert, F. Projected changes in mineral soil carbon of European croplands and grasslands, 1990–2080. *Glob. Chang. Biol.* **2005**, *11*, 2141–2152. [CrossRef]
45. Frank, S.; Schmid, E.; Havlík, P.; Schneider, U.A.; Böttcher, H.; Balkovič, J.; Obersteiner, M. The dynamic soil organic carbon mitigation potential of European cropland. *Glob. Environ. Chang.* **2015**, *35*, 269–278. [CrossRef]
46. Freibauer, A.; Rounsevell, M.D.A.; Smith, P.; Verhagen, J. Carbon sequestration in the agricultural soils of Europe. *Geoderma* **2004**, *122*, 1–23. [CrossRef]
47. Aertsens, J.; De Nocker, L.; Gobin, A. Valuing the carbon sequestration potential for European agriculture. *Land Use Policy* **2013**, *31*, 584–594. [CrossRef]
48. Bellassen, V.; Angers, D.; Kowalczewski, T.; Olesen, A. Soil carbon is the blind spot of European national GHG inventories. *Nat. Clim. Chang.* **2022**, *12*, 324–331. [CrossRef]
49. Lugato, E.; Bampa, F.; Panagos, P.; Montanarella, L.; Jones, A. Potential carbon sequestration of European arable soils estimated by modelling a comprehensive set of management practices. *Glob. Chang. Biol.* **2014**, *20*, 3557–3567. [CrossRef] [PubMed]
50. Panagea, I.S.; Berti, A.; Čermak, P.; Diels, J.; Elsen, A.; Kusá, H.; Piccoli, I.; Poesen, J.; Stoate, C.; Tits, M.; et al. Soil Water Retention as Affected by Management Induced Changes of Soil Organic Carbon: Analysis of Long-Term Experiments in Europe. *Land* **2021**, *10*, 1362. [CrossRef]
51. Piccoli, I.; Seehusen, T.; Bussell, J.; Vizitu, O.; Calciu, I.; Berti, A.; Börjesson, G.; Kirchmann, H.; Kätterer, T.; Sartori, F.; et al. Opportunities for Mitigating Soil Compaction in Europe. Case Studies from the SoilCare Project Using Soil-Improving Cropping Systems. *Land* **2022**, *11*, 223. [CrossRef]
52. Willett, W.; Rockström, J.; Loken, B.; Springmann, M.; Lang, T.; Vermeulen, S.; Garnett, T.; Tilman, D.; DeClerck, F.; Wood, A.; et al. Food in the Anthropocene: The EAT–*Lancet* Commission on healthy diets from sustainable food systems. *Lancet* **2019**, *393*, 447–492. [CrossRef]
53. Kirkby, M.J.; Irvine, B.J.; Jones, R.J.A.; Govers, G.; Team, P. The PESERA coarse scale erosion model for Europe. I.—Model rationale and implementation. *Eur. J. Soil Sci.* **2008**, *59*, 1293–1306. [CrossRef]
54. Abdelwahab, O.M.M.; Ricci, G.F.; De Girolamo, A.M.; Gentile, F. Modelling soil erosion in a Mediterranean watershed: Comparison between SWAT and AnnAGNPS models. *Environ. Res.* **2018**, *166*, 363–376. [CrossRef]

55. Karamesouti, M.; Petropoulos, G.P.; Papanikolaou, I.D.; Kairis, O.; Kosmas, K. Erosion rate predictions from PESERA and RUSLE at a Mediterranean site before and after a wildfire: Comparison & implications. *Geoderma* **2016**, *261*, 44–58. [CrossRef]
56. Berberoglu, S.; Cilek, A.; Kirkby, M.; Irvine, B.; Donmez, C. Spatial and temporal evaluation of soil erosion in Turkey under climate change scenarios using the Pan-European Soil Erosion Risk Assessment (PESERA) model. *Environ. Monit. Assess.* **2020**, *192*, 491. [CrossRef]
57. Meusburger, K.; Konz, N.; Schaub, M.; Alewell, C. Soil erosion modelled with USLE and PESERA using QuickBird derived vegetation parameters in an alpine catchment. *Int. J. Appl. Earth Obs. Geoinf.* **2010**, *12*, 208–215. [CrossRef]
58. Cornes, R.C.; van der Schrier, G.; van den Besselaar, E.J.M.; Jones, P.D. An Ensemble Version of the E-OBS Temperature and Precipitation Data Sets. *J. Geophys. Res. Atmos.* **2018**, *123*, 9391–9409. [CrossRef]
59. Hargreaves, G.H.; Samani, Z.A. Reference Crop Evapotranspiration from Temperature. *Appl. Eng. Agric.* **1985**, *1*, 96–99. [CrossRef]
60. Dosio, A. Projections of climate change indices of temperature and precipitation from an ensemble of bias-adjusted high-resolution EURO-CORDEX regional climate models. *J. Geophys. Res. Atmos.* **2016**, *121*, 5488–5511. [CrossRef]
61. Panagos, P.; Meusburger, K.; Ballabio, C.; Borrelli, P.; Alewell, C. Soil erodibility in Europe: A high-resolution dataset based on LUCAS. *Sci. Total Environ.* **2014**, *479–480*, 189–200. [CrossRef] [PubMed]
62. Abbaspour, K.; Ashraf Vaghefi, S. Harmonized World Soil Database in SWAT Format. *PANGAEA* **2019**. [CrossRef]
63. Andrews, T.; Gregory, J.M.; Webb, M.J.; Taylor, K.E. Forcing, feedbacks and climate sensitivity in CMIP5 coupled atmosphere-ocean climate models. *Geophys. Res. Lett.* **2012**, *39*, L09712. [CrossRef]
64. Kotlarski, S.; Keuler, K.; Christensen, O.B.; Colette, A.; Déqué, M.; Gobiet, A.; Goergen, K.; Jacob, D.; Lüthi, D.; van Meijgaard, E.; et al. Regional climate modeling on European scales: A joint standard evaluation of the EURO-CORDEX RCM ensemble. *Geosci. Model Dev.* **2014**, *7*, 1297–1333. [CrossRef]
65. Van Delden, H.; Fleskens, L.; Muro, M.; Tugran, T.; Vanhout, R.; Baartman, J.; Nunes, J.P.; Vanermen, I.; Salputra, G.; Verzandvoort, S.; et al. *Report on the Potential for Applying Soil-Improving CS across Europe*; Research Institute For Knowledge Systems: Maastricht, The Netherlands, 2021.
66. Sacks, W.J.; Deryng, D.; Foley, J.A.; Ramankutty, N. Crop planting dates: An analysis of global patterns. *Glob. Ecol. Biogeogr.* **2010**, *19*, 607–620. [CrossRef]
67. Rötzer, T.; Chmielewski, F.-M. Phenological maps of Europe. *Clim. Res.* **2001**, *18*, 249–257. [CrossRef]
68. Brunel, S.; Suffert, M.; Petter, F.; Baker, R. Interface between pest risk science and policy: The EPPO perspective. *NeoBiota* **2013**, *18*, 9–23. [CrossRef]
69. Gloning, P.; Estrella, N.; Menzel, A. The impacts of climate change on the winter hardiness zones of woody plants in Europe. *Theor. Appl. Climatol.* **2013**, *113*, 683–695. [CrossRef]
70. Beck, H.E.; Zimmermann, N.E.; McVicar, T.R.; Vergopolan, N.; Berg, A.; Wood, E.F. Present and future Köppen-Geiger climate classification maps at 1-km resolution. *Sci. Data* **2018**, *5*, 180214. [CrossRef] [PubMed]
71. Metzger, M.J.; Bunce, R.G.H.; Jongman, R.H.G.; Mücher, C.A.; Watkins, J.W. A climatic stratification of the environment of Europe. *Glob. Ecol. Biogeogr.* **2005**, *14*, 549–563. [CrossRef]
72. Boons-Prins, E.R.; De Koning, G.H.J.; Van Diepen, C.A.; Penning de Vries, F.W.T. *Crop-Specific Simulation Parameters for Yield Forecasting across the European Community*; Wageningen University & Research: Wageningen, The Netherlands, 1993.
73. Corlouer, E.; Gauffreteau, A.; Bouchet, A.-S.; Bissuel-Bélaygue, C.; Nesi, N.; Laperche, A. Envirotypes Based on Seed Yield Limiting Factors Allow to Tackle G × E Interactions. *Agronomy* **2019**, *9*, 798. [CrossRef]
74. Irvine, B.J.; Kosmas, C. *Pan-European Soil Erosion Risk Assessment—Deliverable 15: PESERA Users Manual*; European Commission: Brussels, Belgium, 2003.
75. De Brogniez, D.; Ballabio, C.; Stevens, A.; Jones, R.J.A.; Montanarella, L.; van Wesemael, B. A map of the topsoil organic carbon content of Europe generated by a generalized additive model. *Eur. J. Soil Sci.* **2015**, *66*, 121–134. [CrossRef]
76. European Environmental Agency Variations in Topsoil Organic Carbon Content across Europe. Available online: https://www.eea.europa.eu/data-and-maps/figures/variations-in-topsoil-organic-carbon (accessed on 30 November 2020).
77. Aagaard Kristensen, J.; Balstrøm, T.; Jones, R.J.A.; Jones, A.; Montanarella, L.; Panagos, P.; Breuning-Madsen, H. Development of a harmonised soil profile analytical database for Europe: A resource for supporting regional soil management. *SOIL* **2019**, *5*, 289–301. [CrossRef]
78. Cerdan, O.; Govers, G.; Le Bissonnais, Y.; Van Oost, K.; Poesen, J.; Saby, N.; Gobin, A.; Vacca, A.; Quinton, J.; Auerswald, K.; et al. Rates and spatial variations of soil erosion in Europe: A study based on erosion plot data. *Geomorphology* **2010**, *122*, 167–177. [CrossRef]
79. Borrelli, P.; Van Oost, K.; Meusburger, K.; Alewell, C.; Lugato, E.; Panagos, P. A step towards a holistic assessment of soil degradation in Europe: Coupling on-site erosion with sediment transfer and carbon fluxes. *Environ. Res.* **2018**, *161*, 291–298. [CrossRef]
80. Rietra, R.; Heinen, M.; Oenema, O. A Review of Crop Husbandry and Soil Management Practices Using Meta-Analysis Studies: Towards Soil-Improving Cropping Systems. *Land* **2022**, *11*, 255. [CrossRef]
81. Jian, J.; Du, X.; Reiter, M.S.; Stewart, R.D. A meta-analysis of global cropland soil carbon changes due to cover cropping. *Soil Biol. Biochem.* **2020**, *143*, 107735. [CrossRef]
82. Luo, Z.; Wang, E.; Sun, O.J. Can no-tillage stimulate carbon sequestration in agricultural soils? A meta-analysis of paired experiments. *Agric. Ecosyst. Environ.* **2010**, *139*, 224–231. [CrossRef]

83. Hamza, M.A.; Anderson, W.K. Soil compaction in cropping systems: A review of the nature, causes and possible solutions. *Soil Tillage Res.* **2005**, *82*, 121–145. [CrossRef]
84. Ministerio Para la Transicion Ecologica y el reto Demografico Inventario Nacional de Erosion de Suelos. Available online: https://www.miteco.gob.es/es/biodiversidad/temas/inventarios-nacionales/inventario-nacional-erosion-suelos/estado_actual.aspx (accessed on 12 April 2022).
85. Umwelt Bundesamt Mittlerer Landjaehriger Sedimenteintrag in Die Oberflaechengewaesser Durch Wassererosion. Available online: https://www.umweltbundesamt.de/themen/boden-landwirtschaft/bodenbelastungen/bodenerosion/bodenerosion-durch-wasser#schutz-vor-erosion-ist-auch-gewasserschutz (accessed on 12 April 2022).
86. BGR Gehalte an Organischer Substanz in Oberboeden Deutschlands 1:1,000,000. Available online: https://www.bgr.bund.de/DE/Themen/Boden/Bilder/Bod_Humuskarte_g.jpg?__blob=normal&v=3 (accessed on 12 April 2022).
87. Poeplau, C.; Don, A. Carbon sequestration in agricultural soils via cultivation of cover crops—A meta-analysis. *Agric. Ecosyst. Environ.* **2015**, *200*, 33–41. [CrossRef]
88. Verheijen, F.G.A.; Jones, R.J.A.; Rickson, R.J.; Smith, C.J.; Bastos, A.C.; Nunes, J.P.; Keizer, J.J. Concise overview of European soil erosion research and evaluation. *Acta Agric. Scand. Sect. B Soil Plant Sci.* **2012**, *62*, 185–190. [CrossRef]
89. Lessmann, M.; Ros, G.H.; Young, M.D.; de Vries, W. Global variation in soil carbon sequestration potential through improved cropland management. *Glob. Chang. Biol.* **2022**, *28*, 1162–1177. [CrossRef] [PubMed]
90. Debeljak, M.; Trajanov, A.; Kuzmanovski, V.; Schröder, J.; Sandén, T.; Spiegel, H.; Wall, D.P.; de Broek, M.; Rutgers, M.; Bampa, F.; et al. A Field-Scale Decision Support System for Assessment and Management of Soil Functions. *Front. Environ. Sci.* **2019**, *7*, 115. [CrossRef]
91. Zwetsloot, M.J.; van Leeuwen, J.; Hemerik, L.; Martens, H.; Simó Josa, I.; de Broek, M.; Debeljak, M.; Rutgers, M.; Sandén, T.; Wall, D.P.; et al. Soil multifunctionality: Synergies and trade-offs across European climatic zones and land uses. *Eur. J. Soil Sci.* **2021**, *72*, 1640–1654. [CrossRef]
92. Vrebos, D.; Jones, A.; Lugato, E.; O'Sullivan, L.; Schulte, R.; Staes, J.; Meire, P. Spatial evaluation and trade-off analysis of soil functions through Bayesian networks. *Eur. J. Soil Sci.* **2021**, *72*, 1575–1589. [CrossRef]
93. Sattari, S.Z.; van Ittersum, M.K.; Bouwman, A.F.; Smit, A.L.; Janssen, B.H. Crop yield response to soil fertility and N, P, K inputs in different environments: Testing and improving the QUEFTS model. *F. Crop. Res.* **2014**, *157*, 35–46. [CrossRef]
94. UNCCD Land Degradation Neutrality. Available online: https://www.unccd.int/land-and-life/land-degradation-neutrality/overview#:~{}:text=WedefineLDNas\T1\textquotedbllefta,andspatialscalesandecosystems.\T1\textquotedblright (accessed on 12 April 2022).
95. Van Delden, H.; Fleskens, L.; Vanhout, R.; Nunes, J.P.; Baartman, J.; Lesschen, J.P.; Verzandvoort, S.; Hessel, R.; All Study Site Partners. *Report on the Integration and Synthesis of Study Site Results and Their Potential for Upscaling*; Wageningen University & Research: Wageningen, The Netherlands, 2021.

land

Article

A New Framework to Assess Sustainability of Soil Improving Cropping Systems in Europe

Abdallah Alaoui [1,*], Moritz Hallama [2], Roger Bär [3], Ioanna Panagea [4], Felicitas Bachmann [3], Carola Pekrun [5], Luuk Fleskens [6], Ellen Kandeler [2] and Rudi Hessel [7]

1 Institute of Geography, University of Bern, Hallerstrasse 12, 3012 Bern, Switzerland
2 Soil Biology Department, Institute of Soil Science and Land Evaluation, University of Hohenheim, Emil-Wolff-Str. 27, 70599 Stuttgart, Germany; moritz.hallama@uni-hohenheim.de (M.H.); ellen.kandeler@uni-hohenheim.de (E.K.)
3 Centre for Development and Environment (CDE), University of Bern, Mittelstrasse 43, 3012 Bern, Switzerland; roger.baer@unibe.ch (R.B.); felicitas.bachmann@unibe.ch (F.B.)
4 Department of Earth and Environmental Sciences, KU Leuven, 3001 Leuven, Belgium; ioanna.panagea@kuleuven.be
5 Agronomy Section, Institute of Applied Agriculture, Nuertingen-Geislingen University, Neckarsteige 6-10, 72622 Nürtingen, Germany; carola.pekrun@hfwu.de
6 Soil Physics and Land Management Group, Wageningen University, Droevendaalsesteeg 3, 6708 PB Wageningen, The Netherlands; luuk.fleskens@wur.nl
7 Wageningen Environmental Research, Droevendaalsesteeg 3, 6708 PB Wageningen, The Netherlands; rudi.hessel@wur.nl
* Correspondence: abdallah.alaoui@giub.unibe.ch; Tel.: +41-31-684-89-22

Abstract: Assessing agricultural sustainability is one of the most challenging tasks related to expertise and support methodologies because it entails multidisciplinary aspects and builds on cultural and value-based elements. Thus, agricultural sustainability should be considered a social concept, reliable enough to support decision makers and policy development in a broad context. The aim of this manuscript was to develop a methodology for the assessment of the sustainability of soil improving cropping systems (*SICS*) in Europe. For this purpose, a decision tree based on weights (%) was chosen because it allows more flexibility. The methodology was tested with data from the SoilCare Horizon 2020 study site in Germany for the assessment of the impact of the integration of cover crops into the crop rotation. The effect on the environmental indicators was slightly positive, but most assessed properties did not change over the short course of the experiment. Farmers reported that the increase in workload was outweighed by a reputation gain for using cover crops. The incorporation of cover crops reduced slightly the profitability, due to the costs for seeds and establishment of cover crops. The proposed assessment methodology provides a comprehensive summary to assess the agricultural sustainability of *SICS*.

Keywords: sustainability framework; overall sustainability; costs and benefits; cover crops

Citation: Alaoui, A.; Hallama, M.; Bär, R.; Panagea, I.; Bachmann, F.; Pekrun, C.; Fleskens, L.; Kandeler, E.; Hessel, R. A New Framework to Assess Sustainability of Soil Improving Cropping Systems in Europe. *Land* **2022**, *11*, 729. https://doi.org/10.3390/land11050729

Academic Editor: Evan Kane

Received: 14 April 2022
Accepted: 10 May 2022
Published: 12 May 2022

1. Introduction

Assessing agricultural sustainability is one of the most complex exercises related to appraisal methodologies because it entails not only multidisciplinary aspects (environmental, economic and social dimensions), but also builds on cultural and value-based elements [1]. Thus, agricultural sustainability should be considered as a social concept that can be modified in response to the requirements of society as a whole and the individuals constituting this society [2,3].

According to the current definitions policy-oriented sustainability assessment is a methodology that can help decision- and policymakers decide what actions they should or should not take to make society more sustainable [4]. For this purpose, sustainability assessment practitioners have developed a large number of tools [5]. Finding the appropriate assessment instrument is critical to match theory with practice, and to have successful outcomes in improving sustainability. More specifically, although many methods exist for monitoring and evaluating the environmental dimension of agricultural management practices, no single method has been widely accepted for assessing it, perhaps due to the complexity and variability of agricultural systems [6]. Though the meanings and uses of the term sustainability remain diverse, it is now widely accepted that sustainability is the path to balancing social, economic, and environmental needs [7–9].

There is broad scientific agreement on the fact that sustainable agriculture is defined as the management and the use of the agricultural ecosystem in a way that allows reaching economic (e.g., income growth or economic stability), social (e.g., equity or the cover of basic needs), and ecological objectives (e.g., ecosystem protection or natural resources regeneration) [7,10]. These objectives need to be continuously evaluated with scientific criteria, acknowledging uncertainty and safety margins.

Many frameworks with various combinations of indicator sets aimed at describing farming and cropping systems exist, from simple ones to complex multi-dimensional assessment tools [11–13]. The choice of the indicators depends on the objective of the study. In many studies, indicators are chosen to characterise the sustainability of the system or the intensity of management and practices (i.e., land-use intensity) [14–16]. However, the collection of the data needed to implement such frameworks is tedious and time consuming, and thus simple and reliable indicators, based on data that are reasonably easy to obtain, are required [5].

A review study related to sustainable agriculture revealed that the social dimension is the most difficult to assess in a quantifiable way when compared to the environmental and economic dimensions due to its inherently more subjective nature [17,18]. Research looking into the social sustainability of farming systems deals with issues and indicators related to (subjective) well-being and quality of life of the farming population, working conditions (workload, working time), gender equality, on-farm and off-farm incomes, access to services (education, advisory services), social relations (family, community), social security, finding work meaningful, life satisfaction, physical and mental health, etc. [18–20]. Hence, socio-cultural acceptability is a prerequisite for the adoption of new agricultural practices.

In their review paper, Alaoui et al. [5] selected frameworks based on the following criteria: (1) are validated through a peer-review process, (2) consider a farm-level assessment, (3) cover universal agricultural sectors including livestock and arable farms, (4) include the three dimensions of sustainability, and (5) are suitable both for Europe and countries worldwide. Based on the selected criteria, the following frameworks were identified: RISE (Response-Inducing Sustainability Evaluation [21]), MASC (Multi-attribute Assessment of Sustainability of Cropping Systems [22]), LADA (Land Degradation Assessment in Drylands [3]), SMART (Sustainability Monitoring and Assessment RouTine, [23]), SAFA (Sustainability Assessment of Food and Agriculture systems [24]) and PG (Public Goods [25]).

The EU Horizon 2020 project SoilCare, aimed "to assess the potential of soil-improving cropping systems (*SICS*), to identify and test these *SICS* to determine their impacts on profitability and sustainability in Europe". This required an assessment framework based on an evaluation of environmental, sociocultural, and economic dimensions of crop production. The methodology needed to allow flexibility; it needed to be applicable to all study sites (SS) across Europe to allow comparison and upscaling and at the same time to be flexible enough to consider site-specific circumstances.

Taking into account the above considerations, none of the reviewed frameworks was suitable for SoilCare because they did not include the indicators needed to evaluate *SICS* and/or did not provide results to evaluate the key terms of SoilCare (such as sustainability, profitability, soil quality) in combination.

The main aim of this research was to develop a comprehensive methodology for assessing the overall sustainability of the farm with special attention to the benefits, drawbacks, profitability, and soil quality of the *SICS* as compared to the *control*-conventional system. To set up a tool for the assessment of the overall sustainability, we chose a decision tree based on weights (%). This is because it allows simple aggregation to assess the three dimensions of sustainability and provides flexibility [22].

In this manuscript, we provide the general concept of the assessment tool developed to calculate sustainability of the *SICS* under consideration. We provide information on the indicators, their weighting factors, their threshold values, and their scores. An application with data from the German SoilCare study site is provided to explain how the tool is used for conservation agricultural techniques and serves as a first critical evaluation to document lessons learned. In this study, we assess the sustainability of the farm/field where the *SICS* is implemented.

2. Materials and Methods

2.1. Assessment Tool

For the evaluation of the overall sustainability of a farm and to facilitate the assessment of the performance of cropping systems three dimensions of sustainability were considered, i.e., environmental, sociocultural and economic. A decision tree was chosen for the aggregation. It breaks the sustainability assessment decisional issue down into simpler units that comprise quantitative as well as qualitative elementary criteria to rate cropping systems. Such aggregation is needed as the data for the three dimensions include various kinds of quantitative and qualitative data, obtained in different ways, including monitoring and questionnaires [22].

Within the decision tree, weights (%) were assigned to adjust the relative importance of the different indicators used within the three dimensions of sustainability. These weighting factor values were established from expert knowledge based on the literature review and can be modified to fit specific conditions and decision makers [5].

In the SoilCare project, the study sites selected to test the sustainability impact of *SICS* were grouped into 4 key topics to improve sustainability, namely, compaction, soil-improving crops, fertilization/amendments, and soil cultivation. The experiments implemented in the SoilCare project were short-term since the project was a 5-year project. To assess the sustainability of a given farm using the tool developed here, input data is needed. The tool calculates sustainability by assigning a higher score to the key topic considered in comparison to the others. This was the reason why we developed a new tool for the assessment of sustainability.

For the assessment of the sustainability of a farm or a field, we have selected plots with the *SICS* and plots without (*controls*) that best characterise the farm or field under consideration. The assessment was carried out by comparing the *SICS* plot with the *control* plot. Figure 1 provides an overview of the three dimensions considered and the related properties. For future use, the users can adapt the weighing to their specific case. This flexibility would help improve the assessment tool for various purposes.

2.1.1. Environmental Dimension

- Monitoring variables

To assess the sustainability of a farm, in situ measurements of the variables were carried out. For this purpose, a monitoring plan was established to harmonize the monitoring including instructions on the treatment replication, randomization, and sampling in which each experiment within a field/location is composed of 3 blocks (corresponding to 3 replicates). Each block contains two experimental units or plots where sampling is carried out for composite or undisturbed samples [26].

- Selection of the indicators and weights

Based on a literature review [5] and considering the *SICS*-related key topics in the SoilCare project, a list of variables for the evaluation of the environmental dimension of the implemented systems was established. Each key topic is defined by a set of indicators with high weight, e.g., soil compaction is assessed by bulk density and penetration resistance with a value of 0.20, and by infiltration capacity and aggregate stability with weight values of 0.15 and 0.10, respectively (refer to the grey boxes in Table 1).

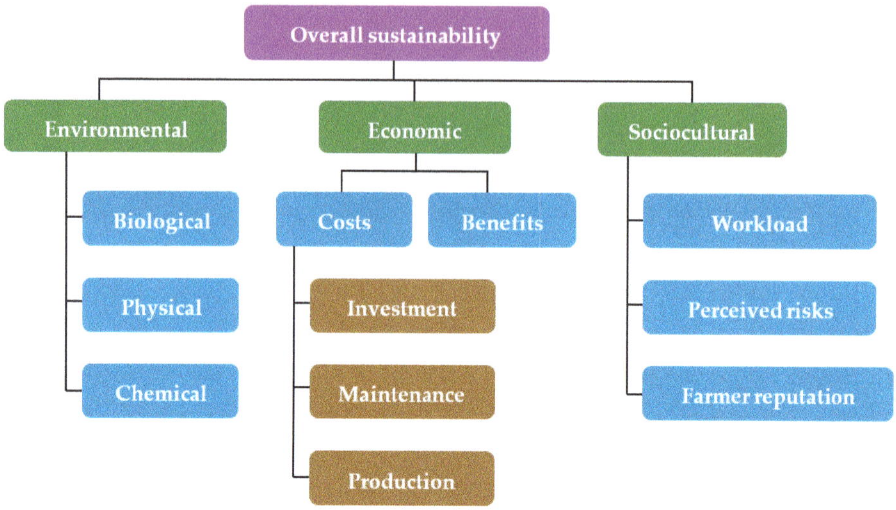

Figure 1. Structure of the aggregation of sustainability dimensions and assessment units.

Table 1. Weighing factors attributed to variables as related to the four key topics (soil cultivation, soil fertilization, soil-improving crops, and soil compaction).

Variable	Weight			
	Soil Cultivation	Fertilisation	Soil Improving Crops	Compaction
Infiltration capacity	0.05	0.01	0.05	0.15
Aggregate Stability	0	0.01	0.05	0.10
Bulk Density	0.08	0.01	0.05	0.20
Penetration Resistance	0	0	0	0.20
Mineral Nitrogen	0.05	0.22	0.05	0.05
SOC	0.05	0.30	0.05	0.05
pH	0.02	0.05	0.05	0.05
Earthworm Density	0.05	0.05	0.05	0.05
Crop Yield	0.20	0.05	0.10	0.05
Yield Quality	0.10	0.10	0.10	0.05
Crop Cover Characteristics	0.25	0.05	0.10	0.05
Pests	0.05	0.05	0.20	0
Root Diseases	0.05	0.05	0.10	0
Weed	0.05	0.05	0.05	0

The selection of variables and their assigned weights was based on the literature review [5] and is presented in Table 1. For more explanation, refer to File S1.

The assessment tool was designed to assess the change in the environmental dimension resulting from the implementation of the *SICS* compared to the *control* cropping system. Prior to inputting data into the assessment tool, the quantitative change of each variable as measured/estimated in the field is transformed into a qualitative score: positive change, no change, or negative change. For more details, refer to the "Metadata sheet" in File S1 using a statistically based approach.

In this tool, an additional option suggests the appropriate methods to be used for the evaluation of the variables listed in Table 1. The aim was to harmonize the methods across all study sites (refer to File S1, input datasheet). Based on the type of method used, the accuracy of the evaluation is evaluated.

- Statistical analysis

To quantify the difference between the variables of soil with the *SICS* and the ones of the *control*, a statistical approach was used. Mixed-effects models were used to estimate if statistically significant differences exist between the *SICS* and related *control* treatments. Mixed effect models were chosen as they allow a larger variety of designs and implementations [27] enabling a better identification and interpretation of interactions and repeated measurements. The statistical data analysis was performed using-R-Studio, R version 3.6.1 [28]. Differences between treatments or dates were analyzed using the full factorial statement "Treatment x Date", or the factor "Treatment" for variables with repeated measurements and only once measured variables, respectively. The experimental design structure effect (block, whole plot, main-plot, etc.) was introduced in all models as a random effect, using the statement "1 | structure" (in the German case study this was 1 | Block). The model's optimum fixed structure was selected for the best fit attaining the lowest value of Akaike's Information Criterion (AIC) using a maximum likelihood function (ML). A visual inspection was performed of the residuals' Q-Q plots and the normalised residuals' plots against the fitted values. The final models were fitted using REML estimation. Estimated marginal means by factors were computed by the least square method and contrasted by the Tukey group comparison method ($p < 0.05$).

The comparison between the variables of *SICS* and the ones of the *control* used for scoring the impact of *SICS* was scored by attributing a value of "1" for the statistically significant positive change (PC) (*SICS* is better than the *control*), "0" for no statistically significant change (NoC) and "−1" for statistically significant negative change (NC) (*control* is better than *SICS*).

2.1.2. Sociocultural Dimension

In contrast to the environmental indicators, most factors that determine the sociocultural acceptability of a *SICS* cannot be easily measured or quantified, which is due to their inherently subjective nature. Therefore, a qualitative approach was applied, and a short questionnaire (summarized in Table 2) was used to grasp the land users' assessment of the tested *SICS* in terms of three key topics: changes in workload, perceived risks, and influence on farmers' reputation.

Three requirements for *SICS* to be socially/culturally acceptable were identified on the basis of a literature review.

- Requirement 1 (Workload): *SICS* should not result in a considerable increase in workload, especially in periods where labour demand is already high.

In agriculture, working hours are generally long, considerably longer than in other professions. Therefore, it is not surprising that farmers are sensitive to an increase in workload, especially when it occurs during periods in which labour input is already high, e.g., in spring (field preparation and sowing) or summer respective autumn (harvesting) [29]. Additionally, an increase in workload not only means long working days, but also leads to higher production costs [29,30].

- Requirement 2 (Risk): In the perception of farmers, a *SICS* should not be a (too) risky practice.

A survey from north-eastern Germany showed that associated risks are among the main drivers when decisions are made to adopt new conservation measures [29]. Trujillo-Barrera et al. [31] concluded from their in-depth interviews with Dutch hog farmers that perceived risk is a barrier to the adoption of sustainable production practices.

- Requirement 3 (Reputation): Applying a *SICS* should not impair the farmer's reputation.

Much evidence exists [29,32–34], that farmers base their decision to adopt or reject conservation practices not exclusively on economic, agronomic, and ecological grounds. To be adopted, practices need to be compatible with land owners' values, and perceptions as to what makes a good farmer, or an aesthetic agricultural landscape (e.g., keeping fields nice and tidy).

Within each requirement, different questions were asked. The possible combinations of the responses and their output scores are reported in the "Metadata sheet" of File S1.

To calculate the final score of the sociocultural dimension, a specific weight was attributed to each requirement listed above (Table 2).

Table 2. Variables considered in the sociocultural dimension based on a questionnaire completed by the land user and their weighing factors.

Topic	Weight	Variable	Range of Answers
1. Workload	0.4	1.1. Increase/decrease in workload	Strongly increased, slightly increased, remained the same, slightly decreased, strongly decreased
		1.2. Workload increase during already existing work peaks	Yes/no
2. Perceived risks	0.4	2.1. Health risk 2.2. Economic risk 2.3. Risk of crop failure 2.4. Risk of conflicts 2.5. Other risk	Yes/no Yes/no Yes/no Yes/no Yes/no
3. Farmer's reputation	0.2	3.1. The (positive/negative) effect *SICS* application has on the reputation of the farmer	Strongly improved, slightly improved, remained the same, slightly worsened, strongly worsened

Given the fact that the two topics of risk perception and workload increase are both crucial for the adoption or rejection of new farming practices, they both have double weight compared to the topic of farmer reputation. In addition, the effect on a farmer's reputation is much more difficult to grasp and verify. Therefore, it was deemed appropriate to give this topic less weight in the assessment.

Study site researchers conducted the interviews at the end of the growing season with those farmers involved in the *SICS* trials. The questionnaire was kept as simple as possible, in order to avoid any limitations related to the implementation of the questionnaire, such as an adequately trained audit team, long duration for the training and implementation.

2.1.3. Assessment of the Economic Dimension

The economic dimension was assessed by evaluating the costs and benefits of the farm using a spreadsheet formatted questionnaire to ensure ease of use. This questionnaire, adapted from [35], contained the different types of costs, such as investment costs, maintenance costs, production costs, and benefits related to both the *control* and the *SICS* fields. Details on costs and benefits should refer to the same area/unit that can be defined in the Overview worksheet (refer to File S2 for more details). A summary of the costs and benefits is directly calculated and provided at the end of the questionnaire to allow the comparison between the *control* and the *SICS*. The details of each cost category are described below.

Investment costs: The assessor should list all one-off investment costs, structured according to the activities and inventorying labour, agricultural inputs, construction material, wood, earth, and other costs.

An activity refers to a defined task needed to establish the *SICS* and may consist of multiple inputs:

- Labour costs indicate total person days, either paid or voluntary.
- Equipment includes tools, machine hours, etc. Cost calculation for machine hours should be based on hiring costs–even if the machinery is owned by the land user.

- Agricultural inputs include seeds, seedling, fertilizer, biocides, compost/manure, etc. and indicate costs and quantities needed.
- Construction material includes stones, wood, earth, sand, etc. and indicate costs and quantities needed.

Maintenance costs: List of maintenance/recurrent activities and their associated costs for the *SICS* and the *control* cropping. It contains the same cost categories as listed for the investment costs above.

Production costs: List of changes pertaining to activities or inputs related to activities that have changed as a consequence of introducing a *SICS* (or a crop). Recurrent costs related to the technology itself should be recorded under maintenance costs.

Benefits: The benefits are considered at the farm level and consequently are defined as "on-site benefits". They can include: (i) Products harvested (cash and food crops, timber, fuelwood, fruits and nuts, animal fodder, etc.), (ii) Grazing/browsing, (iii) Recreation/tourism, (iv) Subsidies (e.g., for agri-environmental measures), and (v) Protection against natural hazards.

Calculation: The cost–benefit score represents the difference between the weighted (see below) relative change in benefits and in costs. A positive score means an improved cost–benefit ratio, a negative score means an impaired cost–benefit ratio. The score was calculated as follows:

$$Cost\ -Benefit\ score = \left(\Delta \text{RC}_{benefits} \times Benefit\ weight\right) - (\Delta \text{RC}_{costs} \times Cost\ weight) \quad (1)$$

With $\Delta\text{RC}_{benefits}$ = Relative change in benefits, calculated as follows:

$$\Delta \text{RC}_{benefits} = \frac{\sum Costs_{SICS}}{\sum Costs_{Control}} \quad (2)$$

ΔRC_{costs} = Relative change in costs, calculated as follows:

$$\Delta \text{RC}_{Costs} = \frac{\sum Benefits_{SICS}}{\sum Benefits_{Control}} \quad (3)$$

The type of costs for both *SICS* and *Control* are: Investment costs + Maintenance costs + production costs.

The benefits of both *SICS* and *Control* include all benefits listed in File S2 and are calculated as follows: Products harvested (cash and food crops, timber, fuelwood, fruits and nuts, animal fodder, etc., +Grazing/browsing + Recreation/tourism + Subsidies (e.g., for agri-environmental measures) + Protection against natural hazards.

In both cases (i.e., change in benefits and change in costs) the relative change has been capped to +/−100%. At the cost end, the positive extreme is theoretically solid as it means that costs involved with the *control* system can be reduced to 0 and that 0 means a perfect (+1) score. A doubling of the cost (and anything above) is regarded as the most negative outcome (−1). At the benefit side, a score of −1 is attributed to any decrease.

The *Benefit weight* and *Cost weight* account for the amplitude of changes in benefits and cost. The *Benefit weight* represents the ratio between the absolute difference in benefits as compared to the absolute difference in costs. The *Cost weight* is the counterpart of the *Benefit weight* and represents the ratio between the absolute difference in costs as compared to the absolute difference in benefits. The weights are calculated as follows:

$$Benefit\ weight = \frac{|Benefits_{SICS} - Benefit_{Control}|}{|Costs_{SICS} - Costs_{Control}| + |Benefits_{SICS} - Benefit_{Control}|} \quad (4)$$

$$Cost\ weight = 1 - Benefit\ weight \quad (5)$$

This allows us to appropriately consider cases with minimal absolute changes at the cost side but large changes at the benefit side, or vice versa. For instance, the doubling of cost from EUR 10 to EUR 20 will have double the weight (0.66) compared to a doubling of benefits from EUR 5 to EUR 10 (0.33).

2.2. The Case Study

In the SoilCare project, panel meetings including scientists, farmers and other stakeholders identified a number of threats to soil quality and fertility and proposed management techniques for the mitigation of these threats [36]. At the study site in Germany, the stakeholder panel suggested focusing on conservation agriculture and to investigate, among other techniques, specifically cover crop management. Therefore, a field experiment was set up at a research farm in Tachenhausen, Germany (48.649800° N, 9.387500° E, 330 m a.s.l.). In the present study, the assessment methodology was applied to the comparison between cover cropping and bare fallow treatments.

The soil is heavy, and loess derived, with a very fine sandy loam texture. The soil profile is characterized as Cambisol (IUSS Working Group WRB, 2015) with four horizons and with a ploughing pan at 40 cm. The climate is temperate with a mean annual temperature of 8.8 °C and 809.3 mm precipitation (monitoring weather station Tachenhausen HfWU, 150 m from the site, 1961–1990). The field has a history of conventional agriculture, with a crop rotation consisting mainly of cereals and sugar beet and winter oilseed rape as alternate break crops. The crop rotation for the experiment was spring barley (2018)–cover crop mixture/bare fallow–silage maize (2019)–spring barley (2020). The main crop for the 2019 cropping season was Zea mays (var. Figaro) sown on 6 May 2019 with 10 plants m−2 and harvested on 17 September 2019.

The experimental layout was a randomized complete block design with eightreplicates with plots of 2.4 × 3 m. The treatments included bare fallow and cover cropped plots. Originally, the field experiment was set up as a full factorial experiment, including also two herbicide treatments. However, as no significant interaction between the cover crop and herbicide treatments could be detected, the measurements could be averaged over the two cover crop treatments. For establishment of the field experiment, a commercial cover crop mixture consisting of 55% Vicia sativa, 20% Trifolium alexandrinum, 16% Phacelia tanacetifolia and 9% Helianthus annus was sown at 25 kg ha^{-1} in rows of 20 cm in the beginning of August. The field was tilled with a rotary harrow in a depth of 10 cm shortly before sowing the cover crops in a regime of non-inversion tillage. Mineral fertilizer was applied in 2018 at the rates of 90 kg ha^{-1} N, 17.5 kg ha^{-1} P, 53.1 kg ha^{-1} K, 8.1 kg ha^{-1} Mg and 20 kg ha^{-1} S. The maize in 2019 was not given any fertilizer. The following spring barley received mineral N-fertilizer at a rate of 89 kg ha^{-1} on 17 April 2020. Herbicides were applied as necessary.

The sampling and measurement of the indicators of the assessment methodology was carried out in spring after the cover crop in 2018–2019 following a monitoring plan with standardized methods for biological, physical and chemical properties of soils [26]. In the case study of Germany, the economic assessment was made possible by taking the values from publicly available tables of agricultural economics. For the calculation of the cost–benefit analysis, a sequence similar to the field experiment consisting of cereal-cover crop-silage-maize-cereal was used, but with winter wheat instead of spring barley. The sociocultural dimension was assessed by conducting semi-structured interviews based on the abovementioned questionnaires with five different farmers and a consultant of the public extension service of the region.

3. Results

3.1. Environmental Dimension

The assessment methodology was applied at the study site in Germany to compare the *SICS* integrating cover crops with the *control* treatment with bare fallow over winter. The environmental performance of the *SICS*, measured as the response of selected soil quality indicators, showed mixed results. Some indicators improved with cover crops (i.e., bulk density and soil cover) or showed a positive trend (number of earthworms) (Table 3). On the contrary, water-stable aggregates and infiltration were higher in the fallow plots, while weeds, tended to be lower than in the cover crop treatments. Mineral nitrogen tended to be lower under cover crops. Most of the other soil quality indicators showed no variation. This slight improvement indicates the positive effect of certain cover crop species on soil quality, especially on soil structure expressed by the reduced bulk density/increase in total porosity [37]. The resulting figures are presented in Supplementary Material File S3, the error bars represent the SE of the model.

Table 3. Impact of cover crops on the variables at least two years after the implementation.

Indicators	Impact of *SICS*	Score	Weight In *SICS*
Infiltration	No change	0	0.05
Aggregate stability	Negative change	−1	0.05
Bulk density	Positive change	+1	0.05
Penetration resistance	No data	0	0.00
Mineral nitrogen	No change	0	0.05
SOC	No change	0	0.05
pH	No change	0	0.05
Earthworm density	No change	0	0.05
Crop yield	No change	0	0.10
Yield quality	No change	0	0.10
Crop cover characteristics	Positive change	+1	0.10
Pests	No data	0	0.20
Root diseases	No data	0	0.10
Weed diseases	No change	0	0.05

Concerning the key topic addressed here, namely soil improving crops, there was a slight increase with an impact index of 0.10 (Table 4).

Table 4. Outcomes of the assessment of the environmental dimension with regard to the key topics.

Properties	Impact Index
Soil cultivation	0.33
Fertilisation	0.05
Soil improving crops	0.10
Compaction	0.15
Environmental dimension	0.18

3.2. Economic Dimension

In order to assess the economic dimension, the benefits of *SICS* were calculated in relation to the costs for the crop sequence of three years. Since the cereal straw was left on the field, the benefit is based on the pure grain yields, respective silage maize yield multiplied by the average market price in the respective year.

When comparing the benefits with the costs for both *control* and *SICS*, there is a loss in both cases (File S2), but less loss for the *control* than for the *SICS*.

The benefit of the *SICS* is higher than that of the *control* (Table 5). The cause of the loss is due to the higher costs for cover crop seeds and sowing that outweigh the slight increase in benefits (yield).

Table 5. Impact index of the economic dimension of *SICS* as compared to *control* considered in the CSS of Germany.

Cropping System	Impact Index
Cost	0.09
Benefits	0.06
Economic dimension	−0.03

3.3. Sociocultural Dimension

The assessment of the sociocultural dimension shows slight positive impact due to the improvement of farmer reputation, although the moderate increase in the workload due to the short time window left after harvest to perform sowing reduces the perceived overall benefit for the farmers (Table 6). The problem with the workload at harvest time could be mitigated by using the technique of harvest–sowing, but this is only possible in some combinations of main and cover crop.

Table 6. Assessment of the changes of the sociocultural dimension with cover crops as soil improving cropping system. The Impact index was calculated using the responses of practitioners in structured interviews at the study region in Germany.

Sociocultural Data	Impact Index
Workload	−0.33
Perceived risks	0.00
Farmer reputation	1.00
Sociocultural dimension	0.07

3.4. Overall Sustainability

The field study in Germany provides an example of the application of the tool (Table 7). In this case, the environmental and the sociocultural dimension improved slightly under cover cropping (*SICS*) compared to bare fallow (*control*). The economic dimension showed a negative scoring, because of a slight increase in costs. Further assessment in the coming years is necessary to confirm these results.

Table 7. Synthesis of the impact of applied *SICS*.

	Impact of Applied SICS
Sociocultural dimension	0.07
Economic dimension	−0.03
Environmental dimension	0.18
Overall sustainability	0.08

4. Discussion and Recommendations

4.1. Outcomes of the CSS of Germany

In order to evaluate the applicability of the assessment tool, it was applied to the dataset resulting from a field experiment at the German study site comparing cover crops and bare fallow as agricultural practices in a common crop rotation with cereals and silage maize. The proposed set of soil quality indicators was used to assess the environmental dimension. The effects on the economic dimension were evaluated by assessing the costs and benefits of the two systems, while the sociocultural dimension was studied using structured interviews with farmers. Generally, statistically observable effects of the *SICS* treatment on the measured soil properties in the field experiment were limited to a few indicators. Reports of positive effects of cover cropping on main crop yield and soil quality are abundant in the literature, but results vary [38,39]. The costs of the *SICS* with the inclusion of cover crops were slightly higher than in the conventional treatment, resulting in a slightly negative score of the economic dimension. The farmers that were interviewed for the assessment of the sociocultural dimension had consistently a positive opinion of cover crops, but also acknowledged management difficulties and a certain dependence on a favourable climate for cover crop establishment and performance.

Regarding the environmental dimension, the positive effect of cover crops on soil cover in spring was significant. Especially under conservation tillage management, cover crop litter constitutes a protective layer on the soil surface and provides important benefits for the agroecosystem, such as erosion protection, reduced evaporation and habitat for soil fauna [40]. Cover cropping had also a positive effect on bulk density, which is related to pore connectivity. As compaction is an increasingly acknowledged soil threat [41], cover crops provide an interesting opportunity to increase porosity, especially in systems with no or decreased mechanical soil loosening [42]. A previous study showed that cover crops reduced soil penetration resistance or compaction by 0–29% (average, 5%). They improve wet aggregate stability by 0–95% (average, 16%) and cumulative infiltration by 0–190% (average, 43%) [43]. In our case, soil biological properties tended to improve, with earthworm numbers showing a positive trend with cover crops although not significant (data not shown here), as well as the potential extracellular activity of β-glucosidase, an enzyme involved in the breakdown of cellulose (not shown). These results clearly showed that the micro-habitat provides more substrate and energy for microbial life under cover crops [38,44]. While most measured soil chemical attributes were not changed.

Despite these positive changes with cover crops, the *SICS* seemed to have also undesired effects on some soil variables: aggregate stability decreased significantly while bulk density decreases (or increase in total porosity). This last observation can be attributed to the dominance of the structural porosity created by earthworms. The unexpected decrease in aggregate stability indicates that positive effects of cover crops on different aspects of soil structure might require time and multiple growing cycles to develop [45]. More on the management side, weed pressure tended to be higher in the cover cropped plots compared to the bare fallow due probably to missing herbicide application, although weed suppression is another expected benefit from cover cropping. Maximising the cover crop biomass and an optimized termination and residue management can improve the weed-suppression capacity of cover crops [46]. The obtained scoring of the environmental dimension of the assessment methodology provided a quite accurate resume of the slight improvement of soil quality with cover crops compared to the bare fallow *control*, but with an uneven response of the different soil quality indicators.

Similarly, the farmers' rather positive opinion about cover crops was reflected by the improvement of the sociocultural dimension. The modest increase in workload was greatly offset by the improved reputation. This underscores the importance of prestige for decision making for practitioners [47], especially since the farmers in the region are increasingly worried about their public image, some of them even mentioned feeling attacked by media. The farmers also acknowledged potential positive benefits of the cover crops being especially interesting when considering the necessary adaption to climate change [48].

The slight negative scoring of the economic dimension matches the reality, as the adoption of *SICS* and other sustainable farming techniques frequently implies higher production and maintenance costs which are not covered economically due to the inability of the market to integrate externalities into pricing [49]. Potential benefits of cover crops were not included in the economic assessment, such as SOC increase, erosion reduction, N input by leguminous cover crops or an increased biodiversity. Nor could external benefits for society be included, such as reduced sediment runoff, C sequestration and positive effects on water quality or landscape, among others [50]. The complexity of management techniques based on (agro-)ecosystem functions means they frequently require a substantial amount of experience to yield satisfying results. Even worse, although management can alleviate many reasons for the underperformance of *SICS*, in some cases significant trade-offs between environmental performance and productivity remain and call for a paradigm shift [51]. Until then, in absence of effective market mechanisms, potential losses can be only compensated by increasing the subsidies for environmentally friendly farming practices.

The overall scoring of the SoilCare assessment methodology of cover cropping is therefore possibly partially biased by an overly negative score of the economic dimension, but seems to provide an acceptable resume of the effects of the adoption of this *SICS* for the sustainability of the system. When evaluating the assessment methodology in workshops at the German study site, the stakeholders provided a heterogeneous rating of the assessment methodology and made some suggestions that could likely improve the applicability and power of the tool.

Regarding the measured soil properties, some farmers suggested to include methods of visual assessment of soil structure, e.g., as in a shovel test, as this method is easy to perform for practitioners and it gives relevant information for practitioners [52]. Participants with a more academic background suggested the measurement of greenhouse gases to cover this important aspect of sustainability. Another possibility would be to integrate methods to assess the soil microbial community into the tool. Creamer et al. [53] explain in detail how soil life could be integrated in the concept of multifunctionality of agricultural ecosystems. It is clear that the selection of adequate methods for judging soil microbiological properties is context dependent. The authors give in their article three different possible contexts where soil microbiological properties could have an additional value: Mechanistically understanding of multifunctionality, optimising sustainable land management and soil quality monitoring over time. Further investigations are needed to include all the above aspects in the here presented tool.

4.2. Strengths and Challenges of the Assessment Tool

This paper describes a new assessment tool to assess the overall sustainability of soil-improving cropping systems illustrated by an example from Germany. An overview of results from SoilCare study sites obtained with the tool under various conditions within SoilCare is provided in [36]. Therefore, from the outset, our intention was to develop a practical and flexible tool to assess the overall sustainability that can be adapted for other purposes and contexts.

In our assessment methodology, we included environmental/soil quality, economic and socio-cultural aspects in order to take into account all factors that are relevant for the success of *SICS*. Nevertheless, it should be realised that our assessment remains a simplification of reality. To be able to develop an assessment methodology for *SICS* that could be used in SoilCare, some assumptions were necessary, and some limitations exist:

- Overall Sustainability assessed in this study has been defined within the three dimensions, environmental, sociocultural, and economic dimensions. The last dimension was restricted to economic benefits to the farmer during the assessment period considered in this study and does not take into account the benefits at larger spatial and temporal scales, e.g., benefits to society, off-site effects, long-term benefits. In reality, an agricultural system is sustainable when the trade-offs between the objectives considered for public evaluation of its performance, economic objectives, social objectives, and ecological objectives reach acceptable values for society as a whole [1].
- An economic assessment at farm level should include the whole rotation that is used, which was not always possible, as rotations are often longer than the 3 years of monitoring that was possible in SoilCare.
- Another limitation of the method, partially solved by considering rental costs, was the lack of detailed costs and benefits related to the equipment that should include depreciation costs occurring at longer time scale than the one considered in this study as well as the use of such equipment for other purposes than the ones related to the *SICS* considered here.
- In general, the SoilCare experiments were too short to show significant effects on the overall sustainability, e.g., soil organic carbon content, mineral nitrogen, pH, earthworm density) (Table 3). Some benefits of the *SICS* may require a longer time period to become detectable [54,55]. Besides, hydraulic conductivity and bulk density have a large spatial and temporal variability in the field, which makes it more difficult to detect significant differences without increasing dramatically the number of measurements [36].
- It should be kept in mind that monitoring was carried out for 2–4 years, and that specific conditions during the years of monitoring can have an impact on the outcomes. For example, the weather conditions during the short-term experiments were quite

specific, especially in 2018 occurred droughts at several study-sites, resulting in a drastic decrease in yield [41]. Moreover, all the years had high, sometimes record-breaking, temperatures.

- Considering the existing distortions of the market and the large dependence of European agriculture on subsidies, it could be debated whether the weight given to the economic dimension in the calculation of the overall sustainability score is biased by ideology instead of a true interest in the well-being of future generations. This societal benefit effect can be captured by an extension of the indicators and extensive data collection. The semi-quantitative nature of the sustainability index would allow for an extension considering the direction of the impact of *SICS* (positive, no change, negative) on different ecosystem services for society even if valuation is not possible.

5. Conclusions

The aim of this paper was to establish a tool for the assessment of the sustainability of the *SICS* at the farm/field scale. For this purpose, a decision tree based on weights (%) was chosen because it allows simple aggregation to assess the three dimensions of sustainability, namely, environmental, economic and sociocultural, and provides flexibility. The decision tree allowed us to set up a comprehensive and standardized methodology that could be further improved and used for different purposes. The methodology was tested with data from the SoilCare Horizon 2020 study site of Tachenhausen, Germany, for the assessment of the effect of integration of cover crops into the crop rotation. The effect on the environmental indicators was slightly positive, but most assessed properties did not change during the short time of implementation (two crop seasons). Regarding the social dimension, farmers reported that the increase in workload was outweighed by an improved reputation for using cover crops. Regarding the economic dimension, the incorporation of cover crops reduced slightly the profitability, due to the costs for seeds and establishment of cover crops, which were greater than the increased income from higher yields. Further development and refinement by considering various pedo-climatic and land management conditions, as well as long-term assessments, are needed to strengthen the predictions.

Supplementary Materials: The following are available online at https://www.mdpi.com/article/10.3390/land11050729/s1, File S1: Excel Tool for the evaluation of the sustainability, File S2: Excel Tool to assess economic dimension, File S3: Results of the statistical analysis–study site in Tachenhausen, Germany.

Author Contributions: Conceptualization: A.A., R.H. and M.H.; methodology: A.A., R.B., F.B., I.P., R.H., M.H. and L.F.; data acquisition: C.P., E.K., M.H. and A.A.; data analysis: I.P., A.A., F.B. and C.P.; writing—review—editing: A.A., M.H., R.B., F.B., I.P., R.H., L.F. and E.K. All authors have read and agreed to the published version of the manuscript.

Funding: This research has received funding from the European Union's Horizon 2020 research and innovation programme under grant agreement No 677407 (SoilCare project). RH also received funding from Dutch Ministry of LNV, via Kennis Basis programma 34, project KB-34-008-005.

Institutional Review Board Statement: Not applicable in this study.

Informed Consent Statement: Not applicable.

Data Availability Statement: The data presented in this study are available in the Supplementary Material.

Acknowledgments: We thank the stakeholders in the SoilCare study sites. Without their collaboration this research would not have been possible.

Conflicts of Interest: The authors declare no conflict of interest.

References

1. Sala, S.; Ciuffo, B.; Nijkamp, P. A systematic framework for sustainability assessment. *Ecol. Econ.* **2015**, *119*, 314–325. [CrossRef]
2. Dufour, D.L. Use of Tropical Rainforests by Native Amazonians. *Bioscience* **2009**, *40*, 652–659. [CrossRef]
3. Nachtergaele, F.; Biancalani, R.; Bunning, S.; George, H. Land Degradation Assessment: The LADA approach. In Proceedings of the 19th World Congress of Soil Science, Brisbane, Australia, 1–6 August 2010; pp. 72–75.
4. Devuyst, D. Linking impact assessment and sustainable development at the local level: The introduction of sustainability assessment systems. *Sustain. Dev.* **2000**, *8*, 67–78. [CrossRef]
5. Alaoui, A.; Barão, L.; Ferreira, C.S.; Hessel, R. Assessment of sustainability in agricultural areas: A review of available indicators, frameworks and approaches. *Land* **2022**, *11*, 537. [CrossRef]
6. Glover, J.D.; Reganold, J.P.; Andrews, P.K. Systematic method for rating soil quality of conventional, organic, and integrated apple orchards in Washington State. *Agric. Ecosyst. Environ.* **2000**, *80*, 29–45. [CrossRef]
7. Gomez-Limon, J.; Riesgo, L. Alternative approaches to the construction of a composite indicator of agricultural sustainability: An application to irrigated agriculture in the Duero basin in Spain. *J. Environ. Manag.* **2009**, *90*, 3345–3362. [CrossRef]
8. Binder, C.R.; Feola, G.; Steinberger, J.K. Considering the normative, systemic and procedural dimensions in indicator-based sustainability assessments in agriculture. *Environ. Impact Assess. Rev.* **2010**, *30*, 71–81. [CrossRef]
9. Bitsch, V. Sustainability as innovation: Challenges and perspectives in measurement and implementation. In *Diffusion and Transfer of Knowledge in Agriculture*; Huyghe, C., Bergeret, P., Svedin, U., Eds.; Matière à Débattre et Decider: Quae, France, 2016; Chapter 2.
10. Van Cauwenbergh, N.; Biala, K.; Bielders, C.; Brouckaert, V.; Franchois, L.; Garcia Cidad, V.; Hermy, M.; Mathijs, E.; Muys, B.; Reijnders, J.; et al. SAFE-A hierarchical framework for assessing the sustainability of agricultural systems. *Agric. Ecosyst. Environ.* **2007**, *120*, 229–242. [CrossRef]
11. Bockstaller, C.; Guichard, L.; Makowski, D.; Aveline, A.; Girardin, P.; Plantureux, S. Agri-environmental indicators to assess cropping and farming systems. A review. *Agron. Sustain. Dev.* **2008**, *28*, 139–149. [CrossRef]
12. Bockstaller, C.; Feschet, P.; Angevin, F. Issues in evaluating sustainability of farming systems with indicators. *Oilseeds Fats Crops Lipids* **2015**, *22*, D102. [CrossRef]
13. Pelzer, E.; Fortino, G.; Bockstaller, C.; Angevin, F.; Lamine, C.; Moonen, C.; Vasileiadis, V.; Guérin, D.; Guichard, L.; Reau, R.; et al. Assessing innovative cropping systems with DEXiPM, a qualitative multi-criteria assessment tool derived from DEXi. *Ecol. Indic.* **2012**, *18*, 171–182. [CrossRef]
14. Bechini, L.; Castoldi, N. On-Farm Monitoring of Economic and Environmental Performances of Cropping Systems: Results of a 2-Year Study at the Field Scale in Northern Italy. *Ecol. Indic.* **2009**, *9*, 1096–1113. [CrossRef]
15. Gaudino, S.; Reidsma, P.; Kanellopoulos, A.; Sacco, D.; van Ittersum, M.K. Integrated assessment of the EU's greening reform and feed self-sufficiency scenarios on dairy farms in Piemonte, Italy. *Agriculture* **2018**, *8*, 137. [CrossRef]
16. Smith, A.; Snapp, S.; Chikowo, R.; Thorne, P.; Bekunda, M.; Glover, J. Measuring sustainable intensification in smallholder agroecosystems: A review. *Glob. Food Secur.* **2017**, *12*, 127–138. [CrossRef]
17. Bond, A.J.; Morrison-Saunders, A.; Pope, J. Sustainability assessmenet: The state of the art. *Impact Assess. Proj. Apprais.* **2012**, *301*, 53–62. [CrossRef]
18. Röös, E.; Fischer, K.; Tidåker, P.; Nordström Källström, H. How well is farmers' social situation captured by sustainability assessment tools? A Swedish case study. *Int. J. Sustain. Dev. World Ecol.* **2019**, *26*, 268–281. [CrossRef]
19. Lynch, J.; Donnellan, T.; Finn, J.A.; Dillon, E.; Ryan, M. Potential development of Irish agricultural sustainability indicators for current and future policy evaluation needs. *J. Environ. Manag.* **2019**, *230*, 434–445. [CrossRef]
20. Mottet, A.; Bicksler, A.; Lucantoni, D.; De Rosa, F.; Scherf, B.; Scopel, E.; Lopez-Ridaura, S.; Gemmill-Herren, B.; Bezner Kerr, R.; Sourisseau, J.M.; et al. Assessing transitions to sustainable agricultural and food systems: A tool for agroecology performance evaluation (TAPE). *Front. Sustain. Food Syst.* **2020**, 252. [CrossRef]
21. Häni, F.; Braga, F.; Stämpfli, A.; Keller, T.; Fischer, M.; Porsche, H. RISE, a tool for holistic sustainability assessment at the farm level. *Int. Food Agribus. Manag. Rev.* **2003**, *6*, 78–90. [CrossRef]
22. Sadok, W.; Angevin, F.; Bergez, J.E.; Bockstaller, C.; Colomb, B.; Guichard, L.; Reau, R.; Messéan, A.; Doré, T. MASC, a qualitative multi-attribute decision model for ex ante assessment of the sustainability of cropping systems. *Agron. Sustain. Dev.* **2009**, *29*, 447–461. [CrossRef]
23. Landert, J.; Pfeifer, C.; Carolus, J.; Schwarz, G.; Albanito, F.; Muller, A.; Smith, P.; Sanders, J.; Schader, C.; Vanni, F.; et al. Assessing agro-ecological practices using a combination of three sustainability assessment tools. *Landbauforschung* **2020**, *70*, 129–144. [CrossRef]
24. *FAO Safa Guidelines—Sustainability Assessment of Food and Agriculture Systems*; FAO: Rome, Italy, 2014; ISBN 9789251084854.
25. Gerrard, C.L.; Smith, L.G.; Pearce, B.; Padel, S.; Hitchings, R.; Measures, M.; Cooper, N. Farming for Food and Water Security. In *Farming for Food and Water Security*; Springe: Dordrecht, The Netherlands, 2012; Volume 10, pp. 1–22, ISBN 978-94-007-4499-8.
26. Alaoui, A.; Schwilch, G.; Bachmann, F.; Panagea, I.; Diels, J.; Wyseure, G.; Hessel, R. Deliverable 4.2. Monitoring Plan for Study Sites. SoilCare Project 2019, Report. Available online: https://www.soilcare-project.eu/ (accessed on 10 April 2022).
27. Smith, A.; Cullis, B.; Thompson, R. The analysis of crop cultivar breeding and evaluation trials: An overview of current mixed model approaches. *J. Agric. Sci.* **2005**, *143*, 449–462. [CrossRef]
28. RStudio Team. *RStudio: Integrated Development Environment for R*; RStudio, Inc.: Boston, MA, USA, 2016.

29. Sattler, C.; Nagel, U.J. Factors affecting farmers' acceptance of conservation measures—A case study from north-eastern Germany. *Land Use Policy* **2010**, *27*, 70–77. [CrossRef]
30. Fantappiè, M.; Lorenzetti, R.; De Meo, I.; Costantini, E.A.C. How to improve the adoption of soil conservation practices? Suggestions from farmers' perception in western Sicily. *J. Rural. Stud.* **2020**, *73*, 186–202. [CrossRef]
31. Trujillo-Barrera, A.; Pennings, J.M.E.; Hofenk, D. Understanding producers' motives for adopting sustainable practices: The role of expected rewards, risk perception and risk tolerance. *Eur. Rev. Agric. Econ.* **2016**, *43*, 359–382. [CrossRef]
32. Schneider, F.; Ledermann, T.; Fry, P.; Rist, S. Soil conservation in Swiss agriculture—Approaching abstract and symbolic meanings in farmers' life-worlds. *Land Use Policy* **2010**, *27*, 332–339. [CrossRef]
33. Schneider, F.; Rist, S. The significance of aesthetics for the adoption of no-tillage farming. *Agrar. Schweiz* **2012**, *3*, 216–223.
34. Sutherland, L.A.; Darnhofer, I. Of organic farmers and 'good farmers': Changing habitus in rural England. *J. Rural. Stud.* **2012**, *28*, 232–240. [CrossRef]
35. Fleskens, L.; Gaaff, M.; Verzandvoort, S.; Irvine, B. Model Assessments of the Cost- Effectiveness of Prevention/ Remediation/ Restoration Measures for All CASE STUDY Areas. RECARE Deliverable 7.2. 2018. Available online: https://www.recare-hub.eu/images/WP7/RECARE_REPORT_29_D72_final.pdf (accessed on 10 April 2022).
36. Hessel, R.; Wyseure, G.; Panagea, I.; Alaoui, A.; Reed, M.S. Soil Improving cropping systems for sustainable and profitable farming in Europe. *Land* 2022, *submitted*.
37. Hallama, M.; Pekrun, C.; Pilz, S.; Jarosch, K.A.; Frąc, M.; Uksa, M.; Marhan, S.; Kandeler, E. Interactions between cover crops and soil microorganisms increase phosphorus availability in conservation agriculture. *Plant Soil* **2021**, *463*, 307–328. [CrossRef]
38. Maltais-Landry, G.; Scow, K.; Brennan, E.; Vitousek, P. Long-term effects of compost and cover crops on soil phosphorus in two California agroecosystems. *Soil Sci. Soc. Am. J.* **2015**, *79*, 688–697. [CrossRef]
39. Cardoso, D.P.; Silva, M.L.N.; de Carvalho, G.J.; de Freitas, D.A.F.; Avanzi, J.C. Potential of cover plant species in reconditioning of physical and chemical properties of soil. *Revista Brasileirade Ciencias Agrarias* **2013**, *8*, 375–382. [CrossRef]
40. Junge, S.M.; Storch, J.; Finckh, M.R.; Schmidt, J.H. Developing Organic Minimum Tillage Farming Systems for Central and Northern European Conditions. In *No-Till Farming Systems for Sustainable Agriculture: Challenges and Opportunities*; Dang, Y.P., Dalal, R.C., Menzies, N.W., Eds.; Springer International Publishing: Cham, Switzerland, 2020; pp. 173–192, ISBN 978-3-030-46409-7.
41. Piccoli, I.; Sartori, F.; Polese, R.; Berti, A. Crop yield after 5decades of contrasting residue management. *Nutr. Cycl. Agroecosystems* **2020**, *117*, 231–241. [CrossRef]
42. Bacq-Labreuil, A.; Crawford, J.; Mooney, S.J.; Neal, A.L.; Ritz, K. Cover crop species have contrasting influence upon soil structural genesis and microbial community phenotype. *Sci. Rep.* **2019**, *9*, 7473. [CrossRef]
43. Blanco-Canqui, H.; Ruis, S.J. Cover crop impacts on soil physical properties: A review. *Soil Sci. Soc. Am. J.* **2020**, *84*, 1527–1586. [CrossRef]
44. Hallama, M.; Pekrun, C.; Mayer-Gruner, P.; Uksa, M.; Abdullaeva, Y.; Pilz, S.; Schloter, M.; Lambers, H.; Kandeler, E. The Role of Microbes in the Increase of Organic Phosphorus Availability in the Rhizosheath of Cover Crops. *Plant Soil* **2022**. [CrossRef]
45. Brennan, E.B.; Acosta-Martinez, V. Cover Cropping Frequency Is the Main Driver of Soil Microbial Changes during Six Years of Organic Vegetable Production. *Soil Biol. Biochem.* **2017**, *109*, 188–204. [CrossRef]
46. Osipitan, O.A.; Dille, J.A.; Assefa, Y.; Radicetti, E.; Ayeni, A.; Knezevic, S.Z. Impact of Cover Crop Management on Level of Weed Suppression: A Meta-Analysis. *Crop Sci.* **2019**, *59*, 833–842. [CrossRef]
47. Mills, J.; Ingram, J.; Dibari, C.; Merante, P.; Karaczun, Z.; Molnar, A.; Sánchez, B.; Iglesias, A.; Ghaley, B.B. Barriers to and opportunities for the uptake of soil carbon management practices in European sustainable agricultural production. *Agroecol. Sustain. Food Syst.* **2020**, *44*, 1185–1211. [CrossRef]
48. Kaye, J.P.; Quemada, M. Using cover crops to mitigate and adapt to climate change. A review. *Agron. Sustain. Dev.* **2017**, *37*, 4. [CrossRef]
49. Bergtold, J.S.; Ramsey, S.; Maddy, L.; Williams, J.R. A Review of Economic Considerations for Cover Crops as a Conservation Practice. Renew. *Agric. Food Syst.* **2019**, *34*, 62–76. [CrossRef]
50. Blanco-Canqui, H.; Shaver, T.M.; Lindquist, J.L.; Shapiro, C.A.; Elmore, R.W.; Francis, C.A.; Hergert, G.W. Cover Crops and Ecosystem Services: Insights from Studies in Temperate Soils. *Agron. J.* **2015**, *107*, 2449–2474. [CrossRef]
51. Wittwer, R.; Bender, S.; Hartman, K.; Hydbom, S.; Lima, R.; Loaiza, V.; Nemecek, T.; Oehl, F.; Olsson, P.; Petchey, O.; et al. Organic and Conservation Agriculture Promote Ecosystem Multifunctionality. *Sci. Adv.* **2021**, *7*, eabg6995. [CrossRef] [PubMed]
52. Alaoui, A.; Barão, L.; Ferreira, C.S.; Schwilch, G.; Basch, G.; Garcia-Orenes, F.; Morugan, A.; Mataix-Solera, J.; Kosmas, C.; Glavan, M.; et al. Visual assessment of the impact of agricultural management practices on soil quality. *Agron. J.* **2020**, *112*, 2608–2623. [CrossRef]
53. Creamer, R.E.; Barel, J.M.; Bongiorno, G.; Zwetsloot, M.J. The Life of Soils: Integrating the Who and How of Multifunctionality. *Soil Biol. Biochem.* **2022**, *166*, 108561. [CrossRef]
54. Kandeler, E.; Tscherko, D.; Spiegel, H. Long-term monitoring of microbial biomass, N-mineralisation and enzyme activities of a Chernozem under different tillage management. *Biol. Fertil. Soils* **1999**, *28*, 343–351. [CrossRef]
55. Kandeler, E.; Stemmer, M.; Klimanek, E.M. Response of soil microbial biomass, urease and xylanase within particle size fractions to long-term soil management. *Soil Biol. Biochem.* **1999**, *31*, 261–273. [CrossRef]

land

MDPI

Article

Do Agricultural Advisory Services in Europe Have the Capacity to Support the Transition to Healthy Soils?

Julie Ingram [1,*], Jane Mills [1], Jasmine E. Black [1], Charlotte-Anne Chivers [1], José A. Aznar-Sánchez [2], Annemie Elsen [3], Magdalena Frac [4], Belén López-Felices [2], Paula Mayer-Gruner [5], Kamilla Skaalsveen [6], Jannes Stolte [6] and Mia Tits [3]

1 Countryside and Community Research Institute, University of Gloucestershire, Cheltenham GL504AZ, UK; jmills@glos.ac.uk (J.M.); jblack2@glos.ac.uk (J.E.B.); cchivers@glos.ac.uk (C.-A.C.)
2 Department of Economy and Business, Research Centre on Mediterranean Intensive Agrosystems and Agrifood Biotechnology (CIAIMBITAL), University of Almería, 04129 Almería, Spain; jaznar@ual.es (J.A.A.-S.); blopezfelices@ual.es (B.L.-F.)
3 Soil Service of Belgium (Bodemkundige Dienst van België) W. de Croylaan 48, 3001 Leuven, Belgium; aelsen@bdb.be (A.E.); mtits@bdb.be (M.T.)
4 Institute of Agrophysics, Polish Academy of Sciences, Doświadczalna 4, 20-290 Lublin, Poland; m.frac@ipan.lublin.pl
5 Soil Biology Department, Institute of Soil Science and Land Evaluation, University of Hohenheim, 70599 Stuttgart, Germany; p.mayer-gruner@uni-hohenheim.de
6 Department of Soil and Land Use, Division of Environment and Natural Resources, Norwegian Institute of Bioeconomy Research, NO-1431 Ås, Norway; kamilla.skaalsveen@nibio.no (K.S.); jannes.stolte@nibio.no (J.S.)
* Correspondence: jingram@glos.ac.uk

Citation: Ingram, J.; Mills, J.; Black, J.E.; Chivers, C.-A.; Aznar-Sánchez, J.A.; Elsen, A.; Frac, M.; López-Felices, B.; Mayer-Gruner, P.; Skaalsveen, K.; et al. Do Agricultural Advisory Services in Europe Have the Capacity to Support the Transition to Healthy Soils? *Land* **2022**, *11*, 599. https://doi.org/10.3390/land11050599

Academic Editor: Amrakh I. Mamedov

Received: 9 March 2022
Accepted: 12 April 2022
Published: 19 April 2022

Abstract: The need to provide appropriate information, technical advice and facilitation to support farmers in transitioning towards healthy soils is increasingly clear, and the role of the Agricultural Advisory Services (AAS) in this is critical. However, the transformation of AAS (plurality, commercialisation, fragmentation, decentralisation) brings new challenges for delivering advice to support soil health management. This paper asks: To what extent do agricultural advisory services have the capacity to support the transition to healthy soils across Europe? Using the 'best fit' framework, analytical characteristics of the AAS relevant to the research question (governance structures, management, organisational and individual capacities) were identified. Analysis of 18 semi-structured expert interviews across 6 case study countries in Europe, selected to represent a range of contexts, was undertaken. Capacities to provide soil health management (SHM) advice are constrained by funding arrangements, limited adviser training and professional development, adviser motivations and professional cultures, all determined by institutional conditions. This has resulted in a narrowing down of access and content of soil advice and a reduced capacity to support the transition in farming to healthy soils. The extent to which emerging policy and market drivers incentivise enhanced capacities in AAS is an important area for future research.

Keywords: agricultural advisory services; soil health; governance; agricultural advisers; sustainable soil management; soil policy; advice

1. Introduction

Soil health has emerged as a priority for high level and national policy makers and for agricultural communities. This is linked to the recognition of the multiple functions that soils fulfil and the soil degradation processes closely linked to agriculture: erosion, organic carbon decline, soil biodiversity decline, compaction, contamination, salinisation and acidification [1,2]. Indeed, soil health is seen as "a key solution for our big challenges" in the newly launched European Union (EU) Soil Strategy, which builds on the European Green Deal and the Farm to Fork Strategy [3,4]. For agricultural soils, soil health and managing soil sustainably are regarded as central to food system transition pathways such as agroecology and regenerative agriculture, managing carbon for climate change mitigation and adaptation and mitigating pollution for human wellbeing.

However, the soil governance landscape (formal and informal institutions related to soil-related decision-making processes) continues to be highly fragmented. It is characterised by multi-level and multi-actor decision making, with no single body responsible at the EU or national levels [5] and could be described as a network mode of governance [6]. A number of public and private mechanisms are applied that influence agricultural soil management decisions (directly or indirectly), reflecting the multiple functions (provisioning, filtering of nutrients, carbon storage, flood mitigation) and private and public good that soils provide [7]. These include public cross-sectoral policy instruments (regulatory and voluntary) at the EU, national and regional levels; market-led (food assurance schemes in the supply chain); and measures which are led by the farming industry and non-governmental organisations (NGO) (voluntary initiatives, partnerships and networks).

This emphasis on soil health and its complex governance arena brings new challenges both for land managers and those that support them. The need to provide appropriate information, technical advice and facilitation to support farmers in transitioning towards sustainable soil management [8] has been identified by a number of researchers and policy makers at the international, European and national levels [8–13].

Agricultural Advisory Services (AAS) have always constituted an important part of farmer decision making with respect to soil management [14,15]. However, the increasing complexities of managing different soil functions and soil health at the farm scale [9] places new demands on these services.

Soil health has been defined as the capacity of soil to function as a vital living system [16]; however, the concept, and how it is operationalised, is still evolving [17]. Consequently, there are many understandings of what constitutes good soil health management (SHM). There are multiple practices embodied in the soil health concept, such as the use of cover crops and residues and reduced tillage; however, there is no single message or set of advice that is relevant to all contexts. Emerging interest in soil health indicators, soil biological processes and soil carbon dynamics [18] and new farming approaches (e.g., regenerative agriculture, agroecology), requires increasingly specialist knowledge and understanding (metrics, sampling techniques and analysis, interpretation) [19–21] beyond the traditional territory of soil fertility and agronomy. Meeting famers' knowledge needs, building their capacity and facilitating shared learning for SHM presents new imperatives for advisers [22]. These challenges exist against a backdrop of a changing farming population operating in a volatile, competitive marketplace negotiating multiple drivers in the agri-food system.

AAS have themselves been in transition, with privatisation and decentralisation occurring to different extents across Europe over the past 30 years [23]. The diversity of actors, intermediaries and organisations from the private (The private farm advisory sector includes profit and non-profit enterprises. Prager et al. [24] distinguishes 'private' as the status of an organisation and 'commercial' as the activities carried out by the organisation (e.g., offering advisory services for a fee)) and public sectors and NGOs engaged in some way in offering advice that influences soil management has grown. In particular, there has been an increase in the number of private advisers (These include: commercial agronomists offering services as part of farm input sales; farm management consultants; independent advisers or technicians within the supply-chain, sector or industry body or employed by farmer-owned groups) [25] and those with commercial links to farmers [26,27]. There is debate about the impact of such diversity on governance with respect to the integration and fragmentation of advice, competition and cooperation and how on access to quality advice [24,28,29]. Arguments about the advantages and disadvantages of privatisation have also been well rehearsed [30–32]. The potentially negative impact of commercialisation on public goods advice [33] and the limited investment in updating environmental knowledge for advisers has been highlighted [34]. The powerful effect of new economic actors, such as those in the supply chain, on environmental objectives has also been demonstrated [26,27,35] and noted specifically for soil management [10,36,37]. Private sector providers who support production goals can promote practices detrimental to soil health (e.g., multiple field operations with heavy machinery, a reliance on inorganic fertilizer, poor budgeting of organic inputs, harvesting in unsuitable conditions) [38]. Meanwhile, resources for public sector advice to farmers on the mitigation of soil degradation processes have also been shown to be inadequate [39].

Although there has been a requirement for all EU member states (since 2007) to establish a Farm Advisory System (FAS) (according to FAS, Regulation (EC) N° 73/2009) to support farmers in meeting cross-compliance requirements, including soil management though Good Agricultural and Environmental Conditions (GAEC), the singular advisory focus on compliance has been to the detriment of other soil functions and soil health outcomes [40].

The role of the individual adviser is also shifting, demanding greater professionalisation in increasingly specialised sectors, technical expertise (subject-matter knowledge), facilitation skills and awareness of a number of policy instruments, innovations, industry demands, certifications and environmental objectives. In such an environment, the assumption is that advisers will pursue different knowledge and strengthen and broaden their suite of professional practices to suit the 'new farming paradigm' [41,42]. At the same time advisers need to stay abreast of the farming community's growing informal soil knowledge networks, [43,44] and the different ways they negotiate their own microAKIS [27]. However, a body of evidence has been accumulating [10,14] suggesting a lack of specialist soil technical support and expertise in the advisory community, a poor understanding of the impact and externalities of their advice for soil, as well as varying motivations. Although studies show that farmers are deferring to advisers for their soil testing, largely in arable sectors [45,46], the lack of meaningful guidance for advisers regarding interpretation of these tests for soil health, especially for soil organic matter, and for specific farm conditions and management, is a concern [22,47]. There are a number of examples of effective soil advice across Europe [39]; however, it is clear that there are variable skill sets [11].

These insights raise questions about the capacity of advisory organisations and the constituent advisers for supporting SHM. This paper asks: To what extent do agricultural advisory services have the capacity to support the transition to healthy soils across Europe?

This addresses a recognised research gap, since understanding how the economic resources and strategies of advisory organisations determine the content of advice has received little (particularly for soil health) attention [26,30]. Equally, although soil literacy and societal engagement are central to the EU Soil Strategy and the implementation of the Mission for Soil and the European Soil Partnership, little has been done to understand the level of knowledge and expertise about soil health management in AAS.

This question is addressed using an analytical framework which positions AAS within the wider Agricultural Knowledge and Innovation System (AKIS), in which AAS are a subsystem. The framework implies that a range of organisations and stakeholders are involved in agricultural innovations along agricultural value chains, as well as agricultural research, agricultural extension and agricultural education [48]. Work conducted in the EU-funded SoilCare project underpins this analysis.

2. Concepts and Framework

AAS can be defined as sets of organisations that support and facilitate people engaged in agricultural production to solve problems and to obtain information, skills and technologies by enabling farmers to co-produce farm-level solutions by establishing service relationships with advisers [30,32,49]. AAS comprise traditional advice providers (chambers of agriculture, public bodies, research institutes), farmer-based organisations (FBOs) (unions, associations, cooperatives), non-governmental organisations (NGOs), independent consultants as well as advisers working in upstream or downstream industries, supply chains and high-tech sectors. However, these distinct categories do not fully capture the different arrangements and the new actors and roles emerging [23,26,27]. The term 'pluralistic' is used to describe the diversity of institutional options in providing and financing AAS [50]. AAS are characterised by a range of approaches, including one-to-one advice, facilitated interactive group approaches to foster peer-to-peer learning and mass dissemination.

We have adapted the 'best fit' (Birner's [49] framework was proposed for identifying modes of providing and financing advisory services that 'best fit' the specific conditions and development priorities of specific countries) framework developed by Birner, Davis, Pender, Nkonya, Anandajayasekeram, Ekboir, Mbabu, Spielman, Horna and Benin [50] to analyse the capacity of AAS for supporting SHM (Figure 1). Due to the multiple interacting dimensions of the AKIS and the AAS, it is difficult to collect data to capture the full complexity and interdependence of the system [51]. This framework provides a means of disentangling the different dimensions within the system. Here, we use selected key analytical categories in the framework relevant to the research questions. We define SHM as 'where management maintains or enhances (and does not impair) the capacity of soil to function as a vital living system, and to provide supporting, provisioning, regulating cultural services'. SHM is underpinned by the following management principles identified in the SoilCare project: integrate crop rotation, maintain continuous soil cover, build organic matter, minimise soil disturbance, prevent soil compaction, manage water for soil, use soil-friendly weed/pest control and consider landscape-scale management. These were derived by scientific review [52] and experimentation [53], as documented in this Special Issue, and have proven soil health benefits [54,55].

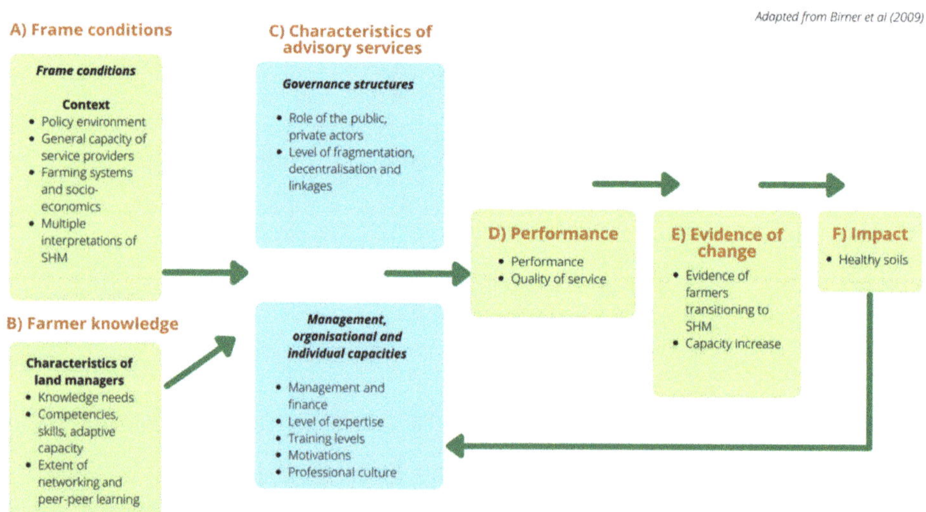

Figure 1. Framework—Blue shading represents the analytical characteristic investigated [50].

The focus of this study is on Characteristics of the system of agricultural advisory services (C in blue in Figure 1) and the implications for SHM. Specifically, according to Birner, Davis, Pender, Nkonya, Anandajayasekeram, Ekboir, Mbabu, Spielman, Horna and Benin [50], we include:

- Governance structures: the roles of the public and private sectors and civil society in providing and financing advisory services, the level of decentralisation and the linkages and partnerships among agents in the innovation system.
- Management, organisational and individual capacities: this refers to the expertise, training, motivations of the members of the advisory service as well as their incentives, professional and organisational culture.

These were translated into characteristics relevant to SHM and framed the data collection. We understand that *Governance* (Usually defined as the systems of institutional rules, policies and processes which govern how roles and responsibilities are delegated, managed and coordinated) *structures* enable and constrain organisational activities, in particular, the institutional options available for financing and the extent of coordination, fragmentation and integration [56]. Regarding the AAS M*management, organisational and individual capacities* to deliver SHM advice, organisational capacity is usually defined in terms of the capacity to perform effectively or to fulfil a goal (e.g., as the set of processes, management practices or attributes that assist an organisation in fulfilling its mission [57]). This capacity also affects the individual. Previous work has shown, for example, that the back-office support such as training and the organisation's knowledge management impacts advisers' capabilities [58].

Regarding individuals, AAS studies recognise that certain individual competencies are necessary for organisations and their advisers to operate effectively, where competence is: to have sufficient knowledge and skills that enable a person to act in a wide variety of situations and the ability to perform something efficiently and effectively (i.e., successfully) [59,60]. These skills include technical skills, which relate to specialist understanding (knowledge, expertise), and process skills, which are soft skills and refer to the collective skills necessary for effective performance of the individual and their organisation. Technical skills with respect to SHM are foregrounded in this study; however, we recognise that the 'soft' skills of facilitators, intermediaries and network builders are important [61].

This infrastructural approach to assessing AAS, which focuses on the presence and interaction of actors and the infrastructures that govern the behaviour of actors [62] also draws on selected criteria that Prager, Creaney and Lorenzo-Arribas [30] identified for evaluating a functional advisory system.

This paper focuses its analysis on C to address the research question, and because these characteristics can be influenced directly by policy makers. However, data for Frame conditions (A), Characteristics of farmers/land managers and their knowledge needs (B), Characteristics in Performance (D), Evidence (E) and Impact (F) (Figure 1) were also collected and analysed. Frame conditions (A) are important contextual factors in shaping the AAS, particularly as these have implications for SHM, and these include: Policy environment; Capacity of potential service providers; Farming systems and socio-economic conditions. Equally, we acknowledge that Farmers' knowledge needs (B) (which we inserted into the framework) are important for assessing the adviser's role. It is not the intention here to follow an impact chain approach to analyse the performance of agricultural advisory services (D,E,F). Assessing (Performance, D) and the quality of advice is challenging, since its measure is the outcome for the farmer and there are multiple factors that affect this [23,50,63].

3. Methods

Countries in Europe are highly diversified in terms of their AAS and AKIS, reflecting the structure of agriculture, farming systems, soils and productivity [64] and the extent to which AKIS are embedded in national institutions, laws and cultures [27]. Six case studies (drawn from country partners in the SoilCare project) were selected to represent a range of AAS approaches and contexts: Norway, Belgium (Flanders) (Belgium (Flanders was the case study for Belgium) as Wallonia operates a different system), Spain, the UK (England) (As the UK's four countries have different political structures and agricultural policies, the focus was on England), Germany and Poland. The selection was based on three broad criteria: firstly, AAS organisations (to ensure the dominant ones were represented); secondly, characterisation of the AKIS, according to strength of national influence and level of integration/fragmentation (based on the the PROAKIS project); and thirdly, to include a range of biogeographical and pedoclimatic zones, as already determined in the SoilCare case study selection process [65].

This selection was informed by detailed AKIS descriptions for each of these countries from a range of sources, including PROAKIS [66] and i2connect [67] project country reports [23,30,32,68–70], and previous reviews for soil [39]. The dominant AAS are represented in the case studies: Spain FBO; England (Private); Germany (Public/Private/FBO); Poland (Public); Norway (Private); and Flanders (Public) according to previous studies [23]. The pedoclimatic zones represented are: Atlantic Central (Flanders, Germany, England), Nemoral/Boreal (Norway), Continental (Poland) and Mediterranean South (Spain).

Semi-structured interviews (2–4) were conducted in each case study (a total of 18). Selection procedures and interviews were carried out in each case study by project partners using standardised guidance and protocols. Respondents were selected were: representatives of decision/policy makers at national and regional level or of AAS organisations who were knowledgeable about SHM. As this was a very limited pool of experts, a purposive sampling strategy was employed. Table 1 lists the respondents and their roles and affiliation and shows the range of AAS organisations represented.

Table 1. The case study respondents and their roles and affiliations.

Norway	Flanders (Belgium)	Spain	England (UK)	Germany	Poland
N1 Representative of NLR (Norsk Landbruksrådgivning) Norwegian Agricultural Advisory *Private independent/ FBO*	BE1 Researcher and extension worker at Flemish Research Station *Public*	ES1 Technical director of agriculture and research at association of farmers and livestock breeders *FBO*	UK1 Independent agricultural consultant *Private independent*	GR1 Representative from the district administration, Agricultural Office Baden-Württemberg *Public*	PL1 Professor of Agriculture Science (fruit and veg sector) *Public*
N2 Representative of NLR *Private independent/ FBO*	BE2 Adviser at the Soil Service of Belgium *Private independent*	ES2 Professor of soil science and agricultural chemistry *Public*	UK2 National farm advice manager for a consultancy *Private commercial*	GR2 Representative from the district administration, Agricultural office, Baden-Württemberg *Public*	PL2 Company producing micro-organisms and organic grower *Private commercial*
N3 Representative of NLR *Private independent/ FBO*	BE3 Representative of Flemish Land Agency *Public*	ES3 Research coordinator at research and transfer centre *Private commercial*		GR3 Board member of agricultural cooperative, Brandenburg *FBO*	P3 Company adviser for horticultural sector *Private commercial*
		ES4 Researcher in agricultural research and transfer centre *Private commercial*			

The analytical categories (Figure 1) were translated into interview questions and topics as shown in Table 2. Interviews were recoded, transcribed and translated into English, then analysed thematically using Nvivo 12. The coding structure followed the analytical categories of the interview but was extended where other themes emerged inductively. In total, 18 interviews provided in-depth analysis of AAS capacities for supporting SHM. A list of abbreviations is provided. A full interview schedule is provided as Supplementary Materials.

Table 2. Analytical categories translated into interview topics: example questions.

Characteristics of AAS for Supporting SHM
Governance Structures
With respect to advice that supports or impacts soil management: • the key actors and organisations providing advice; the key influencers; • the roles of the public and private sectors and civil society in providing and financing advisory services; • the level of diversity, decentralisation, coordination, integration or fragmentation of these services; • the extent of linkages and partnerships among actors.
Management, organisational and individual capacities
• the extent of organisation/management/resourcing of advisory services for delivering advice on soil and the impact on advisers' ability to provide soil advice; • different advisers' expertise for delivering SHM advice, quality of advice, level of soil management training; • the attitudes and motivations of the different sorts of advisers and organisational cultures.

4. Results

Where quotes are provided, the code refers to the notation in Table 1. A summary of the results is provided in Table 3.

Table 3. Summary of AAS characteristics for case studies.

Governance Structure	Norway	Flanders (Belgium)	Spain	England (UK)	Germany	Poland
Integration/ fragmentation	Private or FBOs cooperate and obtain support from public bodies for SHM	Research institutes collaborate to address soil topics	Synergies exist, but some conflict and tensions at farm level	Horizontally fragmented, but partnerships work with shared goal	Synergies exist, but some tensions between individuals	Synergies between public ODRs and private sector but some tensions
AAS capacity						
Management and organisation for SHM advice	Good competence and capacity to deliver SHM advice in NRL Staff recruitment and retention is a problem	Reliance on short-term project funding reduces continuity in SHM advice	Good organisation and management in FBOs but limited in others Culture of short-term projects limits outlook	Absence of planning for the necessary SHM skills and staff	Consultation services are well equipped, good resources; public provision has staffing limitations (Brandenburg)	Public sector under-resourced

Table 3. *Cont.*

Governance Structure	Norway	Flanders (Belgium)	Spain	England (UK)	Germany	Poland
Level of advisers' SHM knowledge *Knowledge and practical experience* *Soil fertility focus* *Environmental shift* *Heterogeneity*	Public bodies not very competent or up to date, main role in supporting fertilizer plans and subsidies, but high standard of soil management in the independent NRL	Good knowledge where advice linked to research institutes and independent services but commercial advisers have fertilizer focus	Unequal distribution of quality SHM advice Commercial and technical advisers emphasise fertilizers Big range in SHM advice quality, poor quality linked to new untrained advisers, good quality in organic sector	Some shift in private adviser activities to environmental and soil advice Large range, some excellent advisers (independent and in non-commercial initiatives) but commercial advisers have limited SHM knowledge	Good quality advice in consultation services, but scope could be wider Commercial advisers' emphasis on fertilization conflicts with advice for other soil functions but some shift to supplying environmental advice	Commercial advisers more active than public advisers but 'locked-in' by company goals Knowledgeable public advisers move to private sector Organic sector provides high-quality SHM advice
Advisers' training for SHM advice	Time and resources for soil training are often limited	Good attendance at dissemination events Limited time and resources for soil training in all sectors	Limited SHM training of technical advisers Need continuing education, as college training inadequate No unified certification	Good attendance at dissemination events Good quality CPD courses but could be more integrated	Large range of training courses, with more offered in recent years	In-service training in ODRs All sectors need continuing education to update college training
Attitudes and motivations of advisers and AAS	Positive NRL adviser attitudes towards the soil	High level of personal commitment to SHM needed	Horticulture advisers' commercial motivations can lead to low social value	A range of attitudes linked to advisers' objectives	High level of personal commitment to SHM needed	Balancing commercial advantage and farmer respect is important

4.1. Framing Conditions

The case studies represent a range of biophysical, political, socio-economic and farming contexts which determine the nature of the agri-food system, the distribution and intensity of production systems, the risk to the soil under agricultural management and public/private support. For example, in Norway, limited areas of arable land coupled with heavy rainfall, constrain timely tillage operations and has led to a national policy prioritising the reduction in the area under autumn ploughing in regions susceptible to soil erosion. In contrast, in Spain, low rainfall areas present challenges for farmers dealing with droughty soils. In Germany (Brandenburg), weather extremes mean water storage capacity, and water-saving cultivation methods are a priority. In both Flanders and Spain, specialised horticultural production systems put pressure on farmers' businesses and, consequently, the soil, while elsewhere, extensive crops such as 'soil friendly' wheat have lower profit margins. In England, arable farmers have expanded with more powerful machinery often implemented by contractors who do not always take account of soil conditions.

With respect to the political context, in Norway and England, there are, to some extent, shared goals between the government and the farming industry (farmer unions and cooperatives) which allow farms to deliver on a range of policies including food production and environmental goals (water, biodiversity, climate, soil). In Spain, a dual system results where intensive horticulture is mainly driven by commercial interests while political interests of soil conservation are more present in extensive agriculture. In Germany, public district and regional offices identified a lack of direction about soil management from the federal government.

From a socio-economic perspective, labour is expensive in Norway, and this affects farm profitability; in Flanders, seasonal land leases hamper any investment in SHM, while strong manure regulations have implications for managing organic matter. Land leasing in Poland leads to exhausted soils, while in Germany (Brandenburg), the large number of cooperatives are well managed by expert agricultural scientists, although the farms themselves are struggling with liquidity. In family-based, non-horticultural farms in Spain, traditional knowledge about and habits concerning soil management continue to be passed on through generations. These variable contextual factors act as framing conditions for AAS for SHM.

4.2. Governance Structures

4.2.1. Governance Arrangements

The six case-study countries have each evolved distinct AAS (and AKIS) in response to a range of framing conditions, with a different mix of public, private and farmer-based organisations (FBOs); non-governmental organisations (NGOs); and research institutes delivering advice that influences and impacts soil management in each (summarised in Box 1 from analysis of interviewee responses). The hybrid and dynamic nature of partnering and funding arrangements is notable across all the case studies. Consequently, there is a diversity and complexity of influencers on farmers' decisions about soil management.

The role of the public support varies across case studies. For most countries, the regional or district agricultural offices have been re-oriented away from technical advice towards administration of subsidies and regulations, where the emphasis is on cross-compliance (GAEC) or supporting scheme applications. For example, in Baden-Württemberg (Germany), the soil service from district administration indirectly controls handling of soils according to the law. Other advice which directly or indirectly impacts SHM tends to be offered through a number of channels; it is often contracted out by the government to private companies, independent companies, FBOs and NGOs and focuses on aspects such as nutrient management and cross-compliance. Only the Soil Service of Belgium, an independent non-profit organisation, is specifically dedicated to soil management. Notably, a public face-to-face advisory service for soil is largely absent or very limited across the case studies. FBOs are significant in Spain, where they are linked with technical soil advice in the production of high-value crops; in Germany, farmer associations are strong, and in Norway, which has a large independent membership organisation (NRL), soil advice is demand-led.

The emergence and influence of the private sector is notable across the case studies. This encompasses a range of advisers working for input suppliers or independently. These advisers play an important on-farm role, where they support day-to-day farming operations. The powerful advisory role of private companies linked to input sales was characterised in some countries as the 'commodification of knowledge', as one Polish respondent (PL3) remarked, "*advice becomes more and more important, and knowledge becomes a commodity that can be bought or sold*". The role of FBOs and the private sector has implications for SHM advice as they respond to farmers' production-oriented needs rather than public goods per se.

Box 1. A summary of the main AAS governance arrangements relevant to SHM in each case study.

Norway's pluralistic advisory system comprises FBOs, public and commercial services. The Norwegian Advisory Service (Norsk landbruksrådgiving NLR) is a decentralised service which provides independent farm, phone and group advice through membership (most large cereal producers are members). NLR also receives subsidies for the organisation's regional and local units to support public good objectives such as soil management and widening access. Other advice comes from advisers working for input companies, independent private consultants and agricultural business cooperatives (input and buyers). Governmental bodies, especially at the local and county levels, have a role in supporting fertilizer plans and subsidies, but there is limited governmental support and responsibility for advisory services.

In **Flanders**, there are different forms of AAS and different sources of funding (regional and provincial funds, farmers' contributions, industry). Public support is still important through funding of regionally embedded Research Stations (RS) which focus on physical and biological soil aspects and act as practical advisory centres, with group dissemination events linking research to farmers and advisors. The Soil Service of Belgium, an independent research and advisory institution, is the main RS for soil. Advisory services subsidised by the government include the CVBB (Coordination centre for education and guidance to sustainable fertilisation), with a focus on nutrient management, now replaced by B3W (Coaching service for a better soil and water quality), with a focus on improvement of soil and water quality. Provincial and regional offices manage administrative issues. FBOs (unions and associations, cooperatives), private consulting companies, Dutch advisors and upstream and downstream industries are a main AAS component and their attention is mainly on crop nutrition and fertilizers.

In **Spain**, there are no public services that specifically provide soil advice on farm, although Agricultural and Fisheries Research and Training Centres hold field events for crop nutrition advice, and regional agricultural offices offer technical advice and training to farmers but are mainly concerned with managing subsidies. Agricultural unions, universities, RDP and operational groups are also involved in advice initiatives. The dominant type of AAS in Spain is the FBOs, the OPAs and the Agro-Food Cooperatives, which are linked to high-value crops and hire their own agricultural technicians, supply companies, certification bodies and have large and established structures. They also have innovation and development centres and provide training to farmers. Farmers with extensive low profit systems (cereals and woody crops) have less access to technical soil advice at farm level.

In **England**, the AAS is diversified and highly fragmented following privatisation. For on-farm advice, agronomists/consultants (independent or commercial) tend to dominate. Where there are commercial interests, historically the emphasis has been on fertilizer recommendations; however, consultants also provide agri-environmental services. Levy bodies (independent/FBOs) offer knowledge exchange for sector production support. Public supported advice has been linked to agri-environment schemes and catchment-based initiatives (soil management to manage diffuse pollution), where cross compliance was a key objective, delivered in partnership with government agencies, water companies and contractors through on-farm and group advice. The government is prioritising supporting public goods (with an emphasis on soil) post-Brexit. A range of NGOs have become increasingly important in facilitating initiatives relevant to all soil health functions.

In **Germany**, there is a heterogeneous and decentralised governance structure where the Federal Government and the 16 Länder take an active role. Due to limited funds, most state services are becoming privatised. These are: (i) the state agricultural offices (free public extension providers) that engage in rural development and regulatory issues, and they also attend to local soil issues; (ii) the chambers of agriculture that offer free and charged advice, education and training; (iii) private consulting and advisory companies offer fee-based advice on specialised topics such as production and business management; (iv) numerous upstream and downstream companies also contribute, as do a broad range of actors who belong to FBOs (boundaries between private organisations and FBO are often fluid). Privatised advisory companies play a key role in the eastern German states.

In **Poland**, advisory services are provided by the state (Agricultural Advice Centres (ODRs)), agricultural chambers, private advisory organisations, companies and NGOs. The ODRs are in Brwinów (centre), branches and Voivodships and are responsible for the education, certification and registration of advisers in Poland. They offer financial and economic advice, while technological advice is limited, as well as organise training courses for farmers. Private agricultural organisations operate in the scope of the publicly funded measures under RDP and other national programmes. Commercial firms, which are extensive, supply advice as part of inputs sales and interact with ODRs. There are a large number of certified individual agricultural advisers who work for various institutions, private companies and farming communities under contract. There are also a large number of active FBOs, and Poland has a long history of agricultural production cooperatives.

4.2.2. Integration/Fragmentation

None of the case studies could be described as having an integrated framework for delivering soil advice. They exhibit different extents of integration and fragmentation in the AAS, which can be characterised by both cooperation and competition (for farmer clients and for project funds) between organisations.

With respect to inter-organisational cooperation, in Norway, private or FBOs cooperate and receive support from public bodies for topics relevant to SHM which do not lend themselves to commercial services. In Flanders, increasing collaboration between the CVBB and B3W advisory services provides a good example of the joint effort of several research institutes to address soil topics. Meanwhile, in England, although the AAS is horizontally fragmented, with multiple uncoordinated actors, organisations and delivery activities concerned with advice for different soil functions, there are a number of partnerships and initiatives where organisations work together towards a shared goal for SHM and water quality (for example, Catchment Sensitive Farming initiative). Synergies were identified in Poland, where the public ODRs host training events which bring together large numbers of farmers and invite private-sector companies, who are knowledgeable about the technologies or products, to participate. However, in Spain, a duality of advice was described with a clear distinction between public and private services, which has implications for soil advice.

There were different perspectives in Germany depending on experiences in the respective states: One respondent described few links between providers and competition between the different consultants and large companies. However, for another respondent (for Baden-Württemberg), the interaction between the state and private consultants at the agriculture office level was a strong point, and they agreed that synergies definitely exist, while there may be tensions between individuals.

In line with this viewpoint, the fragmented landscape and different objectives of public and private providers can have consequences for SHM at the farm level. In Spain, although most respondents did not identify tensions or conflicts in advisory service delivery, one respondent acknowledged that contradictions arise when there are commercial interests:

> *The system is not fully integrated, this affects sustainable soil management negatively because conflicting advice is given, or conflicting objectives are pursued [. . . .] when there are commercial interests, we do find contradiction.* ES2

As with Spain, in England, while advice is *"theoretically joined up"* (for example, a partnership will have shared goals), what actually matters is at an individual farm level, where farmers can be contacted by a number of advisers or projects officers. One respondent (UK2) said, *"I wouldn't say that there's contradictory advice now, but duplication"*, and also noted that farmers have been advised to do things by a commercial company which are questionable with respect to SHM.

A Polish respondent (PL3) also described tension and competition between companies providing agricultural products. Although, as another respondent explained, this depends on the company:

> *There are companies whose approach is to sell their products, and there are companies that act for example together with associations promoting the welfare of the natural environment recommending the use a range of suitable products.* PL2

Regarding vertical research–practice linkages in the soil context, these are considered strong for NLR in Norway which has good links with research; forexample, it is quite common for NLR and the research institute (NIBIO) to be cooperating in projects. This ensures good dissemination but also that projects are relevant to farmers. Researchers, farmers and advisers are also linked in Flanders, where research stations have strong outreach programmes, and in some states in Germany, where district agricultural office carry out practical trials with farmers. In Spain, in the horticulture cooperative sector, there are strong links from research to farmers providing a comprehensive service to these particular farmers. In England, the perception is that research and practice are disconnected, and as the respondent UK2 said, *"It's actually the translation of that [research] into current farming practices, which is where the gap is"*.

4.3. *Advisory Services Capacity*

4.3.1. Management and Organisation for Delivering Advice on SHM

Some respondents considered that there is good organisation and management of AAS but that other limitations prevent effective delivery of SHM advice. For example, in Germany (Brandenburg), the AAS are thought to be well equipped and prepared in terms of technical capacity, with an excellent research infrastructure around Berlin, but lacking political guidance about soil from the federal government. In Spain, some respondents agreed that despite good organisation and management of advice, more information and knowledge transfer are needed for effective SHM advice to be achieved. In Norway, there was consensus from all three NRL (farmer membership organisation) respondents that they have both competence and capacity to deliver advice on SHM. Furthermore, they were optimistic that advice will improve as public funding is now available to increase the focus on soils.

The capacity of public services, where they are provided, tend to be limited by resource constraints, namely, staff and financial. In Germany, there was a sense of good capacity and resourcing in the consultation services hosted by the state agricultural office in Germany (e.g., Baden-Württemberg); however, respondents noted the staffing limitations of public provision and the need for strong personal commitment. This is reiterated later in the analysis:

> *From the public side, we in the agricultural administration are mostly limited by the staff capacities. That is an aspect, which has deteriorated dramatically everywhere in recent years, so if we want to work towards [soil] sustainability, it's only possible through increased commitment beyond the actual working hours.* GR1

Furthermore, the emphasis on inspection and regulation by state bodies in Germany limits their time and scope of work with a focus on inspection. As a consequence, farmers supplement public advice with consultations by private companies.

The Polish state Agricultural Advisory Centres were described as working well to provide advisory services but not yet properly prepared to advise on soil protection, still being stuck in the *"old structures and treatments"* (PL1). They are also constrained by funding and often lose their best advisers to the private sector. The potential of private companies to fill the gap left by public services was identified in Poland. There was consensus that private companies are more visible and accessible and able to meet market demand. Referring to horticultural crops and crop- and soil-borne diseases, this respondent (PL3) remarked:

> *There are private companies that have appeared in the market and provide these services at a good level [. . . . . . .].. The institutes [public] have the potential, equipment, experience and knowledge, but it seems that due to financial and personnel constraints as well as other obligations, they are unable to respond to the very high market demand, and it is very large, while possibilities for conducting research are limited. Private companies, which are more and more visible on the market, are trying to fill this gap, which is good, because such companies can provide services as part of, for example, soil or plant research projects.* PL3

However, for private services, the business model is not always commensurate with building capacity. In England, privatisation of the advisory services has introduced a profit incentive which impacts resourcing, as one respondent, who works for a consultancy, explained:

> *We have to be a profitable organisation, which means that we haven't the luxury of an infinite amount of time [. . . ..] we do the very best we can with the resources we've got, but that some of the expectations of what it actually costs to deliver service are unrealistic.* UK2

This is also a factor in Germany, where dealing with new issues, such as supporting the necessary transition to new cultivation systems or meeting the state policy requirements for environmental programmes, represents an added effort for the consultation services in terms of costs, time and energy. However, adaptation is seen to be essential to ensure future services:

> *And every consultation service is required to adapt, to continuously improve, and to be up to date with the latest science and technology. I think that's actually a very positive development [. . . .]. But it is clear to them that if they do not consult their farms in the direction of sustainability, they will lose them completely in 10–20 years.* GR1

This need to build capacity for the future is reiterated by a respondent (GR3) who works with large cooperatives south of Berlin, where the long-term nature of soil health has become the focus of attention amongst the scientists who advise on the farms.

Private organisations also find that they have to compete for project funding. In Flanders, although the quality of advice is good for soil in the government research institutes and the independent Soil Service, the resourcing of activities is seen to be constrained by a reliance on short-term project funding, reducing the chance to build strong and enduring relationships with farmers. The remark "*True sustainable soil management advice does not exist to my knowledge, the Soil Service provide such integrated advices only as part of projects*" (BE3) is insightful in that it indicates poor continuity, as well as a dependence of projects for funding.

Staff recruitment and retention has implications for advisers' expertise and experience in SHM and was mentioned across a number of countries. In Norway, it can be difficult to recruit advisers who possess sufficient knowledge about soil if, for example, an experienced adviser retires. High turnover of advisers due to a lack of job satisfaction or progression and financial motivations exacerbates this. In Spain, advisers who belong to technical departments in FBOs (companies/associations) are seen to have more room for manoeuvre and are more organised and professionalised compared to commercial advisers. The absence of planning for the necessary skills and staff which may be needed in 2–4 years' time was also raised as a limitation for SHM advice in England.

Regarding an organisation's culture, there was also recognition that advisory organisations themselves have some responsibility to rethink how they advise farmers who are overburdened, face severe economic pressures and are constrained in terms of investing in new equipment, new crop rotations or new fertilization methods. In this respect, the culture of the organisation is seen to be important in Germany, where every consultation service has a specific philosophy that is shaped by the organisations' decision makers.

4.3.2. Level of Advisers' Knowledge about SHM

In the pluralistic advisory systems described here, it is difficult to characterise the expertise or the quality of advice for soil overall and SHM specifically, as this can depend on the sector and systems they support. However, the following provides some insights.

Knowledge and Practical Experience

Practical experience is seen as indicative of good quality advice and private advisers are more likely to acquire this, compared to public advisers, due to their regular on-farm activities. For example, the quality of advice is considered high in consultation services in Germany, although the focus is limited, and wider aspects of SHM advice are not covered:

> *I think, the quality of consultation is high [. . . .] many of the consultants are running agricultural businesses themselves, so they have a certain practical background, or they have simply been working at an agricultural office for many years, so they have a very high level of knowledge [. . . .] so far,[this] has mostly been on crop protection and, I think, especially in terms of sustainability, sustainable soil management, crop rotation, intercropping, things like that—there is still room for improvement.* GR2

Similarly, in Poland, private advisers were regarded as more effective and active than state advisers who, although knowledgeable, lack practical experience and the ability to follow up on advice to keep farmers up to date:

> *A strong point of commercial services is that they have capable advisers. With regard to government institutions, their strong point is certainly their infrastructure and the preparation of speakers, i.e., advisors, who are very knowledgeable, but then somehow nothing happens. And this is the weak point, that there is a lack of continuity, on-site continuity, during on-site workshops. Often these advisers lack practical experience and [. . . .] are unable to keep up with these new solutions and products.* PL2

Another Polish respondent (PL3) noted that, with the loss of good quality government advisers to the private sector, their expert knowledge now only reaches farmers who are customers of private companies. This unequal distribution of quality advice (including SHM) was also identified in Spain, where technicians with a good level of specialist expertise in horticultural production support intensive crop growers, but family-based businesses with extensive systems in other sectors in Spain have limited access to good quality advice on soil. Furthermore, pockets of high-quality SHM advice were described for advisers in the organic sector, as mentioned in Poland and in Spain, and for advisers selling products related to, for example, organic or sustainable management who *"provide information about the nature of living soil, biodiversity or soil quality"* (PL2).

In Norway, governmental and public bodies, especially at the local and county levels, were described as not very competent or up to date, with a main role in supporting fertilizer plans and subsidies. However, the respondents all agreed that the standard of advice for soil management is high in the independent NRL, where the advisers are knowledgeable and have an increasing focus on soil health and environment.

Soil Fertility Focus

In all case studies, there was agreement that private advisers (working for input companies or as independents) are generally trained to advise on soil from the perspective of fertilization and crop nutrition and tend to look at crop management in the shorter term. This emphasis was noted by a respondent (PL3) in Poland who said, *"My impression is that most advisers focus only on the composition of the soil, on just the chemical factors, but they ignore and totally undervalue the importance of soil microbiology"*.

A number of respondents called for a change in the mindset of advisers away from production-orientated to more holistic advice, with a shift in thinking from soil chemistry to a microbiological approach required, to show that *"living soil can achieve more"*.

This focus on soil fertility and crop nutrition can have some negative implications. For example, in Flanders, commercial advisers were known to advise maximum fertiliser recommendations irrespective of crop requirements, which is contrary to good practice recommended by research organisations. This was also noted in Germany, where an emphasis on fertilization as part of an overall crop care package can lead to conflicts with advice for other soil functions. This respondent in Poland highlighted how some advisers are 'locked-in' by their company's commercial imperatives despite being knowledgeable:

> *Many advisers are enslaved by receiving payment from the company, so they have to advise according to the company's offer, and this limits their freedom to act; they have the knowledge but they will necessarily be focused on bonuses, on a raise, on finances, and this restricts them.* PL3

However, respondents did not think commercial advisers purposely provide negative advice, although they may be slightly less inclined to look at the environment or at soil quality, soil biology, etc. In Spain, where consultants are often influenced by their employers, one respondent suggested that there is no intention to damage soil; however, they may not be aware of the externalities of their advice:

> *I don't think there is one main advisory service that has either a positive or negative impact. Normally advisers have the objective of increasing overall production. The adviser does*

> *not go against soil sustainability or soil quality, but [. . .] the use of these technologies continuously without other guidelines in the end leads to an overall degradation of the system, mainly of the soil.* ES2

Environmental Shift

The Green Deal and the demands from supply chain companies and retailers to meet food and farming standards and gain a market advantage were considered by many respondents to be driving advisers towards SHM advice. However, there is some cynicism in Poland that advisers and input-sellers are using slogans related to environmental and soil protection issues but, fundamentally, are still largely dependent on the producers of chemical agents for their income.

In England and Germany, there has been a shift in commercial adviser activities towards supplying environmental advice (supporting agri-environment scheme applications, as well as practices for good soil management), and for agronomists linked to input sales to sell cover crop and, pollinator seeds and biosolutions. A German respondent described the growing demand:

> *In recent years companies have emerged that strive towards sustainability, selling crop fortifiers, soil additives and so on. Active local consultants and some farms use these products in their cultivation. This is of course due to the fact that, in the last few years, little has been done in terms of soil fertility and sustainability on many farms. They are now reaching their limits in terms of plant cultivation, they have problems with diseases, with the soil, etc., and companies, which offer the appropriate products, have been in greater demand in recent years.* GR1

This situation is replicated in Poland where more companies are entering the market with 'natural products' aiming to meet farmer demands.

Heterogeneity

One common factor across all case studies was the heterogeneity in the quality of advisers with respect to soil advice, with a spectrum of very good to very bad commonly being described. In Spain, a range from very good agricultural technicians to others who do not have the necessary knowledge was linked to the number of untrained advisers emerging to meet the increasing demand for sustainability and ecological advice. Similarly, in England, a respondent (UK1) referring to agronomists said, *"I think the good are very good, but I don't think we've got many very good ones, I think a lot of us are in the category of willing triers"*. However, he acknowledged that there are excellent pockets of SHM advice amongst independent advisers and non-commercial initiatives. This range is echoed in another comment by a respondent from England who described the value of long-term experience:

> *Some of them are extremely knowledgeable and interested [about soil] and have been in their post for quite a long time. Some of them are on short term contracts. And some of them who are less good than others, in terms of their understanding of the technicalities of what they're talking about, and what they're being asked to do.* UK2

The same sentiment was expressed by respondents in other case studies, where advisers develop a very good reputation because they have been in the profession for many years. A range of abilities and interests was also described in Germany, where the ease of substituting SHM principles with agrochemicals was blamed on a lack of attention to soil by some advisers:

> *There are consultation services, or even individual consultants on the part of the industry, who attach importance to the topic [soil]. But there are also people who have never bothered with the subject, because it is still possible to achieve good yields with the use of mineral fertilizers or chemical-synthetic pesticides.* GR1

The distinction between the role of the advisers as generalists or experts was widely discussed. There are very few agricultural advisers across the case studies who focus specifically on the soil or get the opportunity to become experts. Some take the view that soil experts can be consulted when necessary, but that wider skills are needed at farm level, as this respondent explained:

> *Rather than being experts on that particular aspect, we reflect the farming community, in the sense that we are people with a wide range of skills, but an expert in nothing. An expert—he's talking purely about the soil, and the health of the soil, we will be talking about it on the profitability of the rotation, the control of various injurious weeds, diseases and pests, and then looking at a rotation that is sustainable, which then comes back to the soil. However, we know where to go to get expert [soil] advice.* UK1

Differing perspectives on the value of experts versus generalists were picked up in the Spanish interviews. One respondent agreed that a historical focus on supporting production has led to fragmentation where an agricultural technician may know a lot about tillage or agricultural equipment but does not have a general vision of sustainable soil management. The other two respondents in Spain, however, argued that advice to farmers on soil management is too general and the level of expertise low; one (ES4) identified *"A strong need for the participation of people who are soil specialists—soil scientists, biotechnologists with application to soil microbiology"*.

4.3.3. Advisers' Training for Delivering SHM Advice

There are a number of opportunities through multiple talks and events for all advisers to expand and update their SHM knowledge, mentioned for all case studies. In Flanders and England, for example, large numbers of advisers reportedly attend dissemination events and demonstration days, and for many, this is important for networking. In addition, there is now comprehensive information about soil topics on the internet and social media and opportunities for peer-to-peer learning and exchange. However, as noted already, poor attention by advisers to SHM has been attributed to the absence of good training.

Time and resources for soil training are a concern for some. As one respondent (N2) in Norway noted, *"unfortunately we have to prioritise covering our hourly rate as employees, so that can affect how much time we have to educate ourselves, go to conferences, seminars"*, illustrating the fact that advisers (from all organisations) are often under financial targets and pressures to the detriment of their training and upskilling in SHM. This imperative steers organisations' decisions about training as well. In Flanders, for example, obtaining a certificate (Flemish Land Agency certification or other quality control procedures) is costly both in terms of time and money, and as a consequence, certification is profitable for only a few advisory institutes/services.

There are a large number of options for in-service training in Poland with ODRs taking on a key role for farmers and advisers. Advisers within the commercial sector in Poland are also considered well trained but only within the sphere of their operations and products:

> *It seems to me that every commercial business tries to train its advisers so that they do at least have this information as regards their own products, how they affect the soil and therefore they must have prior knowledge or learn about the soil, its quality, the processes that take place in the soil environment.* PL3

In Spain, the nature of skills and training depends on the type of agricultural technicians (cooperative, input company or independent). Most respondents agreed that the level of SHM training of technical advisers in Spain is low overall, as one remarked (ES2): *"Advisers do not have sufficient skills and experience to give advice on sustainable soil management [, ...] because they have not had sufficient training during their studies"*. As such, these agricultural technicians need to seek out further training to enable them to meet changing demands. These points were reiterated for Poland, where the notion of continuing education was raised:

> *Every adviser needs to participate in continuing education, as the knowledge gained when graduating from college is not enough [....]. It is necessary to educate, educate and*

> *again educate advisers and farmers, and to provide this new knowledge about sustainable soil management, which is completely different from the information provided before.* PL1

Another respondent from Poland (PL3) supported this, remarking that studies are only the basis and a good adviser has to train for the 'rest of their life', otherwise, they quickly lose touch with reality. In Spain, respondents noted that there is no unified certification validating the agricultural technicians' knowledge. In Germany, large differences in the range of training courses were described, with more being offered in recent years. In England, there is an established continuing professional development (CPD) scheme for advisers (FACTS, BASIS) which offers courses on soil and water management. While acknowledged to be outstanding compared to other European countries, a respondent pointed to the inadequacy of these courses in terms of SHM:

> *In terms of sustainability, I think they're both useless. They've evolved out of a commercial requirement. So it wasn't evolved to deliver good, independent, impartial information [...] they do provide a level of professionalism.* UK1

Another respondent from England thought that a FACTS-qualified adviser would understand about nutrient management but argued that BASIS is too technical and academic and that the modular training does not prepare advisers to deliver integrated advice, considering soils, nutrients, water management together, nor help them understand the underlying principles of SHM:

> *So as far as, is the training fit for purpose for the next generation of advisors? One of the problems that we and the whole industry has got to know that there's plenty of advisers who are qualified, but not necessarily have a good understanding of the principles[...] you need to be able to understand what you're doing. And why are you doing it.* UK2

There was also agreement that capabilities need to be expanded to beyond a focus on production objectives and soil fertility and crop nutrition advice, to meet new demands, reinforcing the points made earlier. This respondent from Spain noted that this was a key limitation for SHM advice:

> *From my point of view, there is enough organisation to provide advice on sustainable soil management and there are enough people capable of providing basic guidelines for sustainable soil management but there is a lack of general training on what is the true nature of soil quality beyond nutrient fertility.* ES3

The extent of informal learning through adviser networking was mentioned by some respondents but did not emerge as a particularly strong aspect in the interviews.

4.3.4. Attitudes and Motivations of Different Advisers and AAS

Positive adviser attitudes towards the soil were described by a number of respondents, however, there is still a range of attitudes linked to economic motivations. Fundamental differences in motivations between advisers were identified in Spain and this aligned to their organisations' objectives:

> *An adviser who belongs to a trade union or a regional agricultural office has a different vocation than an adviser who belongs to a commercial company or to a research centre; their motivations are very different, which means that their inclinations are also very different.* ES1

This can have implications for advisers' reputation and credibility. According to respondents, for example, in the horticulture sector in Almeria, agricultural technicians do not always have high social value and may even start to have a bad reputation. This is echoed in Poland, where the balance between commercial advantage and gaining farmers' respect was seen to be important: *"There is no doubt that an advisor's motivation is influenced both by economics and by the desire to be respected by farmers, it really depends on the person"* (PL1). Many agreed that farmers are able to quickly discern any 'shortermism' and the commitment and motivation of advisers.

For many respondents the different motivations and attitudes of the individual adviser were regarded as more important than the type of organisation they belong to. The high level of personal commitment required by some advisers to pursue their interests in, and deliver, SHM advice was mentioned by respondents. Consultants in Spain, for example, are often limited by specific short-term projects or task forces. When these are finished, if they want to continue with the topic, this has to be done in their own time. Similarly, in Germany, personal effort is linked to quality advice:

> *Yes, well, there are advisers who are just all-around good advisers who really give their best and try to constantly educate themselves in order to be able to provide the best possible consultation to the farmers . . . I think most of the vocational counsellors actually—and yes, I think that of the ones that I know, most really put their full effort into it.* GR2

5. Discussion

5.1. Governance Structures

This analysis confirms the picture painted by previous researchers of considerable AAS diversity between, and pluralism within, European countries [71]. This translates into a diverse landscape for SHM advice with different governance, funding and delivery mechanisms and no evidence of any integrated advisory frameworks for delivering advice for soil management. The analysis shows that institutional options available for financing and the level of coordination are limited with respect to delivering advice for soil management, as observed elsewhere for AAS more generally [50,56]. A reduced central organisational role of government agencies in AAS and an emerging 'knowledge market' [33] has led to a continued decline in the public sector's role in delivering on-farm soil advice for all case studies, with the diversion of their resources and staff towards compliance regulation and scheme/grant administration. Conversely, the prominent role of the private sector, independent organisations, FBOs and NGOs is apparent in filling the gap in delivering on-farm advice that influences and impacts soil management, either through contracts (projects) to fulfil government objectives (e.g., FAS, grants) or commercially in a more market-led environment, as described in other AAS studies [33]. When state and private advisors are incentivised to administer regulations and grant applications, this narrows down choices and limits broader understanding of 'know-why' soil processes [14,72]. New services are also emerging, and overall, the number of advisers with commercial links to economic actors (input suppliers, consultants) is increasing [26].

Fragmentation means competition for clients and project funding, and soil advice at farm level can be compromised by conflicting delivery or duplicating advice in multi-partner approaches, as reported by others [73]. However, many hybrid and dynamic arrangements for partnering and funding for delivering SHM advice are notable. These 'creative alliances' provide opportunities for the effective integration of delivery of soil advice at programme level. This ability of pluralistic advisory services to overcome constraints (shortages in funding, staffing, etc.) through increased cooperation, collaboration and partnerships has been observed elsewhere [29,71,74,75]. Individual relationships of both competition and cooperation, described by Compagnone and Simon [24], were not shown in this analysis.

5.2. Advisory Services Capacity

These governance arrangements provide a backdrop to understanding different organisational arrangements and capacity to provide SHM.

5.2.1. Management and Organisation Capacities for SHM Advice

The analysis identified organisational constraints in resources, funding and staffing, notably in the public services, which are not always able to meet demands, and this impacts the capacity to deliver SHM advice. There are inherent frustrations concerning reliance on short-term project funding for developing and continuing with advice streams, as previously described for environmental advice [31,76]. This often means only committed advisers continue with the SHM advice when the project ends. Poor staff retention [24], with the loss of advisers' knowledgeable about SHM to the private sector, reduces farmer access to SHM advice. Although farmers might look outside of formal advice in such circumstances [72], their options for benefiting from high-quality soil advice are diminished. Other commentators have noted that commercialisation threatens the extension capacity of government agencies [77], however, technical expertise has not been considered.

Investment in staff capacity for SHM advice (training and field days) is restricted in both public and private sector organisations by limited time and resources and the competitive business environment. Small firms also struggle to meet new environmental requirements, corresponding with previous observations [34,42]. Furthermore, in some commercial organisations, economic drivers can lead to an organisational culture that values input sales over expertise in SHM.

These organisational capacities affect individual advisers' capacity to operate effectively, their objectives and motivations, their professional culture and the support they are offered to deliver SHM. As observed by Klerkx and Jansen [34], this wider set of institutional conditions, and the relationship with the 'back-offices' which supports them professionally, is critical for enabling advisers to develop and deliver specialist and professional advice. Furthermore, maintaining a stable or increasing workforce as well as diversifying the expertise and increasing the competence of staff are seen to be critical for AAS [29].

5.2.2. Individual Capacities for SHM Advice

Individual capacity results from a combination of attributes: quality of advice; training; and motivations in relation to SHM. Firstly, regarding advice quality, heterogeneity in levels of advisers' soil knowledge was observed across all cases and across all AAS types, with few advisers considered to be delivering all-round high-quality advice to support SHM. This adds to the emerging body of evidence showing that advice on soil management is suboptimal. What constitutes 'good quality advice' with respect to soil management was understood differently due to advisers' varying goals and their clients' needs. It was generally characterised by, not only extensive on-farm practical experience and a good level of subject-matter knowledge or expertise [67], but also critically by an understanding of soil chemical, biological and physical processes and principles [78]. Private advisers (commercial consultants, technicians and agronomists), while being credible with respect to providing high-quality technical advice, are limited in scope to soil fertility and crop nutrition. This observation is supported by studies showing the predominance of advice based on nutrient testing and interpretation to support farmers' short-term production decisions, e.g., [45]. This limits opportunities to incorporate soil health perspectives into advice, which are critical to understanding the capacity of soil to function as a vital living system [16,17]. Only very few advisers are taking a holistic approach, accounting for non-linear mechanistic relationships between various physical, chemical and biological soil properties considered important for soil health [19].

The significance attributed to practical experience, however, should not be overlooked. This allows advisers to provide localised advice and meet the fine resolution of soil information and data that farmers require [9]. This highlights the value of experiential learning (and co-learning with farmers), which has a particular significance for soil management due to the in-field observations and sensory experiences required [79,80] and is highly appreciated by peers and the practitioner community [78].

Equally, whilst expertise in soil science and management (demonstrated by some individual advisers) is valued, the role for the generalist agronomic adviser who takes a whole farm perspective is seen as important. Interestingly, advisers have been shown to be capable systems thinkers [74] and positioning SHM within the wider farm business and environment is in itself an important skill. Further specialisation, in, for example, soil microbiology was called for by some respondents, in line with emerging farmer interest in soil health, but how such specialists would position themselves in the AAS landscape was not elaborated. Landini [41] suggested that not all individual advisers can hold the same knowledge and capabilities but instead can act in groups to enrich their work. This professional distribution of advisers' SHM roles, skills and specialisms and the way they interact, complement and learn from each other, is an interesting area for future research [24,79]. Furthermore, the changing role of the technical 'expert' needs consideration [81].

Secondly, with respect to training, poor investment in training and particularly in continuing education in SHM in both private and public spheres was seen to be a key reason for the limited scope of advisers' expertise. Training and professional development courses on soil topics, whilst considered to be at a high standard in certain countries, do not always provide an understanding of the principles and processes of SHM. A number of studies have shown that advisers are increasingly relying on each other for sharing soil expertise through professional networking [82]; however, this was not identified in this analysis as an alternative to training. These findings are inconsistent with previous studies [30], although the focus was not SHM.

Thirdly, regarding motivations, personal intrinsic interest in soil was a further facet demonstrated by a few public and independent advisers. The economic motivations of private sector advisers' (linked to input sales) observed here are widely reported in studies that concern soil [10,36,83]. The image of advisers as 'locked into' supporting intensive agriculture pathways has been also described for high-input production systems [26,84], as has the power of supply chain actors [36]. However, analysis here suggests a more nuanced picture, with many private advisers balancing economic motivations with the need to retain respect, social value and trust in the farming community. This loyalty dilemma between private good (what the farmer demands and pays for) and public good (issues of broader importance for society as a whole) [29], may need to be re-examined in a future context when incentives for providing SHM become available (e.g., carbon farming, Environmental Land Management Schemes in England). Organisations are already responding to the market and offering a range of environmental services, and supporting sales of 'natural' biological products. However, the depth of understanding and commitment that accompanies these was queried, and there were calls for a more fundamental shift in advisers' mindsets.

Professional culture is closely connected to individual advisers' motivations and mindsets, accepted norms and values, how they perceive and execute their tasks [34], and their performance rationale and economic strategies [26]. However, adviser roles are not set: Nettle, Crawford and Brightling [42] describe the fluid nature of adviser professional identities and opportunities for evaluating their roles through reflective practice [41,85], which, if organisations were more flexible, could lead to reorientation of soil management advice.

5.2.3. Narrowing Down

Although it was not the intention here to assess the performance characteristic of the framework, some observations can be made. The needs and opportunities, which characterise performance [50] that have been steering advice in relation to soil are: policy (cross-compliance regulation and grant administration support) and markets (farmer demands for crop production advice). As a result, there has been a narrowing down of soil advice, both with respect to content and access, as depicted in Figure 2.

However, the increasing interest in soil health from both farmers, in part due to the recognition of soil degradation [18,86], and policy makers, will provide the new drive and opportunities to widen the scope of advice to cover physical and biological, as well as chemical, processes. To achieve this, AAS organisations will need to invest in adviser training and capacity building and aim to shift professional cultures and mindsets at organisation and individual level. This will require incentivisation, and Dhiab, Labarthe and Laurent [26] identified a need for public policy intervention to support this. This could be through, for example, strengthening national FAS with requirements for member states to provide standardised and certified adviser training and continuing professional development in SHM. Ultimately, however, AAS are shaped by the framing conditions, the priorities within the agricultural sector strategies (high-value commodities or environmental sustainability) which are beyond the direct influence of policy makers and advisory services managers [50]. In turn, these determine the governance structures and the relative capacities of public, private or NGO AAS and the services offered. As Knierim, Labarthe, Laurent, Prager, Kania, Madureira and Ndah [23] point out, the historically grown, path-dependent institutions and institutional constellations in each EU member state play an important role in AAS.

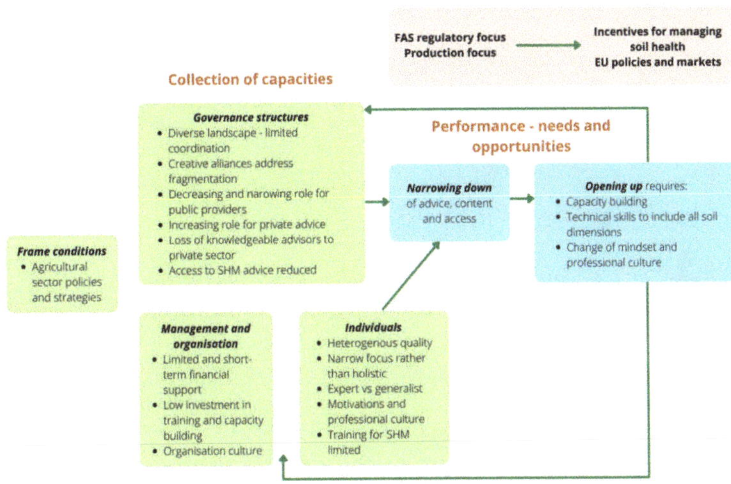

Figure 2. This figure shows how the collection of capacities act to narrow down the nature and extent of advice for soil. Changing needs and opportunities will open up the scope of advice for delivering SHM.

There have been calls for capacity building in knowledge systems at individual, organisational and AKIS levels [42]. This encompasses adviser training and professional development and more back-office support [28,85] as well as the need to understand the varying roles of professional advisers [87]. However, the focus has often been on process skills, the (new) intermediary, advisory styles and facilitatory skills that advisers should master to support and empower farmers in networks of interactive learning [88]. Adviser technical or specialist roles have received less attention, notably for soil, despite the growing demands placed on them for understanding and supporting land managers in the management of complex soil functions.

6. Conclusions

The framework employed allows the collective capacities (governance structures; organisational and individual capacities) of AAS for SHM advice to be revealed. It shows that advisers' competences and skills should not be seen in isolation. As such, the recommendations for expanding the scope of content and access to SHM advice include addressing deficiencies in training and capacity building, shifting professional culture as well as addressing more deep-seated institutional conditions and governance structures. Incentivising such changes will require changes in both policy and market drivers. These insights show that AAS can play a central role in the transformation of food systems more widely [89].

The method based on in-depth interviews (18 experts) provides insights for a cross-section of European countries offering a range of perspectives, as well as common themes with respect to capacities which affect the nature and extent of SHM advice. However, the results can only be indicative for Europe as a whole and further qualitative and quantitative research will be needed to provide a more comprehensive picture. In particular the results show how different advisory services that influence and impact soil evolve in specific country contexts. This suggests that the model of identifying systems that best fit context-specific conditions is suitable for future support of national AAS with respect to SHM. Critically, the methodology did not explore the complexities of the relationship between advisers and farmers/land managers, nor capacities in terms of the soft skills required for co-producing technical soil knowledge or the changing mature of the 'expert' role of advisers.

With the accelerated move towards the integration of soil health issues in a number of European Commission strategies and the actions and ambitious targets set for soil health within the Soil Mission 'A Soil Deal for Europe', the requirements for building capacities and a knowledge base for soil health enhancing practices in agriculture will increase [13]. This will require member states to significantly enhance their AAS capacities to achieve this desired transition, with implications for both European and national level policies.

Supplementary Materials: The following supporting information can be downloaded at: https://www.mdpi.com/article/10.3390/land11050599/s1, SoilCare advice review Interview schedule.

Author Contributions: Conceptualization, J.I., J.M. and C.-A.C.; formal analysis, J.I. and C.-A.C.; investigation, J.M., J.E.B., C.-A.C., J.A.A.-S., A.E., M.F., B.L.-F., P.M.-G., K.S., J.S. and M.T.; methodology, J.I. and J.M.; project administration, J.M.; resources, J.M.; supervision, J.M.; writing—original draft, J.I.; writing—review and editing, J.I., J.M., J.E.B., C.-A.C., J.A.A.-S., A.E., M.F., B.L.-F., P.M.-G., K.S., J.S. and M.T. All authors have read and agreed to the published version of the manuscript.

Funding: This work was part of SoilCare (Soil care for profitable and sustainable crop production in Europe). Grant Agreement 677407 funded by the European Union's Horizon 2020 research and innovation programme. www.Soilcare-project.eu (accessed on 20 September 2021). 2016–2021. This funder had no role to play in the design, data collection, analysis or interpretation of results presented here.

Informed Consent Statement: Ethical standards and guidelines have been applied to the collection, processing and storage of data about persons, in accordance with the agreed project ethical statement. All subjects gave their informed consent for inclusion before they participated in the study.

Data Availability Statement: Data are confidential and not available.

Acknowledgments: We acknowledge the support given by other partners and stakeholders in the SoilCare project and in particular the interview respondents.

Conflicts of Interest: The authors declare no conflict of interest. The funders had no role in the design of the study; in the collection, analyses, or interpretation of data; in the writing of the manuscript, or in the decision to publish the results.

Abbreviations

AAS	Agricultural Advisory Service
AKIS	Agricultural Knowledge and Innovation System
FAS	Farm Advisory System
FBO	Farmer based organisation
NGO	Non-governmental organisation
EU	European Union
GAEC	Good Agricultural and Environmental Condition
PROAKIS	Prospects for Farmers' Support: Advisory Services in European AKIS
NRL	Norwegian Advisory Service (Norsk landbruksrådgiving)
SHM	Soil Health Management
SSM	Sustainable Soil Management
CVBB	Coordination centre for education and guidance to sustainable fertilisation, Belgium
B3W	Coaching service for a better soil and water quality, Belgium
OPAs	Professional farmer's organisations and Agri Food cooperatives, Spain
ODRs	Agricultural Advice Centres, Poland
NIBIO	Norwegian Institute of Bioeconomy Research

UK United Kingdom
CPD Continuing Professional Development

References

1. European Commission. *Communication From the Commission to the Council, the European Parliament, the European Economic and Social Committee and the Committee of the Regions—Thematic Strategy for Soil Protection (COM(2006) 231 Final)*; European Commission: Brussels, Belgium, 2006.
2. Stolte, J.; Tesfai, M.; Øygarden, L.; Kværnø, S.; Keizer, J.; Verheijen, F.; Panagos, P.; Ballabio, C.; Hessel, R. *Soil Threats in Europe*; Publications Office Luxembourg: Luxembourg City, Luxembourg, 2015.
3. Montanarella, L.; Panagos, P. The relevance of sustainable soil management within the European Green Deal. *Land Use Policy* **2021**, *100*, 104950. [CrossRef]
4. European Commission. *The European Green Deal*; European Commission: Brussels, Belgium, 2019.
5. Juerges, N.; Hansjürgens, B. Soil governance in the transition towards a sustainable bioeconomy–A review. *J. Clean. Prod.* **2018**, *170*, 1628–1639. [CrossRef]
6. Prager, K.; McKee, A. Co-production of knowledge in soils governance. *Int. J. Reg. Rural. Remote Law Policy* **2015**, *1*, 81–97. [CrossRef]
7. Vrebos, D.; Bampa, F.; Creamer, R.E.; Gardi, C.; Ghaley, B.B.; Jones, A.; Rutgers, M.; Sandén, T.; Staes, J.; Meire, P. The Impact of Policy Instruments on Soil Multifunctionality in the European Union. *Sustainability* **2017**, *9*, 407. [CrossRef]
8. FAO. *Voluntary Guidelines for Sustainable Soil Management Food and Agriculture Organization of the United Nations Rome, Italy*; FAO: Rome, Italy, 2017.
9. Campbell, G.A.; Lilly, A.; Corstanje, R.; Mayr, T.R.; Black, H. Are existing soils data meeting the needs of stakeholders in Europe? An analysis of practical use from policy to field. *Land Use Policy* **2017**, *69*, 211–223. [CrossRef]
10. McNeill, A.; Bradley, H.; Muro, M.; Merriman, N.; Pederson, R.; Tugran, T.; Lukacova, Z. Inventory of opportunities and bottlenecks in policy to facilitate the adoption of soil-improving techniques. *SoilCare, Scientific Report Milieu*, 2 March 2018. Available online: https://www.soilcare-project.eu/resources/deliverables(accessed on 20 September 2021).
11. Lobry de Bruyn, L.; Ingram, J. Soil information sharing and knowledge building for sustainable soil use and management: Insights and implications for the 21st Century. *Soil Use Manag.* **2019**, *35*, 1–5. [CrossRef]
12. Aznar-Sánchez, J.A.; Velasco-Muñoz, J.F.; López-Felices, B.; del Moral-Torres, F. Barriers and Facilitators for Adopting Sustainable Soil Management Practices in Mediterranean Olive Groves. *Agronomy* **2020**, *10*, 506. [CrossRef]
13. European Commission. *A Soil Deal for Europe Implementation Plan*; European Commission: Brussels, Belgium, 2022.
14. Ingram, J. Are farmers in England equipped to meet the knowledge challenge of sustainable soil management? An analysis of farmer and advisor views. *J. Environ. Manag.* **2008**, *86*, 214–228. [CrossRef]
15. Baumgart-Getz, A.; Prokopy, L.S.; Floress, K. Why farmers adopt best management practice in the United States: A meta-analysis of the adoption literature. *J. Environ. Manag.* **2012**, *96*, 17–25. [CrossRef]
16. Doran, J.W.; Zeiss, M.R. Soil health and sustainability: Managing the biotic component of soil quality. *Appl. Soil Ecol.* **2000**, *15*, 3–11. [CrossRef]
17. Lehmann, J.; Bossio, D.A.; Kögel-Knabner, I.; Rillig, M.C. The concept and future prospects of soil health. *Nat. Rev. Earth Environ.* **2020**, *1*, 544–553. [CrossRef] [PubMed]
18. Mattila, T.J.; Hagelberg, E.; Söderlund, S.; Joona, J. How farmers approach soil carbon sequestration? Lessons learned from 105 carbon-farming plans. *Soil Tillage Res.* **2022**, *215*, 105204. [CrossRef]
19. Hou, D.; Bolan, N.S.; Tsang, D.C.; Kirkham, M.B.; O'Connor, D. Sustainable soil use and management: An interdisciplinary and systematic approach. *Sci. Total Environ.* **2020**, *729*, 138961. [CrossRef] [PubMed]
20. Karlen, D.L.; Veum, K.S.; Sudduth, K.A.; Obrycki, J.F.; Nunes, M.R. Soil health assessment: Past accomplishments, current activities, and future opportunities. *Soil Tillage Res.* **2019**, *195*, 104365. [CrossRef]
21. Rinot, O.; Levy, G.J.; Steinberger, Y.; Svoray, T.; Eshel, G. Soil health assessment: A critical review of current methodologies and a proposed new approach. *Sci. Total Environ.* **2019**, *648*, 1484–1491. [CrossRef] [PubMed]
22. Briggs, S.; Eclair-Heath, G. Helping UK farmers to choose, use, and interpret soil test results to inform soil management decisions for soil health. *Asp. Appl. Biol. Crop Prod. South. Br.* **2017**, *134*, 161–168.
23. Knierim, A.; Labarthe, P.; Laurent, C.; Prager, K.; Kania, J.; Madureira, L.; Ndah, T.H. Pluralism of agricultural advisory service providers–Facts and insights from Europe. *J. Rural. Stud.* **2017**, *55*, 45–58. [CrossRef]
24. Compagnone, C.; Simon, B. Cooperation and competition among agricultural advisory service providers. The case of pesticides use. *J. Rural. Stud.* **2018**, *59*, 10–20. [CrossRef]
25. Kania, J.; Vinohradnik, K.; Knierim, A. WP3—AKIS in the EU: The Inventory Final Report Volume I—Summary Findings; Krakow, Poland, 2014. Available online: https://proakis.webarchive.hutton.ac.uk/sites/proakis.hutton.ac.uk/files/FINAL_REPORT_08_07_2014_VOL_I.pdf (accessed on 20 September 2021).
26. Dhiab, H.; Labarthe, P.; Laurent, C. How the performance rationales of organisations providing farm advice explain persistent difficulties in addressing societal goals in agriculture. *Food Policy* **2020**, *95*, 101914. [CrossRef]
27. Laurent, C.; Nguyen, G.; Triboulet, P.; Ansaloni, M.; Bechtet, N.; Labarthe, P. Institutional continuity and hidden changes in farm advisory services provision: Evidence from farmers' microAKIS observations in France. *J. Agric. Educ. Ext.* **2021**, 1–24. [CrossRef]

28. Klerkx, L.; Proctor, A. Beyond fragmentation and disconnect: Networks for knowledge exchange in the English land management advisory system. *Land Use Policy* **2013**, *30*, 13–24. [CrossRef]
29. Garforth, C.; Angell, B.; Archer, J.; Green, K. Fragmentation or creative diversity? Options in the provision of land management advisory services. *Land Use Policy* **2003**, *20*, 323–333. [CrossRef]
30. Prager, K.; Creaney, R.; Lorenzo-Arribas, A. Criteria for a system level evaluation of farm advisory services. *Land Use Policy* **2017**, *61*, 86–98. [CrossRef]
31. Klerkx, L.; De Grip, K.; Leeuwis, C. Hands off but strings attached: The contradictions of policy-induced demand-driven agricultural extension. *Agric. Hum. Values* **2006**, *23*, 189–204. [CrossRef]
32. Prager, K.; Labarthe, P.; Caggiano, M.; Lorenzo-Arribas, A. How does commercialisation impact on the provision of farm advisory services? Evidence from Belgium, Italy, Ireland and the UK. *Land Use Policy* **2016**, *52*, 329–344. [CrossRef]
33. Labarthe, P.; Laurent, C. Privatization of agricultural extension services in the EU: Towards a lack of adequate knowledge for small-scale farms? *Food Policy* **2013**, *38*, 240–252. [CrossRef]
34. Klerkx, L.; Jansen, J. Building knowledge systems for sustainable agriculture: Supporting private advisors to adequately address sustainable farm management in regular service contacts. *Int. J. Agric. Sustain.* **2010**, *8*, 148–163. [CrossRef]
35. Pedersen, A.B.; Nielsen, H.Ø.; Christensen, T.; Ørum, J.E.; Martinsen, L. Are independent agricultural advisors more oriented towards recommending reduced pesticide use than supplier-affiliated advisors? *J. Environ. Manag.* **2019**, *242*, 507–514. [CrossRef]
36. Kik, M.; Claassen, G.; Meuwissen, M.P.; Smit, A.; Saatkamp, H. Actor analysis for sustainable soil management–A case study from the Netherlands. *Land Use Policy* **2021**, *107*, 105491. [CrossRef]
37. Madureira, L.; Barros, A.B.; Fonseca, A.F. *Deliverable D2. 5: Synthesis Report Innovation, Farm Advice, and Micro-AKIS in Europe*; AgriLink: Castanet, France, 2021.
38. Louwagie, G.; Gay, S.H.; Sammeth, F.; Ratinger, T. The potential of European Union policies to address soil degradation in agriculture. *Land Degrad. Dev.* **2011**, *22*, 5–17. [CrossRef]
39. Ingram, J.; Mills, J. Are advisory services "fit for purpose" to support sustainable soil management? An assessment of advice in Europe. *Soil Use Manag.* **2019**, *35*, 21–31. [CrossRef]
40. House of Commons. *Environmental Audit Committee on Soil Health*; House of Commons: London, UK, 2016.
41. Landini, F. How to be a good rural extensionist. Reflections and contributions of Argentine practitioners. *J. Rural. Stud.* **2016**, *43*, 193–202. [CrossRef]
42. Nettle, R.; Crawford, A.; Brightling, P. How private-sector farm advisors change their practices: An Australian case study. *J. Rural. Stud.* **2018**, *58*, 20–27. [CrossRef]
43. Mills, J.; Reed, M.; Skaalsveen, K.; Ingram, J. The use of Twitter for knowledge exchange on sustainable soil management. *Soil Use Manag.* **2019**, *35*, 195–203. [CrossRef]
44. Skaalsveen, K.; Ingram, J.; Urquhart, J. The role of farmers' social networks in the implementation of no-till farming practices. *Agric. Syst.* **2020**, *181*, 102824. [CrossRef]
45. Daxini, A.; O'Donoghue, C.; Ryan, M.; Buckley, C.; Barnes, A.P.; Daly, K. Which factors influence farmers' intentions to adopt nutrient management planning? *J. Environ. Manag.* **2018**, *224*, 350–360. [CrossRef]
46. Prager, K.; McKee, A. Use and awareness of soil data and information among local authorities, farmers and estate managers. *James Hutton Inst. Intern. Rep.* **2014**. Available online: https://www.hutton.ac.uk/sites/default/files/files/Use%20of%20soil%20information_final%20report_14Jan2014.pdf (accessed on 20 September 2021).
47. Rhymes, J.M.; Wynne-Jones, S.; Williams, A.P.; Harris, I.M.; Rose, D.; Chadwick, D.R.; Jones, D.L. Identifying barriers to routine soil testing within beef and sheep farming systems. *Geoderma* **2021**, *404*, 115298. [CrossRef]
48. Poppe, K. 2. Agricultural Knowledge and Innovation Systems in transition: Findings of the SCAR Collaborative Working Group on AKIS. In *Improving Agricultural Knowledge and Innovation Systems*; OECD Publishing: Paris, France, 2012; p. 41.
49. Labarthe, P.; Caggiano, M.; Laurent, C.; Faure, G.; Cerf, M. *Concepts and Theories Available to Describe the Functioning and Dynamics of Agricultural Advisory Services. Learning for the Inventory (PRO AKIS WP3): Deliverable WP2-1 (Pro AKIS: Prospect for Farmers' Support: Advisory Services in European AKIS; WP2: Advisory Services within AKIS: International Debates)*; European Union: Brusssels, Belgium, 2013.
50. Birner, R.; Davis, K.; Pender, J.; Nkonya, E.; Anandajayasekeram, P.; Ekboir, J.; Mbabu, A.; Spielman, D.J.; Horna, D.; Benin, S. From best practice to best fit: A framework for designing and analyzing pluralistic agricultural advisory services worldwide. *J. Agric. Educ. Ext.* **2009**, *15*, 341–355. [CrossRef]
51. Faure, G.; Desjeux, Y.; Gasselin, P. New challenges in agricultural advisory services from a research perspective: A literature review, synthesis and research agenda. *J. Agric. Educ. Ext.* **2012**, *18*, 461–492. [CrossRef]
52. Rietra, R.; Heinen, M.; Oenema, O. A Review of Crop Husbandry and Soil Management Practices Using Meta-Analysis Studies: Towards Soil-Improving Cropping Systems. *Land* **2022**, *11*, 255. [CrossRef]
53. Piccoli, I.; Seehusen, T.; Bussell, J.; Vizitu, O.; Calciu, I.; Berti, A.; Börjesson, G.; Kirchmann, H.; Kätterer, T.; Sartori, F.; et al. Opportunities for Mitigating Soil Compaction in Europe—Case Studies from the SoilCare Project Using Soil-Improving Cropping Systems. *Land* **2022**, *11*, 223. [CrossRef]
54. Jian, J.; Lester, B.J.; Du, X.; Reiter, M.S.; Stewart, R.D. A calculator to quantify cover crop effects on soil health and productivity. *Soil Tillage Res.* **2020**, *199*, 104575. [CrossRef]

55. Nunes, M.R.; Karlen, D.L.; Veum, K.S.; Moorman, T.B.; Cambardella, C.A. Biological soil health indicators respond to tillage intensity: A US meta-analysis. *Geoderma* **2020**, *369*, 114335. [CrossRef]
56. Nettle, R.; Klerkx, L.; Faure, G.; Koutsouris, A. Governance dynamics and the quest for coordination in pluralistic agricultural advisory systems. *J. Agric. Educ. Ext.* **2017**, *23*, 189–195. [CrossRef]
57. Eisinger, P. Organizational Capacity and Organizational Effectiveness among Street-Level Food Assistance Programs. *Nonprofit Volunt. Sect. Q.* **2002**, *31*, 115–130. [CrossRef]
58. Labarthe, P.; Laurent, C. The Importance of the Back-office for Farm Advisory Services. *EuroChoices* **2013**, *12*, 21–26. [CrossRef]
59. Davis, K. The new extensionist: Core competencies for individuals. *GFRAS Brief* **2015**, *3*. Available online: https://www.g-fras.org/en/gfras/652-the-new-extensionist-core-competencies-for-individuals.html (accessed on 20 September 2021).
60. Suvedi, M.; Ghimire, R.; Channa, T. Examination of core competencies of agricultural development professionals in Cambodia. *Eval. Program Plan.* **2018**, *67*, 89–96. [CrossRef]
61. Kilis, E.; Adamsone-Fiskovica, A.; Šūmane, S.; Tisenkopfs, T. (Dis)continuity and advisory challenges in farmer-led retro-innovation: Biological pest control and direct marketing in Latvia. *J. Agric. Educ. Ext.* **2021**, 1–18. Available online: https://www.tandfonline.com/doi/full/10.1080/1389224X.2021.1997770 (accessed on 20 September 2021). [CrossRef]
62. Klerkx, L.; Van Mierlo, B.; Leeuwis, C. Evolution of systems approaches to agricultural innovation: Concepts, analysis and interventions. In *Farming Systems Research into the 21st Century: The New Dynamic*; Springer: Berlin/Heidelberg, Germany, 2012; pp. 457–483.
63. Herrera, B.; Gerster-Bentaya, M.; Tzouramani, I.; Knierim, A. Advisory services and farm-level sustainability profiles: An exploration in nine European countries. *J. Agric. Educ. Ext.* **2019**, *25*, 117–137. [CrossRef]
64. Eurostat. *Key Figures on the European Food Chain*; Publications Office of the European Union: Luxembourg, 2021.
65. Metzger, M.J.; Bunce, R.G.H.; Jongman, R.H.; Mücher, C.A.; Watkins, J.W. A climatic stratification of the environment of Europe. *Glob. Ecol. Biogeogr.* **2005**, *14*, 549–563. [CrossRef]
66. PROAKIS Project. Prospects for Farmers' Support: Advisory Services in European AKIS (PRO AKIS). Available online: https://proakis.hutton.ac.uk/ (accessed on 1 March 2020).
67. i2connect. i2onnect Project AKiS Country Reports. Available online: https://i2connect-h2020.eu/resources/akis-country-reports/ (accessed on 2 February 2021).
68. Knuth, U.; Knierim, A. Interaction with and governance of increasingly pluralistic AKIS: A changing role for advisory services. In *Knowledge and Innovation Systems towards the Future*; Publications Office of the European Union: Luxembourg, 2016; p. 104.
69. Klerkx, L.; Petter Stræte, E.; Kvam, G.-T.; Ystad, E.; Butli Hårstad, R.M. Achieving best-fit configurations through advisory subsystems in AKIS: Case studies of advisory service provisioning for diverse types of farmers in Norway. *J. Agric. Educ. Ext.* **2017**, *23*, 213–229. [CrossRef]
70. Vuylsteke, A.; De Schepper, S. Agricultural knowledge and innovation systems. In *The Case of Flanders*; Deparment of Agriculture and Fisheries, Division for Agricultural Policy Analysis and Institute for Agricultural and Fisheries Research: Brussels, Belgium, 2011.
71. Knierim, A.; Boenning, K.; Caggiano, M.; Cristóvão, A.; Dirimanova, V.; Koehnen, T.; Labarthe, P.; Prager, K. The AKIS concept and its relevance in selected EU member states. *Outlook Agric.* **2015**, *44*, 29–36. [CrossRef]
72. Sutherland, L.-A.; Madureira, L.; Dirimanova, V.; Bogusz, M.; Kania, J.; Vinohradnik, K.; Creaney, R.; Duckett, D.; Koehnen, T.; Knierim, A. New knowledge networks of small-scale farmers in Europe's periphery. *Land Use Policy* **2017**, *63*, 428–439. [CrossRef]
73. Vrain, E.; Lovett, A. The roles of farm advisors in the uptake of measures for the mitigation of diffuse water pollution. *Land Use Policy* **2016**, *54*, 413–422. [CrossRef]
74. Ingram, J.; Chiswell, H.; Mills, J.; Debruyne, L.; Cooreman, H.; Koutsouris, A.; Alexopoulos, Y.; Pappa, E.; Marchand, F. Situating demonstrations within contemporary agricultural advisory contexts: Analysis of demonstration programmes in Europe. *J. Agric. Educ. Ext.* **2021**, *27*, 615–638. [CrossRef]
75. Creaney, R.; McKee, A.; Prager, K. Designing, implementing and maintaining (rural) innovation networks to enhance farmers' ability to innovate in cooperation with other rural actors. Monitor Farms in Scotland, UK. Report for AKIS on the Ground: Focusing Knowledge Flow Systems (WP4) of the PRO AKIS Project; February 2015. Available online: www.proakis.eu/publicationsandevents/pubs (accessed on 20 September 2021).
76. Knuth, U.; Knierim, A. Characteristics of and challenges for advisors within a privatized extension system. *J. Agric. Educ. Ext.* **2013**, *19*, 223–236. [CrossRef]
77. Marsh, S.P.; Pannell, D. Agricultural extension policy in Australia: The good, the bad and the misguided. *Aust. J. Agric. Resour. Econ.* **2000**, *44*, 605–627. [CrossRef]
78. Chivers, C.-A.; Collins, A.L. (Un)willingness to contribute financially towards advice surrounding diffuse water pollution: The perspectives of farmers and advisors. *J. Agric. Educ. Ext.* **2022**, 1–24. Available online: https://www.tandfonline.com/doi/abs/10.1080/1389224X.2022.2043917?journalCode=raee20 (accessed on 20 September 2021). [CrossRef]
79. Compagnone, C.; Hellec, F. Farmers' Professional Dialogue Networks and Dynamics of Change: The Case of ICP and No-Tillage Adoption in Burgundy (France). *Rural. Sociol.* **2015**, *80*, 248–273. [CrossRef]
80. Ingram, J. Technical and social dimensions of farmer learning: An analysis of the emergence of reduced tillage systems in England. *J. Sustain. Agric.* **2010**, *34*, 183–201. [CrossRef]

81. Krzywoszynska, A. Making knowledge and meaning in communities of practice: What role may science play? *The case of sustainable soil management in England. Soil Use Manag.* **2019**, *35*, 160–168. [CrossRef]
82. Phillipson, J.; Proctor, A.; Emery, S.B.; Lowe, P. Performing inter-professional expertise in rural advisory networks. *Land Use Policy* **2016**, *54*, 321–330. [CrossRef]
83. Rust, N.A.; Stankovics, P.; Jarvis, R.M.; Morris-Trainor, Z.; de Vries, J.R.; Ingram, J.; Mills, J.; Glikman, J.A.; Parkinson, J.; Toth, Z.; et al. Have farmers had enough of experts? *Environ. Manag.* **2022**, *69*, 31–44. [CrossRef] [PubMed]
84. Vanloqueren, G.; Baret, P.V. Why are ecological, low-input, multi-resistant wheat cultivars slow to develop commercially? A Belgian agricultural 'lock-in'case study. *Ecol. Econ.* **2008**, *66*, 436–446. [CrossRef]
85. Gorman, M. Becoming an agricultural advisor—The rationale, the plan and the implementation of a model of reflective practice in extension higher education. *J. Agric. Educ. Ext.* **2019**, *25*, 179–191. [CrossRef]
86. Carlisle, L. Factors influencing farmer adoption of soil health practices in the United States: A narrative review. *Agroecol. Sustain. Food Syst.* **2016**, *40*, 583–613. [CrossRef]
87. Cerf, M.; Guillot, M.-N.; Olry, P. Acting as a change agent in supporting sustainable agriculture: How to cope with new professional situations? *J. Agric. Educ. Ext.* **2011**, *17*, 7–19. [CrossRef]
88. Schneider, F.; Fry, P.; Ledermann, T.; Rist, S. Social learning processes in Swiss soil protection—The 'from farmer-to farmer'project. *Hum. Ecol.* **2009**, *37*, 475–489. [CrossRef]
89. Klerkx, L. Advisory services and transformation, plurality and disruption of agriculture and food systems: Towards a new research agenda for agricultural education and extension studies. *J. Agric. Educ. Ext.* **2020**, *26*, 131–140. [CrossRef]

land

Article

An Overview of Sustainability Assessment Frameworks in Agriculture

Abdallah Alaoui [1,*,†], Lúcia Barão [2,†], Carla S. S. Ferreira [3,4,5] and Rudi Hessel [6]

1 Institute of Geography, University of Bern, Hallerstrasse 12, 3012 Bern, Switzerland
2 Center for Ecology, Evolution and Environmental Changes (cE3c), University of Lisbon, Campo Grande, 1749-016 Lisbon, Portugal; albarao@fc.ul.pt
3 Department of Physical Geography and Bolin Centre for Climate Research, Stockholm University, 11419 Stockholm, Sweden; carla.ferreira@natgeo.su.se
4 Navarino Environmental Observatory, Costa Navarino, Navarino Dunes, 24001 Messinia, Greece
5 Research Centre of Natural Resources, Environment and Society (CERNAS), Polytechnic Institute of Coimbra, Coimbra Agrarian Technical School, Bencanta, 3045-601 Coimbra, Portugal
6 Wageningen Environmental Research, 6708 WG Wageningen, The Netherlands; rudi.hessel@wur.nl
* Correspondence: abdallah.alaoui@giub.unibe.ch
† These authors contributed equally to this work.

Abstract: Recent research established a link between environmental alterations due to agriculture intensification, social damage and the loss of economic growth. Thus, the integration of environmental and social dimensions is key for economic development. In recent years, several frameworks have been proposed to assess the overall sustainability of farms. Nevertheless, the myriad of existing frameworks and the variety of indicators result in difficulties in selecting the most appropriate framework for study site application. This manuscript aims to: (i) understand the criteria to select appropriate frameworks and summarize the range of those being used to assess sustainability; (ii) identify the available frameworks to assess agricultural sustainability; and (iii) analyze the strengths, weaknesses and applicability of each framework. Six frameworks, namely SAFA, RISE, MASC, LADA, SMART and public goods (PG), were identified. Results show that SMART is the framework that considers, in a balanced way, the environmental, sociocultural and economic dimensions of sustainability, whereas others focused on the environmental (RISE), environmental and economic (PG) and sociocultural (SAFA) dimension. However, depending on the scale assessment, sector of application and the sustainability completeness intended, all frameworks are suitable for the assessment. We present a decision tree to help future users understand the best option for their objective.

Keywords: agriculture; sustainability frameworks; socio-economic and environmental indicators; soil land management

Citation: Alaoui, A.; Barão, L.; Ferreira, C.S.S.; Hessel, R. An Overview of Sustainability Assessment Frameworks in Agriculture. *Land* **2022**, *11*, 537. https://doi.org/10.3390/land11040537

Academic Editors: Guido Wyseure, Julián Cuevas González and Jean Poesen

Received: 16 March 2022
Accepted: 1 April 2022
Published: 7 April 2022

1. Introduction

Agricultural land covers over a third of the earth's surface [1] and 41% of the European Union's 28 member states [2]. Agriculture uses and affects natural resources, such as soil and water, shaping the landscape and contributing to establishing and maintaining semi-natural habitats [3]. Over the last decades, agricultural management practices have changed considerably to enhance crop yields and productivity to ensure food security [4]. This has been achieved through (i) technological developments, particularly by improving and adapting machinery to different management requirements, the genetic improvement of seeds and development of new agrochemicals [5], (ii) the plantation of extensive areas of monocultures [6] and (iii) the high use of mineral fertilizers and phytopharmaceuticals (e.g., pesticides and herbicides) [7–9].

The pressure on the agriculture sector will continue to rise due to global challenges, such as an increasing population and food requirements, and climate change [10]. To meet the world's projected food demands by 2050, food production must increase by 60–100% [11]. Furthermore, global agricultural production will be affected by increasing competition with certain non-food crops for several economic sectors (e.g., energy for bio-fuels production), a reduction in market prices due to globalization and limited natural resources driven by, e.g., land degradation and water scarcity [12,13] exacerbated by climate change [14].

Agricultural intensification is often associated with environmental degradation, including soil erosion, water, and soil contamination, and biodiversity loss [15–18]. By the end of the 20th century, the consequences of the intensive agriculture approach, especially in developed countries, were thoroughly investigated and frequently reported. As a result, agriculture had been highlighted as one of the main activities worldwide contributing to water depletion [19], soil degradation/pollution [20,21], biodiversity loss [22] and climate change [23]. According to the EU Soil Thematic Strategy [24], the erosion and loss of organic matter are some of the major soil threats affecting agricultural areas, along with compaction, contamination, salinization and loss of soil biodiversity.

Besides environmental problems, intensive agriculture also causes social damage and the loss of economic growth itself in the medium/long term [25]. Thus, the integration of environmental and social dimensions is key for economic development itself, and sustainable agriculture is therefore seen as the only approach towards a successful future [26]. When assessing the sustainability of different agricultural land-uses and land management practices, it is therefore important to consider not only the immediate economic benefit but also how they compromise the overall environmental quality and affect the rural communities, since these factors are relevant to sustaining future economic growth in the short and long-term [27].

As stated in the literature "Sustainability is a multidimensional concept [28] of a dignified life for the present without compromising a dignified life in the future or endangering the natural environment and ecosystem services" [29],. Its evaluation process plays an important role in the development and promotion of sustainable agricultural systems [30]. To investigate the transition towards more sustainable production, various frameworks have been proposed to gain knowledge about the sustainability performance of such production systems [31,32]. Some of these frameworks are based on indicators, whereas others are based on indices (e.g., [33]). Indicator-based sustainability assessment frameworks combining environmental, economic and social issues require the processing of a wide range of information (qualitative vs. quantitative), parameters and uncertainties [34]. They also differ in scope, target audience, indicator selection, aggregation, weighting and scoring methods, as well as the time required to complete the assessment [35]. Although many frameworks emphasize the necessity of including socio-economic and environmental aspects in sustainability assessment, many others focus only on environmental indicators to investigate the short- and long-term effects of different agricultural management practices [36] or are applied to a specific context [37]. In addition, existing assessment methodologies to investigate agricultural sustainability are scattered, focusing on single, complicated and demanding aspects regarding time, cost and required skills.

The main aim of this paper is to identify and summarize the indicators and frameworks used to assess sustainability in agricultural areas. The specific aims are (i) understanding the criteria to select appropriate frameworks and summarize the range of those being used to assess the environmental and the socio-economic themes of agricultural sustainability; (ii) identifying the frameworks available to assess agricultural sustainability; and (iii) understanding the methodological approach and analyzing the strengths, weaknesses and applicability of each framework.

2. General Considerations

The following section summarizes the general considerations about the indicators' importance, and selection criteria to set the context for those commonly used in the selected frameworks to assess sustainability in agriculture.

2.1. Criteria for Selecting Sustainability Indicators

Indicators are set to monitor and highlight the current conditions and enable stakeholders (e.g., farmers, businesses, policymakers) to identify trends and compare performances among specific places, such as farms, regions or countries, concerning their sustainability performance [38]. They should present the results in a way that is understandable by people with different occupations and sociocultural and educational backgrounds, since they are a powerful public communication tool [39].

The selection of indicators is crucial since it influences conclusions. Thus, the purpose of the assessment, the system boundaries (e.g., aims, scope and temporal and spatial scales) and the end-users should be clearly identified [40]. The assessment should also establish a baseline or reference value (starting point to measure change from a certain state or date) or target (usually established by policymakers). The comparison and contextualization helps to understand the current state or trend [41] and to support the interpretation/significance of the results [39]. Criteria to select indicators include: (i) coverage of environmental, economic and sociocultural dimensions of sustainability [1]; (ii) practicability and simplicity considering field measurements and data availability (e.g., historical data), which should consider spatial and temporal data coverage, reliability, accuracy and consistency [38,42]; (iii) the meaningful use of the indicators to take into consideration the differences in culture and geography to match them to locally relevant problems [39]; (iv) the system's sensitivity to both anthropogenic and natural stresses [1]; (v) meaningfulness to end-users in order to respond to stakeholders' expectations and support policy decisions [40]; and (vi) cost-effectiveness, since the costs to produce the information should justify the benefits of the knowledge produced [40].

Selected indicators can be assessed by qualitative or quantitative techniques [41]. Qualitative techniques are typically based on visual evaluations applied at the field scale and have been increasingly used to evaluate the soil quality (e.g., soil structure and texture, rooting depth and slope) and farm management information [42]. Ball et al. [43] summarized the visual assessment techniques that can be used to monitor soil structure, soil quality and fertility as impacted by land management. Quantitative techniques include: (i) direct measurements via field data collection (e.g., crop yields); (ii) a compilation of secondary data based on a literature review; (iii) statistical correlations of the existing data (e.g., soil compaction); (iv) modeling approaches based on empirical models (e.g., biophysical and economic); or (v) sensing approaches, such as spectroscopic techniques and remote sensing [1].

2.2. Indicators Typically Used

Table 1 summarizes chronologically some relevant studies assessing the sustainability of different agricultural practices using indicators. These studies acknowledge the need for a coherent and consistent methodology to successfully evaluate the agricultural management practices and the adoption of three-dimensional indicators. They demonstrate that an oversimplification of the evaluation does not provide a comprehensive overview of the sustainability potential of the different farming practices. These studies also show the myriad of indicators/methodologies that can be used when assessing agriculture sustainability, namely when different farming systems, practices and geographical locations are considered.

Due to the growing concern for environmental issues, numerous indicators devoted to the environmental dimension have been used, and relatively little integration of social and economic aspects on farm assessments has been considered [40]. Environmental indicators reflect the complex interaction between agriculture and environment, providing a cause-and-effect relationship. They tend to include the number and type of crops in the farm, since it links to agricultural biodiversity; soil cover, which is linked to soil erosion; water use; nutrient balance (particularly of nitrogen); and the use of pesticides [44,45], given their toxicity to the environment.

Since soil is nowadays seen as one of the most valuable resources on Earth, given its essential elements to sustain and maintain life, it has received increasing attention under the environmental indicators, and typically includes physical, chemical and biological aspects. Bünemann et al. [46] identified the most frequently proposed soil quality indicators and summarized the measured soil properties that have been used for assessing the environmental dimension in agricultural land uses from 65 soil quality assessment approaches.

Economic indicators aim to address the economic context, focusing on the economic viability defined by profitability, stability, liquidity and productivity, based on input and output prices and yields [47].

Profitability is calculated by cost and revenue, and includes variable and fixed costs (e.g., land rent), whereas liquidity measures the ability of an enterprise/farm to meet short- and long-term obligations and stability is determined by the equity share and equity development [39]. Another important indicator is productivity, which measures the ability of production systems to generate output [48]. Typical economic indicators also consider public subsidies for the farmers, since they provide protection regarding their agricultural activities. GDP is sometimes considered as an indicator of the difference between producers' income and transfers to other economy sectors (variable costs, subsidies) [44].

Most social indicators focus on the following: (i) the sustainability of the farming community, which involves the welfare of the relevant actors and communities; and (ii) the sustainability of society as a whole. The first type of indicators focuses mainly on working conditions, education and the quality of life defined by physical well-being and psychological well-being [40]. Social sustainability is linked to society's demands, with regards to its values and concerns [49], and may be grouped in: multifunctionality (e.g., quality of rural life, contribution to local employment and to ecosystem services) [50], sustainable agricultural practices (e.g., animal welfare, environmental impacts) and product quality (e.g., quality processes, food safety) [40]. These indicators also tend to measure the socio-economic implications of agriculture in the rural income, and may be measured by the total labor generated, as well as by the seasonal variations linked to individual crop requirements, often associated with peaks in agricultural employment (e.g., sowing and harvesting) [44]. Measuring indicators of a sociocultural dimension is challenging, since they are based on a qualitative assessment and are therefore subjective. Farm-community-based indicators are usually based on farmers' self-evaluation gained from surveys or interviews [40].

Table 1. Most relevant papers on sustainability providing relevant information on the indicators and their relevance for case study applications and different conditions. The papers in relation to the frameworks considered in this manuscript are not reported, except those comparing different frameworks.

Reference	Summary of the Study
[50]	Presents the farmer sustainability index (FSI), relying on sustainability scores for diverse agricultural management practices to avoid an oversimplification of the reality. The study focuses on 33 production practices implemented by [51] Malaysian farmers to assess the FSI scores.
[39]	The sustainability of the agricultural systems is assessed based on different points and levels, considering the need to improve the assessment methods used for some agricultural sustainability subthemes. The limited availability of tools to evaluate qualitative aspects, such as landscapes and animal welfare, was identified as a major shortcoming. It also highlights the need to couple economics and social sciences with environmental processes for a better understanding of the overall agricultural system.
[52]	By analyzing the impact of agri-environmental indicators (AEIs) on policy outcomes, the paper examines the potential impacts of Agri-environmental Regulation EC 2078/92 on European agricultural landscapes. It discusses the frameworks divided in policy outcomes and policy performances and analyzes the obstacles to measuring policy outcomes directly. The study focuses on intensification and abandonment problems in extensive agricultural areas of Spain and Denmark.

Table 1. *Cont.*

Reference	Summary of the Study
[53]	The environmental impacts of agriculture are investigated through life cycle assessment (LCA). The LCA framework was adapted in terms of functional unit and impact categories of the agricultural production process. The framework was applied in 18 grassland dairy farms managed under different intensity levels in southern Germany.
[54]	Investigates a method for evaluating the environmental impacts of arable farming systems. The method is based on agro-ecological indicators (AEI) to rank or classify the cropping systems. The agro-ecological indicators tested include phosphorus and nitrogen fertilization, irrigation, pesticides, organic matter, cropping pattern, crop succession and covering, ecological structures, soil management and energy.
[55]	Environmental impacts, economic viability and social acceptability are investigated in two production systems. The sustainability of the system is based on 12 indicators assessed through empirical data from household survey, soil samples, field observations and information supplied by key informants. Management of soil fertility, pests and diseases, the use of agro-chemicals and crop diversification were significantly different between both systems. In turn, indicators, including crop yield and stability, land-use pattern, food security and risk and uncertainties, showed similar results.
[56]	The use of pesticides, nutrients and energy in 55 farming systems was compared using input–output accounting systems (IOA) covering the topics of the farm's use of nutrients, pesticides and energy. The indicators and approach used varies from systems using physical input–output units to systems based on good agricultural practices (GAP).
[57]	Proposes the Sustainability Assessment of Farming and the Environment (SAFE) framework, aiming to assess the sustainability of agricultural systems through several criteria and indicators. The framework can be applied at different spatial scales, including parcel, farm, landscape, region or state. This is a hierarchical framework, comprising structured principles, criteria and indicators. SAFA serves as an assessment tool for identifying, developing and evaluating the overall sustainability of agricultural systems, techniques and policies.
[58]	It presents the Indicateurs de Durabilité des Exploitations Agricoles or Farm Sustainability Indicators method (IDEA) tool, which includes 41 sustainability indicators, and is devoted to supporting farmers and policy makers. The study reveals that the IDEA method requires adaptation of indicators to local farming.
[48]	Based on an irrigated agriculture area in Spain, authors perform a comparative analysis of different methods for developing composite indicators to analyze agricultural sustainability. The study uses indicators calculated from several farms and policy scenarios.
[59]	Develops a methodology to evaluate the sustainability of two agricultural systems in Spain (rain-fed vs. irrigated) through composite indicators. It reveals farm heterogeneity in each individual agricultural system in terms of sustainability, and analyzes the influencing variables to support decision making.
[60]	Proposes a framework for an integrated assessment of sustainability in European regions and policy options. The framework is used in ex ante assessment of land use policy scenarios and includes environmental, economic and social aspects in different sectors (forestry, agriculture, tourism, transport and energy). The conceptual framework can be applied at different scales (regional, European), and considers the variability of the European regions.
[61]	Presents a project funded by the UK government to develop a methodology for assessing the sustainability of both conventional and organic farming systems. The project includes 40 environmental, social and economic indicators. Data were collected to support the chosen indicators. The selected set of indicators assesses the advantages and disadvantages of the different farming systems, and the results can be useful to improve the sustainability of the farming systems.
[62]	Provides a review of current management tools to address sustainability in small and medium-sized enterprises (SMEs) and highlight the advantages of such tools for SMEs. Results show that most tools are not implemented by the majority of SMEs, and summarize the barriers for this. The paper also suggests criteria to facilitate future implementation.
[63]	The MASC framework is used to evaluate the performance of 31 agriculture cropping systems. Conservation agriculture displayed a greater sustainability performance, especially regarding the environmental criteria. However, conservation agriculture systems revealed several weaknesses, namely regarding those of technical or social nature.

Table 1. *Cont.*

Reference	Summary of the Study
[64]	Four sustainability assessment tools (RISE, SAFA, PG and IDEA) were compared regarding the indicators used for perceiving practical requirements, procedures and the complexity of their application on five Danish farms. The scoring and aggregation method used in each tool vary widely, as well as the data input and time requirements. RISE was considered as the most relevant tool. However, farmers seem hesitant in applying the outcomes of the tools to support decision making and management.
[65]	Develops a set of indicators based on generally available data to assess the sustainability of urban food systems. Through a participatory process, an assessment method considering 97 indicators for evaluating 51 of the 58 subthemes was considered developed. The method was tested in Basel city, Switzerland, and revealed that it was useful to improve the sustainability of the tested investigated food system.
[66]	By using a set of environmental, social and economic indicators, the sustainability of an agricultural sites in Italy was assessed. The indicators were identified based on IDEA, RISE, SAFE, SOSTARE and MOTIFS methodologies. The framework developed provides easy-to-read results relevant for different scales assessment, and relies on balanced features of data availability and reliability.
[67]	The environmental sustainability of the ornamental plant production sector (including both nurseries growing plants in container production (CP) and in open field (FP)) is assessed through impact indicators. The results exposed the higher environmental impacts of the CP comparing with the FP due to their peculiar production structure, which, thus, must be improved to assure an acceptable environmental performance.
[68]	The social sustainability of the Swedish (livestock) farming system is investigated using the social indicators considered in existent sustainability assessment tools (RISE, SAFA, IDEA). From these three tools, RISE seems best at capturing the social situation of the farmers, although not fully addressing the finding work aspect. Both SAFA and IDEA fail to capture several aspects relevant to describing the situation of the farmers.
[69]	Investigates how existent sustainability assessment tools support decision making regarding management practices by farmers. It shows that farmers need more basic and rapid overviews of the complexity dimension, whereas the management dimension is useful to develop and implement new farm strategies.
[70]	An ex ante evaluation of several conventional practices is used to enhance the sustainability of cropping systems. The sustainability of five diversified cropping systems is compared with less diversified systems in several arable areas of France. The diversified systems revealed fewer greenhouse gas emissions, improved water and air quality and a high biodiversity. Nevertheless, diversification can cause negative impacts in some indicators, such as NH_3 volatilization, NO_3^- lixiviation, pesticide use and gross margin.
[71]	A multi-criteria analysis (MCA) tool is developed to assess the sustainability of four Italian organic farms with durum-wheat-based crop rotations. The best sustainability scores were noticed in both ex ante and ex post analysis by diversified cereal farming systems with short supply chain mechanisms to sell their products.
[72]	A sustainability assessment of the flowering potted plants (FPP) value chain was performed, including all of the phases from breeding to distribution. The selected indicators relied on SAFA and RISE sustainability assessment tools. The study shows that SAFA and RISE tools do not cover the overall sustainability subthemes, and emphasizes the need for a system-specific view in unique systems, such as the FPP.
[73]	The relationship between agricultural sustainability and economic resilience is investigated through an empirical analysis of Northern European countries. Composite indicators are settled based on decision-making criteria. Results highlight that sustainability indicators cannot be replaced by economic resilience ones, and that the latter should be considered in addition to the economic sustainability indicators.

3. Methodology

During the past 20 years, various approaches and tools have been proposed for assessing the overall sustainability in the agricultural production system and food sector [31,74,75]. However, these methods have many limitations. As an example, life cycle assessment tools quantify many aspects of the environmental dimension in a narrow way, need a high amount of data and do not consider the impacts on soil quality and biodiversity [76] and economic and socio-cultural impacts [77], or can only be applied to agricultural enterprises [32]. Eco-management and audit schemes, as well as sustainability reporting systems, include procedures accounting for the sustainability of a company, but do not enable comparison between the outcomes of different ones since they are not science-based assessments [78].

In this study, we selected indicators and frameworks based on the following criteria: (1) went through a peer review process, (2) have a farm assessment level, (3) cover universal agricultural sectors, (4) include the three dimensions of sustainability, (5) suitable for Europe and countries worldwide and (6) present transparency of information allowing for an informed assessment as well as solid cultural and value-based elements.

For the search of the frameworks, we considered literature including at least one peer reviewed publication, reports and presentations available online by searching on scientific web platforms.

Each framework selected was therefore described by stating information on the type of tool used (software, database, etc.) and where it can be found available, requisites for running the tool, type of input data required, time needed for the assessment and number and description of indicators (environmental, socio-cultural and economic) used.

The six sustainability assessment frameworks were also compared according to their ability to cover the main themes of environmental, economic and sociocultural dimensions, and their themes were reported. We compared their strengths and weaknesses and developed a decision tree based on possible scales, sectors of applicability and the completeness of sustainability dimensions required to help stakeholders decide which framework is the most suitable for their sustainable assessment purposes.

4. Results

Based on selected criteria, the following frameworks were identified: SAFA, RISE, MASC, LADA, SMART and PG. Below, each framework is briefly described, as are the environmental (Table 2), sociocultural (Table 3) and economic (Table 4) indicators included in each one of them. In the next section, their strengths and weaknesses are highlighted individually.

Table 2. Environmental themes, sustainability objectives, indicators and measured parameters for each framework considered in this study.

Theme	Sustainability Objectives	Indicators	Framework	Parameters
Water use	Water conservation	Water management	RISE	Water consumption monitoring and measures for water saving
			PG	Irrigation, flooding defences, pollution reduction, water management plan
			SAFA	Reduction in water consumption/water withdrawals
		Dependency of water	MASC	Irrigation, water availability and crop water requirements
	Water security (supply without compromising available water resources)	Water Supply	RISE	Assessment at watershed scale
	Availability of water resources for irrigation, salinization	Irrigated areas	LADA	Water availability
Water quality	Water resources degradation	Overexploitation of water resources, salinization	LADA	Groundwater level, salinity of water, arsenic contamination
		Clean water target	SAFA	Concentration of water pollutants, wastewater quality
Water pollution	Water pollution risks	Pesticides losses in water	SMART	NO_3 losses, phosphorus losses

Table 2. *Cont.*

Theme	Sustainability Objectives	Indicators	Framework	Parameters
Soil quality/land degradation	Providing the best conditions for plant growth and soil health, preventing land degradation	Physical and chemical properties	SMART	Compaction, erosion, SOC, phosphorus fertility,
			PG	Cultivation, winter grazing, NPK management, cropland diversity, livestock diversity
			RISE	Soil reaction
			SAFA	Soil chemical and biological quality, soil structure and SOM
	Identification of soil and terrain resource degradation	Erosion, compaction, nutrient and soil biodiversity decline, salinization (regional)	LADA	Texture, structure, pH, organic matter, water infiltration/drainage, salinity, soil depth, landslides, gullies
			RISE	Soil erosion, soil compaction
			SAFA	Soil health, soil degradation, net loss/gain of productive land
		Soil resources (local)	LADA	Heavy metals, earthworms (and others), root development, soil color
Air quality	Prevention of air pollutant emissions and elimination of ozone-depleting substances	GHG, air quality	SMART RISE	Air pollution, ozone substances, GHG
			SAFA	Emission of air pollutants, number of days of the year with exceedance of air pollution values, GHG emission, net direct GHG emission
Climate	*Climate resources:* Identification drought/desertification and water erosion	Aridity, soil moisture, variability of rainfall	LADA	Aridity index, soil moisture change, inter-annual and trends of rainfall
	Extreme events: Tsunami, heavy rains, long drought, dust storms, volcanic eruption, water erosion	Extreme events, disasters, slope/land use	LADA	Salinization, landslides, loss of land cover and biodiversity, sedimentation
Plant and fertility	Fertilizer conservation: Prevent nutrient losses through runoff	Wastewater quality	SAFA	Nitrate and orthophosphate concentrations
			RISE	Material flows, fertilisation Environment pollution
	Abiotic resources conservation	Phosphorus conservation	SMART	Crop phosphorus needs, phosphorus use autonomy
			PG	Manure management
	Reduce plant protection: Reduce application of chemicals and avoid environmental exposure	Plant protection Practices [1]	RISE	Agreement with integrated plant protection principles

Table 2. *Cont.*

Theme	Sustainability Objectives	Indicators	Framework	Parameters
Biodiversity	Preserve diversity of ecosystem, species and generic	Species conservation practices	SMART	Conservation of functional integrity, agrifood ecosystem, wild and domesticated species
			PG	Conservation plan, habitats, rare species
			SAFA	Rare and endemic species, wild animals, threatened or vulnerable wild species
		Functioning and connectivity of ecosystem services	SAFA	Ecosystem services, connectivity, structural diversity of ecosystems, land-use and land-cover change
	Preserve vegetation resources	Changes in land cover	LADA	Loss of biodiversity/loss of nutrient
		Genetic diversity	SAFA	Wild genetic diversity, agro-biodiversity, locally adapted varieties/breeds, rare and traditional varieties and breeds
	Pesticide use intensity	Number of doses	MASC	Sprayed area, insecticides, fungicides, herbicides
	Infrastructure and production	Management and production	RISE	Management of biodiversity, ecological infrastructure, distribution of ecological infrastructures, diversity and intensity of agricultural production
Energy use (temperature control/heating storage and transport)	Reduce GHG emissions and energy consumption	Measures to save energy	SAFA	Implementation of energy-saving practices
			PG	GHG emissions
		Energy conservation	MASC	Energy consumption, energetic efficiency
			PG	Energy balance, benchmarking
			RISE	Energy management, energy intensity, greenhouse gas balance
	Reduce non-renewable energy sources' dependency	Renewable energy	SAFA	Net of energy use and share of sustainable energy transports
	Waste reduction and disposal		SMART	Prevention of waste generation
			PG	Disposal of farm waste
Energy use Substrate and containers	Reduce non- renewable materials (e.g., plastic, peat)	Material consumption practices	SAFA	Replacement of non-renewable materials by renewable and recycled materials
	Reduce non-degradable waste such as plastic or substrate (perlite)	Waste reduction practices	SAFA	Reducing the generation and hazardousness of waste, food loss and waste reduction

Table 2. *Cont.*

Theme	Sustainability Objectives	Indicators	Framework	Parameters
Animal welfare	Animal health and freedom from stress	Animal health	SMART	No thirst, hunger, injury and disease
			PG	Housing, bio security, ability to perform natural behaviors
			RISE	Animal production management, productivity of animal production, possibility of species-appropriate behavior, living conditions, animal health
			SAFA	Reduce pain and injury risk of animals, condition of animal husbandry

[1] Originally "Plant protection" in the RISE framework.

Table 3. Social themes, sustainability objectives, indicators and measured parameters for each framework considered in this study.

Theme	Sustainability Objectives	Indicators	Framework	Parameters
Employment contract/agreement	Workers' stability and secure workplace through legal contracts	Employment relations; ability to cover the costs of production, right of suppliers	SAFA	Written agreements with employees
		No forced labor, no child labor, freedom of association and right to bargaining	SMART	Fair prices, rights of suppliers are respected, labor rights
Workload	Allows overtime compensation and quality of life	Working hours	RISE	Working hours and vacations recorded and following the standards
Wages	Wages provide reasonable life quality for workers and their families	Wage level	SAFA SMART	Living wage paid to employees
		Profession and education, financial situation, social relations, personal freedom and values, health	RISE	Education, economic and social situation, health
Health safety	*Occupational health and operational difficulties:* Employees trained for health and safety issues/complexity of implementation	Safety and health trainings/health risks	SAFA MASC	Existence and effectiveness of employees' health and safety training/physical constraints, number of specific operations, number of crops
	Safe working environment	Safety of workplace	SAFA SMART	Determining safe, clean and healthy workplace
	Medical care: Access to affordable medical care for employees;	Health coverage and access to medical care	SAFA	Employees' access to medical care; and health provisions
Job satisfaction	Attract and retain employees	Capacity development	SAFA	Opportunities for employees' capacity development and advancement
			PG	Skills and knowledge
Decent livelihood	Enjoy a livelihood, time for culture and nutritionally adequate diet, training and education, access to means of production	Life quality, development capacity, fair access to production income	SMART	Adequate livelihood, possibilities for education and training, access to production means

Table 3. *Cont.*

Theme	Sustainability Objectives	Indicators	Framework	Parameters
Gender equality/equity	No gender discrimination, including support of working mothers through provision of maternity leave; non discrimination, support to vulnerable people	Gender equality equity, non-discrimination	SAFA SMART	Resources to provide women's pregnancy rights; equity and non-discrimination policies are taken into account; disadvantaged groups are promoted and supported.
Cultural diversity	Freedom of choice and ownership in regards to production means	Indigenous knowledge, food sovereignty	SMART	Intellectual property right, choice and ownership in regards to production means
Benefits to/investment in local communities	Support of/invest in local communities	Community investment	SAFA	Investment to meet local community needs
Employment	Contribution to local/ regional employment	Regional workforce	SAFA MASC	History of preferential hiring of local employees when possible,
			PG	Community engagement
Consumer safety	Product free of highly hazardous pesticides	Hazardous pesticides	SAFA	Any highly hazardous and other pesticides used (safety to consumers and pollinators)
Transparency	Consumer informed of product quality through a reliable labeling system	Product labeling	SAFA	Products are labeled in compliance with standards

Table 4. Economic themes, sustainability objectives, indicators and measured parameters for each framework considered in his study.

Theme	Subtheme	Sustainability Objectives	Indicators	Framework	Parameters
Profitability	Net income/autonomy	Maintain short- and long-term profitability of the business/autonomy	Net income	SAFA MASC SMART	Total revenue in the last five years associated with producing goods and services exceeds the totalprofitability, independency, efficiency, specific equipment needs
			Liquidity, stability, profitability, indebtedness, livelihood	RISE	Liquidity, stability, indebtedness, livelihood
	Profitability per unit product	Costs of unit production are lower than the price per unit of product sold	Cost of production	SAFA RISE	Cost of the products sold per unit of production, break-even point
				PG	Financial viability
Vulnerability	Stable production	Mitigating production risk such as unpredictable weather conditions and pathogen infestation	Production risk [1]	SAFA	Implementation of mechanisms to prevent disruption of volume or quality
				SMART	Stable business relationships and accessibility to alternative procurement channels
				SAFA	Procurement channels to reduce the risk of having input supply shortages, stability of supplier relationships
	Assortment	Diversified products to ensure market growth, product differentiation and reduced risk (market, weather, price)	Product diversification	SAFA	Number and type of products, as well as development of new products
	Diversified income	Diversified income structure (marketing channels and buyers) and production contract with buyers	Stability of market	SAFA SMART	Activities to diversify marketing channels and stabilize prices
	Risk management	Internal and external risks (e.g., demand uncertainty, shortage in workforce)	Risk management	SAFA SMART RISE	Existence of a plan or a strategy to reduce risks and adapt [3]
				PG	Farm resilience
	Liquidity	Financial liquidity to withstand shocks	Financial liquidity [2] /independence	RISE MASC SMART	Cash flow plus available credit lines divided by average weekly expenditure
				SAFA	Net cash flow, safety nets

Table 4. *Cont.*

Theme	Subtheme	Sustainability Objectives	Indicators	Framework	Parameters
Accountability	Product traceability, food safety and quality	Products can be traced along the value chain	Traceability system	SMART	Share of production that can be traced along the value chain, food safety and quality
				SAFA	Product labeling, traceability system, certified production, food quality, control measures, hazardous pesticides, food contamination
				PG	Food quality certification
Investment	Internal, community, long-ranging investment	Sustainable performance and development of a community aiming at long-term sustainability	Resilience	SMART	Enhancing sustainability performance, sustainable development of a community, long-term sustainability
				SAFA	Long-term profitability, business plan
			Internal investment	SAFA	Improved social, economic, environmental and governance performance
			Community Investment	SAFA	Balance between the community needs and efficient use of environmental resource
Local economy	Value creation, local procurement	Benefit of the local economies through procurement from local suppliers	Local economy	SMART	Benefit to local economies through employment and payment of local taxes,
				PG	Local food, production of fresh produce
				SAFA	Regional workforce, fiscal commitment, local procurement
Economic risk	Loss of land	Identification of the risk related to the loss of profit	Frequency of forest fires, presence of land mines, under-management resource, urbanization, livestock pressure, human-induced disasters	LADA	Deforestation, complete loss of land, nutrient loss/erosion, sealing, compaction, loss of land cover, isotope fall out (radio nuclear)
				PG	Landscape features, management of boundaries

[1] Originally "guaranty of production levels" in the SAFA framework. [2] Originally "liquidity" in the RISE framework. [3] Addressed by operational management with the indicators: goals, strategy and implementation, information availability, risk management and sustainable relationships. SMART has a 4th dimension "Good Governance", with the following themes: corporate ethics, accountability, participation, rule of law and holistic management (not included here).

4.1. SAFA

The Sustainability Assessment of Food and Agriculture systems (SAFA) is a framework developed and proposed by FAO to assess the environmental and social impacts of food and agricultural operations [79]. It offers a comprehensive reference framework for assessing sustainability in agricultural, forestry and fishery chain systems. The framework is designed hierarchically starting with four dimensions: environmental integrity, social well-being, economic resilience and good governance [72].

The available software (https://www.fao.org/nr/sustainability/sustainability-assessments-safa/safa-tool/en/) (accessed on 4 March 2022) calculates 116 indicators that target the principles of sustainable development. Measured and/or calculated data from production sites with defined unit processes of a system include a wide diversity of sources, including literature or available databases, and public and other independent sources of information. Additionally, interviews are carried out with local employees in the sector considered. Data analyses should be conducted by an expert in sustainability. SAFA-Tool assists users with setting system boundaries and scoring ranges, and selecting targets, practices or performance indicators from qualitative or quantitative information. The latest software version 2.4.1 allows the user to add their own indicators. Depending on the complexity level of the analysis, determined by the choices made by the user, data collection may range from ±2 h to weeks, and the total assessment from 0.5 days to months [69].

Environmental indicators established in SAFA cover a broad range of themes including water use, wastewater quality, soil quality, air quality, species conservation practices and ecosystem diversity, energy-saving practices, material consumption and reduction practices, energy use and animal welfare, all linked to the food and agriculture processes (Table 2). The social angle of the evaluation process is also very well represented in SAFA, with the rating of indicators covering themes such as employment contracts, the wage level of employees, safety and health environment, job satisfaction, gender equality, cultural diversity or even transparency in the labeling, safety for the consumer and the impact of using a regional workforce (Table 3).

Finally, economic indicators figuring in SAFA cover both profitability and vulnerability topics, such as the net income, production cost and risk and stability of the market or risk management, among others. It also includes indicators related to accountability, such as the existence of system traceability, the investment potential and the will to invest in local economy (Table 4).

LADA data are extracted from the LADA indicators' toolbox developed for LADA (see [80]); the indicators of LADA are divided into two types: those describing the state of the resources+ and those describing direct pressure on the resources++; thus, the indicators used are those that indicate the degradation type

4.2. RISE

The framework RISE (Response-Inducing Sustainability Evaluation) was developed by Hafel, in Switzerland, for evaluating the environmental, sociocultural and economic sustainability of farm operations [80]. Currently, the RISE version 3.0 software can be found online (RISE 3.0 - Software Manual (bfh.ch)) (accessed on 4 March 2022) or offline (Microsoft SilverlightTM plug-in required) to analyze the data. It includes a total of 50 indicators addressing environmental, social, economic and land management aspects. The data are collected with a questionnaire-based methodology, where farmers are interviewed for 3 to 5 h, which, with the additional time for data computation, requires a total assessment time of 5–9 h [80]. The framework should be used by agronomists or specialists in agricultural advisory. The results are thoroughly discussed with farmers and used to support the continuing improvement of farm sustainability. The environmental indicators included are mainly related to water use and plant protection (Table 2), whereas the social dimension is focused on the workload and the economic dimension mainly tackles the business vulnerability by assessing the financial liquidity (Table 4).

4.3. MASC

INRA (Institut National de la Recherche Agronomique) developed MASC (Multi-attribute Assessment of Sustainability of Cropping Systems) to assess how cropping systems contribute to sustainability at the farm level [13]. The tool that is currently available (http://wiki.inra.fr/wiki/deximasc/Main/) (accessed on 4 March 2022) uses a decision tree to break down the sustainability assessment decisional issue into 32 input criteria. Indicators used to assess these basic input criteria can be chosen by the user depending on their accuracy and the context of their study, as well as the available data [63].

Qualitative and quantitative information is collected through questionnaires and reported results. Methods such as MASC that are suited for the analysis of qualitative data may be more relevant for sorting and categorizing technical solutions when considering a wide range of performances [13,81]. The tool should be managed by a researcher/professional, who then interprets the results obtained.

The indicators included in this framework deal with the evaluation of environmental aspects such as water use, biodiversity and energy use through indicators of water dependency, number of pesticides doses and energy conservation (Table 2). Social indicators are also included, especially targeting the safety and health trainings of employees and the priority to employ a regional workforce. The economic dimension is assessed through indicators of net income and financial liquidity (Table 4).

4.4. LADA

The LADA tool (Land Degradation Assessment in Drylands) framework was developed by FAO (Food and Agriculture Organization of the United Nations) for assessing and quantifying the nature, severity, impact and extent of land degradation on ecosystem services across different spatial and temporal scales. In order to support policy decisions to combat land degradation, the framework aims to identify hotspots and bright spots [82]. It is available as a tool-kit (https://www.fao.org/nr/kagera/tools-and-methods/lada-local-level-assessment-manuals/en/) (accessed on 4 March 2022) that identifies the state of the land resources through different indicators, the pressures and driving forces that caused this status and the impacts on ecosystem services and on livelihoods. The data required are collected through agricultural and other national surveys and censuses and maps of soil and natural resources, as well as digital and computer-assisted methods.

LADA environmental indicators focus on water quality and water use, soil quality and the soil degradation status. It includes an assessment of the irrigation area and the over-exploitation of water resources, as well as the salinization process, and includes indicators focused on general soil threats, including erosion, compaction and nutrient loss. Biodiversity is also tackled through indicators of land cover (Table 2). Additionally, LADA also includes economic indicators related to the economic risk caused by land degradation problems, through the assessment of land loss by fires, urbanization and livestock pressure, among others (Table 4). The sociocultural dimension is represented by the pressures on the resources that will impact society as a whole. The change in land users' life is not investigated. The LADA framework considers climate components illustrated by climate resources and climate extreme events.

4.5. SMART

The SMART (Sustainability Monitoring and Assessment RouTine) framework was developed by FiBL (Research Institute of Organic Agriculture) to assist farms and enterprises in the food sector for assessing their sustainability level in a credible and transparent manner [83]. The specific software (https://www.fibl.org/en/themes/smart-en/smart-method) (accessed on 4 March 2022) is used to compute context-specific indicators (up to 200) that are compiled individually for each case study. Data needed for the assessment are semi-quantitative and collected using a standardized interview procedure [84]. The time for data collection is 2–3 h [64]. The software should be handled by scientists and/or field practitioners. The extensive list of indicators includes transversal environmental topics from water pollution to soil quality and degradation, air quality, fertilizer consumption, biodiversity, energy use and even animal welfare. Examples of the broad list of environmental indicators in the framework include pesticide presence in water, greenhouse gas emissions, phosphorus crops content, conservation of species and the use of renewable energy (Table 2). Social indicators are also included in the framework, assessing employees' rights and their wage level for a dignified life. The social dimension also includes gender equality and non-discrimination, cultural diversity, health coverage and access to medical care (. Finally, economic indicators cover a set of themes, from profitability to vulnerability, accountability, the resilience of the investment and the value of local economy (Table 4).

4.6. PG

PG (public goods) is a framework developed by the Organic Research Centre in the United Kingdom for assessing the provision of a broad range of public goods from farming activities [84]. It is based on the premise that agriculture produces many by-products that are deemed public goods [85].

Information related to the farming activity is gathered and computed in an excel sheet (https://www.organicresearchcentre.com/our-research/research-project-library/public-goods-tool/) (accessed on 4 March 2022), where 11 individual public goods are scored. Information is collected using questionnaires with several key "activities" and includes qualitative and quantitative data. The analysis is normally undertaken by famers and/or sustainability experts. The time of data collection varies between 2 and 4 h [84].

Environmental indicators from PG framework include water management and soil quality through the assessment of the irrigation method used, flooding defenses implemented and the existence of water and nutrients management plans, cultivation types and cropland and livestock diversity. Biodiversity and energy use are also tackled extensively through the screening of conservation plans, the presence of habitats and rare species, GHG emissions, energy balance and the correct disposal of farm waste. The animal welfare is accounted through parameters such as housing, biosecurity and their ability to behave naturally (Table 2). The social indicators are basically represented in the job satisfaction through the skills and knowledge of the employees and the contribution to local/regional employment assessed by the level of community engagement. Economic indicators range from financial viability and farm resilience to others, such as accountability by food quality certification, the local economy value through assessing the production of local products and the economic risk by checking landscape features and the management of boundaries (Table 4).

5. Discussion

5.1. Strengths and Weaknesses of the Frameworks

5.1.1. SAFA

The study by Landert et al. [83] aimed to transform intensive livestock farming in 15 European countries with a high impact on the environment, society and economy in sustainable livestock farming, which reduces emissions and the costs associated with this. The authors showed that farms with an optimized governance component can improve sustainability in general and that the farmers should learn about this context and improve their production and economic performance within each individual farm. In this context, SAFA is an important tool to provide recommendations for future actions to support achieving sustainability [86].

The study by [87] in the central Sicily Mountains showed that a growing economy would also require more resources to reduce environmental impacts, modernize animal shelters and use renewable energy sources to make them more sustainable. It illustrates how, on the one hand, the sustainability areas that are discussed in SAFA are interconnected, and, on the other, that there are many open pathways for Sicilian organic farms to improve their performance. Although SAFA is a valid asset for addressing the sustainability potential of food in urban system contexts, two main weaknesses related to some subthemes have been pointed out by Landert et al. [83]: (i) the subtheme Remedy, Restoration and Prevention would need a specific adaptation to become food-focused, and (ii) the subtheme Rights of Suppliers does not include the full web of existing relations and processes normally present in these systems. In addition, the subthemes Long-Ranging Investment, Profitability, Stability of Supply, Stability of Market and Liquidity are not flexible for use in this system [65]. In this context, by setting the boundaries of the system, the majority of the indicators became less responsive to drivers or pressures. In turn, this led to poorer analyses of the cost-effectiveness and political and societal acceptance.

5.1.2. RISE

Grenz et al. [88] showed that RISE is an effective tool for field production since it measures fertilizer application relative to soil nutrients and crop requirements for optimum crop growth and calculates the non-renewable energy percentage, as well as the farm financial security (e.g., diversifying income sources, securing access to land, maintaince of infrastructure). Röös et al. [68] observed the potential of this framework to integrate the social dimension of the farm, although some modifications would be necessary to enhance its relevance for the specific context of the study. The authors perceived the results of RISE as highly solid because they are based on quantitative data input and integrate experts on the subject.

RISE becomes complex due to complicated calculations and the elevated number of data required. However, regarding the tool, farmers consider it as relatively simple to understand [68] because of the language adopted compared to the more general one found in in SAFA (e.g., rule of law) [88] and IDEA (e.g., organization of space) [68]. Regarding the relevance of RISE, the farmers recognize that the obtained outcomes reflect the positive and negative points of their farming activities well. Therefore, in comparison to other frameworks (e.g., IDEA, PG), farmers consider RISE as one of the most appropriate frameworks to use [72]. However, it was shown that the time investment and time required for learning RISE are relatively long in comparison to other frameworks [68], while also not being highly transparent as other frameworks due to the complexity of the calculations that complicates the computation rationale behind it [64]. In addition, using standardized quantitative measures makes it hard to capture the specific situation (e.g., farmers' financial situation and working situation), since farming activities will always endorse high variability from one case study to another [68].

Havardi-Burger et al. [72] showed that the process of selecting indicators in RISE becomes difficult since, on the one hand, one must include all of the significant indicators that represent the system well, but, on the other, the number of indicators cannot be too high otherwise it compromises the application of the tool. This aspect is observed for all frameworks except for SAFA, which includes a relatively high number of indicators. In describing this difficulty, Binder et al. [31] refer to parsimony as a principle in order to strive for the system representation under consideration and the sufficiency to address its complexity. Overcoming this difficulty by setting different indicators from different sustainable dimensions and themes is not an easy task since one becomes easily lost on what is actually under study. One possible example is the indicator stability used in RISE to address how financially stable a farm is (e.g., farm infrastructure, long-term access to land, the number of customers and main source of income). The authors showed that covering more aspects would be a benefit, as also shown in the indicator liquidity combining two SAFA indicators (safety nets and net cash flow). This allows the adoption of concrete measures to improve the business performance, even when under financial stress [31].

5.1.3. MASC

MASC can be described as an objective and broad tool. Its ability to incorporate qualitative data in addition to its ease-of-use in terms of the necessary input becomes very helpful for real situations and enables a high comprehensibility of the outputs. Quantitative values can be processed as qualitative information by simply using thresholds, and, thus, MASC integrates both measurements (e.g., yields), calculated data (e.g., semi-net margin) and empirical knowledge (e.g., physical difficulties of crop interventions) into the indicators. This ensures that the best available information is used and that there is a high participation approach, since, as an example, the users' point of view can be integrated in the framework, since normally it would be difficult to address them by using quantitative indicators [63].

Graheix et al. [62] applied MASC to evaluate 31 cropping systems previously chosen to study different management practices, from conventional tillage systems to other systems where conservation agriculture principles were incorporated. In this study, the integrative approach of the MASC framework provided a benefit for the understanding of how the different cropping systems behave when considering, at the same time, (i) the multiple objectives of the dimensions (economic, social and environmental); (ii) various time scales and (iii) the objective worries and goals of the farmers, and generally also the society, raised by different stakeholder groups with various interests. While the results of many studies have highlighted advantages of MASC for adapting cropping systems through conservation agriculture [63], they also identify a weakness in terms of MASCs' inability to properly evaluate the agronomic effects of biodiversity (e.g., normally, a higher biodiversity is an advantage, but decreasing the soil tillage may also contribute to a higher diversity of pests and weeds) from a simple description of the practices employed. The diversification systems may have many advantages (e.g., lower GHG emissions) in comparison with the conventional reference system. They may improve both the air and water quality and contribute to a higher biodiversity [70]. The indicators were initially determined based on scientific knowledge and the context available at the time of the development of MASC, with the aim of keeping its use relatively simple [25]. This probably led to a too generalized meaning of the indicators that cannot highlight the specific context found in different pedoclimatic conditions and under different agricultural management practices of the different studies. As reported by Médière et al. [89], "we still have little scientific information concerning the responses of biological process to agricultural practices in a given pedoclimatic context". The balance between benefits from the services provided and the negative effects that are often observed when tillage is reduced is still unknown, and crop rotation is included, which results in a higher biodiversity [90]. Al Shamsi et al. [91] showed that the best practice reduces the need for off-farm inputs while increasing the product range. However, it is also reported that this diversification can cause negative impacts, i.e., NO_3 leaching, NH_3 volatilization or pesticide use [70]. When assessing the effect of a combination of different practices in one single indicator, some complexity is added, since this will also be dependent on the pedoclimatic conditions, the intrinsic performance of the system and the goals set for the sustainability performance [70]. Thus, using such frameworks and interpreting its results should be carried out carefully, since there is a high level of subjectivity that cannot be erased [25].

5.1.4. LADA

LADA is a framework that is focused on the following items: biomass production, yearly biomass increments, soil health, water quality and quantity, biodiversity, economic value of the land use and social services of the land and its use [82]. It is also very solid in providing baseline data for improving the land degradation status, offering valid assets to plant, prioritizing and monitoring [92]. The cost-effectiveness is reasonable, i.e., the mapping activity, which includes the land use systems classification, costs approximately USD 250,000 for a country the size of South Africa [92]. This framework also operates with both local and national scales when assessing the land degradation and sustainable land management, cooperating with different stakeholders and proving applicable in at least 18 countries [93]. This is seen as a strength, since the contribution given by different stakeholders (locally and/or nationally) contributes significantly to equilibrated responses and results. For instance, the same status of a land may be classified differently depending on the stakeholder value system [82]. The LADA framework differs from others in its integration of climate factors, which may account for the long-term performance under climate change conditions.

The use of the framework, however, is still rather limited to people with multi-sectoral expertise [92]. This is linked to the need to build a comprehensive database to store both the quantitative and qualitative data obtained during the assessment operations. The assessment should provide a fixed baseline to monitor future changes and trends, and to feed more in-depth knowledge and understanding into the findings of the national assessment for the area in question [94,95]. Reed et al. [93] also states that, in this framework, land degradation assessment and the impact of the soil management practices that could be applicable in each specific situation should be tighter.

5.1.5. SMART

The tool has the advantage of having a high number of indicators to assess the trade-off and synergy analysis. It operationalizes the SAFA guidelines by including indicators that are based on scientific procedures and extensive literature revision. SMART is distinguished from all sustainable assessment frameworks studied by Landert et al. [83] because it integrates the contribution of the stakeholders in its development, which strengthens the acceptance by the end-users while also being specific to local situations [94,96], whereas the others typically involve stakeholders in the application of the framework, but only partly in its development [31]. Therefore, there is a compromise in the intended global applicability of the sustainable assessment tools and the incorporation of a local context.

SMART can be combined with other available tools to improve items such as the system boundary definitions and cut-off criteria when assessing farming activities. The study by Landert et al. [83] used three tools when assessing farm sustainability: COMPAS (an economic farm assessment tool); Cool Farm Tool (a greenhouse gas inventory, water footprint and biodiversity assessment tool, CFT); and the SMART Farm Tool. The results showed that SMART results can be used in combination with quantitative data from COMPAS and CFT. This study was a pioneer in showing the sustainability outcome for 15 different farms in Europe at different stages of their agro-ecological transition. The interdisciplinary tune of this research is characterized by its quantitative contributions and the plurality of view [96]. However, this framework proved to be too time consuming for all of the stakeholders involved, as well as for the interviewers. The combination of SMART with different tools and an improved standard method to incorporate data between the frameworks would facilitate this in the future.

Ssebunya et al. [97] used SMART to assess the sustainability performance of certified organic and fair-trade coffee when compared to non-certified in Uganda. The farm scores were included in the study, which enable analyses of synergies and the trade-off between different sustainable themes. Results showed a link between the certification and the improvement of the sustainable performance of the coffee farms. The framework was also used to enhance the governance objectives by suggesting alterations in group organizations and collective capacities, which, circularly, would also impact other sustainable dimensions. The authors pointed out three main limitations and specific requirements for credible and more consistent outcomes. One of these limitations is related to the comprehensiveness, which is related to the necessary trade-offs for the analysis specificity of some sun-themes. For example, 'Energy Use' and 'Greenhouse Gases' might be more accurately quantified through life cycle assessment methods. Profitability can also be calculated from detailed data from farm incomes and expenditures, whereas this is impossible for other sub-themes. Another limitation is related to the implementation, since the use of SMART requires an adequately trained audit team, involving very time-consuming practice activities to properly understand the functioning of the framework, its indicators and application range. Finally, the team also requires an expertise background on agronomy.

5.1.6. PG

PG is a user-friendly tool, with scores of the indicators coming directly from farmers' answers. One of the strengths of this framework is, therefore, its ease of application. On the one hand, data needed to compute the sustainability assessment are easy to obtain from simple interviews with farmers [85], and the questions include accessible data from the farm accounts and management. On the other, the framework was specifically designed to be simple, which means that input data requirements are modest, and are easily translated in the calculation methods and results [84]. This also implies that relatively little time is required for an assessment, since both manuals are simple to use and questions and calculations are easy to follow. This framework was specifically developed for agri-environmental schemes, making it the best option for policy makers wanting to address questions on whether suggested schemes/subsidies will significantly impact the different sustainability dimensions. Famers also have a direct answer on the impact that future improvements will have on the provision of public goods [84].

Other strengths of PG include the high level of transparency and the opportunity to transform the results directly into understandable outcomes of public goods in agriculture. Additionally, its user friendliness integrates better farmers and provides a useful tool for them to gain awareness on their sustainability farming activities, which is the first step to adopt better practices [96]. The main weaknesses, however, are also related to the simplicity of the tool, based on qualitative data collection and the lack of quantitative indicators, which allows for subjectivity in the scoring and results. Other more minor weaknesses are related to the presence of terminology related to nature conservation, which can be unfamiliar to farmers, the lack of the possibility to select indicators and the impossibility of including indicators within the framework [96].

Scoring the Frameworks

For the environment dimension, RISE, SMART and SAFA show a higher number of indicators covered (seven of eight themes), whereas MASC includes only three themes (water, soil and biodiversity). PG and LADA cover six and four themes, respectively, with water, soil and biodiversity as common themes (Table 5). Although an important subject, climate change seems to be missing in most of the frameworks studied, except in the LADA framework.

In the sociocultural dimension, SAFA is the strongest framework, including nine indicators of a total of twelve themes, followed by SMART covering seven, whereas RISE, MASC and PG cover two themes. SMART and SAFA cover the most important aspects of the sociocultural dimension, whereas RISE assesses only two (workload and wages) and LADA does not assess the sociocultural dimension at the individual level, but rather through land degradation that affects the society as whole (Table 3). In addition to the sociocultural advantages of SMART and SAFA, they enable us to engage stakeholders in different steps in order to increase their acceptance by end-users.

In the economic dimension, SMART, SAFA and PG all cover five themes out of six, followed by RISE and MASC with two themes each (profitability and vulnerability), and LADA with one theme (economic risk). SMART and SAFA assess all themes of the economic dimension except economic risks, whereas PG excludes only the investment theme. Despite the low number of economic themes included, farmers perceive RISE and SMART as the most indicated frameworks for understanding the level of sustainability achieved in their farm because they are based on quantitative data, which are then used for specific contexts [63,64].

In summary, SAFA is the framework with more focus on sociocultural aspects, while still covering some environmental and economic themes. SMART is also homogenous, and covers all three dimensions, but with fewer themes in each one in comparison to SAFA. In contrast, LADA does not include the sociocultural dimension at the individual level and is focused on the environmental dimension. The same is true to some extent for RISE and PG, which include few themes of the sociocultural dimension, while being focused on the environment and /or economy, respectively (Figure 1).

5.2. Which Frameworks Should Farmers Select?

To help stakeholders decide which framework is the most suitable for their sustainable assessment, we have developed a decision tree based on possible scales, sectors of applicability and the completeness of sustainability dimensions required (Table 6). For global assessments, there are both SAFA and LADA, but SAFA differs from LADA in assessing food systems in addition to land degradation. In addition, SAFA covers all dimensions, whereas LADA excludes the sociocultural dimension at the individual level, and it includes only a few economic themes.

Table 5. Framework's scoring based on their environmental (**A**), sociocultural (**B**) and economic (**C**) assessment themes.

(A)	Water	Soil	Air	Climate	Plant and Fertility	Biodiversity	Energy Use	Animal Well Being	Total
RISE	X	X	X	-	X	X	X	X	7
MASC	X	-	-	-	-	X	X	-	3
LADA	X	X	-	X	-	X	-	-	4
SMART	X	X	X	-	X	X	X	X	7
SAFA	X	X	X	-	X	X	X	X	7
PG	X	X	-	-	X	X	X	X	6

(B)	Employment Agree-ment	Workload	Wages	Health Safety	Job Satis-faction	Decent liveli-hood	Gender Equality	Cultural Diversity	Investment in Local Communities	Employment	Consumer Safety	Transparency	Total
RISE	-	X	X	-	-	-	-	-	-	-	-	-	2
MASC	-	-	-	X	-	-	-	-	-	X	-	-	2
LADA	-	-	-	-	-	-	-	-	-	-	-	-	0
SMART	X	-	X	X	-	X	X	X	-	-	-	-	7
SAFA	X	-	X	X	X	-	X	-	X	X	X	X	9
PG	-	-	-	-	X	-	-	-	-	X	-	-	2

(C)	Profitability	Vulnerability	Accountability	Investment	Local Economy	Economic Risk	Total
RISE	X	X	-	-	-	-	2
MASC	X	X	-	-	-	-	2
LADA	-	-	-	-	-	X	1
SMART	X	X	X	X	X	-	5
SAFA	X	X	X	X	X	-	5
PG	X	X	X	-	X	X	5

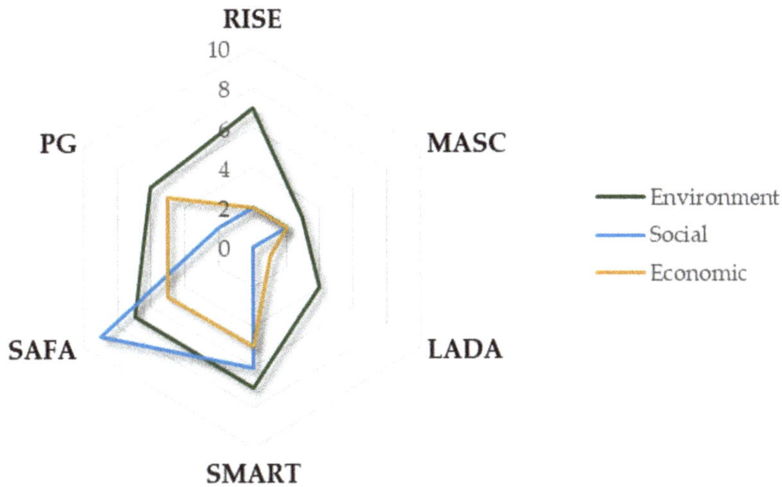

Figure 1. Total number of environmental, sociocultural and economic indicators used in each framework under study: RISE, MASC, LADA, SMART, SAFA and PG.

Table 6. Decision tree according to the framework scale assessment (global/local), sector of application (cropping system, livestock system, forestry system, urban system and food sector) and completeness of sustainability assessment (environmental, economic and sociocultural dimensions). Icons in black represent a higher number of themes whereas grey represent a lower number of themes in each dimension. Strengths and weaknesses related to the user-friendliness of the tool and the use of qualitative/quantitative data are also mentioned.

Scale Assessment	Sector of Application	Completeness Assessment	Framework	Strengths (+) and Weaknesses (−)
Global 			SAFA	• Qualitative and quantitative data (+) • Complex framework, requires expert in sustainability (−)
			LADA	• Qualitative and quantitative data (+) • Limited to people with multi-sectoral expertise (−)
Farm 			RISE	• Quantitative and qualitative data (+) • High number of input data, requires specialist (−)
			PG	• Only qualitative data used (−) • Scores of the indicators coming directly from farmers answers (+)
			MASC	• Highly adaptable for qualitative and quantitative data (+) • Requires researcher/professional (−)
			SMART	• Uses semi-quantitative data (+) • Very time-demanding and limited to scientists (−)

When the stakeholder intends to perform a sustainability assessment on a farm level, he/she has four choices: RISE, PG, MASC and SMART, but the latest only covers the cropping sector and is rather limited in the number of themes covered. The other three frameworks include cropping and livestock systems, whereas SMART also includes the food sector, which is the only possible choice if that is the user's goal. The selection between RISE, PG and SMART depends on the level of completeness intended for the analysis. SMART covers all dimensions, but with fewer themes in each dimension, whereas the other two include more themes in the environment and economy, respectively. However, MASC and RISE are more complex frameworks, whereas PG is the most user friendly and accessible for farmers.

6. Summary and Conclusions

The comparison between the six sustainability assessment frameworks (SAFA, RISE, MASC, LADA, SMART and PG) showed that they have different characteristics with regard to their assessment methodologies, time and data requirements to operate, and different outcomes with a different accuracy and level of complexity. Balancing all of these aspects in the development of the sustainability frameworks in order to meet the expectations of the main actors has proven to be a challenging task.

The high variety of characteristics of each sustainability frameworks derives from the fact that they were developed to serve different end-users: (i) farmers for assessing their farm performance; (ii) advisories and technicians for advising farmers on how they can improve their sustainability; (iii) researchers who conduct comprehensive regional and local assessments adaptable for context-specific conditions by combining, for example, different indicators from different frameworks.

The six sustainability assessment frameworks were compared according to their ability to cover the main themes of environmental, economic and sociocultural dimensions, and their themes were reported. We have also developed a decision tree based on possible scales, sectors of applicability and the completeness of sustainability dimensions required to help stakeholders decide which framework is the most suitable for their sustainable assessment purposes.

This overview study reveals that a multi-actor approach is necessary to enable the acceptance of the outcomes and their adoption by the main actors (i.e., farmers). When a value judgement is incorporated into a framework without involving farmers (e.g., assuming that organic farming will be more sustainable), the results may become irrelevant and are not considered useful by them [58,98,99].

It might be difficult to include alterations occurring in climatic, environmental, socio-economic or technological dimensions, in both the short- and/or long-term in the agricultural and societal aspects, but it may also offer new opportunities for more sustainable development [100]. Therefore, assessing the long-term performance under climate change conditions should be addressed further while assessing agricultural sustainability. For this purpose, realistic climate scenarios should be included.

Author Contributions: Conceptualization and writing—original draft preparation, A.A., L.B. and C.S.S.F.; review and editing, R.H. All authors have read and agreed to the published version of the manuscript.

Funding: This work is a part of the EU—H2020 project entitled 'Soil Care for profitable and sustainable crop production in Europe' (SoilCare), contract number 677407, which aims to identify and evaluate promising soil improving CS and agronomic techniques, increasing profitability and sustainability across scales in Europe.

Institutional Review Board Statement: Not applicable.

Informed Consent Statement: Not applicable.

Data Availability Statement: Not applicable.

Acknowledgments: We would like to thank the handling editor as well as the three reviewers for their constructive comments which considerably improved this paper.

Conflicts of Interest: The authors declare no conflict of interest.

References

1. Kanter, D.R.; Musumba, M.; Wood, S.L.R.; Palm, C.; Antle, J.; Balvanera, P.; Dale, V.H.; Havlik, P.; Kline, K.L.; Scholes, R.J.; et al. Evaluating agricultural trade-offs in the age of sustainable development. *Agric. Syst.* **2018**, *163*, 73–88. [CrossRef]
2. Palmieri, A.; Martino, L.; Dominici, P.; Kasanko, M. Land Capability Classification to Plan for a Changing Climate. *Land Qual. Land Use Inf. Eur. Union* **2011**, 289–296. Available online: https://www.researchgate.net/project/Land-Capability-and-Agricultural-Suitability-in-a-Changing-Climate (accessed on 4 April 2022).
3. EEA. *Climate Change Adaptation and Disaster Risk Reduction in Europe, Enhancing Coherence of the Knowledge Base, Policies and Practices*; EEA: Brussels, Belgium, 2017.
4. FAO. The Future of Food and Agriculture—Alternative Pathways to 2050, Rome, p. 224, Licence: CC BY-NC-SA 3.0 IGO. 2018. Available online: https://www.fao.org/3/I8429EN/i8429en.pdf (accessed on 4 April 2022).
5. EEA. *Agriculture and Environment in EU-15—the IRENA Indicator Report*; European Environment Agency: Copenhagen, Denmark, 2005; ISBN 9291677795.
6. Chavarria, D.N.; Pérez-Brandan, C.; Serri, D.L.; Meriles, J.M.; Restovich, S.B.; Andriulo, A.E.; Jacquelin, L.; Vargas-Gil, S. Response of soil microbial communities to agroecological versus conventional systems of extensive agriculture. *Agric. Ecosyst. Environ.* **2018**, *264*, 1–8. [CrossRef]
7. Urruty, N.; Deveaud, T.; Guyomard, H.; Boiffin, J. Impacts of agricultural land use changes on pesticide use in French agriculture. *Eur. J. Agron.* **2016**, *80*, 113–123. [CrossRef]
8. FAOSTAT Food and Agriculture Organization of the United Nations. Pesticides indicators for the period 1990 to 2019. Available online: https://www.fao.org/faostat/en/#data/EP (accessed on 4 April 2022).
9. Carvalho, F.P. Pesticides, environment, and food safety. *Food Energy Secur.* **2017**, *6*, 48–60. [CrossRef]
10. Evans, D.L.; Janes-Bassett, V.; Borrelli, P.; Chenu, C.; Ferreira, C.S.S.; Griffiths, R.I.; Kalantari, Z.; Keesstra, S.; Lal, R.; Panagos, P.; et al. Sustainable futures over the next decade are rooted in soil science. *Eur. J. Soil Sci.* **2021**, *73*, e13145. [CrossRef]
11. Valin, H.; Sands, R.D.; van der Mensbrugghe, D.; Nelson, G.C.; Ahammad, H.; Blanc, E.; Bodirsky, B.; Fujimori, S.; Hasegawa, T.; Havlik, P.; et al. The future of food demand: Understanding differences in global economic models. *Agric. Econ.* **2014**, *45*, 51–67. [CrossRef]
12. Ferreira, C.S.S.; Seifollahi-Aghmiuni, S.; Destouni, G.; Ghajarnia, N.; Kalantari, Z. Soil degradation in the European Mediterranean region: Processes, status and consequences. *Sci. Total Environ.* **2022**, *805*, 150106. [CrossRef]
13. Sadok, W.; Angevin, F.; Bergez, J.E.; Bockstaller, C.; Colomb, B.; Guichard, L.; Reau, R.; Messéan, A.; Doré, T. MASC, a qualitative multi-attribute decision model for ex ante assessment of the sustainability of cropping systems. *Agron. Sustain. Dev.* **2009**, *29*, 447–461. [CrossRef]
14. Rahman, S.; Islam, M.S.; Khan, M.N.H.; Touhiduzzaman, M. Climate change adaptation and disaster risk reduction (DRR) through coastal afforestation in South-Central Coast of Bangladesh. *Manag. Environ. Qual. An. Int. J.* **2019**, *30*, 498–517. [CrossRef]
15. Arif, M.S.; Riaz, M.; Shahzad, S.M.; Yasmeen, T.; Ashraf, M.; Siddique, M.; Mubarik, M.S.; Bragazza, L.; Buttler, A. Fresh and composted industrial sludge restore soil functions in surface soil of degraded agricultural land. *Sci. Total Environ.* **2018**, *619–620*, 517–527. [CrossRef] [PubMed]
16. Chen, Y.-H.; Wen, X.-W.; Wang, B.; Nie, P.-Y. Agricultural pollution and regulation: How to subsidize agriculture? *J. Clean. Prod.* **2017**, *164*, 258–264. [CrossRef]
17. Ferreira, C.S.S.; Keizer, J.J.; Santos, L.M.B.; Serpa, D.; Silva, V.; Cerqueira, M.; Ferreira, A.J.D.; Abrantes, N. Runoff, sediment and nutrient exports from a Mediterranean vineyard under integrated production: An experiment at plot scale. *Agric. Ecosyst. Environ.* **2018**, *256*, 184–193. [CrossRef]
18. Nacinovic, M.G.G.; Mahler, C.F.; de Avelar, A.S. Soil erosion as a function of different agricultural land use in Rio de Janeiro. *Soil Tillage Res.* **2014**, *144*, 164–173. [CrossRef]
19. Smidt, S.J.; Haacker, E.M.K.; Kendall, A.D.; Deines, J.M.; Pei, L.; Cotterman, K.A.; Li, H.; Liu, X.; Basso, B.; Hyndman, D.W. Complex water management in modern agriculture: Trends in the water-energy-food nexus over the High Plains Aquifer. *Sci. Total Environ.* **2016**, *566–567*, 988–1001. [CrossRef] [PubMed]
20. Gomiero, T. Soil degradation, land scarcity and food security: Reviewing a complex challenge. *Sustainability* **2016**, *8*, 281. [CrossRef]
21. Zalidis, G.; Stamatiadis, S.; Takavakoglou, V.; Eskridge, K.; Misopolinos, N. Impacts of agricultural practices on soil and water quality in the Mediterranean region and proposed assessment methodology. *Agric. Ecosyst. Environ.* **2002**, *88*, 137–146. [CrossRef]
22. Newbold, T.; Hudson, L.N.; Hill, S.L.; Contu, S.; Lysenko, I.; Senior, R.A.; Börger, L.; Bennett, D.J.; Choimes, A.; Collen, B.; et al. Global effects of land use on local terrestrial biodiversity. *Nature* **2015**, *520*, 45–50. [CrossRef]
23. Lenka, S.; Lenka, N.K.; Sejian, V.; Mohanty, M. Contribution of Agriculture Sector to Climate Change. In *Climate Change Impact on Livestock: Adaptation and Mitigation*; Springer: New Delhi, India, 2015; pp. 37–48. ISBN 9788132222651.
24. European Commission. *Towards a Thematic Strategy for Soil Protection*; European Commission: Brussels, Belgium, 2002.

25. Bell, S.; Morse, S. Sustainability Indicators Measuring the Immeasurable? Earthscan: London, UK, 2018; ISBN 9781844072996.
26. Tilman, D.; Balzer, C.; Hill, J.; Befort, B.L. Global food demand and the sustainable intensification of agriculture. *Proc. Natl. Acad. Sci. USA* **2011**, *108*, 20260–20264. [CrossRef]
27. Barão, L.; Alaoui, A.; Ferreira, C.; Basch, G.; Schwilch, G.; Geissen, V.; Sukkel, W.; Lemesle, J.; Garcia-Orenes, F.; Morugán-Coronado, A.; et al. Assessment of promising agricultural management practices. *Sci. Total Environ.* **2019**, *649*, 610–619. [CrossRef]
28. Valizadeh, N.; Hayati, D. Development and validation of an index to measure agricultural sustainability. *J. Clean. Prod.* **2021**, *280*, 123797. [CrossRef]
29. Dufour, D.L. Use of Tropical Rainforests by Native Amazonians. *BioScience* **2009**, *40*, 652–659. [CrossRef]
30. Ebel, R. Are small farms sustainable by nature?-review of an ongoing misunderstanding in agroecology. *Chall. Sustain.* **2020**, *8*, 17–29. [CrossRef]
31. Binder, C.R.; Feola, G.; Steinberger, J.K. Considering the normative, systemic and procedural dimensions in indicator-based sustainability assessments in agriculture. *Environ. Impact Assess. Rev.* **2010**, *30*, 71–81. [CrossRef]
32. Schader, C.; Grenz, J.; Meier, M.S.; Stolze, M. Scope and precision of sustainability assessment approaches to food systems. *Ecol. Soc.* **2014**, *19*, 42. [CrossRef]
33. Talukder, B.; Blay-palmer, A.; Gary, W.; Hipel, K.W. Environmental and Sustainability Indicators Towards complexity of agricultural sustainability assessment: Main issues and concerns. *Environ. Sustain. Indic.* **2020**, *6*, 100038. [CrossRef]
34. Talukder, B.; Blay-Palmer, A. Comparison of Methods to Assess Agricultural Sustainability. In *Sustainable Agriculture Reviews*; 2017; pp. 149–168. ISBN 9783319586793.
35. Marchand, F.; Debruyne, L.; Triste, L.; Gerrard, C.; Padel, S.; Lauwers, L. Key characteristics for tool choice in indicator-based sustainability assessment at farm level. *Ecol. Soc.* **2014**, *19*, 46. [CrossRef]
36. Bai, Z.; Caspari, T.; Gonzalez, M.R.; Batjes, N.H.; Mäder, P.; Bünemann, E.K.; de Goede, R.; Brussaard, L.; Xu, M.; Ferreira, C.S.S.; et al. Effects of agricultural management practices on soil quality: A review of long-term experiments for Europe and China. *Agric. Ecosyst. Environ.* **2018**, *265*, 1–7. [CrossRef]
37. Schindler, J.; Graef, F.; König, H.J. Methods to assess farming sustainability in developing countries A review. *Agron. Sustain. Dev.* **2015**, *35*, 1043–1057. [CrossRef]
38. Reytar, K.; Hanson, C.; Henninger, N. *Indicators of Sustainable Agriculture: A Scoping Study*; Working Paper, Installment 6 of Creating a Sustainable Food Future; World Resources Institute: Washington, DC, USA, 2014; pp. 1–20. Available online: http://www.worldresourcesreport.org (accessed on 4 April 2022).
39. Yli-Viikari, A. Indicators for sustainable agriculture—A theoretical framework for classifying and assessing indicators. *Agric. Food Sci. Finl.* **1999**, *8*, 265–283. [CrossRef]
40. Latruffe, L.; Diazabakana, A.; Bockstaller, C.; Desjeux, Y.; Finn, J.; Kelly, E.; Ryan, M.; Uthes, S. Measurement of sustainability in agriculture: A review of indicators. *Stud. Agric. Econ.* **2016**, *118*, 123–130. [CrossRef]
41. Moldan, B.; Janoušková, S.; Hák, T. How to understand and measure environmental sustainability: Indicators and targets. *Ecol. Indic.* **2012**, *17*, 4–13. [CrossRef]
42. Alaoui, A.; Basch, G.; Kosmas, C.; Barão, L.; Ferreira, C.S.S.; Schwilch, G.; Garcia-orenes, F.; Morugan, A.; Mataix-solera, J.; Glavan, M.; et al. Visual assessment of the impact of agricultural management practices on soil quality. *Agron. J.* **2020**, *12*, 2608–2623. [CrossRef]
43. Ball, B.C.; Guimarães, R.M.L.; Cloy, J.M.; Hargreaves, P.R.; Shepherd, T.G.; McKenzie, B.M. Visual soil evaluation: A summary of some applications and potential developments for agriculture. *Soil Tillage Res.* **2017**, *173*, 114–124. [CrossRef]
44. Gómez-Limón, J.A.; Riesgo, L. Alternative approaches to the construction of a composite indicator of agricultural sustainability: An application to irrigated agriculture in the Duero basin in Spain. *J. Environ. Manage.* **2009**, *90*, 3345–3362. [CrossRef] [PubMed]
45. Panagos, P.; Borrelli, P.; Poesen, J.; Ballabio, C.; Lugato, E.; Meusburger, K.; Montanarella, L.; Alewell, C. The new assessment of soil loss by water erosion in Europe. *Environ. Sci. Policy* **2015**, *54*, 438–447. [CrossRef]
46. Bünemann, E.K.; Bongiorno, G.; Bai, Z.; Creamer, R.E.; De Deyn, G.; de Goede, R.; Fleskens, L.; Geissen, V.; Kuyper, T.W.; Mäder, P.; et al. Soil quality—A critical review. *Soil Biol. Biochem.* **2018**, *120*, 105–125. [CrossRef]
47. Lebacq, T.; Baret, P.V.; Stilmant, D. Sustainability indicators for livestock farming. A review. *Agron. Sustain. Dev.* **2013**, *33*, 311–327. [CrossRef]
48. Castoldi, N.; Bechini, L. Integrated sustainability assessment of cropping systems with agro-ecological and economic indicators in northern Italy. *Eur. J. Agron.* **2010**, *32*, 59–72. [CrossRef]
49. Christen, O.; O'Halloranetholtz, Z. *Indicators for a Sustainable Development in Agriculture*; Vol. I—Sustainable Development of Agriculture, Fisheries, and Forestry—Olaf Christen and Wolfgang Heyer; ©Encyclopedia of Life Support Systems (EOLSS); UNESCO: Paris, France, 2022.
50. Gayatri, S.; Gasso-tortajada, V.; Vaarst, M. Assessing Sustainability of Smallholder Beef Cattle Farming in Indonesia: A Case Study Using the FAO SAFA Framework. *J. Sustain. Dev.* **2016**, *9*, 236. [CrossRef]
51. Taylor, D.C.; Mohamed, Z.A.; Shamsudin, M.N.; Mohayidin, M.G.; Chiew, E.F. Creating a farmer sustainability index: A Malaysian case study. *Am. J. Altern. Agric.* **1993**, *8*, 175–184. [CrossRef]
52. Oñate, J.J.; Andersen, E.; Peco, B.; Primdahl, J. Agri-environmental schemes and the European agricultural landscapes: The role of indicators as valuing tools for evaluation. *Landsc. Ecol.* **2000**, *15*, 271–280. [CrossRef]

53. Haas, G.; Wetterich, F.; Geier, U. Life cycle assessment framework in agriculture on the farm level. *Int. J. Life Cycle Assess.* **2000**, *5*, 345–348. [CrossRef]
54. Girardin, P.; Bockstaller, C.; Van Der Werf, H. Assessment of potential impacts of agricultural practices on the environment: The AGRO*ECO method. *Environ. Impact Assess. Rev.* **2000**, *20*, 227–239. [CrossRef]
55. Rasul, G.; Thapa, G.B. Sustainability of ecological and conventional agricultural systems in Bangladesh: An assessment based on environmental, economic and social perspectives. *Agric. Syst.* **2004**, *79*, 327–351. [CrossRef]
56. Halberg, N.; Verschuur, G.; Goodlass, G. Farm level environmental indicators; are they useful? An overview of green accounting systems for European farms. *Agric. Ecosyst. Environ.* **2005**, *105*, 195–212. [CrossRef]
57. Van Cauwenbergh, N.; Biala, K.; Bielders, C.; Brouckaert, V.; Franchois, L.; Garcia Cidad, V.; Hermy, M.; Mathijs, E.; Muys, B.; Reijnders, J.; et al. SAFE-A hierarchical framework for assessing the sustainability of agricultural systems. *Agric. Ecosyst. Environ.* **2007**, *120*, 229–242. [CrossRef]
58. Zahm, F.; Viaux, P.; Vilain, L.; Girardin, P.; Mouchet, C. Farm Sustainability Assessment using the IDEA Method: From the concept of farm sustainability to case studies on French farms. In *1. INFASA Symposium (p. np)*; International Institute for Sustainable Development: Winnipeg, MB, Canada, 2006; Volume 14, pp. 77–110.
59. Gómez-Limón, J.A.; Sanchez-Fernandez, G. Empirical evaluation of agricultural sustainability using composite indicators. *Ecol. Econ.* **2010**, *69*, 1062–1075. [CrossRef]
60. Paracchini, M.L.; Pacini, C.; Jones, M.L.M.; Pérez-Soba, M. An aggregation framework to link indicators associated with multifunctional land use to the stakeholder evaluation of policy options. *Ecol. Indic.* **2011**, *11*, 71–80. [CrossRef]
61. Lillywhite, R.; Keeling, C.; Courtney, P.; Lampkin, N.; Pearce, B.; Rayns, F.; Reed, M.; Schmutz, U.; Watson, C.; Williams, A. Assessing the economic, environmental and social characteristics of UK farming systems. In Proceedings of the 10th European IFSA symposium, Aarhus, Denmark, 1–4 April 2012.
62. Johnson, M.P.; Schaltegger, S. Two Decades of Sustainability Management Tools for SMEs: How Far Have We Come? *J. Small Bus. Manag.* **2016**, *54*, 481–505. [CrossRef]
63. Craheix, D.; Angevin, F.; Doré, T.; de Tourdonnet, S. Using a multicriteria assessment model to evaluate the sustainability of conservation agriculture at the cropping system level in France. *Eur. J. Agron.* **2016**, *76*, 75–86. [CrossRef]
64. De Olde, E.M.; Oudshoorn, F.W.; Sørensen, C.A.; Bokkers, E.A.; De Boer, I. Assessing sustainability at farm-level: Lessons learned from a comparison of tools in practice. *Ecol. Indic.* **2016**, *66*, 391–404. [CrossRef]
65. Landert, J.; Schader, C.; Moschitz, H.; Stolze, M. A holistic sustainability assessment method for Urban food system governance. *Sustainability* **2017**, *9*, 490. [CrossRef]
66. Gaviglio, A.; Bertocchi, M.; Demartini, E. A tool for the sustainability assessment of farms: Selection, adaptation and use of indicators for an Italian case study. *Resources* **2017**, *6*, 60. [CrossRef]
67. Lazzerini, G.; Merante, P.; Lucchetti, S.; Nicese, F.P. Assessing environmental sustainability of ornamental plant production: A nursery level approach in Pistoia District, Italy. *Agroecol. Sustain. Food Syst.* **2018**, *42*, 911–932. [CrossRef]
68. Röös, E.; Fischer, K.; Tidåker, P.; Nordström Källström, H. How well is farmers' social situation captured by sustainability assessment tools? A Swedish case study. *Int. J. Sustain. Dev. World Ecol.* **2019**, *26*, 268–281. [CrossRef]
69. Coteur, I.; Wustenberghs, H.; Debruyne, L.; Lauwers, L.; Marchand, F. How do current sustainability assessment tools support farmers' strategic decision making? *Ecol. Indic.* **2020**, *114*, 106298. [CrossRef]
70. Viguier, L.; Cavan, N.; Bockstaller, C.; Cadoux, S.; Corre-Hellou, G.; Dubois, S.; Duval, R.; Keichinger, O.; Toqué, C.; Toupet de Cordoue, A.L.; et al. Combining diversification practices to enhance the sustainability of conventional cropping systems. *Eur. J. Agron.* **2021**, *127*, 126279. [CrossRef]
71. Iocola, I.; Colombo, L.; Guccione, G.D.; De Vita, P.; Palumbo, M.; Ritunnano, V.; Sciacca, F.; Virzì, N.; Canali, S. A multi-criteria qualitative tool for the sustainability assessment of organic durum wheat-based farming systems designed through a participative process. *Ital. J. Agron.* **2021**, *16*, 1–15. [CrossRef]
72. Havardi-Burger, N.; Mempel, H.; Bitsch, V. Framework for sustainability assessment of the value chain of flowering potted plants for the German market. *J. Clean. Prod.* **2021**, *329*, 129684. [CrossRef]
73. Volkov, A.; Morkunas, M.; Balezentis, T.; Streimikiene, D. Are agricultural sustainability and resilience complementary notions? Evidence from the North European agriculture. *Land Use Policy* **2022**, *112*, 105791. [CrossRef]
74. Grimm, C. *Nachhaltige Landwirtschaft: Indikatoren, Bilanzierungsansätze*; Erich-Schmidt Verlag: Berlin, Germany, 2009.
75. Singh, R.K.; Murty, H.R.; Gupta, S.K.; Dikshit, A.K. An overview of sustainability assessment methodologies. *Ecol. Indic.* **2009**, *9*, 189–212. [CrossRef]
76. Schader, C.; Stolze, M.; Gattinger, A. Environmental performance of organic farming. In *Green Technologies in Food Production and Processing*; Boye, J., Arcand, Y., Eds.; Springer US: New York, NY, USA, 2012; pp. 183–210. ISBN 9781461415879.
77. Finkbeiner, M.; Schau, E.M.; Lehmann, A.; Traverso, M. Towards life cycle sustainability assessment. *Sustainability* **2010**, *2*, 3309–3322. [CrossRef]
78. Dumay, J.; Guthrie, J.; Farneti, F. GRI sustainability reporting guidelines for public and third sector organizations: A critical review. *Public Manag. Rev.* **2010**, *12*, 531–548. [CrossRef]
79. FAO. *SAFA GUIDELINES—Sustainability Assessment of Food and Agriculture Systems*; FAO: Rome, Italy, 2014; ISBN 9789251084854.
80. Häni, F.; Braga, F.; Stämpfli, A.; Keller, T.; Fischer, M.; Porsche, H. RISE, a tool for holistic sustainability assessment at the farm level. *Int. Food Agribus. Manag. Rev.* **2003**, *6*, 78–90. [CrossRef]

81. Figueira, J.; Greco, S.; Ehrgott, M. *Multiple Criteria Decision Analysis—State of the Art*; Springer: Boston, MA, USA, 2005; ISBN 0387230815.
82. Nachtergaele, F.; Biancalani, R.; Bunning, S.; George, H. Land Degradation Assessment: The LADA approach. In Proceedings of the 19th World Congress of Soil Science, Brisbane, Australia, 1–6 August 2010; pp. 72–75. Available online: https://www.iuss.org/19th%20WCSS/Symposium/pdf/1730.pdf (accessed on 4 April 2022).
83. Landert, J.; Pfeifer, C.; Carolus, J.; Schwarz, G.; Albanito, F.; Muller, A.; Smith, P.; Sanders, J.; Schader, C.; Vanni, F.; et al. Assessing agro-ecological practices using a combination of three sustainability assessment tools. *Landbauforschung* **2020**, *70*, 129–144. [CrossRef]
84. Gerrard, C.L.; Smith, L.G.; Pearce, B.; Padel, S.; Hitchings, R.; Measures, M.; Cooper, N. Farming for Food and Water Security. In *Farming for Food and Water Security*; Springer: Dordrecht, The Netherlands, 2012; Volume 10, pp. 1–22. ISBN 978-94-007-4499-8.
85. Cooper, T.; Hart, K.; Baldock, D. *Provision of Public Goods through Agriculture in the European Union*; Institute for European Environmental Policy: London, UK, 2009.
86. Soldi, A.; Meza, M.J.A.; Guareschi, M.; Donati, M.; Ortiz, A.I. Sustainability assessment of agricultural systems in Paraguay: A comparative study using FAO's SAFA framework. *Sustainability* **2019**, *11*, 3745. [CrossRef]
87. Cammarata, M.; Timpanaro, G.; Scuderi, A. Assessing sustainability of organic livestock farming in Sicily: A case study using the Fao Safa framework. *Agriculture* **2021**, *11*, 274. [CrossRef]
88. Grenz, J.; Mainiero, R.; Schoch, M.; Sereke, F.; Stalder, S.; Thalmann, C.; Wyss, R. *RISE 3.0—Manual: Sustainability Themes and Indicators*; Bern University of Applied Sciences: Bern, Switzerland, 2021.
89. Médiène, S.; Valantin-Morison, M.; Sarthou, J.P.; De Tourdonnet, S.; Gosme, M.; Bertrand, M.; Roger-Estrade, J.; Aubertot, J.N.; Rusch, A.; Motisi, N.; et al. Agroecosystem management and biotic interactions: A review. *Agron. Sustain. Dev.* **2011**, *31*, 491–514. [CrossRef]
90. Palm, C.; Blanco-Canqui, H.; DeClerck, F.; Gatere, L.; Grace, P. Conservation agriculture and ecosystem services: An overview. *Agric. Ecosyst. Environ.* **2014**, *187*, 87–105. [CrossRef]
91. Al Shamsi, K.B.; Compagnoni, A.; Timpanaro, G.; Cosentino, S.L.; Guarnaccia, P. A sustainable organic production model for "food sovereignty" in the United Arab Emirates and Sicily-Italy. *Sustainability* **2018**, *10*, 620. [CrossRef]
92. Schwilch, G.; Bestelmeyer, B.; Bunning, S.; Critchley, W.; Herrick, J.; Kellner, K.; Liniger, H.; Nachtergaele, F.; Ritsema, C.; Schuster, B.; et al. Experiences in monitoring and assessment of sustainable land management. *Land Degrad. Dev.* **2011**, *22*, 214–225. [CrossRef]
93. Reed, M.S.; Buenemann, M.; Atlhopheng, J.; Bachmann, F.; Bastin, G.; Bigas, H.; Chanda, R.; Dougill, A.J.; Essahli, W.; Evely, A.C.; et al. Cross-Scale Monitoring And Assessment Of Land Degradation And Sustainable Land Management: A Methodological Framework For Knowledge Management. *Land Degrad. Dev.* **2011**, *271*, 261–271. [CrossRef]
94. Bunning, S.; McDonagh, J.; Rioux, J. *Land Degradation Assessment in Drylands*; Food and Agricultural Organization of the United Nations: Rome, Italy, 2016; Volume 1.
95. Arulnathan, V.; Heidari, M.D.; Doyon, M.; Li, E.; Pelletier, N. Farm-level decision support tools: A review of methodological choices and their consistency with principles of sustainability assessment. *J. Clean. Prod.* **2020**, *256*, 120410. [CrossRef]
96. De Olde, E.M.; Moller, H.; Marchand, F.; McDowell, R.W.; MacLeod, C.J.; Sautier, M.; Halloy, S.; Barber, A.; Benge, J.; Bockstaller, C.; et al. When experts disagree: The need to rethink indicator selection for assessing sustainability of agriculture. *Environ. Dev. Sustain.* **2017**, *19*, 1327–1342. [CrossRef]
97. Ssebunya, B.R.; Schader, C.; Baumgart, L.; Landert, J.; Altenbuchner, C.; Schmid, E.; Stolze, M. Sustainability Performance of Certified and Non-certified Smallholder Coffee Farms in Uganda. *Ecol. Econ.* **2019**, *156*, 35–47. [CrossRef]
98. Gasparatos, A.; Scolobig, A. Choosing the most appropriate sustainability assessment tool. *Ecol. Econ.* **2012**, *80*, 1–7. [CrossRef]
99. Vatn, A. Rationality, institutions and environmental policy. *Ecol. Econ.* **2005**, *55*, 203–217. [CrossRef]
100. Baležentis, T.; Liobikienė, G.; Štreimikienė, D.; Sun, K. The impact of income inequality on consumption-based greenhouse gas emissions at the global level: A partially linear approach. *J. Environ. Manag.* **2020**, *267*, 110635. [CrossRef] [PubMed]

land

MDPI

Article

Opportunities for Mitigating Soil Compaction in Europe—Case Studies from the SoilCare Project Using Soil-Improving Cropping Systems

Ilaria Piccoli [1], Till Seehusen [2,*], Jenny Bussell [3], Olga Vizitu [4], Irina Calciu [4], Antonio Berti [1], Gunnar Börjesson [5], Holger Kirchmann [5], Thomas Kätterer [6], Felice Sartori [1], Chris Stoate [3], Felicity Crotty [7], Ioanna S. Panagea [8], Abdallah Alaoui [9] and Martin A. Bolinder [6]

1 Department of Agronomy, Food, Natural Resources, Animals and Environment, University of Padova, Viale dell'Università 16, 35020 Legnaro, Italy; ilaria.piccoli@unipd.it (I.P.); antonio.berti@unipd.it (A.B.); felice.sartori@phd.unipd.it (F.S.)
2 Norwegian Institute of Bioeconomy Research, P.O. Box 115, 1431 Ås, Norway
3 The Game & Wildlife Conservation Trust, Allerton Project, Loddington, Leicestershire LE7 9XE, UK; jbussell@gwct.org.uk (J.B.); cstoate@gwct.org.uk (C.S.)
4 National Research and Development Institute for Soil Science, Agrochemistry and Environmental Protection, 011464 Bucharest, Romania; olga.vizitiu@icpa.ro (O.V.); Irina.calciu@icpa.ro (I.C.)
5 Department of Soil and Environment, Swedish University of Agricultural Sciences, P.O. Box 7014, 75007 Uppsala, Sweden; gunnar.borjesson@slu.se (G.B.); holger.kirchmann@slu.se (H.K.)
6 Department of Ecology, Swedish University of Agricultural Sciences, P.O. Box 7044, 75007 Uppsala, Sweden; thomas.katterer@slu.se (T.K.); martin.bolinder@slu.se (M.A.B.)
7 Centre for Agriculture, Royal Agricultural University, Stroud Rd., Cirencester GL7 6JS, UK; felicity.crotty@rau.ac.uk
8 Department of Earth and Environmental Science, KU Leuven, 3001 Leuven, Belgium; panagea.ioanna@gmail.com
9 Institute of Geography, University of Bern, Hallerstrasse 12, 3012 Bern, Switzerland; abdallah.alaoui@giub.unibe.ch
* Correspondence: till.seehusen@nibio.no; Tel.: +47-93269878

Citation: Piccoli, I.; Seehusen, T.; Bussell, J.; Vizitu, O.; Calciu, I.; Berti, A.; Börjesson, G.; Kirchmann, H.; Kätterer, T.; Sartori, F.; et al. Opportunities for Mitigating Soil Compaction in Europe—Case Studies from the SoilCare Project Using Soil-Improving Cropping Systems. *Land* **2022**, *11*, 223. https://doi.org/10.3390/land11020223

Academic Editor: Amrakh I. Mamedov

Received: 16 December 2021
Accepted: 31 January 2022
Published: 2 February 2022

Abstract: Soil compaction (SC) is a major threat for agriculture in Europe that affects many ecosystem functions, such as water and air circulation in soils, root growth, and crop production. Our objective was to present the results from five short-term (<5 years) case studies located along the north–south and east–west gradients and conducted within the SoilCare project using soil-improving cropping systems (SICSs) for mitigating topsoil and subsoil SC. Two study sites (SSs) focused on natural subsoil (>25 cm) compaction using subsoiling tillage treatments to depths of 35 cm (Sweden) and 60 cm (Romania). The other SSs addressed both topsoil and subsoil SC (>25 cm, Norway and United Kingdom; >30 cm, Italy) using deep-rooted bio-drilling crops and different tillage types or a combination of both. Each SS evaluated the effectiveness of the SICSs by measuring the soil physical properties, and we calculated SC indices. The SICSs showed promising results—for example, alfalfa in Norway showed good potential for alleviating SC (the subsoil density decreased from 1.69 to 1.45 g cm^{-1}) and subsoiling at the Swedish SS improved root penetration into the subsoil by about 10 cm—but the effects of SICSs on yields were generally small. These case studies also reflected difficulties in implementing SICSs, some of which are under development, and we discuss methodological issues for measuring their effectiveness. There is a need for refining these SICSs and for evaluating their longer-term effect under a wider range of pedoclimatic conditions.

Keywords: degree of compaction; soil penetration resistance; relative normalised density; air-filled porosity; tillage; straw incorporation; bio-drilling crops; subsoiling; crop productivity

1. Introduction

Soil compaction (SC) is a form of physical degradation due to the disruption of soil micro- and macro-aggregates, which are deformed, reduced in volume, or destroyed under pressure. Compaction is a "hidden" threat that occurs belowground and is one of eight European soil threats [1], affecting as much as 18 to 36% of croplands [2,3]. There are several consequences of SC because of its influence on many important soil functions. For example, it can negatively affect physical soil properties, such as gas permeability and water infiltration and storage [4,5]. This hampers the ecological function of the soil, leading to reduced soil fertility and crop production [6–8]. Furthermore, SC problems can reduce water infiltration and, in addition to causing problems with runoff and erosion, the soil workability may be reduced due to high water content, and the crops may not be able to explore the entire growing season (e.g., delayed seeding date) [9–11]. In this regard, the climate and the expected climate changes are important; for example, Northern Europe may be subject to increasing precipitations and wetter conditions during the growing season [12]. Indeed, soil compaction is one of the main reasons for stagnating yields [10,13]. A study also showed that even if SC does not necessarily lead to a reduction in yields, it can cause considerable amounts of extra costs not only for the farmers but also for society [14].

There are several common reasons for SC in most European countries. Compaction may occur in both the topsoil (i.e., arable layer) and subsoil layers (i.e., below the arable layer) due to pressure from the passage of machinery and repeated trampling of grazing animals, or occur naturally from previous geological periods during the initial ground formation under land ice. Subsoil compaction is also associated with in-furrow ploughing, during which tractor wheels that are in direct contact with the subsoil transmit the pressure to deeper soil horizons, especially when using heavy machinery under wet and sub-optimal soil conditions [15]. Unlike topsoils, subsoils are not loosened annually, and compaction may become cumulative [16,17]. Another feature regarding the SC of subsoil is the formation of a plough pan layer that results from repeated ploughing and is less permeable for roots and limits water flow and gaseous exchange. Ruser et al. [18] report that compaction can become almost irreversible once it reaches the threshold of the pre-consolidation stress (i.e., the index of soil load-bearing capacity).

Even though certain climatic conditions and processes (i.e., drying/wetting or freezing/thawing and shrinking cycles) can be effective in counteracting the SC of clayey soils [19,20], these processes are mostly absent on silty soils, making them especially susceptible to subsoil compaction [21]. While ploughing is effective for loosening up compaction of the upper soil layers, there is a lack of measures for persistently loosening up the subsoil [22]. There is a need for developing strategies to avoid subsoil SC and to stabilise and improve subsoil structure. For example, plant roots can be effective for loosening up subsoil, a strategy referred to as "bio-drilling" where roots modify the soil structure by pushing aside soil particles, thereby creating large pores that improve both hydraulic conductivity and gas flow [23–26]. Cresswell and Kirkegaard [27] defined bio-drilling as the creation of bio-pores by deeply penetrating taproots as low-resistance pathways for the roots of a succeeding crop. For this purpose, alfalfa, forage radish, or oilseed crops, which are known for having deep taproot systems, may be efficient for improving the soil structure even deeper in the soil profile [24,28,29]. However, the understanding of optimising the effect of bio-drilling crops through appropriate management remains limited, and their effects on crop yields vary with climatic conditions [29].

Mechanical subsoil loosening, referred to as deep loosening, deep ripping, or subsoiling, is a common practice to loosen up dense soil layers below the topsoil [30,31]. Subsoil loosening can decrease penetration resistance and bulk density [32] and increase infiltration [33], root development [34], and crop yield [35–37]. There is a need for loosening subsoil under optimal soil moisture conditions. When the soil is too wet and loose, the soil might be smeared and compacted [38,39]. When the soil is too dry, thick clods are formed [39]. Furthermore, the benefits of subsoiling are often not long-lasting due to re-compaction by the overburden topsoil and field operations [40–42]. However, when combining mechanical subsoil loosening with the addition of organic materials into the subsoil, loosening may last for several years [43,44].

Great efforts have been made to quantify SC, which is needed both for identifying SC problems and for evaluating the effectiveness of mitigating strategies. For instance, Huber et al. [45] suggested the following indicators: soil bulk density, air and water permeability, mechanical resistance, and a visual assessment of the soil structure and rooting. The proposed indicators involve several common measurements, such as bulk density and penetration resistance. However, suitable definitions of critical limit values linked to crop impairment are difficult to define. A number of penetration resistance threshold values above which rootability is impaired [46] can be found. They range between 1 and 2 MPa or higher [47–54] and are strictly linked to pedoclimatic conditions and soil management (e.g., tillage vs. no-tillage). Similarly, for SC characterization, Håkansson [8] suggested an index of the degree of compaction (DC). The DC index represents the bulk density-to-reference density ratio and is considered detrimental for crop development when it exceeds 87% [8]. Although the DC is a fast and easy index, two issues have recently been raised—the identification of the correct reference bulk density is not obvious and the 87% threshold seems to not be applicable for all pedoclimatic conditions [55].

Compaction is one of the threats included in the EU "SoilCare" project (soil care for profitable and sustainable crop production in Europe). This project addressed the use of different soil-improving cropping systems (SICS) involving both the crop type and rotation, as well as specific management techniques aiming to improve soil quality and functions (http://soilcare-project.eu/, accessed on 1 January 2022). In this paper, we present the main outcomes from five case studies within the SoilCare project using different SICSs to counteract compaction. The study sites (SSs) were located in five European countries, where we investigated different innovative strategies for mitigating SC under various soil and climatic conditions. The SICSs involved different types of tillage, including subsoiling and various deep-rooted bio-drilling crops.

2. Materials and Methods

All SICSs had a common objective—to counteract soil compaction. They were located in five countries along the north-to-south and east-to-west gradients from Norway to Romania (Figure 1 and Supplementary Materials, Table S1).

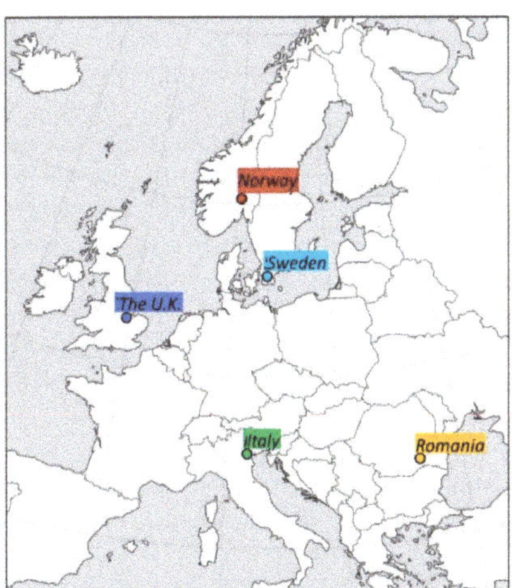

Figure 1. Location of the five study sites involved in the present study.

The SICSs examined in each country for alleviating topsoil and subsoil SC comprised the use of various deep-rooting crops and different types of tillage operations, including subsoiling (Table 1). At all SSs, the SICSs were compared with a reference standard practice, and both topsoil and subsoil samplings were made at different depths according to the soil characteristics for each of the experiments (Supplementary Materials, Table S2). Although exactly the same measurements were not performed at all SSs (as detailed below), some were similar for all SSs. This allowed us to make a generic analysis and identify the relationships with soil properties using the three SC indices described in Section 2.4.

Table 1. Soil-improving cropping systems (SICS) applied at the five study sites (SSs) and the reference standard practice at each site.

Country	Institution	SICS	Standard Practice
Norway	NIBIO	Bio-drilling crop roots	Conventional tillage
Sweden	SLU	Loosening of subsoil with and without straw incorporation	Conventional tillage
United Kingdom	GWCT	Ploughing	Direct drilling
Italy	UNIPD	No-tillage with deep-rooted cover crop	Conventional tillage with bare soil between the main crops
Romania	ICPA	Ploughing, subsoiling, chisel	Disking as main soil tillage

2.1. Norway

2.1.1. Experimental Design

The Norwegian SS investigated soil compaction alleviation by using bio-drilling crops. The soil was characterised by poor natural drainage and medium erosion risk. This field has been under cultivation for several decades and the site was drained. In the early summer of 2015, a multiple wheel-by-wheel approach was used for establishing the initial compaction with a tractor and trailer combination passing across the plots ten times, with a total weight of 17 Mg and resulting in a wheel load of 2.8 Mg for the trailer tandem axles (compacted "C" plot). This is a typical wheel load for small- and medium-sized farms in Norway and representative of other machinery, such as a combine harvester. There was little precipitation the days before the compaction treatment and none during it, resulting in workable conditions and higher soil moisture tension in the topsoil and subsoil (−25 and −63 kPa, respectively) than assumed at the field capacity (−10 kPa) while wheeling. The site was used for yield studies until 2017; for further details, please see the work of Seehusen et al. [56]. Thereafter, four different rotation treatments were applied during a 4-year period—(1 and 2) oilseed (*Brassica rapa* L. ssp. Oleifera) and barley (*Hordeum vulgare* L.) rotation, (3) barley monoculture, (4) alfalfa (*Medicago sativa* L.) monoculture. The experimental design was a split plot with two replicates, with the compaction level as the splitting factor (compacted "C" vs. reference "R" plot) and the rotation treatment (1 to 4) as the main plot factor. Crops were grown in 5 × 1.5 m plots for a total of 16 plots, that is, four rotations × two compaction levels (compacted vs. uncompacted) × two replicates.

All plots were subject to spring ploughing at 25 cm beginning in 2015 (after the compaction) except for the plots with perennial alfalfa. The ploughing was assumed to be effective for alleviating compaction and aligning the root effects, and therefore, only the topsoils from Treatments 3 and 4 were sampled in 2020. Management practices (seeding, fertiliser, and tillage) were done in the same way as the surrounding fields.

2.1.2. Soil Sampling, Field Measurements, and Laboratory Analysis

Undisturbed cylinder cores (100 cm^{-3}) were collected at both 10–20 and 40–50 cm depth in 2015 (n = 4–5 per depth) and 2020 (n = 4 per treatment and depth) for soil physical analysis. The soil bulk density (BD) was determined gravimetrically by weighing the soil before and after drying for 24 h at 105 °C. In 2015, the BD represented the field conditions at sampling, while in 2020, it represented the BD at −3 kPa. The water retention was studied in both years by first saturating the samples and then draining them at different matric potentials (−3, −50, and −1500 kPa in 2015, and −2, −10, −100, and −150 kPa in 2020). In the latter year, the wilting point (−1500 kPa) was calculated using a pedotransfer function [57]. The pore size distribution was derived from the water retention curves (details in Seehusen et al. [56]). The air capacity was measured assuming a field capacity of −10 kPa by measuring the airflow through the soil samples at a pressure of 2 kPa [58]. Saturated hydraulic conductivity (K_s) was determined with the hood permeameter method [59] on saturated soil samples in 2015, while in 2020, it was derived from the air permeability according to Riley's pedotransfer functions [57] (for further details, see Seehusen et al. [60]).

In 2015, the data from the compaction trial were analysed using the R statistical software package (2014) (details in Seehusen et al. [56]). In 2020, the data for the Norwegian SS were analysed using general linear models and Fisher tests in Minitab 19. Comparisons between the values from 2015 and 2020 were done by two-sample t-tests and confidence intervals. The results after Treatments 1–4 from 2020 of the reference plot (no compaction) were compared with the reference values from 2015, and the results after Treatments 1–4 from 2020 of the earlier compacted plot were compared with the values of the compacted plot from 2015.

2.2. Sweden

2.2.1. Experimental Design

The SS in Sweden is located at a farm in southern Sweden, where the subsoil is naturally compacted (1.7–1.9 g cm^{-3}) due to its formation under land ice and the root growth of crops is restricted to the topsoil, with hardly any roots below 30 cm [61,62]. The site has been under cultivation for at least a century and is tile-drained. The experiment consisted of a pilot study starting in September 2018 that investigated the possibility of improving the upper subsoil through the supply of undecomposed organic material in combination with a mechanical subsoil loosening. A randomised block design (12 plots, 6 × 20 m) with four replicates was established, involving three treatments—(a) a control treatment, (b) loosening of the subsoil (to a depth of about 35 cm) without the incorporation of organic material, and (c) loosening of the subsoil with the incorporation of undecomposed straw pellets at amounts of about 25 Mg ha^{-1}. Subsoiling and straw incorporation were performed using adapted HE-VA sub-tiller equipment at a speed of 1 km per hour to 24–35 cm depth. Straw pellets were pumped from a tank mounted on the front of the tractor and injected under pressure into the upper subsoil through oval openings in metal pipes welded behind each vertical bill. The loosening of the subsoil and the addition of straw pellets was performed only once (in 2018). Thereafter, normal tillage practices, including mouldboard ploughing to 25 cm, were applied in all plots. Crop fertilisation followed the local recommendations. Winter wheat (*Triticum aestivum* L.) and sugar beet (*Beta vulgaris* L.) yields were recorded in the 2019 and 2020 growing seasons, respectively.

2.2.2. Soil Sampling, Field Measurements, and Laboratory Analysis

In 2019, at the end of the winter wheat heading (growth stage Z60 according to the Zadoks scale), a soil profile description was conducted in one plot per treatment. The portions of the upper subsoil (24–35 cm) volume and surface affected by subsoiling and the presence of roots were visually evaluated. A more detailed soil sampling was done in 2020 about six weeks before harvest within a small area in the middle of each plot that was kept free from sugar beet plants starting around mid-summer. In this area, a soil pit 65–75 cm long and 25 cm wide was dug, and six undisturbed soil cylinders (7.2 cm diameter, 5 cm height) were taken at 10–15 cm depth, as well as six at 28–33 cm depth, by placing the cylinders one after the other in a row at a spacing of about 5 cm between each. Before removing the cylinders from the 28–33 cm depths, six penetration resistance (PR) tests were collected along the row with cylinders. On the same occasion, a soil profile for the control plots (only) was obtained using an auger sampling with depths divided into increments of

0–20, 20–22.5, 22.5–25, 25–27.5, 27.5–30, 30–35, and 35–40 cm. Each increment was analysed for the total C and N, and the pH and soil texture were measured in the 0–20 cm layer and in a combined sample for the 25–40 cm depth.

The soil moisture content and dry soil bulk density were determined from each of the cylinders. Each of the cylinder samples was also passed through a 2-mm sieve, and the occurrence of gravel and small stones (i.e., particles >2 mm) was determined by measuring both their weight and volume fraction. Thereafter, the total C and N concentrations were measured by dry combustion, and the pH (water) was determined for each sample from the 28–33 cm depth; only a pooled sub-sample for the 0–15 cm depth was retained for these analyses.

All statistical analyses were done with the GLM Procedure in SAS software (SAS Institute, Cary, NC, USA). The means were compared using Fisher's least significant difference (LSD) when the F-value in ANOVA was statistically significant ($p < 0.05$).

2.3. *United Kingdom*

2.3.1. Experimental Design

The United Kingdom SS is located at the Allerton Project—a 300 ha mixed arable and livestock research, demonstration, and education Farm. The experiment aimed to examine the alleviation of compaction by using tillage. The experiment started in October 2017 at the Allerton Project. This SS historically used a wheat–rape (*Brassica napus* L.) rotation with a "break" spring crop, and over the last ten years, had a reduction in tillage, going from a plough-based system to direct drilling. Soil compaction was artificially created by driving a tractor (Massey Ferguson 7720, approx. 8 tons total weight) across the area, ensuring a tractor tyre was running over the whole plot twice. Directly afterwards, measurements with a penetrometer verified the degree of compaction, showing the average compaction was 15% higher to a depth of 45 cm, with the highest compaction (+32%) occurring at 7.5 cm depth. The experimental design was a randomised complete block design with 3 replicates involving a total of 6 plots (9 m wide and 40 m long). The ploughing system (20 cm depth) was compared with a no-cultivation direct-drilled control treatment. Following the fall cultivations in 2017, winter barley grew across all plots and was harvested in July 2018. The compaction and treatments were repeated in October 2018, keeping the same plot structure, and faba beans (*Vicia faba* L.) were planted across all plots and harvested in September 2019. In March 2020, spring wheat was planted across all plots and harvested in October 2020.

2.3.2. Soil Sampling, Field Measurements, and Laboratory Analysis

The measurements of BD and penetration resistance (PR) were split into topsoil (0–25 cm), which was within the cultivation depth of the plough, and subsoil (>25 cm), which was below the depth of cultivation. PR measurements were conducted in 2020 after crop drilling in May using a field penetrometer (Field Scout, SC900) to a depth of 45 cm, with 10 measurements taken per plot and averaged. The bulk density was also measured in May 2020 using a soil cylinder (196 cm^3) in the topsoil and subsoil layers. The soil was dried for 48 h at 105 °C and weighed to calculate the bulk density. Soil samples were also collected from the topsoil layer and the particle size distribution and soil organic carbon were analysed. Infiltration was measured using the double ring method (outer ring diameter of 53 cm, inner ring diameter of 28 cm diameter, water depth of 24 cm). Both rings were partially buried in the soil and the outer ring was kept topped up with water to prevent lateral leaking. Once the water loss reached a stable rate, the water loss from the inner ring was recorded over time and converted to saturated hydraulic conductivity (K_S). The crop yield was measured at harvest each year by taking a reading from the combine after each plot was harvested.

Differences between treatments were analysed using Genstat version 18. A general linear model was used, with blocking treated as a random effect in all analyses. Where topsoil and subsoil measurements were both included in the analysis, a split-plot design was used, with the sample depths of the split-plot and treatment (plough vs. direct-drill) as the main plot effects.

2.4. Italy

2.4.1. Experimental Design

The Italian SS aimed to prevent soil SC by combining no-tillage and cover crops. In the area, the shallow water table ranged from about 0.5–1.5 m in late winter to early spring to 1–2 m in summer. The experiment has been ongoing since 2018 and has a split-plot design (12 plots in total, 12 m wide × 85 m long) with two replicates, two levels of tillage intensity (main plot), and three levels of soil cover (sub-plot). The no-tillage (NT) system based on sod seeding was compared with the conventional practice (CT) based on mouldboard ploughing to 30 cm, followed by disk-harrowing to 15 cm. The main crop was maize (*Zea mays* L.), while during fall, the soil remained bare (BS) or was covered with cover crops, for example, winter wheat (WW) or tillage radish (*Raphanus sativus* L.) (TR), which are characterised by fibrous and taproot root systems, respectively. Subsurface band fertilisation was applied at sowing in NT, while side-dressing fertiliser was followed by hoeing in the CT treatment. Pesticide applications depend on the crop requirements but were the same for all the plots.

2.4.2. Soil Sampling, Field Measurements and Laboratory Analysis

Soil samples were collected before seedbed preparation in spring 2020 from the topsoil (i.e., tilled layer) and subsoil (i.e., below the tilled layer), as reported in the Supplementary Materials, Table S2. Undisturbed soil cores (7 cm in diameter, 60 cm in height) were collected with a hydraulic sampler and cut to extract the 0–20 and 40–60 cm soil layers. Remoulded soil samplings were collected at the same depth for chemical–physical analysis. PR measurements were performed up to 60 cm depth before tillage operation (at the end of February), with a digital cone penetrometer (Eijkelkamp, Giesbeek, The Netherlands) with a base area of 2 cm^2 and an apex angle of 30°. Undisturbed soil cores were oven-dried at 105 °C for 24 h to calculate the volumetric water content (VWC) and BD using the core method [63]. Remoulded soil samples were air-dried, sieved at 2 mm, and analysed for particle size distribution according to the methods by Bittelli et al. [64] and the soil organic carbon concentration (SOC). On-field soil hydraulic properties were measured inside each plot using a double-ring infiltrometer (inner ring diameter of 60 cm, outer ring diameter of 80 cm) according to the methods by Parr and Bertrand [65]. The hydraulic conductivity (K_s) and sorptivity were calculated by applying Philip's infiltration equations [66]. At the end of the growing season, the maize grain yield was collected at the commercial moisture content from four representative areas (2 m^2) in each plot and then dried at 65 °C until a constant weight was obtained to determine the dry weight.

The data were analysed by applying a linear mixed-effect model based on the restricted maximum likelihood estimation method considering tillage, soil covering, and their interaction as fixed and block as random factor. Post-hoc pairwise comparisons of the least-squares means were performed using the Tukey method to adjust for multiple comparisons ($p < 0.05$). Statistical analyses were performed with SAS software (SAS Institute Inc. Cary, NC, USA), 5.1 version.

2.5. Romania

2.5.1. Experimental Design

The Romania SS is located in an area characterised by natural subsoil compaction. The experiment consisted of a pilot study established in March 2018, and its aim was to mitigate natural SC by tillage. The experimental design was a split plot (36 plots, 6 × 33 m) with three blocks and involving four treatments—(TR1) mouldboard ploughing with furrow inversion to 25 cm depth, (TR2) subsoiling to 60 cm by ripping and disking to 12 cm depth, (TR3) a control treatment with 2-times disking, and (TR4) chiselling to 25 cm depth with furrow inversion. All treatments were repeated every year. The testing of tillage treatments also involved three rotations with deep-rooting leguminous crops. Only the main effect of the tillage treatments on the soil physical properties is reported here.

2.5.2. Soil Sampling, Field Measurements and Laboratory Analysis

Soil physical and chemical parameters were measured in all plots during the three years of the experiment. For this, disturbed soil samples were collected in autumn after crop harvesting for soil water-stable aggregates (WSA) >250 um, and undisturbed soil cores (100 cm^3 volume) were sampled at 10–20 cm and 40–50 cm depths for soil physical analyses (Ks and BD).

The content of water-stable aggregates (in % g/g) was measured by the Henin–Feodoroff method based on wet sieving (SR EN ISO 10930:2012). The Ks was determined according to the steady-state falling head method (Romanian standard: STAS 7184/15–91). The BD was gravimetrically determined by weighing the soil core samples before and after drying for 24 h at 105 °C (SR EN ISO 11272:2017).

The data obtained for the soil properties measured at the Romanian SS were analysed by one-way repeated measure ANOVA considering either the soil tillage or year as the tested factor. Post-hoc pairwise comparisons of the least-squares means were performed using the Tukey method to adjust for multiple comparisons ($p < 0.05$). All statistical analyses were performed with OriginLab 6.1 software (Origin Lab Corporation, Northampton MA, USA).

2.6. Soil Compactions Indices

The effects of the SICSs across the different SSs were investigated using three soil compaction indices—degree of compaction (DC), relative normalised density (RND), and air-filled porosity (AFP). The DC was calculated as follows:

$$DC = {}^{BD}/_{BD_{ref}} \times 100 \tag{1}$$

where BD is the bulk density and BDref is the reference bulk density. The BDref was calculated according to Equation (12) reported by Keller and Håkansson [67], as follows:

$$\begin{aligned} BD_{ref} = 1.308 + 0.0119\ \text{clay} + 0.0103\ \text{sand} + 0.00018\text{clay}^2 - 0.00008\text{sand}^2 \\ -0.00062\text{siltOM} - 0 : 00059\text{sandOM} \end{aligned} \tag{2}$$

where OM is the soil organic matter. The RND index was derived from the ratio between the BD and the critical bulk density (BD_{crit}), the latter being 1.6 g cm^{-3} for soils with clay < 16.7% or calculated with the following equation for soils with clay > 16.7% [68]:

$$BD_{crit} = 1.75 - 0.0009 \times \text{clay} \tag{3}$$

The air-filled porosity (AFP) at the sampling was calculated as the difference between the total porosity and the volumetric water content.

3. Results and Discussion

3.1. Norway

In the topsoil, there were no significant differences in the BD, TPV, or AC between the treatments in 2020 or between years, with average values of 1.35 g cm^{-3}, 47.6%, and 10.8%, respectively. Similarly, there were no significant differences between the treatments for Ks and air permeability in 2020. However, for both treatments, the Ks and air permeability were significantly lower in 2020 compared to 2015 (Supplementary Table S3).

In the subsoil, multiple wheeling in 2015 led to a significant increase in BD, with 1.69 g cm^{-3} in C as compared to 1.59 g cm^{-3} in the R plots (Table 2). Five years after the compaction event, the BD was still significantly higher in the C than in the R plots in 2020, with the exception of Treatments 2 and 4. In the uncompacted R plot, the BD in 2020 was significantly decreased compared to 2015 for all treatments, from 1.59 to 1.45 g cm^{-3}. In the

C plot, the BD was also significantly reduced after 5 years for Treatment 4 (1.45 g cm^{-3}), reaching the same level as Treatment 4 in the uncompacted R plot (1.44 g cm^{-3}). There was also a trend towards a reduction in the BD after 5 years for Treatment 2, from 1.69 to 1.55 g cm^{-3}. Compared to the topsoil (Supplementary Materials, Table S3), the BDs in the subsoil for the R plot were about 20 and 10% higher in 2015 and 2020, respectively (Table 2). The subsoil TPV significantly decreased by about 6% in 2015 following the compaction event (Table 2). In 2020, there were no significant treatment effects on the TPV, which was 45.5% on average. However, both Treatment 1 (+5.1%) and Treatment 4 (+7.0%) led to a significant increase in the TPV on the C plots in 2020 compared to 2015.

Table 2. Subsoil (30–40 cm) bulk density (BD), total pore volume (TPV), air capacity (AC), saturated hydraulic conductivity (Ks), and air permeability (Air perm) in uncompacted reference and compacted plots at the Norwegian study site in 2015 and for the different treatments in these plots in 2020. Treatments 1, 2: oilseed rape–barley rotation; Treatment 3: barley monoculture; Treatment 4: alfalfa monoculture.

	BD (g cm^{-3})	TPV (%)	AC (%)	Ks (m day^{-1})	Air Perm (um^2)
2015					
Reference plot	1.59 ± 0.04 a	46.1 ± 1.3 a	3.42 ± 0.82 ns	$7.56 \times 10^{-2} \pm 1.02 \times 10^{-1}$ ns	11.9 ± 13.32 ns
Compacted plot	1.69 ± 0.04 b	40.0 ± 1.7 b	3.38 ± 1.03 ns	$7.30 \times 10^{-3} \pm 5.1 \times 10^{-3}$ ns	26.1 ± 20.90 ns
2020					
Reference plot					
Treatment 1	1.44 ± 0.06 ns¶	48.2 ± 5.5 ns	6.31 ± 1.06 ns¶	$9.68 \times 10^{-2} \pm 1.24 \times 10^{-1}$ ns	1.1 ± 1.30 ns
Treatment 2	1.45 ± 0.05 ns¶	47.0 ± 1.5 ns	6.41 ± 0.53 ns¶	$3.00 \times 10^{-3} \pm 2.7 \times 10^{-3}$ ns	0.1 ± 0.06 ns
Treatment 3	1.48 ± 0.04 ns¶	45.2 ± 2.5 ns	5.19 ± 1.78 ns	$1.5 \times 10^{-3} \pm 8.00 \times 10^{-4}$ ns	0.1 ± 0.02 ns
Treatment 4	1.44 ± 0.03 ns¶	46.7 ± 1.3 ns	5.33 ± 1.12 ns¶	$1.46 \times 10^{-1} \pm 2.18\times 10^{-1}$ ns	1.5 ± 1.97 ns
Compacted plot					
Treatment 1	1.63 ± 0.07 a	45.1 ± 3.3 ns¶	9.68 ± 5.36 a	$1.48 \times 10^{-2} \pm 1.71 \times 10^{-2}$ ns	0.3 ± 0.27 ns¶
Treatment 2	1.55 ± 0.16 ab	42.5 ± 5.1 ns	6.58 ± 1.06 ab¶	$3.13 \times 10^{-1} \pm 3.76 \times 10^{-1}$ ns	2.7 ± 3.16 ns¶
Treatment 3	1.68 ± 0.05 a	42.4 ± 1.7 ns	5.26 ± 0.86 b¶	$1.90 \times 10^{-3} \pm 1.00 \times 10^{-3}$ ns¶	0.1 ± 0.02 ns¶
Treatment 4	1.45 ± 0.07 b¶	47.0 ± 1.8 ns¶	5.71 ± 1.12 ab¶	$60.1\times 10^{-1} \pm 1.20$ ns	3.8 ± 7.49 ns

Mean ± standard deviation (2015 n = 5, 2020 n = 4). For 2015, values followed by different letters are significantly different. For 2020, different letters after values indicate significant differences between the treatments within reference (R) and compacted (C) plots. ns= not significant at $p < 0.05$. ¶ indicates a significant difference between 2020 and the value in 2015 for each treatment (i.e., in R or C plots).

In contrast, multiple wheeling in 2015 had no significant effect on the subsoil AC. In the uncompacted R plot, the AC was an average of 5.81% after 5 years and there were no significant differences between the treatments, while in the C plot, Treatment 3 presented the lowest increase of all treatments in 2020 (from 3.38 to 5.26%). With the exception of Treatment 3 on the R plot and Treatment 1 on the C plot, the AC values significantly increased during the research period, from 3.4 to 5.9% on average across treatments.

Similarly to the TPV, soil compaction in 2015 did not lead to a significant reduction in either the saturated hydraulic conductivity or air permeability (Table 2). In 2020, Ks followed a similar pattern to air permeability since it was estimated using a pedotransfer function based on air permeability. In both the R and C plots, there were no significant differences between the treatments in 2020 in either the Ks or air permeability, which were 0.15 m day^{-1} and 1.2 um^2 on average, respectively. Compared to 2015, there was a significant reduction in Ks by 5.40 $\times 10^{-3}$ m day^{-1} in the C plots for Treatment 3, while there was a significant reduction in the air permeability for Treatments 1–3 in the C plot, with an average of 23.9 um^2. Compared to the topsoil (Table 2), there were very large differences regarding both the Ks and air permeability in the subsoil for the R plot in 2015 and 2020.

There was only a significant difference between the same treatment in the R and C plots in 2020 for BD (Treatments 1 and 3) and not for any of the other soil physical properties (data not shown).

3.2. Sweden

The soil visual assessment showed that the straw was not mixed with the subsoil in rows but located at the bottom of the subsoil rows created by the bills in the subsoiling + straw treatment (Figure 2a). Indeed, the two subsoiling treatments forced the topsoil into the subsoil, forming distinct rows, while the subsoil moved into the topsoil irregularly (Figure 2a). However, subsoiling affected only a portion of the upper subsoil layer (24–35 cm) below the topsoil. We evaluated that the volume of the subsoil affected by the subsoiling treatments varied between 36 and 40% and that the surface of the subsoil affected varied between 42 and 49%. Analysis of the soil profile samples for the control plots that characterised the experimental site more precisely showed that the sand, silt, and clay contents were 62, 27, and 11%, and 64, 27, and 9% in the topsoil and subsoil layers, respectively.

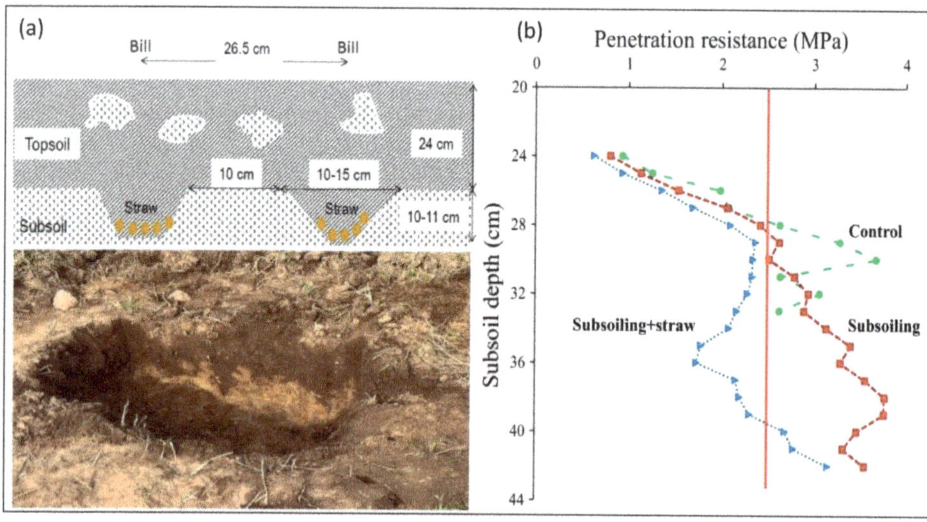

Figure 2. (**a**) Illustration (top) and photo (bottom) of a Swedish soil profile used for evaluating the effects of the subsoiling + straw treatment. (**b**) Changes in the penetration resistance with depth in 2020, a metric used for evaluating the effects of subsoil loosening and loosening + straw incorporation treatments at the Swedish study site. The vertical line (2.5 MPa) indicates the critical limit for root penetration. Data are mean values of six measurements made across treatment stripes covering a width of about 40 cm in each experimental plot.

As shown by the visual assessment for the presence of roots in 2019, which was done by counting the number of roots along a 10-cm line at two depths in the topsoil (10 and 20 cm) and in the subsoil (30 cm), there were more roots present in the subsoiling treatments at the 30 cm depth. Meanwhile, there were almost no roots present in the subsoil for the control treatment, and the subsoiling + straw treatment also appeared to improve the number of roots at all three depths compared to the control (Supplementary Materials, Table S4). The mean maximum root penetrations into the subsoil (>24 cm) were about 4 cm in the control and 11 cm in the subsoiling treatments. The maximum penetrations were more variable for the subsoiling treatments since among the six measurements made within each plot, some presented values similar to those for the control, but some values were much deeper, indicating the measurements were sometimes penetrating the subsoil rows created by the bills (data not shown). The measurements in the control plots almost never exceeded 6 cm. The changes in the soil penetration resistance with depth in 2020 showed a mean maximum (i.e., exceeding the 2.5 MPa critical limit for root penetration) rooting depth of about 28 cm in the control, almost 30 cm in the subsoiling alone, but much deeper at around 40 cm for the subsoiling + straw treatments (Figure 2b).

There were no significant differences between the SOCs in the topsoil (10–15 cm) and the subsoil (28–33 cm) cylinder soil samples (Supplementary Materials, Table S4). The soil total C/N ratios, as well as the pH values in the top- and subsoils, were also not significantly different between treatments at around 10.0 and 6.0, respectively.

Compared to the subsoiling + straw treatment, the BD was significantly higher in the topsoil in the subsoiling treatment. It was higher also in the subsoil compared to the control (Supplementary Materials, Table S4). However, when correcting the BD for the presence of gravel and stones [69], which varied between 6.1 and 8.3%, there were no significant differences between the treatments in either the top- or subsoils. Since this site had a naturally compacted subsoil with high soil densities, we were restricted to using smaller cylinders than usual (i.e., 204 vs. 408 cm^{-3}), which provided less precise and more variable measurements. The experimental site was also heterogeneous, and there was a negative correlation between the SOC contents and the BDs (Supplementary Materials, Figure S1). This may have contributed to the differences because the subsoil SOC content in the control was slightly higher than for the subsoiling treatments, even if not significant.

3.3. United Kingdom

The soil BD was not affected by the compaction alleviation treatment and showed no significant difference between the treatments in the topsoil or subsoil layers. Nevertheless, a trend of lower BDs was observed under the direct-drilling treatment in both the topsoil (1.46 g cm^{-3} $\pm$ 0.073 vs. 1.52 g cm^{-3} $\pm$ 0.086) and the subsoil (1.43 g cm^{-3} $\pm$ 0.17 vs. 1.64 g cm^{-3} $\pm$ 0.029). A significant (p = 0.007) sample depth $\times$ compaction alleviation treatment interaction was observed for the PR results, with the treatments ranked as follows: plough topsoil < direct-drilling topsoil < plough subsoil < direct-drilling subsoil (Figure 3). Overall, the PR was significantly lower in the plough plots ($p < 0.001$) and significantly higher in the subsoil compared to the topsoil ($p < 0.001$). In the subsoil, the PR exceeded the 2.5 MPa limit in about 30% of the measurements but did not exceed it in any in the topsoil.

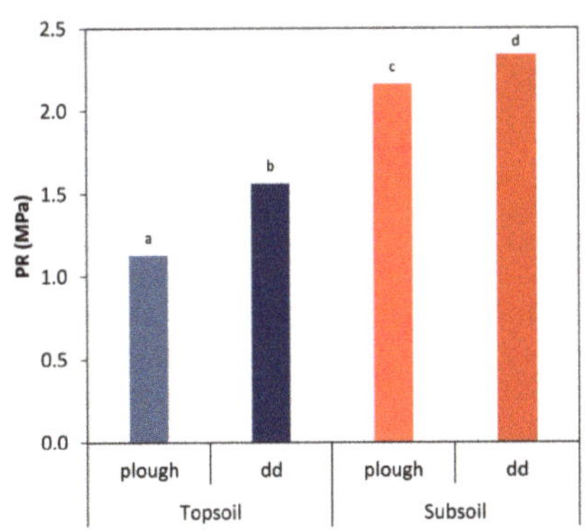

Figure 3. Penetration resistance (PR) in topsoil (0–25 cm) and subsoil (25–45 cm) in ploughed and control direct-drilled (dd) plots at the UK study site.

The measurements of the SOCs in the topsoil showed no significant differences between treatments (plough: 2.85% $\pm$ 0.70; control dd: 2.85% $\pm$ 0.80). The measurements of Ks also showed no significant differences between treatments (plough: 1.28×10^{-2} m s^{-1} $\pm$ 5.22×10^{-3}, control dd: 1.53×10^{-2} m s^{-1} $\pm$ 6.25×10^{-3}).

3.4. Italy

The BD was not affected by agronomic management neither in the topsoil nor in the subsoil despite a tendency for denser topsoil being observed in the NT compared to the CT plot (1.43 vs. 1.35 g cm^{-3}). The topsoil PR was affected by both tillage and soil covering, where the NT was 5% higher than the CT plot (1.05 vs. 0.82 MPa) (Figure 4). At the same depth, the PR had a lower value in the BS (0.85 MPa) compared to the WW (1.06 MPa). Contrarily, in the subsoil, the PR was not affected by the agronomic management. PR observations were always < 2.5 MPa for the topsoil, while 33% of the measurements exceeded that limit for the subsoil. No significant treatment effects were found.

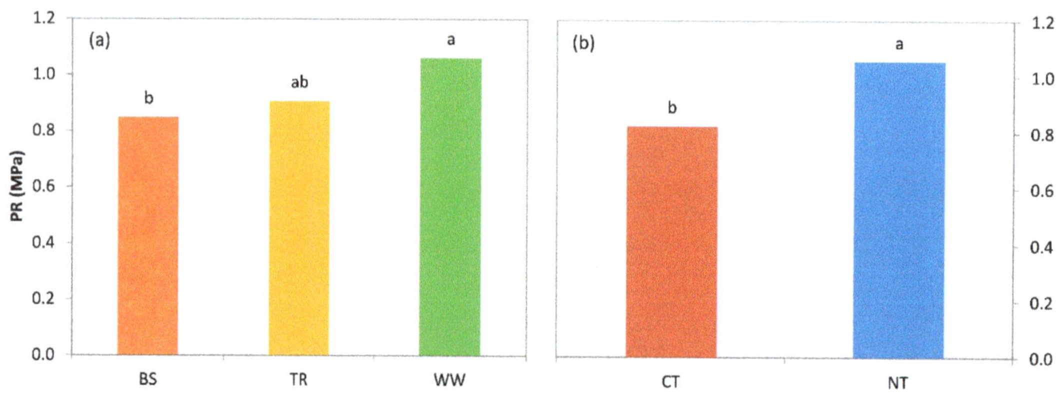

Figure 4. Topsoil (0–20 cm) penetration resistance (PR) as affected by soil cover (**a**) and tillage (**b**) at the Italian study site. Different letters indicate a significant difference according to the Tukey test at $p < 0.05$. BS: bare soil; TR: tillage radish; WW: winter wheat; CT: conventional tillage; NT: no-tillage in VWC, SOC content, or stock.

Tillage affected the hydraulic parameters. Indeed, the sorptivity ($p = 0.05$) increased almost five-fold under NT management. The Ks, despite not significant with a $p < 0.05$, showed a three-fold value for the NT compared to the CT (Supplementary Materials, Figure S2). For further details, see Supplementary Materials, Table S5.

3.5. Romania

For the BD in the topsoil (10–20 cm), throughout all three years of the Romanian experiment, the mean value in the subsoiling treatment (TR2) ranged from 1.28 to 1.32 g cm^{-3} and was always significantly different from the other treatments, which had higher values between 1.42 to 1.48 g cm^{-3} (Figure 5a). With the exception of the control treatment (TR3), where no significant differences in BD occurred between the years, the BD was significantly lower in 2020 compared to 2018 for all treatments (Supplementary Materials, Figure S3a).

For the BD in the subsoil (40–50 cm), the results follow the same trend as for the topsoil regarding the subsoiling treatment and always had significantly lower values over all three years (Figure 5b). The control treatment also presented a significantly higher BD during each year of the study compared to the other three treatments. The bulk density in the subsoil was significantly lower in 2020 compared to 2018 for both the subsoiling and control treatments, but no significant differences were observed for the other two treatments (Supplementary Materials, Figure S3b).

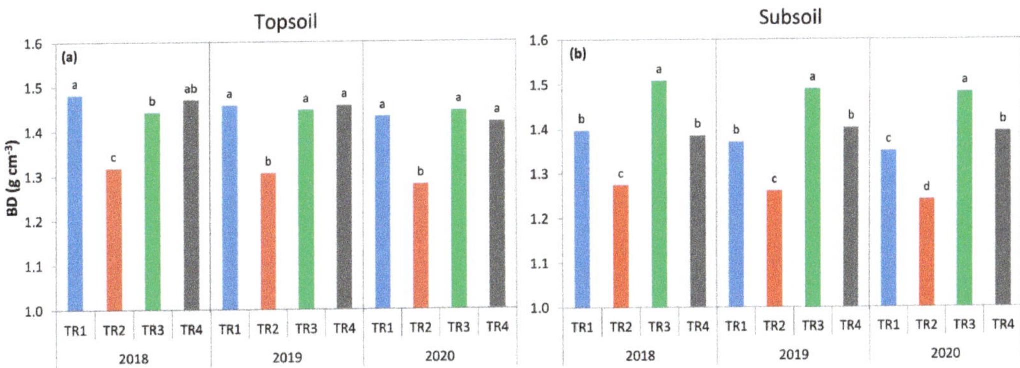

Figure 5. Bulk density (BD) in topsoil (10–20 cm) (**a**) and subsoil (40–50 cm) (**b**) as affected by different tillage systems during the three years of the Romanian experiment. Different letters represent statistically significant differences according to the Tukey post-hoc test at $p < 0.05$. TR1: mouldboard ploughing with furrow inversion to 25 cm depth; TR2: subsoiling to 60 cm + disking to 12 cm depth; TR3: control treatment with 2-times disking; TR4: chiselling to 25 cm depth with furrow inversion.

The tillage treatments significantly affected the WSA in the topsoil throughout the 3 years of the experiment (Supplementary Materials, Figure S4a), with subsoiling (TR2) always exhibiting the highest WSA (23.9, 28.9, and 29.3% for 2018, 2019, and 2020, respectively) with respect to all other treatments (mean values across treatments of 17.0, 17.9, and 17.7% for 2018, 2019, and 2020, respectively). Except for 2018, the control (T3) always had a significantly lower percentage of WSA. The WSA remained the same during the study period for all treatments except for subsoiling, where the aggregation was significantly higher in 2019 and 2020 (Supplementary Materials, Figure S5a). The tillage treatments also significantly affected the Ks over all three studied years and were always 4 to 5 times higher for the subsoiling treatment (Supplementary Materials, Figure S4b). Differences in the Ks for other treatments only occurred in 2018, where TR2 ($202 \times = 10^{-8}$ m s^{-1}) >TR4 (74×10^{-8} m s^{-1}) >TR3 and TR1 (60×10^{-8} m s^{-1}). Only TR1 and TR2 differed between the years, with lower values of Ks in 2018 compared to the other years (Supplementary Materials, Figure S5b).

3.6. Soil Compaction Indices and Crop Yield

Generally, SC is considered to impair crop performance [70]. In the present study, the crop yield was only affected by the adopted SICS for the Romanian SS, a site predisposed to natural subsoil compaction as well as having a plough pan at 30 cm, which restricted the rooting depth. Compared to the other tillage types, the main effect of conducting subsoiling every year always gave the best crop performances after 3 years, with yields of 5.8, 1.6, and 3.4 Mg of dry matter ha^{-1} for maize, soybean, and spring barley, respectively. There were no significant effects of the SICSs on crop yields at the other SSs (data not shown).

In contrast to the Romanian SS, the Swedish SS only applied the subsoiling operation once, and there were no significant differences in yields between the treatments throughout the experimental period. Subsoiling did not affect the whole area but only a portion of it (i.e., distinct subsoil rows), in which roots would theoretically be able to grow deeper and take up more water and nutrients by exploring a greater volume of soil. This trial differed in this respect from the other types of experimental treatments that affected the whole area. Thus, the measured yields of the whole field represent a weighted mean value of the treated and untreated subsoil volumes. Conceptually, calculating yields as the weighted mean of the affected and unaffected subsoil may be a more reasonable indicator of the effect of subsoil loosening. To illustrate this, we recalculated the measured relative winter wheat yield (2019) of the whole area compared to the control for the subsoiling treatments

(Figure 6). This was done by scaling the two subsoiling treatments by factors of 100/38 and 100/45, considering they affected either 38% of the subsoil volume or 45% of the subsoil surface, respectively.

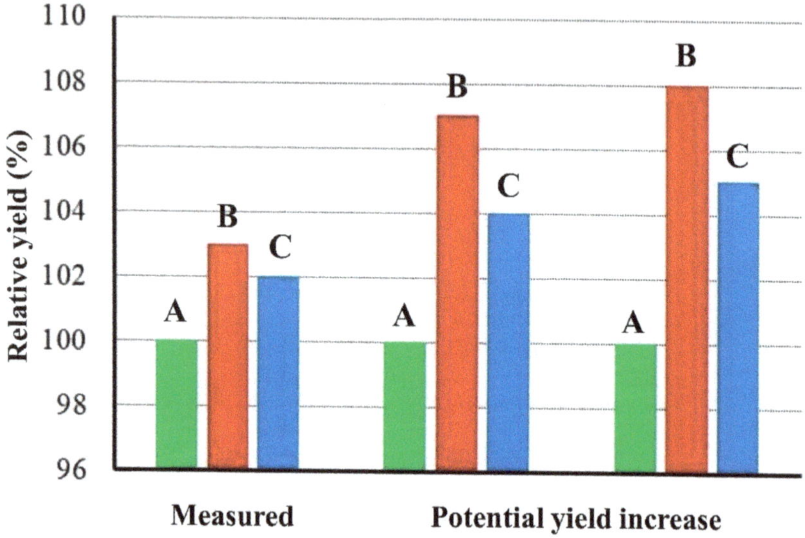

Figure 6. The measured relative winter wheat yield for the subsoiling (B) and subsoiling + straw (C) treatments compared to the control (A) in 2019 (left columns) at the Swedish site. Assuming that the whole (100%) subsoil was affected, and not only a portion of the subsoil surface (45%) or subsoil volume (38%), the potential yield increase is proportionally higher (middle and right columns).

To predict the possible effect of SC on crop performances, a few indices can be adopted [8,67,71,72]. A combined indicator of critical conditions in the soil (e.g., PR, porosity, and gas exchange) is the degree of compactness (DC), defined as the ratio of bulk density-to-reference density [8]. A threshold of 87% has been suggested as critical for root growth and crop development [8,67]. At the five SSs, the DC ranged from a minimum of 56% to a maximum of 124% (Figure 7a). The Norwegian SS exhibited the highest DC, which always exceeded the 87% threshold in both the topsoil and subsoil. On average, values in the subsoil were higher in the compacted compared to the reference plots (113 vs. 101%) (Figure 8b). At the Swedish SS, the DC was always <87%, averaging at 75 and 81% for the topsoil and subsoil, respectively (Figure 7a). For the UK SS, the DC averaged at 81%, with small variation between the topsoil and subsoil (Figure 7a). The DC limit exceeded the threshold of 87% for about one-third of the observations at the UK SS. At the Italian SS, the DC was higher for the no-tillage plots compared to the conventionally tilled plots (88 vs. 81% in the topsoil and 103 vs. 100% in the subsoil) (Figure 8a,b). At the Romanian SS, 47% (topsoil) and 25% (subsoil) of the measurements exceeded the DC limit but with lower magnitudes, where the maximum recorded DC was 90% (Figure 7a). At this site, a DC of >87% was frequently found in the topsoil under ploughing, chiselling, and disking, while this was only the case for the subsoil under disking (Figure 8a,b).

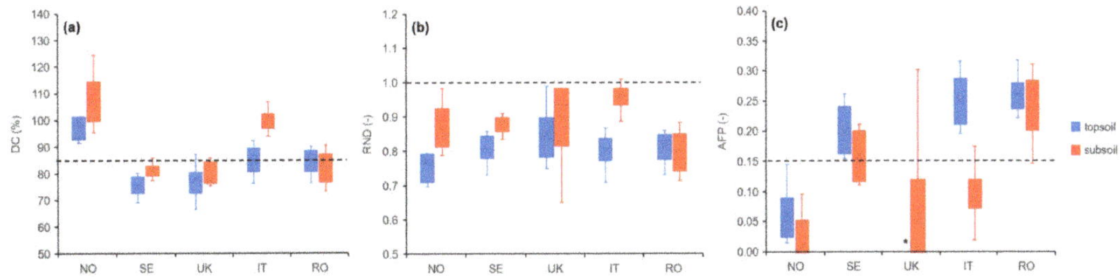

Figure 7. Box and whisker plots of degree of compactness (DC) (**a**), relative normalised density (RND) (**b**), and air-filled porosity (AFP) (**c**) in topsoil and subsoil at the five study sites. N: Norway; SE: Sweden; UK: United Kingdom; IT: Italy; RO: Romania. The box delimits values from low to high (from the 25th percentile to 75th percentile). Inside the box, the median and mean are indicated by a line and an X, respectively. The whiskers represent the minimum and maximum values in the range. * data not shown for the UK topsoil due to high frequencies of zero values.

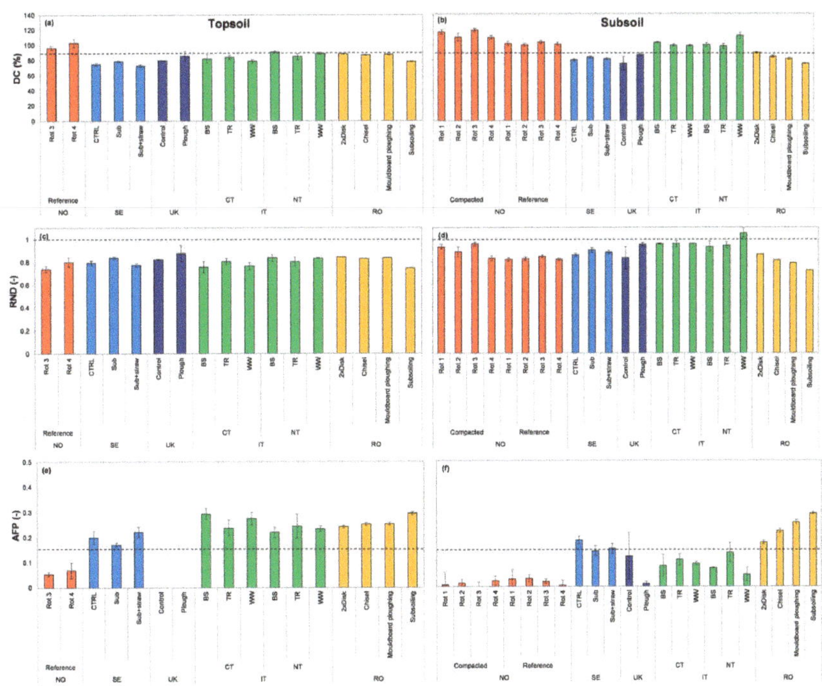

Figure 8. Degree of compaction (**a**,**b**), relative normalised density (**c**,**d**), and air-filled porosity (**e**,**f**) across the study sites (N: Norway; SE: Sweden; UK: United Kingdom; IT: Italy; RO: Romania) for topsoil (**a**,**c**,**e**) and subsoil (**b**,**d**,**f**). The dotted horizontal lines indicate an 87% degree of compactness, relative normalised density = 1, and air-filled porosity = 0.15, which represent suggested limits for good crop growth. The values shown are the mean and standard error. Rot 1, 2: oilseed rape–barley rotation; Rot 3: barley monoculture; Rot 4: alfalfa monoculture (N). CTRL: control treatment; Sub: subsoiling; Sub + straw: subsoiling + straw (SE). Control: direct-drilled treatment; Plough: ploughing system (UK). CT: conventional tillage; NT: no-tillage; BS: bare soil; TR: tillage radish; WW: winter wheat (IT). TR1: mouldboard ploughing with furrow inversion to 25 cm depth; TR2: subsoiling to 60 cm + disking to 12 cm depth; TR3: control treatment with 2-times disking; TR4: chiselling to 25 cm depth with furrow inversion (RO). For further details on adopted treatments, see Section 2. Materials and Methods.

The relative normalised density (RND), sometimes referred to as the degree of "over-compaction", is a texture-modified expression of density that might be useful to compare the state of compactness across differently textured soils [68]. Soil is defined as compacted when the RND > 1. In this study, the RND ranged from 0.65 to 1.09 across all the SSs (Figure 7b). The Norwegian SS always exhibited an RND of < 1.0 in the subsoil, with a higher value for the artificially compacted (0.90) plots compared to the reference plot (0.83) (Figure 8c,d). Similarly, the Swedish SS showed RND values in the subsoil below the suggested limit, being lower in the subsoiling + straw plot (0.78) compared to the subsoiling plot alone (0.84) (Figure 8c,d). At the UK SS, the RND ranged from a minimum of 0.65 to a maximum of 0.99 (Figure 7b). At the Italian SS, the RND was lower in the topsoil than in the subsoil (0.80 vs. 0.96), and the RND in the subsoil was only above 1 in the treatment with a winter wheat cover crop in the no-tillage system (Figure 8c,d). At the Romanian SS, the RND was always <1, with higher values in the topsoil associated with disking, chiselling, and ploughing (0.84, on average) compared to subsoiling (0.75), while the RND values in the subsoil of the different treatments ranked as follows: disking (0.87) < chiselling (0.81) < ploughing (0.79) < subsoiling (0.73) (Figure 8c,d).

In compacted soils, the soil–root contact may be very close, and reduced porosity could result in reduced soil aeration [73]. Therefore, the AFP may also be a useful index to estimate the compaction impact on crop growth. Except for tolerant crops, the ideal AFP is in the 10–15% range [74]. In the present study, the AFP ranged from values close to zero in the UK and some of the Norwegian treatments to a maximum of 0.32 found at both the Italian and Romanian SSs (Figure 7c). Higher topsoil values compared to the subsoil were found at both the Norwegian (0.06 vs. 0.01) and Italian SSs (0.25 vs. 0.09), while small differences between the soil layers were found at the Swedish (0.18 on average) and Romanian SSs (0.25 on average) (Figure 7c). At the latter SS, the AFP was affected by the tillage treatments in both the topsoil and subsoil, following the opposite trend observed for the RND (Figure 8e,f). Lower values in the topsoil were associated with disking, chiselling, and ploughing (0.25 on average) compared to subsoiling (0.30), while the AFP values for the subsoil of the different treatments ranked as follows: subsoiling (0.29) > ploughing (0.25) > chiselling (0.22) > disking (0.18) (Figure 8e,f).

Although roots may benefit from soil cracks and pre-existing bio-macro-pores [75], to fully exploit the soil, matrix roots must be able to explore the intra-aggregate space [76]. It is generally recognised that a root can either penetrate a soil aggregate or be deflected along its surface depending on the soil strength [72]. A total root growth decrease and impaired crop yield are observed when the PR exceeds a soil-specific limit, which typically ranges from 1 MPa [51] to 2 MPa or greater [47–53]. In this study, only the Swedish, UK, and Italian SSs directly measured the PR in the field. At the Swedish SS, the PR measurements in the subsoiling + straw incorporation treatment showed values below 2.5 MPa down to about 40 cm depth, while the control and subsoiling alone treatments showed values >2.5 MPa higher up in the soil profile (Figure 2b). Indeed, visual assessments for the presence of roots and maximum penetration (Supplementary Materials, Table S4) indicated that subsoiling had a positive impact on both the root growth and rooting depth at this site. At the UK and Italian SSs, soils under no-tillage presented higher PR values than the ploughed treatments, with at least 30% of the subsoils exceeding 2.5 MPa, which might impair root-growing conditions.

With the exception of the Romanian SS, there was no relationship (data not shown) between the crop yield and the SC indicators (i.e., DC, RND, AFP, and PR). At this SS, we found a 2% yield reduction for every percentage of DC increase or every unit of AFP decrease (Figure 9). This implies that in passing from a DC of 83% (average of Romanian soils) to 87% (DC limit for crop growth according to Håkansson [8]), a 7% reduction in the crop yield might be a possible scenario for this SS, irrespective of the crop type. The response of the crop yield to different levels of SC is usually considered parabolic, with low production in loosened soil, high yields at an optimal degree of soil compaction, and lower yields for compacted soils [77]. Only the descending part of this parabolic relationship may have been observed at the Romanian SS, and it is possible that the optimal DC for crop production in its fine-textured soil might be located at a DC lower than 87% [6].

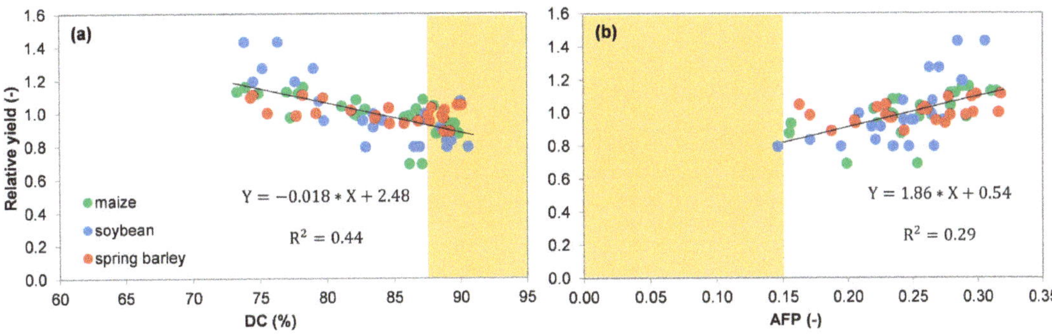

Figure 9. Linear regressions between relative crop yield (yield-to-average yield ratio) and degree of compaction (DC) (**a**) and air-filled porosity (AFP) (**b**) at the Romanian study site. The regression equation and R2 values are shown. Yellow areas represent the zone with DC >87% and AFP < 0.15.

3.7. Soil Compaction and SICS with Tillage

The relationship between SC and tillage has been thoroughly investigated, especially in northern [16,78,79] compared to central [80] and southern Europe [81]. It is generally recognised that topsoil compaction might be mitigated using annual soil loosening with conventional tillage practices such as mouldboard ploughing, while subsoil compaction has proven to be persistent and difficult to recover and sometimes requires more specialised tillage operations [82].

In this study, all countries except Norway involved different intensities of tillage for mitigating topsoil and subsoil compaction. Inversion tillage through mouldboard ploughing was adopted in the UK (to 20 cm depth), Italy (to 30 cm depth), and Romania (to 25 cm depth). The UK and Italian SSs also included no-tillage (i.e., direct-drilling) treatments. Subsoiling treatments were used at the Romanian SS (to 60 cm depth) and in Sweden (to 35 cm depth), the latter with and without the injection of organic materials. Lower-intensity tillage (i.e., reduced or no-tillage) is considered to improve soil structural stability and, therefore, theoretically, tillage practices prevent SC [83]. On the contrary, high-intensity tillage might produce an unstable soil structure more prone to soil compaction. Disking is one of the less conservative soil aggregate tillage practices, often resulting in a greater proportion of micro-aggregates (2–250 μm) but a lower proportion of macro-aggregates (>250 μm) [84]. At the Romanian SS, the treatment with 2-times disking (TR3) decreased the topsoil BD but showed the lowest proportions of WSA and Ks together with the highest DC and RND and the lowest AFP. These findings suggest that despite providing suitable conditions for crop establishment, disking can make the soil more prone to SC due to greater soil structure instability. Mouldboard ploughing inversion tillage is considered responsible for soil aggregate fragmentation [85], although this negative effect on soil structure may be counteracted by increasing organic C inputs to the soil from crop residues or the incorporation of organic amendments [86]. At the Romanian SS, the results suggest that mouldboard ploughing inversion tillage had a less negative effect on the WSA compared to 2-times disking (Supplementary Materials, Figure S4a).

No-tillage or direct drilling is usually considered a more sustainable agronomic practice [87,88] because it is thought to be less harmful to soil biota, and by keeping the crop residues at the soil surface, it reduces the risk of soil erosion [89]. Verhulst et al. [90] found a greater proportion of large macro-aggregates (>2000 μm) and macro-aggregates (250–2000 μm) under no-tillage compared to conventionally-tilled soils, confirming both the positive effect of tillage absence and crop residue retention. The UK and Italian SSs show somewhat opposite results when comparing the effects of no-tillage and direct-drilling against inversion tillage with mouldboard ploughing, with both the DC and RND being higher under no-tillage and direct drilling in Italy but not in the United Kingdom. Derpsch [91] identified four phases after the adoption of no-tillage—an "initial phase" (0–5 years), when crop residues are expected to be low due to lower yields and with no measurable changes in the SOC while the soil starts rebuilding aggregates; a "transition phase" after 5 to 10 years, when crop residues and SOCs are expected to increase, although these changes are accompanied by higher SC; improvements are expected only after 10–20 years during the "consolidation phase" followed by the "maintenance phase", characterised by stabilised agro-ecological conditions. Six et al. [92] found that for drier climates before a positive trend occurs, no-tillage could even have a negative effect on the SOC during the first 5–10 years. According to this classification, the Italian soil was in the initial phase and experienced SC, while the UK soil was reaching the end of the transition phase (around 10 years after the first adoption), showing improved soil conditions. Different soil types and their interactions with agronomic practices may explain the differences between these two SSs. The Italian SS is mainly formed from Calcisols and Cambisols (WRB, 2006), with low SOC content (<1.0%) and far from equilibrium, having a silty texture and poor aggregate stability [93]. Piccoli et al. [81,94] previously postulated that the limited amount of non-complexed SOC available for interaction with clay minerals and the low clay-to-silt ratio could prevent the formation of a resilient structure that goes beyond the adopted agronomic management. In contrast, the higher clay and SOC (2.88%) contents at the UK SS might have fostered an improved soil structure by ensuring high stability of the macro-pores [95], better exploiting the benefits related to no-tillage.

Subsoiling is primarily aimed at counteracting subsoil SC and does not disturb the soil surface unless it is associated with another tillage operation. At the Romanian SS, subsoiling was associated with shallow disking (to 12 cm depth), while the plots with the subsoiling and subsoiling + straw treatments at the Swedish SS were subjected to the same conventional tillage as for the control plots (i.e., mouldboard ploughing to 25 cm depth and normal seedbed preparations). The hypothesis for the Swedish SS, which is naturally compacted, was that the incorporation of organic material, in addition to subsoiling alone, would stimulate biological activity and lead to the stabilisation of the soil structure at a lower density, enabling roots to grow deeper. The results show a positive impact on root growth and rooting depth, particularly for the subsoiling + straw treatment, and partly confirm this hypothesis. However, when corrected for gravel and stones, there was no significant effect on the BD (Supplementary Materials, Table S4). The subsoiling treatment at the Romanian SS was aimed at counteracting the natural compaction and preventing the formation of a plough pan layer. At this site, both the topsoil and subsoil BD were significantly improved with subsoiling, as also reflected in the SC indices, and subsoiling had a positive effect on the topsoil WSA and Ks (Figure 5 and Supplementary Materials, Figure S4).

A stronger response to subsoiling at the Romanian SS compared to the Swedish SS was probably related to the frequency of subsoiling. It occurred only once in Sweden, whereas it was repeated every year during the Romanian experiment. However, it may also relate to the clay, silt, and SOC, as discussed in a meta-analysis by Schneider et al. [89]. They suggested that for many soils with a clay-to-silt ratio of <0.3, subsoiling might result in a complete collapse of the natural soil structure and SC instead of loosening, while for soils with a clay content of >20%, subsoiling may have a better possibility of lowering the BD and increasing the macro-porosity. The clay content was twice as high at the Romanian SS as at the Swedish site and the clay-to-silt ratio was 1.5, while this ratio was 0.34 in the Swedish subsoil. In fact, the response to the SICSs was faster at the UK and Romanian SSs compared to those in Italy and Sweden. This may be related to the high clay contents at the former sites (31 and 44%, respectively), which were much lower at the latter SSs (18 and 10%, respectively). Furthermore, high silt (58%) and sand (63%) contents characterise the Italian and Swedish SSs, and these inherent soil properties may partly explain the lower responsiveness to SICSs. The dynamics also differ between the topsoil and the subsoil;

the former is more often subject to external factors (e.g., meteorological conditions), while the latter mostly follows natural dynamics (e.g., pedofauna activity). The results from the Italian SS agree with this reasoning, as there was a relationship between the PR (as an index of soil strength) and the fine silt + clay particles (0–20 μm), relating to the SOC protection against microbial degradation [96] only in the subsoil ($p < 0.01$ and 0.65 R^2) and not in the topsoil (Figure 10). In this subsoil, a 10% increase in soil fines reflected a PR reduction of 0.6 MPa among the studied range.

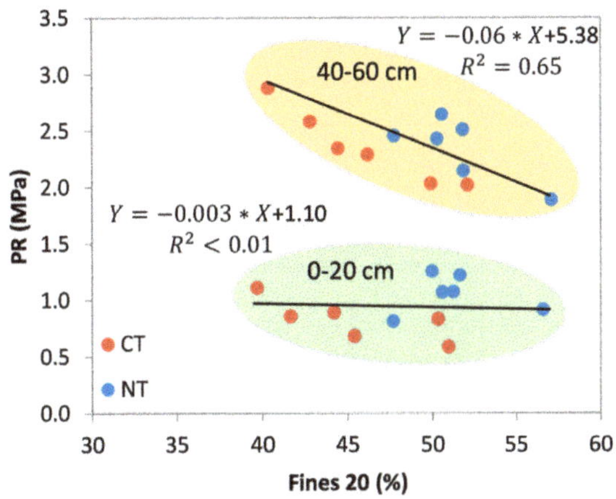

Figure 10. Scatterplot of fines 20 (% of particles below 20 μm) against penetration resistance (PR) at the Italian study site. Linear interpolation equations and their coefficient of determination (R^2) are also indicated for topsoil (0–20 cm) and subsoil (40–60 cm).

3.8. Soil Compaction and SICS with Deep-Rooted Bio-Drilling Crops

Another possibility for mitigating SC at deeper horizons or under no-tillage management is the adoption of deep-rooted crops, which may be either cash (e.g., alfalfa) or cover crops (e.g., tillage radish, mustard). Two SSs tested the effect of crops with a taproot apparatus—Norway (oilseed and alfalfa) and Italy (tillage radish).

Despite a relatively low machinery weight, multiple wheeling in 2015 led to subsoil compaction at the Norwegian SS [56], with increased BD and decreased TPV compared to the uncompacted reference plot (Table 2). In the topsoil, the presence of alfalfa (plots not ploughed) shows comparable results to Treatment 3 (barley monoculture), which was ploughed each year (Supplementary Materials, Table S3), suggesting that the alfalfa was equally effective at loosening the topsoil compared to ploughing. In 2015, the BD in the subsoil often exceeded 1.5–1.6 g cm^{-3}, which represents a threshold for root growth [97]. All subsoil BD observations in 2015 were classified as "very compact" according to Pagliai et al. [98], while in 2020, all values could be classified as "compact", suggesting that the SC mitigation occurred during the five years after the compaction event. In particular, this was evident in Treatment 4 (alfalfa), where the Ks was improved compared to 2015. More specifically, the subsoil Ks under the alfalfa treatment was frequently higher than the proposed limit for good soil functioning (e.g., 0.10 m day^{-1}) [22]. On the contrary, the same Ks threshold was undercut for all the other plots except for Treatment 2 in the compacted plots, confirming how subsoil compaction may be long-lasting [99]. Alfalfa was efficient at reducing the subsoil BD and restoring the TPV in the compacted plots, which were on the same level in 2020 as the uncompacted reference plot in 2015. These results are similar to other studies (e.g., [24,100]), showing that alfalfa, especially if grown over several years, is efficient for restoring soil structure. Effects on other parameters, such as air permeability and water infiltration, were less clear due to both higher data variability and a methodological issue. In fact, during the sampling operations, the alfalfa was still growing, and the living roots blocked the bio-pores (data not shown). Consequently, positive effects on the soil structure (e.g., improvements in AC and water and airflow) might be more recognisable over time, after the roots have decomposed [101].

Furthermore, oilseed crops are known to have deep-growing taproots that efficiently loosen up the soil structure [29], but contrary to that observed for alfalfa, the oilseed established only poorly at the Norwegian SS mostly due to a short growing season at this high latitude. Therefore, the root system was not well established, and this crop was not effective at loosening the soil or mitigating SC.

At the Italian SS, using the deep-rooted tillage radish cover crop during the winter season was not reflected in soil improvements, either in terms of SC mitigation (e.g., BD, DC, and RND) or in terms of soil functioning (e.g., Ks). Only the PR test suggested a trend for higher soil strength in changing from bare to covered soil (Figure 4a). As mentioned for Norway, a methodological issue might have impaired the results for the tillage radish crop in Italy due to incomplete taproot decomposition during sampling since it was necessary to take measurements for the PR and Ks prior to the following field operations (i.e., in early spring before tillage and seedbed preparation for the main crop, maize). The higher temperature during the subsequent maize growing season may have promoted complete root degradation and, in turn, fostered improved soil functioning. Indeed, in higher density soils, the presence of a few vertical macro-pores may dominate structure dynamics and soil functions (e.g., water infiltration and gas exchange) [94,102] and possibly counteract the negative effects of increased BD and soil strength. Moreover, the bio-macro-pores left by tillage radish provide low resistance paths for the subsequent cash crop roots [103]. Bio-drilling with cover crops was previously demonstrated to be more effective for topsoil under no-tillage management than with conventional tillage because bio-pores can persist and function for a longer time without tillage [104]. Nevertheless, in subsoil below the tillage depth, root-derived bio-pores might also persist even if shallower tillage occurred.

Beyond its correlation with compaction, the overall impact of bio-drilling on cash crop yields varies with climate conditions [29], improving crop performances under highly rainy climates (e.g., tropical) [105] and reducing yields in semiarid environments [106]. The response of crop yields to bio-drilling might also be dependent on the number of years since its first adoption [29]. The first year of bio-drilling adoption may not result in a boosted crop yield, while after several years, a more positive effect can be expected [107,108]. Finally, bio-drilling crops may contribute to SOC formation by providing more above- and belowground C inputs to the soil, in addition to the crop residues from the main crop [109]. Particularly because they have an important and deep root system, and compared to aboveground biomass, root-derived C is about twice as efficient in the C input conversion into stable SOC. However, changes in the SOC occur slowly and become measurable only after longer periods (>5 to 10 years) [110,111].

4. Future Prospects and Conclusions

Strategies to avoid SC, stabilise soil structure, and loosen up compacted layers are clearly needed. Many conventional tillage practices are effective in loosening topsoil SC, but measures to counteract subsoil SC are scarce [22]. The use of deep-rooted bio-drilling cash and cover crops showed potential for being applicable in European countries. At the Norwegian SS, a cash crop such as alfalfa had good potential for mitigating both topsoil and subsoil SC. However, in using a cover crop such as tillage radish, the Italian SS did not obtain the expected positive outcome for SC. Both SSs experienced methodological difficulties in evaluating the effectiveness of these mitigating strategies. For practical reasons, some measurements could not be conducted at the optimum time and the presence of actively growing bio-drilling crop roots hampered the evaluation of water infiltration and hydraulic conductivity. Further studies are needed for investigating and identifying suitable crop varieties, as shown at the Norwegian SS, where it was difficult to establish the oilseed bio-drilling crop because of a short growing season at this high latitude. This highlights the need for optimising the management and crop growth of bio-drilling species for different pedoclimatic conditions. There is also a need for policymakers to address the economical dimension, as farmers need financial support for adopting deep-rooted bio-drilling cash or cover crops. For example, alfalfa involves low production costs, but it is difficult to find a profitable market, and all types of relevant cover crops are not necessarily covered by current subsidies.

The two case studies with SICSs involving subsoiling were found to be efficient for improving both SC and crop yields only at the Romanian SS. However, it was also found that applying subsoiling every year was time and energy consuming, and the financial benefit for farmers is questionable. There is a need for further evaluating if the subsoiling at this SS could be done only periodically, such as every 3 or 4 years. At the Swedish SS, subsoiling was only done once and using pilot-scale equipment that is still under development. Although there was an improved rooting depth with subsoiling, and a lowering of the PR with the subsoiling treatment with the incorporation of organic material into the upper subsoil, there was no significant effect on crop production. There is a need for long-time studies with this equipment on other crop and soil types, to test other sources of organic materials and, in particular, to examine the effects of repeated subsoiling treatments over time.

The responsiveness of the SICSs investigated in these case studies appeared to be at least partially influenced by inherent soil properties, such as texture, as illustrated by the different responses to no-tillage observed at the UK and Italian SSs. The effect of climate was not evaluated directly since exactly the same SICS was not present at a sufficiently large number of SSs. However, the effect of climate is not negligible. Furthermore, the effect of future climate change might vary between European regions. In northern Europe, greater precipitation is expected during the growing season [12,82], which will lead to a reduction in workable days for field operations [20,112]. The use of heavy machinery under future sub-optimal conditions may further increase the risk for SC, especially in the subsoil [15]. Soil compaction may reduce water infiltration [6,113], which may shorten the growing season and thereby increase the risk of leaching and erosion [9–11]. It is also expected that wetter growing seasons might give greater yield reductions due to subsoil compaction than the drier seasons [114].

In southern Europe, a higher frequency of dry days during the growing season is predicted [82]. Since compacted soil may suffer from poor rooting conditions during drought [114], this could increase the demand for freshwater for irrigation [115]. Deep tillage can be an effective measure to mitigate drought stress and improve the resilience of crops under climate change scenarios in soils by creating a more stable soil structure and alleviating root-restricting layers [89].

The case studies on different SICSs for mitigating SC showed encouraging results as well as several difficulties relating to their implementation and evaluation. Some were pilot studies and need more technical development, and all were short-term studies. More research is needed to refine these SICSs and evaluate their long-term effects at more SSs covering a wider range of pedoclimatic conditions.

Supplementary Materials: The following are available online at https://www.mdpi.com/article/10.3390/land11020223/s1, Table S1: study sites description, Table S2: sampling depth, Table S3: topsoil results at the Norwegian study site, Table S4: results at the Swedish study site, Table S5: statistics at the Italian study site, Figure S1: relationship between soil organic carbon and dry soil bulk density at the Swedish study site Figure S2: saturated hydraulic conductivity at the Italian study site, Figure S3: bulk density at the Romanian study site, Figure S4: water-stable aggregates and saturated hydraulic conductivity by treatment at the Romanian study site, Figure S5: water-stable aggregates and saturated hydraulic conductivity by year at the Romanian study site.

Author Contributions: Conceptualization, I.P. and M.A.B.; formal analysis, I.P., T.S., J.B., O.V., I.C., A.B., G.B., H.K., T.K., F.S., C.S., F.C. and M.A.B.; investigation, I.P., T.S., J.B., O.V., I.C., A.B., G.B., H.K., T.K., F.S., C.S., F.C. and M.A.B.; data curation, I.P., T.S., J.B., O.V., I.C., A.B., G.B., H.K., T.K., F.S., C.S., F.C. and M.A.B.; writing—original draft preparation, I.P. and M.A.B.; writing—review and editing, I.P., T.S., J.B., O.V., I.C., A.B., G.B., H.K., T.K., F.S., C.S., F.C., I.S.P., A.A. and M.A.B. All authors have read and agreed to the published version of the manuscript.

Funding: The research leading to these results received funding from the European Union HORIZON2020 Programme for Research, Technological Development, and Demonstration under grant agreement No. 677407 (SOILCARE Project). Martin A. Bolinder also received financial support provided by the Swedish Farmers' Foundation for Agricultural Research, grant number O-18-23-141.

Institutional Review Board Statement: Not applicable.

Informed Consent Statement: Not applicable.

Data Availability Statement: The data that support the findings of this study are available from the corresponding author upon reasonable request.

Acknowledgments: Sølve Stiauren, Norwegian Agricultural Extension Service (NLR Øst), Kirkenær Norway took care of all the Norwegian practical fieldwork (except soil sampling) in the research field. The Italian study site wishes to acknowledge Riccardo Polese for his contribution to the field experiment. We wish to thank T. Ingelsson (farmer and consultant) for his contribution in adapting the subsoiling equipment used for the Swedish SS, and the Swedish Rural Economy and Agricultural Society for their help in managing the experiment. We acknowledge the technical and administrative personnel, and the many work package members and stakeholders for their efforts and insights. In particular, we thank Rudi Hessel (coordinator) for his leadership throughout the SoilCare project. More information on other SSs and SICSs, methodologies, and fact sheets as well as on other aspects, such as socio-cultural and economic dimensions, can be found at https://www.soilcare-project.eu, last accessed on 1 January 2022.

Conflicts of Interest: The authors declare no conflict of interest.

References

1. Jones, A.; Panagos, P.; Barcelo, S.; Bouraoui, F.; Bosco, C.; Dewitte, O.; Gardi, C.; Hervás, J.; Hiederer, R.; Jeffery, S.; et al. *The State of Soil in Europe: A Contribution of the JRC to the European Environment Agency's Environment State and Outlook Report—SOER 2010*; Europeran Union: Luxembourg, 2012.
2. Crescimanno, G.; Lane, M.; Owens, P.; Rydel, B.; Jacobsen, O.; Düwel, O.; Böken, H.; Berényi Üveges, J.; Castillo, V.; Imeson, A. *Final Report, Working Group on Soil Erosion, Task Group 5: Links with Organic Matter and Contamination Working Group and Secondary Soil Threats*; European Commission: Brussels, Belgium, 2004.
3. Van Camp, L.; Bujarrabal, B.; Gentile, A.; Jones, R.J.A.; Montanarella, L.; Olazábal, C.; Selvaradjou, S. *Reports of the Technical Working Groups Established under the Thematic Strategy for Soil Protection. Vol. IV: Contamination and Land Management*; Office for Official Publications of the European Communities: Luxembourg, 2004.
4. Lipiec, J.; Hatano, R. Quantification of compaction effects on soil physical properties and crop growth. *Geoderma* **2003**, *116*, 107–136. [CrossRef]
5. Lipiec, J.; Simota, C. Role of soil and climate factors in influencing crop responses to soil compaction in Central and Eastern Europe. *Dev. Agric. Eng.* **1994**, *11*, 365–390.
6. Arvidsson, J.; Håkansson, I. Response of different crops to soil compaction-Short-term effects in Swedish field experiments. *Soil Tillage Res.* **2014**, *138*, 56–63. [CrossRef]
7. Horn, R.; Fleige, H. Risk assessment of subsoil compaction for arable soils in Northwest Germany at farm scale. *Soil Tillage Res.* **2009**, *102*, 201–208. [CrossRef]
8. Håkansson, I. A method for characterizing the state of compactness of the plough layer. *Soil Tillage Res.* **1990**, *16*, 105–120. [CrossRef]
9. Hartmann, P.; Zink, A.; Fleige, H.; Horn, R. Effect of compaction, tillage and climate change on soil water balance of Arable Luvisols in Northwest Germany. *Soil Tillage Res.* **2012**, *124*, 211–218. [CrossRef]
10. Keller, T.; Sandin, M.; Colombi, T.; Horn, R.; Or, D. Historical increase in agricultural machinery weights enhanced soil stress levels and adversely affected soil functioning. *Soil Tillage Res.* **2019**, *194*, 104293. [CrossRef]
11. Rogger, M.; Agnoletti, M.; Alaoui, A.; Bathurst, J.C.; Bodner, G.; Borga, M.; Chaplot, V.; Gallart, F.; Glatzel, G.; Hall, J.; et al. Land use change impacts on floods at the catchment scale: Challenges and opportunities for future research. *Water Resour. Res.* **2017**, *53*, 5209–5219. [CrossRef]

12. Olesen, J.E.; Bindi, M. Consequences of climate change for European agricultural productivity, land use and policy. *Eur. J. Agron.* **2002**, *16*, 239–262. [CrossRef]
13. Colombi, T.; Keller, T. Developing strategies to recover crop productivity after soil compaction—A plant eco-physiological perspective. *Soil Tillage Res.* **2019**, *191*, 156–161. [CrossRef]
14. Graves, A.R.; Morris, J.; Deeks, L.K.; Rickson, R.J.; Kibblewhite, M.G.; Harris, J.A.; Farewell, T.S.; Truckle, I. The total costs of soil degradation in England and Wales. *Ecol. Econ.* **2015**, *119*, 399–413. [CrossRef]
15. Batey, T. Soil compaction and soil management—A review. *Soil Use Manag.* **2009**, *25*, 335–345. [CrossRef]
16. Berisso, F.E.; Schjønning, P.; Keller, T.; Lamandé, M.; Etana, A.; De Jonge, L.W.; Iversen, B.V.; Arvidsson, J.; Forkman, J. Persistent effects of subsoil compaction on pore size distribution and gas transport in a loamy soil. *Soil Tillage Res.* **2012**, *122*, 42–51. [CrossRef]
17. Schjønning, P.; Lamandé, M.; Berisso, F.E.; Simojoki, A.; Alakukku, L.; Andreasen, R.R. Gas Diffusion, non-Darcy air permeability, and computed tomography images of a clay subsoil affected by compaction. *Soil Sci. Soc. Am. J.* **2013**, *77*, 1977. [CrossRef]
18. Ruser, R.; Flessa, H.; Russow, R.; Schmidt, G.; Buegger, F.; Munch, J.C. Emission of N_2O, N_2 and CO_2 from soil fertilized with nitrate: Effect of compaction, soil moisture and rewetting. *Soil Biol. Biochem.* **2006**, *38*, 263–274. [CrossRef]
19. Bamgbopa, O.S. Investigation of shrinkage and cracking in clay soils under wetting and drying cycles. *Int. J. Eng. Res. Technol.* **2016**, *5*, 283–320.
20. Seehusen, T.; Mordhorst, A.; Riggert, R.; Fleige, H.; Horn, R.; Riley, H. Subsoil compaction of a clay soil in South-East Norway and its amelioration after 5 years. *Int. Agrophys.* **2021**, *35*, 145–157. [CrossRef]
21. Horn, R.; Domzzał, H.; Słowińska-Jurkiewicz, A.; van Ouwerkerk, C. Soil compaction processes and their effects on the structure of arable soils and the environment. *Soil Tillage Res.* **1995**, *35*, 23–36. [CrossRef]
22. Lebert, M.; Böken, H.; Glante, F. Soil compaction—Indicators for the assessment of harmful changes to the soil in the context of the German Federal Soil Protection Act. *J. Environ. Manag.* **2007**, *82*, 388–397. [CrossRef]
23. Löfkvist, J.; Whalley, W.R.; Clark, L.J. A rapid screening method for good root-penetration ability: Comparison of species with very different root morphology. *Acta Agric. Scand. Sect. B-Soil Plant* **2007**, *55*, 120–124. [CrossRef]
24. Uteau, D.; Pagenkemper, S.K.; Peth, S.; Horn, R. Root and time dependent soil structure formation and its influence on gas transport in the subsoil. *Soil Tillage Res.* **2013**, *132*, 69–76. [CrossRef]
25. Colombi, T.; Braun, S.; Keller, T.; Walter, A. Artificial macropores attract crop roots and enhance plant productivity on compacted soils. *Sci. Total Environ.* **2017**, *574*, 1283–1293. [CrossRef]
26. Kautz, T.; Lüsebrink, M.; Pätzold, S.; Vetterlein, D.; Pude, R.; Athmann, M.; Küpper, P.M.; Perkons, U.; Köpke, U. Contribution of anecic earthworms to biopore formation during cultivation of perennial ley crops. *Pedobiologia* **2014**, *57*, 47–52. [CrossRef]
27. Cresswell, H.P.; Kirkegaard, J.A. Subsoil amelioration by plant-roots—The process and the evidence. *Soil Res.* **1995**, *33*, 221–239. [CrossRef]
28. Chen, G.; Weil, R.R. Penetration of cover crop roots through compacted soils. *Plant Soil* **2010**, *331*, 31–43. [CrossRef]
29. Zhang, Z.; Peng, X. Bio-tillage: A new perspective for sustainable agriculture. *Soil Tillage Res.* **2021**, *206*, 104844. [CrossRef]
30. Abadi Ghadim, A.; Kingwell, R.; Pannell, D. An economic evaluation of deep tillage to reduce soil compaction on crop-livestock farms in Western Australia. *Agric. Syst.* **1991**, *37*, 291–307. [CrossRef]
31. Díaz-Zorita, M. Effect of deep-tillage and nitrogen fertilization interactions on dryland corn (*Zea mays* L.) productivity. *Soil Tillage Res.* **2000**, *54*, 11–19. [CrossRef]
32. Varsa, E.G.; Chong, S.K.; Abolaji, J.O.; Farquhar, D.A.; Olsen, F.J. Effect of deep tillage on soil physical characteristics and corn (*Zea mays* L.) root growth and production. *Soil Tillage Res.* **1997**, *43*, 219–228. [CrossRef]
33. Raper, R.L.; Bergtold, J.S. In-row subsoiling: A review and suggestions for reducing cost of this conservation tillage operation. *Appl. Eng. Agric.* **2007**, *23*, 463–471. [CrossRef]
34. Getahun, G.T.; Kätterer, T.; Munkholm, L.J.; Parvage, M.M.; Keller, T.; Rychel, K.; Kirchmann, H. Short-term effects of loosening and incorporation of straw slurry into the upper subsoil on soil physical properties and crop yield. *Soil Tillage Res.* **2018**, *184*, 62–67. [CrossRef]
35. Hamza, M.A.; Anderson, W.K. Responses of soil properties and grain yields to deep ripping and gypsum application in a compacted loamy sand soil contrasted with a sandy clay loam soil in Western Australia. *Aust. J. Agric. Res.* **2003**, *54*, 273–282. [CrossRef]
36. Adcock, D.; McNeill, A.M.; McDonald, G.K.; Armstrong, R.D.; Adcock, D.; McNeill, A.M.; McDonald, G.K.; Armstrong, R.D. Subsoil constraints to crop production on neutral and alkaline soils in south-eastern Australia: A review of current knowledge and management strategies. *Aust. J. Exp. Agric.* **2007**, *47*, 1245–1261. [CrossRef]
37. Jakobs, I.; Schmittmann, O.; Schulze Lammers, P. Short-term effects of in-row subsoiling and simultaneous admixing of organic material on growth of spring barley (*H. vulgare*). *Soil Use Manag.* **2017**, *33*, 620–630. [CrossRef]
38. Soane, G.C.; Godwin, R.J.; Marks, M.; Spoor, G. Crop and soil response to subsoil loosening, deep incorporation of phosphorus and potassium fertilizer and subsequent soil management on a range of soil types: Part 2: Soil structural conditions. *Soil Use Manag.* **1987**, *3*, 123–130. [CrossRef]
39. Schulte-Karring, H.; Haubold-Rosar, M. Subsoiling and deep fertilizing with new technique as a measure of soil conservation in ag agriculture, viniculture and forestry. *Soil Technol.* **1993**, *6*, 225–237. [CrossRef]

40. Larney, F.J.; Fortune, R.A. Recompaction effects of mouldboard ploughing and seedbed cultivations on four deep loosened soils. *Soil Tillage Res.* **1986**, *8*, 77–87. [CrossRef]
41. Johnson, B.S.; Erickson, A.E.; Voorhees, W.B. Physical conditions of a lake plain soil as affected by deep tillage and wheel traffic. *Soil Sci. Soc. Am. J.* **1989**, *53*, 1545–1551. [CrossRef]
42. Håkansson, I.; Grath, T.; Olsen, H. Influence of machinery traffic in Swedish farm fields on penetration resistance in the subsoil. *Swedish J. Agric. Res.* **1996**, *26*, 181–187.
43. Gill, J.S.; Clark, G.J.; Sale, P.W.; Peries, R.R.; Tang, C. Deep placement of organic amendments in dense sodic subsoil increases summer fallow efficiency and the use of deep soil water by crops. *Plant Soil* **2012**, *359*, 57–69. [CrossRef]
44. Gill, J.S.; Sale, P.W.G.; Peries, R.R.; Tang, C. Changes in soil physical properties and crop root growth in dense sodic subsoil following incorporation of organic amendments. *Field Crop. Res.* **2009**, *114*, 137–146. [CrossRef]
45. Huber, S.; Prokop, G.; Arrouays, D.; Banko, G.; Bispo, A.; Jones, R.J.A.; Kibblewhite, M.G.; Lexer, W.; Möller, A.; Rickson, R.J.; et al. *Environmental Assessment of Soil for Monitoring: Volume I Indicators and Criteria*; OPOCE: Luxembourg, 2008; ISBN 9789279097089.
46. Colombi, T.; Herrmann, A.M.; Vallenback, P.; Keller, T. Cortical cell diameter is key to energy costs of root growth in wheat. *Plant Physiol.* **2019**, *180*, 2049–2060. [CrossRef] [PubMed]
47. Blanchar, R.W.; Edmonds, C.R.; Bradford, J.M. Root growth in cores formed from fragipan and B2 horizons of hobson soil. *Soil Sci. Soc. Am. J.* **1978**, *42*, 437–440. [CrossRef]
48. Cockroft, B.; Barley, K.; Greacen, E. The penetration of clays by fine probes and root tips. *Soil Res.* **1969**, *7*, 333–348. [CrossRef]
49. de Moraes, M.T.; Debiasi, H.; Carlesso, R.; Franchini, J.C.; da Silva, V.R. Critical limits of soil penetration resistance in a rhodic Eutrudox. *Rev. Bras. Ciência Do Solo* **2014**, *38*, 288–298. [CrossRef]
50. Ehlers, W.; Köpke, U.; Hesse, F.; Böhm, W. Penetration resistance and root growth of oats in tilled and untilled loess soil. *Soil Tillage Res.* **1983**, *3*, 261–275. [CrossRef]
51. Imhoff, S.; Kay, B.D.; da Silva, A.P.; Hajabbasi, M.A. Evaluating responses of maize (*Zea mays* L.) to soil physical conditions using a boundary line approach. *Soil Tillage Res.* **2010**, *106*, 303–310. [CrossRef]
52. Kay, B.D.; Hajabbasi, M.A.; Ying, J.; Tollenaar, M. Optimum versus non-limiting water contents for root growth, biomass accumulation, gas exchange and the rate of development of maize (*Zea mays* L.). *Soil Tillage Res.* **2006**, *8*, 42–54. [CrossRef]
53. Taylor, H.M.; Roberson, G.M.; Parker, J.J. Soil strength-root penetration relations for medium- to coarse-textured soil materials. *Soil Sci.* **1966**, *102*, 18–22. [CrossRef]
54. Groenevelt, P.H.; Grant, C.D.; Semetsa, S. A new procedure to determine soil water availability. *Aust. J. Soil Res.* **2001**, *39*, 577–598. [CrossRef]
55. Piccoli, I.; Lazzaro, B.; Furlan, L.; Berti, A.; Morari, F. Examining crop root apparatus traits in a maize-soybean-winter wheat rotation under conservation agriculture management. *Eur. J. Agron.* **2021**, *122*, 126171. [CrossRef]
56. Seehusen, T.; Riggert, R.; Fleige, H.; Horn, R.; Riley, H. Soil compaction and stress propagation after different wheeling intensities on a silt soil in South-East Norway. *Acta Agric. Scand. Sect. B Soil Plant Sci.* **2019**, *69*, 343–355. [CrossRef]
57. Riley, H. Estimation of physical properties of cultivated soils in south east Norway from readily available information, no. 25. *Norw. J. Agric. Sci.* **1996**, 25–51.
58. Green, R.D.; Fordham, S.J. *A Field Method for Determining Air Permeability in Soil*; Ministry of Agriculture, Fisheries and Food: London, UK, 1975.
59. Hartge, K.H. Saturated hydraulic conductivity measurement at soil core samples and its evaluation. *Soil Technol.* **1993**, *6*, 115–121.
60. Seehusen, T.; Børresen, T.; Rostad, B.I.; Fleige, H.; Zink, A.; Riley, H. Verification of traffic-induced soil compaction after long-term ploughing and 10 years minimum tillage on clay loam soil in South-East Norway. *Acta Agric. Scand. Sect. B* **2014**, *64*, 312–328. [CrossRef]
61. Carlgren, K.; Mattsson, L. Swedish soil fertility experiments. *Acta Agric. Scand. Sect. B Soil Plant Sci.* **2001**, *51*, 49–76. [CrossRef]
62. Kirchmann, H.; Eriksson, J. Properties and Classification of Soils of the Swedish Long-Term Fertility Experiments II. Sites at Örja and Orup. *Acta Agric. Scand. Sect. B Soil Plant Sci.* **1993**, *43*, 193–205. [CrossRef]
63. Grossman, R.B.; Reinsch, T.G. 2.1 Bulk density and linear extensibility. In *Methods of Soil Analysis: Part 4 Physical Methods*; Dane, J.H., Topp, C.G., Eds.; Soil Science Society of America: Madison, WI, USA, 2002; pp. 201–228.
64. Bittelli, M.; Andrenelli, M.C.; Simonetti, G.; Pellegrini, S.; Artioli, G.; Piccoli, I.; Morari, F. Shall we abandon sedimentation methods for particle size analysis in soils? *Soil Tillage Res.* **2019**, *185*, 36–46. [CrossRef]
65. Parr, J.F.; Bertrand, A.R. Water infiltration into soils. *Adv. Agron.* **1960**, *12*, 311–363.
66. Philip, J.R. The theory of infiltration: 4. Sorptivity and algebraic infiltration equations. *Soil Sci.* **1957**, *84*, 257–264. [CrossRef]
67. Keller, T.; Håkansson, I. Estimation of reference bulk density from soil particle size distribution and soil organic matter content. *Geoderma* **2010**, *154*, 398–406. [CrossRef]
68. van den Akker, J.J.H.; Hoogland, T. Comparison of risk assessment methods to determine the subsoil compaction risk of agricultural soils in the Netherlands. *Soil Tillage Res.* **2011**, *114*, 146–154. [CrossRef]
69. Poeplau, C.; Reiter, L.; Berti, A.; Kätterer, T. Qualitative and quantitative response of soil organic carbon to 40 years of crop residue incorporation under contrasting nitrogen fertilisation regimes. *Soil Res.* **2017**, *55*, 1–9. [CrossRef]
70. Arvidsson, J.; Håkansson, I. A model for estimating crop yield losses caused by soil compaction. *Soil Tillage Res.* **1991**, *20*, 319–332. [CrossRef]

71. Munkholm, L.J.; Hansen, E.M.; Olesen, J.E. The effect of tillage intensity on soil structure and winter wheat root/shoot growth. *Soil Use Manag.* **2008**, *24*, 392–400. [CrossRef]
72. Whiteley, G.M.; Dexter, A.R. The behaviour of roots encountering cracks in soil. *Plant Soil* **1984**, *77*, 141–149. [CrossRef]
73. Van Noordwijk, M.; Kooistra, M.J.; Boone, F.R.; Veen, B.W.; Schoonderbeek, D. Root-soil contact of maize, as measured by a thin-section technique—I. Validity of the method. *Plant Soil* **1992**, *139*, 109–118. [CrossRef]
74. Lipiec, J.; Håkansson, I. Influences of degree of compactness and matric water tension on some important plant growth factors. *Soil Tillage Res.* **2000**, *53*, 87–94. [CrossRef]
75. McKenzie, B.M.; Bengough, A.G.; Hallett, P.D.; Thomas, W.T.B.; Forster, B.; McNicol, J.W. Deep rooting and drought screening of cereal crops: A novel field-based method and its application. *Field Crop. Res.* **2009**, *54*, 165–171. [CrossRef]
76. Hirth, J.R.; McKenzie, B.M.; Tisdall, J.M. Ability of seedling roots of *Lolium perenne* L. to penetrate soil from artificial biopores is modified by soil bulk density, biopore angle and biopore relief. *Plant Soil* **2005**, *272*, 327–336. [CrossRef]
77. Håkansson, I.; Lipiec, J. A review of the usefulness of relative bulk density values in studies of soil structure and compaction. *Soil Tillage Res.* **2000**, *53*, 71–85. [CrossRef]
78. Lamandé, M.; Schjønning, P. Transmission of vertical stress in a real soil profile. Part I: Site description, evaluation of the Söhne model, and the effect of topsoil tillage. *Soil Tillage Res.* **2011**, *114*, 57–70. [CrossRef]
79. Schjønning, P.; Thomsen, I.K. Shallow tillage effects on soil properties for temperate-region hard-setting soils. *Soil Tillage Res.* **2013**, *132*, 12–20. [CrossRef]
80. Romero-Ruiz, A.; Linde, N.; Baron, L.; Solazzi, S.G.; Keller, T.; Or, D. Seismic signatures reveal persistence of soil compaction. *Vadose Zone J.* **2021**, *20*, e20140. [CrossRef]
81. Piccoli, I.; Furlan, L.; Lazzaro, B.; Morari, F. Examining conservation agriculture soil profiles: Outcomes from northeastern Italian silty soils combining indirect geophysical and direct assessment methods. *Eur. J. Soil Sci.* **2020**, *71*, 1064–1075. [CrossRef]
82. Schjønning, P.; van den Akker, J.J.H.; Keller, T.; Greve, M.H.; Lamandé, M.; Simojoki, A.; Stettler, M.; Arvidsson, J.; Breuning-Madsen, H. Driver-Pressure-State-Impact-Response (DPSIR) analysis and risk assessment for soil compaction-A European perspective. *Adv. Agron.* **2015**, *133*, 183–237.
83. Chamen, T.; Alakukku, L.; Pires, S.; Sommer, C.; Spoor, G.; Tijink, F.; Weisskopf, P. Prevention strategies for field traffic-induced subsoil compaction: A review: Part 2. Equipment and field practices. *Soil Tillage Res.* **2003**, *73*, 161–174. [CrossRef]
84. Barzegar, A.R.; Hashemi, A.M.; Herbert, S.J.; Asoodar, M.A. Interactive effects of tillage system and soil water content on aggregate size distribution for seedbed preparation in Fluvisols in southwest Iran. *Soil Tillage Res.* **2004**, *78*, 45–52. [CrossRef]
85. Pirmoradian, N.; Sepaskhah, A.R.; Hajabbasi, M.A. Application of fractal theory to quantify soil aggregate stability as influenced by tillage treatments. *Biosyst. Eng.* **2005**, *90*, 227–234. [CrossRef]
86. Alcántara, V.; Don, A.; Well, R.; Nieder, R. Deep ploughing increases agricultural soil organic matter stocks. *Glob. Chang. Biol.* **2016**, *22*, 2939–2956. [CrossRef]
87. Soane, B.D.; Ball, B.C.; Arvidsson, J.; Basch, G.; Moreno, F.; Roger-Estrade, J. No-till in northern, western and south-western Europe: A review of problems and opportunities for crop production and the environment. *Soil Tillage Res.* **2012**, *118*, 66–87. [CrossRef]
88. Sartori, F.; Piccoli, I.; Polese, R.; Berti, A. A multivariate approach to evaluate reduced tillage systems and cover crop sustainability. *Land* **2021**, *11*, 55. [CrossRef]
89. Schneider, F.; Don, A.; Hennings, I.; Schmittmann, O.; Seidel, S.J. The effect of deep tillage on crop yield—What do we really know? *Soil Tillage Res.* **2017**, *174*, 193–204. [CrossRef]
90. Verhulst, N.; Govaerts, B.; Verachtert, E.; Mezzalama, M.; Wall, P.C.; Chocobar, A.; Deckers, J.; Sayre, K.D. Conservation agriculture, improving soil quality for sustainable production systems? In *Advances in Soil Science: Food Security and Soil Quality*; CRC Press: Boca Raton, FL, USA, 2010; pp. 137–208. ISBN 139781439800584.
91. Derpsch, R. No-tillage and conservation agriculture: A progress report. In *No-Till Farming Systems*; Goddard, T., Zoebisch, M.A., Gan, Y.T., Ellis, W., Watson, A., Sombatpanit, S., Eds.; World Association of Soil and Water Conservation: Bangkok, Thailand, 2008; pp. 7–39. ISBN 9789748391601.
92. Six, J.; Ogle, S.M.; Breidt, F.J.; Conant, R.T.; Mosiers, A.R.; Paustian, K. The potential to mitigate global warming with no-tillage management is only realized when practised in the long term. *Glob. Chang. Biol.* **2004**, *10*, 155–160. [CrossRef]
93. Dal Ferro, N.; Cocco, E.; Lazzaro, B.; Berti, A.; Morari, F. Assessing the role of agri-environmental measures to enhance the environment in the Veneto Region, Italy, with a model-based approach. *Agric. Ecosyst. Environ.* **2016**, *232*, 312–325. [CrossRef]
94. Piccoli, I.; Schjønning, P.; Lamandé, M.; Furlan, L.; Morari, F. Challenges of conservation agriculture practices on silty soils. Effects on soil pore and gas transport characteristics in North-eastern Italy. *Soil Tillage Res.* **2017**, *172*, 12–21. [CrossRef]
95. Alaoui, A.; Lipiec, J.; Gerke, H.H. A review of the changes in the soil pore system due to soil deformation: A hydrodynamic perspective. *Soil Tillage Res.* **2011**, *115–116*, 1–15. [CrossRef]
96. Hassink, J. The capacity of soils to preserve organic C and N by their association with clay and silt particles. *Plant Soil* **1997**, *191*, 77–87. [CrossRef]
97. Entrup, N.L.; Oehmchen, J. *Lehrbuch des Pflanzenbaues. Band 2: Kulturpflanzen*; Entrup, N.L., Oehmchen, J., Eds.; AgroCooncept GmbH: Bonn, Germany, 2000; ISBN 378620117X.
98. Pagliai, M.; Vignozzi, N.; Pellegrini, S. Soil structure and the effect of management practices. *Soil Tillage Res.* **2004**, *79*, 131–143. [CrossRef]

99. Etana, A.; Håkansson, I. Swedish experiments on the persistence of subsoil compaction caused by vehicles with high axle load. *Soil Tillage Res.* **1994**, *29*, 167–172. [CrossRef]
100. Pagenkemper, S.K.; Puschmann, D.U.; Peth, S.; Horn, R. Investigation of time dependent development of soil structure and formation of macropore networks as affected by various precrop species. *Int. Soil Water Conserv. Res.* **2014**, *2*, 51–66. [CrossRef]
101. Wahlström, E.M.; Kristensen, H.L.; Thomsen, I.K.; Labouriau, R.; Pulido-Moncada, M.; Nielsen, J.A.; Munkholm, L.J. Subsoil compaction effect on spatio-temporal root growth, reuse of biopores and crop yield of spring barley. *Eur. J. Agron.* **2021**, *123*, 126225. [CrossRef]
102. Piccoli, I.; Schjønning, P.; Lamandé, M.; Zanini, F.; Morari, F. Coupling gas transport measurements and X-ray tomography scans for multiscale analysis in silty soils. *Geoderma* **2019**, *338*, 576–584. [CrossRef]
103. Williams, S.M.; Weil, R.R. Crop cover root channels may alleviate soil compaction effects on soybean crop. *Soil Sci. Soc. Am. J.* **2004**, *68*, 1403–1409. [CrossRef]
104. Çerçioğlu, M.; Anderson, S.H.; Udawatta, R.P.; Alagele, S. Effect of cover crop management on soil hydraulic properties. *Geoderma* **2019**, *343*, 247–253. [CrossRef]
105. Calonego, J.C.; Raphael, J.P.A.; Rigon, J.P.G.; Oliveira Neto, L.d.; Rosolem, C.A. Soil compaction management and soybean yields with cover crops under no-till and occasional chiseling. *Eur. J. Agron.* **2017**, *85*, 31–37. [CrossRef]
106. Nielsen, D.C.; Lyon, D.J.; Higgins, R.K.; Hergert, G.W.; Holman, J.D.; Vigil, M.F. Cover crop effect on subsequent wheat yield in the central great plains. *Agron. J.* **2016**, *108*, 243–256. [CrossRef]
107. Acuña, J.C.M.M.; Villamil, M.B. Short-term effects of cover crops and compaction on soil properties and soybean production in Illinois. *Agron. J.* **2014**, *106*, 860. [CrossRef]
108. Blanco-Canqui, H.; Shaver, T.M.; Lindquist, J.L.; Shapiro, C.A.; Elmore, R.W.; Francis, C.A.; Hergert, G.W. Cover crops and ecosystem services: Insights from studies in temperate soils. *Agron. J.* **2015**, *107*, 2449–2474. [CrossRef]
109. Bolinder, M.A.; Crotty, F.; Elsen, A.; Frac, M.; Kismányoky, T.; Lipiec, J.; Tits, M.; Tóth, Z.; Kätterer, T. The effect of crop residues, cover crops, manures and nitrogen fertilization on soil organic carbon changes in agroecosystems: A synthesis of reviews. *Mitig. Adapt. Strateg. Glob. Chang.* **2020**, *256*, 929–952. [CrossRef]
110. Kätterer, T.; Bolinder, M.A.; Berglund, K.; Kirchmann, H. Strategies for carbon sequestration in agricultural soils in northern Europe. *Acta Agric. Scand. Sect. A—Anim. Sci.* **2012**, *62*, 181–198. [CrossRef]
111. Smith, P. How long before a change in soil organic carbon can be detected? *Glob. Chang. Biol.* **2004**, *10*, 1878–1883. [CrossRef]
112. Seehusen, T.; Waalen, W.; Hoel, B.; Uhlen, A.K.; Persson, T.; Strand, E. *Landbruket i Møte Med Klimaendringen—Effekter av Endret Klima og Behov for Tilpasninger- Norsk Kornproduksjon—Vedlegg til Hovedrapporten: Utredninger om Landbrukets Utfordringer i Møte Med Klimaendringene—Fagnotater Som Underlag for Arbeidsgruppens*; Landbruks og Matdepartement: Oslo, Norway, 2016.
113. Alaoui, A.; Rogger, M.; Peth, S.; Blöschl, G. Does soil compaction increase floods? A review. *J. Hydrol.* **2018**, *557*, 631–642. [CrossRef]
114. Liu, H.; Colombi, T.; Jäck, O.; Keller, T.; Weih, M. Effects of soil compaction on grain yield of wheat depend on weather conditions. *Sci. Total Environ.* **2022**, *807*, 150763. [CrossRef] [PubMed]
115. Vörösmarty, C.J.; Green, P.; Salisbury, J.; Lammers, R.B. Global water resources: Vulnerability from climate change and population growth. *Science* **2000**, *289*, 284–288. [CrossRef] [PubMed]

land

MDPI

Article

Long-Term Effect of Pig Slurry and Mineral Fertilizer Additions on Soil Nutrient Content, Field Pea Grain and Straw Yield under Winter Wheat–Spring Barley–Field Pea Crop Rotation on Cambisol and Luvisol

Lukáš Hlisnikovský *, Ladislav Menšík, Pavel Čermák, Kateřina Křížová and Eva Kunzová

Crop Research Institute, 16106 Prague, Czech Republic; ladislav.mensik@vurv.cz (L.M.); pavel.cermak@vurv.cz (P.Č.); krizovak@vurv.cz (K.K.); kunzova@vurv.cz (E.K.)
* Correspondence: l.hlisnik@vurv.cz; Tel.: +420-233-022-248

Abstract: Different fertilizers have different effects on soil chemistry and crop yields. In this paper, we analyzed how long-term and regular application of mineral fertilizers, pig slurry and their combinations (15 fertilizer treatments totally) affect soil pH, nutrient content and yield of field pea at two sites with different soil (cambisol and luvisol) and climatic conditions. The long-term trials evaluated in this paper were established in 1972 at Pernolec and Kostelec, Czech Republic. Results of the soil analyses (evaluated period) are from the years 2015–2020, covering two sequences of crop rotation (winter wheat–spring barley–field pea). The fertilizer treatments significantly affected the soil reaction; application of mineral fertilizers and their combinations resulted in the lowest pH values. On the other hand, the same treatments provided the highest yields and left the highest pool of nutrients in the soil. Pig slurry can provide the same yields of field pea as mineral NPK fertilizers, without a negative effect on soil reaction. Analyzing the mineral fertilizers only, a reasonable dose of N (according to the linear-plateau model) can range from 73 and 97 kg ha^{-1} N in Pernolec, according to the weather conditions.

Keywords: *Pisum sativum* L.; organic manure; NPK; pH; SOM; macronutrients; nutrient content

Citation: Hlisnikovský, L.; Menšík, L.; Čermák, P.; Křížová, K.; Kunzová, E. Long-Term Effect of Pig Slurry and Mineral Fertilizer Additions on Soil Nutrient Content, Field Pea Grain and Straw Yield under Winter Wheat–Spring Barley–Field Pea Crop Rotation on Cambisol and Luvisol. *Land* **2022**, *11*, 187. https://doi.org/10.3390/land11020187

Academic Editors: Guido Wyseure, Julián Cuevas González, Jean Poesen and Angelinus Franke

Received: 3 December 2021
Accepted: 21 January 2022
Published: 25 January 2022

1. Introduction

The nutrient content of the soil is one of the parameters determining its fertility and quality. It is a parameter influenced by a wide range of natural, anthropogenic and interrelated factors such as soil type [1], farming method (conventional, organic farming) [2], crop rotations and fertilization [3–5], microbial activity in the soil [6], or soil organic matter content [7]. The application of fertilizers represents the main way of supplying nutrients to the soil; for the crops grown, fertilizers thus directly affect soil chemical [8–11], physiological [12,13] and biological [14,15] properties and crop growth.

Fertilizers are divided into three categories, namely mineral and organic fertilizers and organic manures. They differ in origin, composition and nutrient content, speed of nutrient release and availability to farmers. Mineral fertilizers are fast-acting and have a precisely defined composition, which makes it easier to adjust the dose of nutrients delivered. On the other hand, they are costly and, if used unwisely, can pose a significant threat to the environment [16,17] or arable products [18]. In particular, the effect of nitrogen mineral fertilizers on soil pH poses a risk of acidification [19–22] and a risk to elements' availability [23]. Manure fertilizers have a low nutrient content and must be applied in large doses (the classic dose of cow farmyard manure is 40 t ha^{-1} in Czech Republic). The nutrients contained in manure are released gradually, depending on the origin [24] and C:N ratio. Manures with a low C:N ratio (slurries) release nutrients to a greater extent already in the first year of application; on the other hand, manures with a high C:N ratio (farmyard manure) release nutrients over a longer period in smaller doses [25]. According to [26], approximately 11% of the organic N is mineralized during the first year from the application of composted manure and around 20% for non-composted manure. In the case of slurries, approximately 40% of the organic N is mineralized at the same time [25]. The application of organic manures is usually associated with positive effects on soil properties [7,8,27–29], but one has to be careful about the doses and dry matter content. In the case of slurry, the dry matter and nutrient content is very important information in order to correctly adjust the applied dose. Ignorance of this information can easily lead to overdosing, which can significantly damage the crops grown or adversely affect the environment via leaching and volatilization of nitrogen and salinization [30–33]. In addition to directly supplied nutrients, the unifying factor for the positive effects of organic manures and the nutrient content, pH value, physical and biological properties of soils, is organic matter. Soil organic carbon and nutrient content are usually higher after application of solid organic manures [34–36], while the benefit of liquid organic manures, such as slurries, is mainly to increase the nutrient content and the effect on soil organic carbon can be neutral (no changes) [37] or positive [38,39] as the liquid manures contain a lower amount of organic carbon than the solid. As in the case of mineral fertilizers, organic manures can also pose a threat to the environment if not applied judiciously [40,41] or because of the presence of pharmaceuticals [42].

One of the major problems of agriculture in Czech Republic is the disruption between crop and livestock production, reduction of cultivated crops in crop rotations and the fact that most of the arable land is rented [43]. Disruption of the balance between crop and livestock production is manifested by a lack of organic manure and reduced input of organic carbon into the soil. Together with the significant dependence of crop production on mineral nitrogen (and the low level of phosphate and potassium fertilizers applied), we then experience soil erosion (lack of organic carbon), lower content of macronutrients (doses of P and K mineral fertilizers) and soil acidification (due to nitrogen fertilizers application) [44,45]. One way to reduce the negative impact of mineral nitrogen fertilizers on the soil while ensuring good soil nutrient supply and crop yields is to apply mineral and manure fertilizers together. Multiple scientific papers have indicated that joint application of mineral fertilizers and manure has a positive effect on both crop yields and reduction of negative impacts of mineral fertilizers on the soil properties [46–50]. Another problem of Czech agriculture is the reduction of crops in crop rotation. Over the years, there has been a change in the proportion of crops grown, mainly in favor of winter rape. Soil-improving crops such as root crops, forage crops and legumes are grown to a lesser extent than in the past [51]. While root crops (potatoes and sugar beet) are considered as soil-improving plants mainly due to the manure applied to them, legumes have a unique ability to fix airborne nitrogen in the soil, due to their symbiosis with rhizobacteria. Field pea (FP) is the most cultivated legume in Czech Republic (79% of all legumes), yet its representation in the crop rotations of Czech Republic is low (1.2%) (average values from 2015 to 2019 [52]). From the point of view of human nutrition and soil care, it is a valuable crop. Thanks to their symbiosis with rhizobia bacteria, legumes and FP cover a large part of their nitrogen needs from the symbiosis (depending on the type of legumes, they cover their nitrogen requirement from the soil from approximately 15-30%) and leave nitrogen in the soil for use by subsequent crops [53,54]. Although FP can use nitrogen from symbiosis with rhizobia bacteria, fertilizer application significantly affects its yield and quality. Foliar application of phosphorus can significantly improve yield and quality parameters of FP, especially on soils with low phosphorus content [55], but even on soils rich in P content, it has a positive effect on FP yields [56]. N fertilization can also increase yield and quality. Early application of N fertilizers is important, as the actual fixation of airborne nitrogen takes place only in the later stages of growth. Depending on soil and climate conditions, optimum N rates can range from 40 to 80 kg ha^{-1} N and higher doses can provide lower yields as high N doses can reduce bacteria nodule mass [57]. However, under different soil and climate conditions the response of FP to N fertilization may be different as yields can increase up to the dose of 135 kg ha^{-1} N [58].

In 1972, long-term experiments were set up at two sites with different soil and climatic conditions to study the effect of the application of organic manure (pig slurries), mineral fertilizers and their combinations on soil chemistry and yields of wheat, barley and peas. The design of this experiment allows us to analyze the long-term effect of different fertilizer combinations on soil properties, which is currently a hot topic due to the dependence of conventional agriculture on mineral nitrogen, the low rates of applied P and K fertilizers and the limited availability of organic manures (slurries). In other words, our experiment can provide answers on how to take better care of the soil with the help of organic manure and how to avoid undesirable effects of mineral nitrogen applied without organic manure (current situation in Czech Republic). Soil types are represented by Cambisol (about 45% of the soil in Czech Republic) and Luvisol (about 13% of the soil in Czech Republic), representing the two most widespread soil types in Czech Republic. The article includes an analysis of the effect of fertilization on pea yields in 2017 and 2020 in Pernolec and the determination of a reasonable dose (using a linear-plateau model) of mineral nitrogen fertilization.

2. Materials and Methods

2.1. General Information and Sites Description

The results come from two long-term field trials located at Pernolec and Kostelec, Czech Republic, Central Europe. Both trials were established in 1972. The long-term trials aim to analyze the effect of mineral fertilizers (mineral nitrogen–N, phosphorus–P and potassium–K), pig slurry (three different doses), and their combination on the yield of arable crops. The crop rotation of both trials consists of winter wheat (*Triticum aestivum* L., WW), spring barley (*Hordeum vulgare* L., SB) and field pea (*Pisum sativum* L., FP). At the same time, the effect of long-term fertilizer application on basic soil properties is monitored (pH, the concentration of soil P, K, magnesium–Mg, calcium–Ca, the content of soil organic carbon–Cox, the content of soil total N–Ntot). In this paper, we assessed the period from 2015 to 2020 (six years) to analyze how long-term regular application of mineral fertilizers and organic manures affects soil properties and FP yields (yields from the years 2017 and 2020; 2015–WW, 2016–SB, 2017–FP, 2018–WW, 2019–SB, 2020–FP).

According to Köpper–Geiger climate classification [59], both sites are located in warm summer humid continental climate (Dfb). The basic site description of both localities is shown in Table 1. Detailed weather information can be found in Section 2.3. It should be noted here that our team was not the team that established the experiments in 1972 and we have not been able to find the results of soil analyses from the period of the trial establishment.

Table 1. The description of trial sites—Pernolec and Kostelec.

	Pernolec	Kostelec
GPS	N 49°46′ E 12°41′	N 50°12′ E 16°20′
Altitude (m a.s.l.)	530	290
Long-term average total annual precipitation (mm)	557	714
Long-term average annual temperature (°C)	7.5	8.5
Soil type [60]	Sandy loam, gleyiccambisol	luvisol
Top soil layer (cm)	0–28	0–30
Precipitation 2017 (mm)	423	774
Precipitation 2020 (mm)	544	961
Temperature 2017 (°C)	8.4	9.3
Temperature 2020 (°C)	9.0	10.0

Note: the long-term average precipitation and temperature for Pernolec are based on the data from the years 1977–2014 (37 years) and for Kostelec from the years 1982–2014 (32 years).

2.2. Field Trials Description

In both long-term trials, the effect of a total of fifteen fertilization treatments with four replications has been running since 1972. The trial consists of sixty plots (15 × 4) arranged in a completely randomized block design. The plot size is 8 m × 5.5 m (44 m^2). The fertilization treatments are identical in both trials, but the fertilization rates differ slightly (Tables 2 and 3 show fertilizer treatments and rates applied to FP in Pernolec and Kostelec. Tables S1 and S2 show the fertilizer treatments and rates applied over the whole three-year crop rotation—the sum of nutrients applied to all three crops over the three years). Mineral N was applied in two forms. Ammonium sulfate (AS) was applied in the spring, before the planting. Calcium ammonium nitrate (CAN) was applied during the beginning of stem elongation (BBCH 30). Mineral P was applied as superphosphate and mineral K as potassium sulfate. Both mineral fertilizers and PS were applied in the autumn, before tillering. Mineral fertilizers were spread by hand at both sites. Pig slurries were applied by manual sprayer. The average content of dry matter (DM) ranged from 0.68% (Pernolec) to 1.8% (Kostelec). This is a very low value: the amount of DM in slurry usually ranges from 0.7% to 24% [61] and is significantly affected by the season of the year [62]. Quality slurry is considered to have a dry matter content between 6% and 8%. The average pH and concentrations of N, P, K, Ca and Mg (% of DM) in Pernolec were 7.75, 1.79%, 0.52%, 16.77%, 1.11%, and 0.73%, respectively. In Kostelec, the average pH value of PS was 7.68 and the concentrations of N, P, K, Ca and Mg (% of DM) were 1.95%, 1.53%, 14.53%, 3.84% and 0.98%, respectively. Pig slurries were obtained from the nearest livestock farms that were able to supply manure in time. The FP (cul. Eso) was sown at the beginning of April (one million germinating seeds per ha, approximately 270 kg) and harvested in the first half of August.

Table 2. Forms and doses of mineral fertilizers and pig slurry (PS) according to the fertilizer treatments applied to FP in Pernolec.

	N (kg ha^{-1})		PS		P (kg ha^{-1})	K (kg ha^{-1})
	AS	CAN	t ha^{-1}	N (kg ha^{-1})		
Control	0	0	0	0	0	0
PK	0	0	0	0	19.8	132.8
NPK	30	0	0	0	19.8	132.8
PS1	0	0	17	85	0	0
PS1+PK	0	0	17	85	19.8	132.8
PS1+NPK	30	0	17	85	19.8	132.8
PS2	0	0	34	170	0	0
PS2+PK	0	0	34	170	19.8	132.8
PS2+NPK	30	0	34	170	19.8	132.8
PS3	0	0	51	255	0	0
PS3+PK	0	0	51	255	19.8	132.8
PS3+NPK	30	0	51	255	19.8	132.8
NPK E1	70	0	0	0	15	25
NPK E2	70	70	0	0	30	50
NPK E3	45	75	0	0	45	75

Note: AS–ammonium sulphate; CAN–calcium ammonium nitrate; N (kg ha^{-1}) in the PS column represents the content of N applied in PS.

2.3. Weather Information

Weather data (average monthly temperatures and monthly precipitation) were evaluated according to [63], which describes the World Meteorological Organization's recommendations for describing meteorological and climatological conditions of a defined period (text in Czech, tables in English). The weather analysis was based on long-term records. In Pernolec we compared the years 2017 and 2020 with the period from 1977 to 2014 (37 years). In Kostelec, we based our analysis on the period from the years 1982 to 2014 (32 years).

In Pernolec, the year 2017 was evaluated as warm (+0.9 °C in comparison with long–term average) and 2020 as very warm (+1.5 °C). In terms of precipitation, 2017 was a very dry year (76% of the long–term average), while 2020 was normal (98%). In Kostelec, the year 2017 was evaluated as warm (+0.8 °C) and 2020 as very warm (+1.5 °C). In terms of precipitation, 2017 was a normal year (109%), while 2020 was a very wet year (135%, Table 1). Detailed weather information for 2017 and 2020 at both sites, including assessments, is provided in Tables S3 and S4.

Table 3. Forms and doses of mineral fertilizers and pig slurry (PS) according to the fertilizer treatments applied to FP in Kostelec.

	N (kg ha^{-1})		Pig slurry		P (kg ha^{-1})	K (kg ha^{-1})
	AS	CAN	t ha^{-1}	N (kg ha^{-1})		
Control	0	0	0	0	0	0
PK	0	0	0	0	19.8	132.8
NPK	30	30	0	0	19.8	132.8
PS1	0	0	20	70	0	0
PS1+PK	0	0	20	70	19.8	132.8
PS1+NPK	30	30	20	70	19.8	132.8
PS2	0	0	40	140	0	0
PS2+PK	0	0	40	140	19.8	132.8
PS2+NPK	30	30	40	140	19.8	132.8
PS3	0	0	60	210	0	0
PS3+PK	0	0	60	210	19.8	132.8
PS3+NPK	30	30	60	210	19.8	132.8
NPK E1	30	60	0	0	18	28
NPK E2	30	60	0	0	36	56
NPK E3	30	60	0	0	54	84

Note: AS–ammonium sulphate; CAN–calcium ammonium nitrate; N (kg ha^{-1}) in the PS column represents the content of N applied in PS.

2.4. Soil Analyses

Following the harvest of the crops, soil samples were taken using the stainless-steel soil probe sampler. The soil samples were taken from the topsoil layer (0–20 cm). Four samples from each plot were taken. The samples were then mixed and transported to the laboratory, where they were dried and sieved to get fine and dry soil. The soil pH was analyzed potentiometrically using 0.2 mol KCl (inoLab pH 730, WTW, Xylem Analytics, Weilheim, Germany). The concentration of total N (Ntot) was analyzed using sulfuric acid in the heating block (Tecator, Foss Analytics, Hillerød, Denmark), followed by the Kjeldahl method [64]. The concentrations of P, K, Ca and Mg were analyzed using the Mehlich III solution [65], followed by the ICP-OES analysis (Thermo Scientific iCAP 7400 Duo, Thermo Fisher Scientific, Cambridge, UK). The SOC content was analyzed colorimetrically and via oxidimetric titration according to [66,67].

2.5. Data Analyses

One-way and multivariate analysis of variance (ANOVA, MANOVA) was used to compare the results of pH and soil element concentrations as affected by fertilization treatments and to analyze the effect of weather and fertilization treatments on FB yields (Pernolec locality only). Due to the occurrence of certain problems, we have only the summed FP yield values (average values without repeats) from the Kostelec site. For this reason, it was not possible to perform statistical analysis as in the case of the Pernolec. However, the average FP yield values from the Kostelec were suitable for PCA. FP yields from the Kostelec site are shown in Table S5. In this article, we have analyzed a total of fifteen fertilization treatments. Such a large set makes the interpretation of the results difficult and ambiguous (the results of the post hoc analysis overlap widely). For this reason, we proceeded to group the treatments (Control, PK, NPK, PS, PS+PK, PS + NPK) and calculate separate ANOVA for soil parameters where significant differences between the fertilizer treatments were recorded previously. If statistically significant differences were found, we used Tukey's HSD post hoc analysis to separate treatments. Statistical analyses were performed in Statistica 13.3. (Tibco Software Inc. Palo Alto, CA, USA). The nitrogen use efficiency (NUE) was calculated as ((GY_T-GY_C)/N rate) where GY_T represents grain yield from the particular fertilizer treatment and GY_C represents the grain yield from the Control treatment. The NUE was calculated from seven fertilizer treatments (NPK, PS1, PS1+NPK, PS2, PS2+NPK, PS3, and PS3+NPK. To evaluate the relationships between the yields, fertilizer treatments and soil parameters, principal component analysis (PCA) and factor analysis (FA) were used (Statistica 14.0.). MS Excel 2019 was used for weather analyses (Microsoft Corporation, Washington, DC, USA). The linear-plateau model, analyzing the reasonable N dose for FP (calculated from mineral fertilizer treatments), was calculated using R software (R: A language and environment for statistical computing. R Foundation for Statistical Computing, Vienna, Austria, 2020), together with three R packages [68–70].

3. Results

3.1. Comparison of Localities

The two sites are statistically significantly different from each other in all observed soil parameters (results from all fertilizer treatments and for the whole period 2015–2020, Table 6). Compared to Kostelec, the soil in Pernolec is characterized by a higher pH value, and lower mean content of available P, K and Ca. In contrast, the average content of Mg, Cox and Ntot is higher in Pernolec.

3.2. The Effect of Fertilizer Treatments on Soil Chemical Properties

In the following sections the results of the effect of fertilization on pH, nutrient concentration, Cox and Ntot at each site will be presented. A summary description of the relationships between the fertilization treatments and the individual parameters is given in the last section, in which the PCA results are presented.

3.2.1. Soil Reaction

In Pernolec (Cambisol), the soil pH was statistically significantly affected by the fertilizer treatment (d.f. = 14; F = 10.6; $p < 0.001$). Comparing all 15 treatments, the lowest mean pH value (4.73) was recorded in the NPK E3 treatment. The highest mean pH value (5.93) was recorded in PK treatment (Table 4). Comparing the groups of fertilizers, the lowest mean pH was recorded in NPK treatments (5.16), followed by PS+NPK (5.48), while the highest pH was recorded in PS+PK (5.81) and PK (5.93) treatments (Figure 1a).

In Kostelec (Luvisol), the value of the pH was also significantly affected by the fertilizer treatment (d.f. = 14; F = 4.2; $p < 0.001$). The lowest mean pH value was recorded in PS2 + NPK (5.04) treatments, while the highest was in Control (5.75) and PK (5.67) treatments (Table 5). Comparing the groups of fertilizer treatments, similarly to Pernolec, the lowest mean pH values were recorded in treatments with mineral N–NPK (5.33) and PS+NPK (5.14), while the highest value was recorded in Control (5.75) treatment (Figure 1b).

The results show that the application of NPK, either alone or in combination with PS, results in the lowest pH values. In Kostelec, the pH values for the NPK and PS+NPK treatments were comparable and significantly different from the other treatments. In Pernolec, the effect of NPK was most significant, while the combined application of PS+NPK was comparable to PS, yet lower. The negative effect of ammonium nitrogen on pH is particularly noticeable when compared to the PK treatment (Figure 1).

3.2.2. Phosphorus

The concentration of P in the soil was not affected by the long-term application of slurry and mineral fertilizers in Pernolec (d.f. = 14; F = 0.6; $p = 0.84$). The lowest mean concentration was recorded in Control (58 mg kg^{-1}), and the highest in PS3+PK treatment (111 mg kg^{-1})

(Table 4). A different situation occurred in Kostelec, where differences between fertilizer treatments were significant (d.f. = 14; F = 16.47; $p < 0.001$). As in Pernolec, the lowest concentration was recorded in Control (124 mg kg^{-1}), and the highest in PS3+NPK (262 mg kg^{-1}) treatment (Table 5). Comparing the groups of fertilizers, ANOVA separated three groups of fertilizers according to their effect on soil P concentration in Kostelec (Figure 2a). The lowest mean concentration was recorded in Control (124 mg kg^{-1}), followed by NPK, PK and PS treatments. The combined application of PS+NPK and PS+PK resulted in the highest mean P concentrations, ranging from 229 to 235 mg kg^{-1} (Figure 2a).

Table 4. Soil pH value, the concentration of P, K, Ca and Mg (mg kg^{-1}), the content of organic carbon (Cox, %) and total nitrogen (Nt, %) as affected by the fertilizer treatments (2015–2020) in Pernolec.

	pH	P	K	Ca	Mg	Cox	Nt
Control	5.69 ± 0.08 C–E	58 ± 11	120 ± 5 A	1356 ± 36 B–D	114 ± 8	0.89 ± 0.02	0.11 ± 0.01
PK	5.93 ± 0.08 E	87 ± 13	205 ± 10 C	1331 ± 27 B–D	119 ± 11	0.88 ± 0.03	0.11 ± 0.01
NPK	5.52 ± 0.08 B–E	83 ± 15	184 ± 9 BC	1293 ± 49 A–D	111 ± 11	0.96 ± 0.03	0.12 ± 0.01
PS1	5.76 ± 0.08 CD	76 ± 22	147 ± 4 AB	1447 ± 29 D	131 ± 10	0.88 ± 0.04	0.11 ± 0.01
PS1+PK	5.89 ± 0.10 E	97 ± 23	208 ± 10 C	1371 ± 56 CD	128 ± 12	0.94 ± 0.05	0.12 ± 0.01
PS1+NPK	5.73 ± 0.10 CD	98 ± 22	215 ± 11 C	1282 ± 52 A–D	124 ± 9	0.92 ± 0.02	0.12 ± 0.01
PS2	5.66 ± 0.08 B–E	76 ± 14	132 ± 4 A	1288 ± 52 A–D	124 ± 10	0.89 ± 0.04	0.11 ± 0.01
PS2+PK	5.70 ± 0.06 CD	94 ± 16	191 ± 9 C	1287 ± 39 A–D	127 ± 11	0.88 ± 0.04	0.11 ± 0.01
PS2+NPK	5.34 ± 0.10 B–D	94 ± 17	193 ± 10 C	1176 ± 34 A–C	113 ± 9	0.99 ± 0.05	0.13 ± 0.01
PS3	5.68 ± 0.10 C–E	93 ± 15	138 ± 4 A	1329 ± 53 B–D	125 ± 10	0.91 ± 0.05	0.12 ± 0.01
PS3+PK	5.83 ± 0.10 CD	111 ± 19	198 ± 6 C	1366 ± 55 B–D	130 ± 11	0.93 ± 0.05	0.12 ± 0.01
PS3+NPK	5.38 ± 0.12 B–D	101 ± 23	196 ± 11 C	1252 ± 32 A–D	121 ± 10	1.02 ± 0.03	0.12 ± 0.01
NPK E1	5.19 ± 0.10 A–C	76 ± 9	116 ± 6 A	1166 ± 19 AB	115 ± 11	0.94 ± 0.06	0.12 ± 0.01
NPK E2	5.18 ± 0.12 AB	78 ± 10	126 ± 8 A	1191 ± 50 A–C	118 ± 11	0.96 ± 0.05	0.12 ± 0.01
NPK E3	4.73 ± 0.17 A	82 ± 13	125 ± 7 A	1113 ± 29 A	104 ± 11	0.97 ± 0.06	0.12 ± 0.01

Mean values (±SE) followed by the same letter (a vertical comparison of the effect of fertilizer treatment) are not statistically significantly different. Columns without letters (P, Mg, Cox, Nt) represent values without statistically significant differences, where the effect of fertilizer treatments was insignificant.

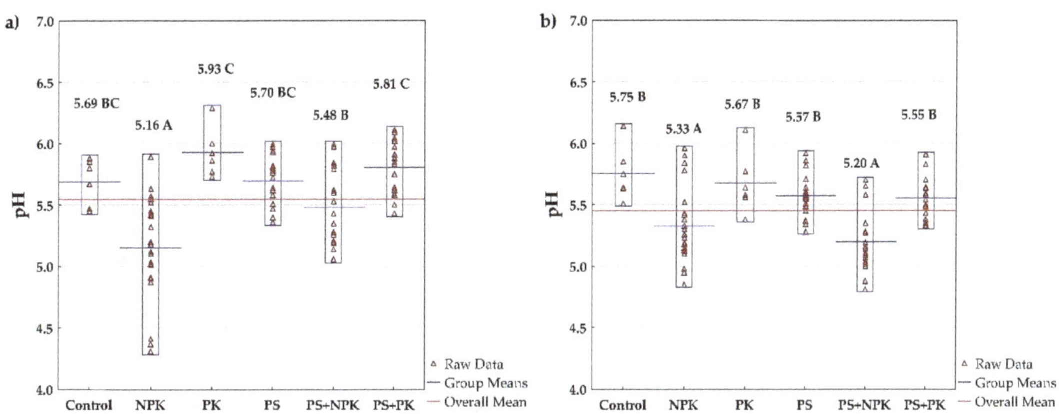

Figure 1. The effect of fertilizer treatments on pH in (**a**) Pernolec and (**b**) Kostelec (2015–2020). Mean values followed by the same letter are not significantly different. Red triangles represent the raw data. The blue lines represent the mean value of the particular treatment, while the red line represents the mean value calculated from all fertilizer treatments.

Table 5. Soil pH value, the concentration of P, K, Ca and Mg (mg kg^{-1}), the content of organic carbon (Cox, %) and total nitrogen (Nt, %) as affected by the fertilizer treatments (2015–2020) in Kostelec.

	pH	P	K	Ca	Mg	Cox	Nt
Control	5.75 ± 0.09 ^{C}	124 ± 7 ^{A}	113 ± 7 ^{A}	1549 ± 31 ^{D}	73 ± 2 $^{A–D}$	0.82 ± 0.03	0.10 ± 0.01
PK	5.67 ± 0.10 ^{C}	170 ± 7 $^{A–D}$	200 ± 8 $^{D–H}$	1430 ± 44 $^{A–D}$	64 ± 3 ^{AB}	0.77 ± 0.03	0.10 ± 0.01
NPK	5.34 ± 0.12 $^{A–C}$	174 ± 8 $^{B–D}$	187 ± 10 $^{C–G}$	1341 ± 64 $^{A–D}$	58 ± 3 ^{A}	0.80 ± 0.03	0.10 ± 0.01
PS1	5.59 ± 0.06 ^{BC}	194 ± 7 $^{C–F}$	168 ± 7 $^{B–E}$	1465 ± 41 $^{B–D}$	80 ± 2 ^{CD}	0.81 ± 0.03	0.11 ± 0.01
PS1+PK	5.51 ± 0.08 $^{A–C}$	225 ± 6 $^{E–H}$	234 ± 13 ^{GH}	1330 ± 46 $^{A–D}$	73 ± 3 $^{B–D}$	0.78 ± 0.03	0.11 ± 0.01
PS1+NPK	5.27 ± 0.09 $^{A–C}$	231 ± 7 $^{F–H}$	221 ± 11 $^{F–H}$	1277 ± 27 $^{A–C}$	68 ± 2 $^{A–C}$	0.81 ± 0.03	0.11 ± 0.01
PS2	5.57 ± 0.10 ^{BC}	174 ± 8 $^{B–D}$	151 ± 9 $^{A–D}$	1468 ± 41 $^{B–D}$	79 ± 2 ^{CD}	0.82 ± 0.03	0.11 ± 0.01
PS2+PK	5.52 ± 0.09 $^{A–C}$	216 ± 12 $^{D–H}$	228 ± 8 ^{GH}	1355 ± 41 $^{A–D}$	72 ± 3 $^{A–D}$	0.79 ± 0.02	0.11 ± 0.01
PS2+NPK	5.04 ± 0.20 ^{A}	214 ± 9 $^{D–G}$	207 ± 10 $^{E–H}$	1213 ± 48 ^{A}	65 ± 3 $^{A–C}$	0.82 ± 0.03	0.11 ± 0.01
PS3	5.57 ± 0.05 ^{BC}	197 ± 10 $^{C–F}$	170 ± 11 $^{B–F}$	1476 ± 52 $^{B–D}$	83 ± 3 ^{D}	0.84 ± 0.04	0.11 ± 0.01
PS3+PK	5.63 ± 0.05 ^{BC}	246 ± 13 ^{GH}	252 ± 14 ^{H}	1486 ± 54 ^{CD}	82 ± 4 ^{D}	0.83 ± 0.02	0.11 ± 0.01
PS3+NPK	5.12 ± 0.10 ^{AB}	262 ± 14 ^{H}	237 ± 18 ^{GH}	1249 ± 29 ^{AB}	75 ± 4 $^{B–D}$	0.87 ± 0.04	0.11 ± 0.01
NPK E1	5.39 ± 0.12 $^{A–C}$	139 ± 7 ^{AB}	112 ± 8 ^{A}	1443 ± 60 $^{A–D}$	68 ± 3 $^{A–C}$	0.80 ± 0.02	0.10 ± 0.01
NPK E2	5.44 ± 0.10 $^{A–C}$	164 ± 9 $^{A–C}$	135 ± 7 ^{AB}	1409 ± 53 $^{A–D}$	64 ± 2 ^{AB}	0.79 ± 0.03	0.11 ± 0.01
NPK E3	5.14 ± 0.14 ^{AB}	183 ± 12 $^{B–E}$	146 ± 10 $^{A–C}$	1303 ± 54 $^{A–C}$	61 ± 3 ^{AB}	0.81 ± 0.01	0.11 ± 0.01

Mean values (±SE) followed by the same letter (a vertical comparison of the effect of fertilizer treatment) are not statistically significantly different. Columns without letters (P, Mg, Cox, Nt) represent values without statistically significant differences, where the effect of fertilizer treatments was insignificant.

Table 6. Average values of soil parameters in Pernolec and Kostelec. The values are based on the results of soil analyses of all fertilization treatments and all analyzed years (2015–2020).

	F	d.f.	*p*	Pernolec	Kostelec
pH	4.33	1	<0.05	5.55 ± 0.04 ^{B}	5.44 ± 0.03 ^{A}
P (mg kg^{-1})	302	1	<0.001	87 ± 4 ^{A}	194 ± 5 ^{B}
K (mg kg^{-1})	6.80	1	<0.01	166 ± 4 ^{A}	184 ± 5 ^{B}
Ca (mg kg^{-1})	26	1	<0.001	1283 ± 13 ^{A}	1386 ± 15 ^{B}
Mg (mg kg^{-1})	312	1	<0.001	120 ± 3 ^{B}	71 ± 1 ^{A}
Cox (%)	76	1	<0.001	0.93 ± 0.01 ^{B}	0.81 ± 0.01 ^{A}
Ntot (%)	25	1	<0.001	0.12 ± 0.01 ^{B}	0.11 ± 0.01 ^{A}

Note: F: F statistic; d.f.: degree of freedom; *p*: level of significance. Mean values ± standard error of the mean (SE) followed by the same letter are not significantly different.

3.2.3. Potassium

Of all the analyzed parameters, potassium was the element most affected by the fertilization treatments (d.f. = 14; F = 22.83; $p < 0.001$ for Pernolec and d.f. = 14; F = 19.11; $p < 0.001$ for Kostelec). In Pernolec, the lowest concentration was recorded in NPK E1 treatment (116 mg kg^{-1}), and the highest in PS1+NPK treatment (215 mg kg^{-1}) (Table 4). In Kostelec, the K mean concentration varied from 112 mg kg^{-1} (NPK E1) to 252 mg kg^{-1} (PS3+PK) (Table 5). If we compare the fertilizer groups, we find that both localities have a comparable pattern. In Pernolec, application of no fertilizers (Control), NPK and PS resulted in lower K soil concentrations without differences between these treatments, while application of PK, PS+NPK and PS+PK resulted in higher K concentrations (Figure 3a). The situation in Kostelec was similar, with one exception, namely for the PK and PS treatments. The differences between these two treatments were not significant, as in Pernolec (Figure 3b).

3.2.4. Calcium

In Pernolec, the mean Ca soil concentrations varied significantly (d.f. = 14; F = 4.83; $p < 0.001$) between the treatments and ranged from 1113 mg kg^{-1} (NPK E3) to 1447 mg kg^{-1} (PS1) (Table 4). Similarly, in Kostelec, the differences between fertilization treatments were significant (d.f. = 14; F = 4.46; $p < 0.001$), and varied from 1213 mg kg^{-1} (PS2+NPK) to 1549 mg kg^{-1} (Control) (Table 5). Comparing the fertilizer groups, we find that the effect of fertilization on soil Ca content is similar in the two sites. The lowest mean Ca concentrations were recorded in NPK and PS+NPK treatments (Figure 4a,b), while application of no fertilizers (Control) and PS resulted in the highest Ca concentrations.

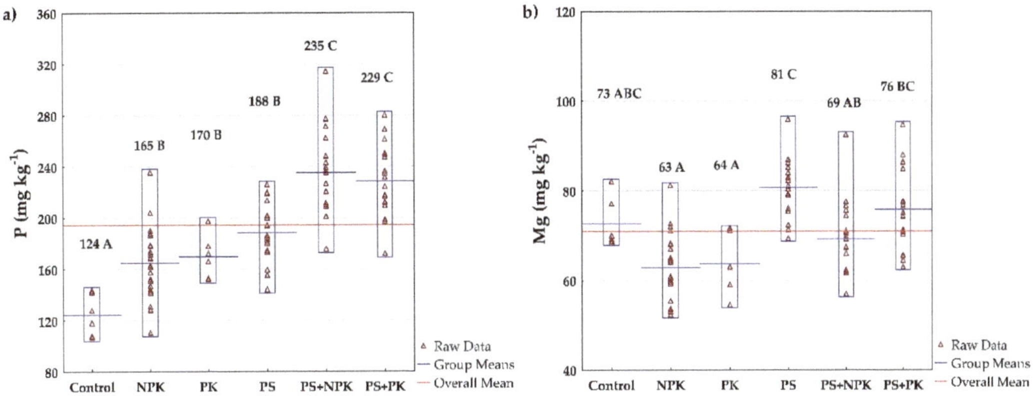

Figure 2. The effect of fertilizer treatments on (**a**) soil P and (**b**) Mg concentration in Kostelec (2015–2020). The differences between P and Mg concentrations as affected by fertilizer treatment were insignificant in Pernolec (Table 6). Mean values followed by the same letter are not significantly different. Red triangles represent the raw data. The blue lines represent the mean value of the particular treatment, while the red line represents the mean value calculated from all fertilizer treatments.

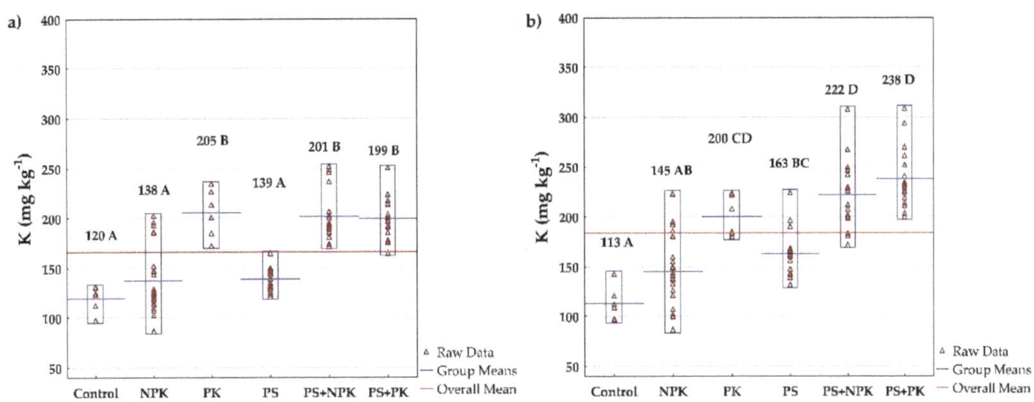

Figure 3. The effect of fertilizer treatments on soil K concentration in (**a**) Pernolec and (**b**) Kostelec (2015–2020). Mean values followed by the same letter are not significantly different. Red triangles represent the raw data. The blue lines represent the mean value of the particular treatment, while the red line represents the mean value calculated from all fertilizer treatments.

3.2.5. Magnesium

Average soil Mg concentrations in Pernolec were not significantly affected by the fertilization treatments (d.f. = 14; F = 0.57; p = 0.88) and ranged from 104 mg kg^{-1} (NPK E3) to 131 mg kg^{-1} (PS1) (Table 4). In Kostelec, on the other hand, the long-term application of slurry and mineral fertilizers had a significant effect on the Mg concentration (d.f. = 14; F = 7.10; p = 0.001), which varied from 58 mg kg^{-1} (NPK) to 83 mg kg^{-1} (PS3) (Table 5). Comparing the groups of fertilizers, the lowest mean concentrations were recorded in NPK and PK treatments, while the highest concentrations occurred in PS and PS+PK treatments (Figure 2b).

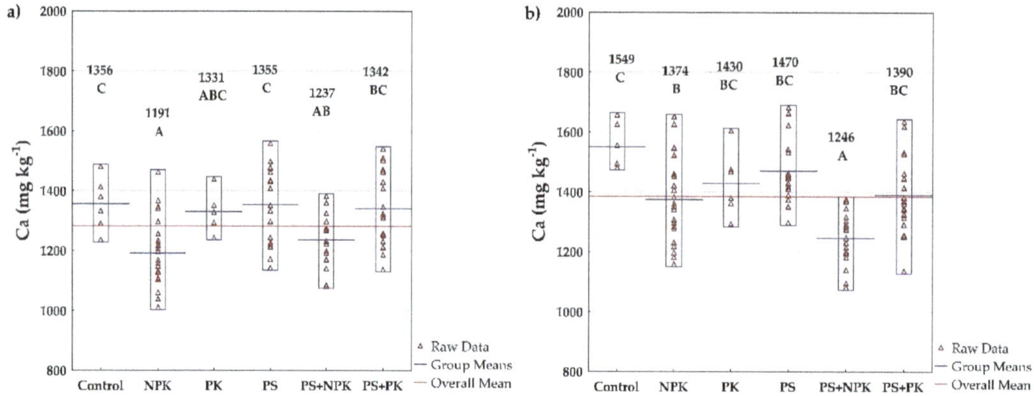

Figure 4. The effect of fertilizer treatments on soil Ca concentration in (**a**) Pernolec and (**b**) Kostelec (2015–2020). Mean values followed by the same letter are not significantly different. Red triangles represent the raw data. The blue lines represent the mean value of the particular treatment, while the red line represents the mean value calculated from all fertilizer treatments.

3.2.6. Soil Organic Carbon Content

Long-term and regular application of slurry, mineral fertilizers and their combinations did not significantly affect the soil organic carbon content in either Pernolec (d.f. = 14; F = 0.91; p = 0.56) or Kostelec (d.f. = 14; F = 0.77; p = 0.70). In Pernolec, the Cox content in the soil varied from 0.88% (PK, PS1) to 1.02% (PS3+NPK) (Table 4). In Kostelec, the Cox ranged from 0.77% (PK) to 0.87% (PS3+NPK) (Table 5).

3.2.7. Total Nitrogen Content

Similar to soil organic carbon, long-term and regular application of manure, mineral fertilizers and their combinations did not significantly affect total soil nitrogen content at either of the two sites (Pernolec: d.f. = 14; F = 0.52; p = 0.91; Kostelec: d.f. = 14; F = 0.64; p = 0.83). In Pernolec, the Ntot content ranged from 0.11% to 0.13% (Table 4), in Kostelec from 0.10% to 0.11% (Table 5).

3.2.8. Principal Component Analysis (PCA)

Based on the PCA results (Figure 5a), we can classify the fertilizers in Pernolec (cambisol) into four categories according to their effect on yield and soil properties (Figure 5b). (1) The unfertilized treatment (Control) gives lower crop yields and has low P and K concentrations due to no external supply of nutrients. (2) Pig slurry (PS) applied alone, application of mineral P and K (PK), and combination of PS+PK (generally the fertilizers without mineral N): these fertilizers have a positive relationship with pH and Ca and Mg content, and there is no decrease in pH compared to other treatments. On the other hand, the absence of mineral N puts this group at a disadvantage in terms of low grain and straw yields and the soils have a low organic matter content (no organic matter in the PK treatment and low organic matter in the slurry). (3) The third group is represented by PS+NPK treatments. The joint application of PS and mineral NPK represents a kind of golden mean ensuring relatively high grain and straw yields, nutrient and soil organic matter content. However, the presence of the ammonium form of mineral N negatively affects soil pH. (4) The fourth group consists of separately applied mineral fertilizers (NPK, without manure supplement). Mineral fertilizers are clearly closely and positively associated with yield, followed by soil organic carbon and nitrogen. On the other hand, the presence of the ammonium form of nitrogen, accompanied by the absence of slurry, accentuates the negative effect on pH even more significantly (compared to PS+NPK combinations).

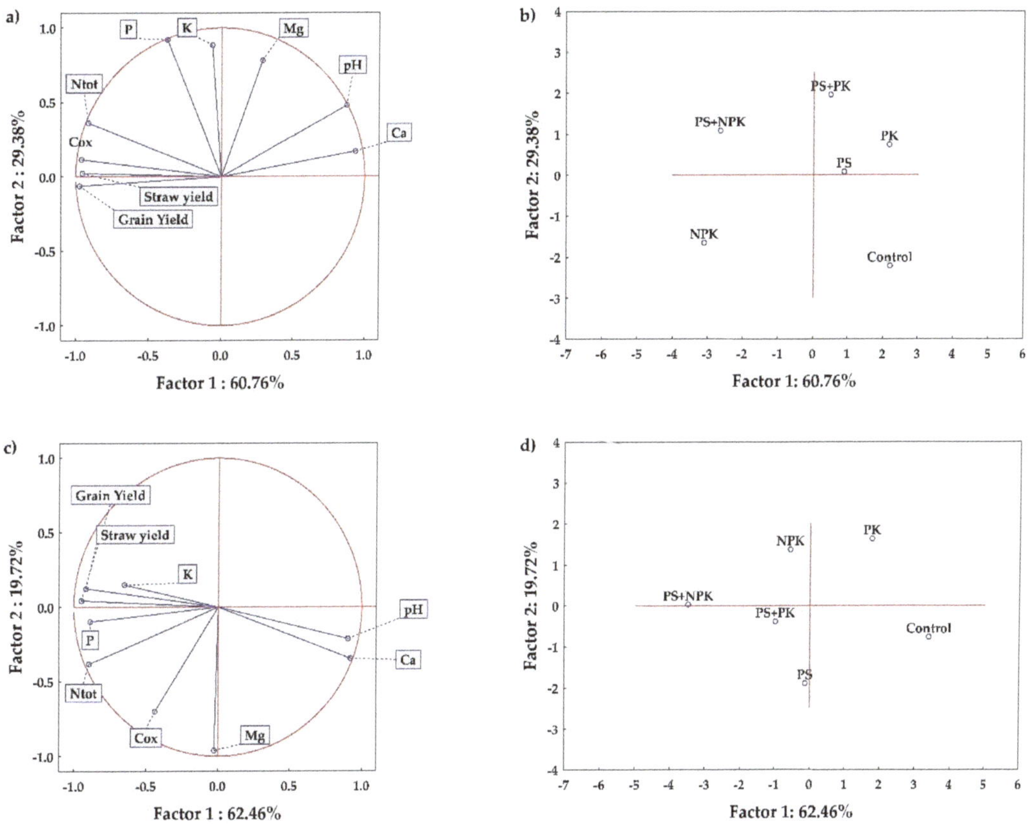

Figure 5. Results of the PCA—relationships between soil chemical parameters and grain and straw yields as affected by the fertilization treatments: (**a**,**b**) Pernolec, (**c**,**d**) Kostelec. Grain and straw yields are based on the average WW, SB and FP yield from 2015 to 2020.

With a change in soil type (Kostelec, luvisol), we can see a different response to the long-term application of manure and mineral fertilizers on yield and soil properties (Figure 5c). The separation of fertilizers in Kostelec (Figure 5d) is not as clear-cut as in Pernolec, which means that the differences between fertilizer treatments are not as pronounced. As in Pernolec, unfertilized Control is strongly and positively correlated with pH and soil Ca content. On the other hand, treatment without external nutrient inputs (Control) is associated with low grain and straw yields and also with low concentrations of soil P and K (soil depletion). The PK group (mineral P and K fertilizers) has a completely different status than PS (in Pernolec these two fertilizer groups were together in one cluster). PK has a strong negative relationship with soil organic carbon and total nitrogen. This treatment highlights the need for nitrogen, either supplied in mineral form or the form of manure. In contrast, the application of pig slurry (PS) is strongly and positively associated with soil organic carbon content combined with a neutral relationship to both yield and pH. Mineral fertilizers (NPK) occupy a similar position to PS in terms of yield and pH, with the exception that they are closer to higher yields and lower pH. Quite different (compared to Pernolec) is their relationship to soil organic carbon, with which it is moderately and rather negatively correlated. Similarly to Pernolec, the PS+NPK fertilizer group is dominant. It is associated with high yields, high soil nutrient content, a relatively neutral relationship to soil organic matter and a significantly (strongly) negative relationship to soil pH (stronger negative relationship to pH than in Pernolec).

3.2.9. FP Grain and Straw Yields

As mentioned in Section 2.5, the results of the FP grain and straw yields from Kostelec cannot be statistically analysed. The average grain and straw yields in 2017 and 2020 are shown in Table S5. In 2017, the grain yields varied from 2.8 t ha^{-1} (PS1, PS1+PK) to 3.7 t ha^{-1} (PS3+NPK), while in 2020 the grain yields varied from 2.5 t ha^{-1} (NPK E2) to 3.3 t ha^{-1} (PS1+PK). Straw yields varied from 2.1 t ha^{-1} (PS1) to 3.2 t ha^{-1} (PS3+NPK) in 2017 and from 3.0 t ha^{-1} (PS1+NPK, NPK E1) to 3.6 t ha^{-1} (PS2) (Table S5).

According to MANOVA results, the FP grain yields were significantly affected by year (d.f. = 1; F = 71.55; p < 0.001), fertilizer treatment (d.f. = 14; F = 6.76; $p < 0.001$), and their interaction (d.f. = 14; F = 2.33; $p < 0.01$) in Pernolec. The effect of year was dominant (89%), while the effect of fertilizer treatment influenced yields by 8%. If we look at the weather in a particular year, we find that 2017 in Pernolec was marked by drought in May and June. Moreover, 2017 was significantly marked by very high temperatures in June and July (Table S3). These were factors that caused significantly lower yields compared to 2020, which was characterized by both higher precipitation and milder temperatures (Table 1). Straw yields were comparable in 2017 and 2020 as the differences were insignificant in Pernolec (d.f. = 1; F = 0.40; $p = 0.53$), while the effect of the fertilizer treatment was significant (d.f. = 14; F = 4.32; $p < 0.001$). The interaction between the factors of year and treatment was insignificant (d.f. = 14; F = 1.90; $p < 0.07$).

In 2017, the grain yields were significantly affected by the fertilization (d.f. = 14; F = 3.18; $p < 0.01$) and varied from 1.2 t ha^{-1} (Control) to 2.3 t ha^{-1} (PS3+PK and NPK E3) (Table 7). Significantly different were Control and PS1+PK against PS3+PK, PS3+NPK and NPK E3. Grain yield slightly increased with increasing nitrogen rate (Figure 6a). According to the linear-plateau model, calculated from the mineral fertilizer treatments (NPK, NPK E1, NPK E2, NPK E3), the FP yield response to different rates of mineral N plateaued at 97 kg ha^{-1} N, with a corresponding yield of 2.08 t ha^{-1} (Figure 7, left). Comparing the nitrogen use efficiency (NUE), the highest NUE was recorded in NPK treatment (23.3 kg per 1 kg of N applied), followed by PS1 (5.9 kg), PS1+NPK (4.3 kg), PS2, PS3 and PS3+NPK (3.5 kg), the lowest NUE was recorded in PS2+NPK treatment (3.0 kg per 1 kg of N applied). This calculation shows that mineral fertilizers, compared to organic manures (slurries), supply nutrients very quickly and, even in small quantities can significantly and efficiently promote growth. On the other hand, their effectiveness is offset by their negative effect on the soil environment.

In 2020, the grain yields varied from 1.8 t ha^{-1} (PS1 and PS1+PK) to 2.8 t ha^{-1} (NPK E3). As we can see, the response to the fertilization was a little different as the weather conditions changed (Figure 6b, the red line representing a quadratic model). We can see that grain yield slightly increased with increasing N dose, as in 2017. The course of the function indicates the attainment of a local maximum, which, according to the quadratic model, is located at an N rate of 400 kg ha^{-1}. At this rate, the maximal average yield of 2.4 t ha^{-1} would be achieved, which is actually lower than the yields already obtained with lower inputs (Figure 6b). According to the linear-plateau model, the response of FP yields to different rates of N doses plateaued at 73 kg ha^{-1}, corresponding with the yield 2.71 t ha^{-1} (Figure 7, right), showing better weather conditions for yield development in 2020. Comparing the NUE, the highest efficiency was again recorded in NPK treatment (10.0 kg per 1 kg N applied), followed by PS2+NPK (3 kg), PS1+NPK (2.6 kg), PS3+NPK (2.4 kg), PS3 (0.4 kg) and PS1 and PS2 (−2.4 and −0.5, respectively), where the efficiency was negative as the mean yield was lower than in the Control treatment.

Table 7. The effect of the year (2017, 2020) and fertilizer treatment on FP grain and straw yield (t ha^{-1}) in Pernolec.

Fertilizer Treatment	Grain Yield (t ha^{-1}) 2017	Grain Yield (t ha^{-1}) 2020	Mean	Straw Yield (t ha^{-1}) 2017	Straw Yield (t ha^{-1}) 2020	Mean
Control	1.2 ± 0.2^{A}	2.0 ± 0.1^{AB}	1.6 ± 0.2^{A}	1.3 ± 0.1^{A}	1.5 ± 0.1^{A}	1.4 ± 0.1^{A}
PK	1.4 ± 0.2^{ABC}	2.1 ± 0.2^{ABCD}	1.8 ± 0.2^{AB}	1.5 ± 0.1^{A}	1.9 ± 0.2^{AB}	1.7 ± 0.2^{ABC}
NPK	1.9 ± 0.3^{ABC}	2.3 ± 0.1^{ABCDE}	2.1 ± 0.2^{ABCD}	2.3 ± 0.6^{A}	1.9 ± 0.2^{AB}	2.1 ± 0.3^{ABC}
PS1	1.7 ± 0.1^{ABC}	1.8 ± 0.2^{A}	1.8 ± 0.1^{AB}	1.7 ± 0.4^{A}	1.7 ± 0.1^{AB}	1.7 ± 0.2^{ABC}
PS1+PK	1.3 ± 0.1^{A}	1.8 ± 0.1^{A}	1.6 ± 0.1^{A}	1.4 ± 0.1^{A}	1.7 ± 0.1^{AB}	1.5 ± 0.1^{AB}
PS1+NPK	1.7 ± 0.1^{ABC}	2.3 ± 0.1^{ABCDE}	2.0 ± 0.1^{ABC}	1.7 ± 0.1^{A}	1.8 ± 0.1^{AB}	1.8 ± 0.1^{ABC}
PS2	1.8 ± 0.2^{ABC}	1.9 ± 0.2^{A}	1.9 ± 0.1^{AB}	1.8 ± 0.4^{A}	1.7 ± 0.1^{AB}	1.7 ± 0.2^{ABC}
PS2+PK	1.7 ± 0.1^{ABC}	2.2 ± 0.1^{ABCDE}	2.0 ± 0.1^{ABC}	2.0 ± 0.1^{A}	1.9 ± 0.1^{AB}	2.0 ± 0.1^{ABC}
PS2+NPK	1.8 ± 0.3^{ABC}	2.6 ± 0.1^{BCDE}	2.2 ± 0.2^{BCD}	2.2 ± 0.2^{A}	2.1 ± 0.2^{AB}	2.1 ± 0.1^{ABC}
PS3	2.1 ± 0.2^{ABC}	2.1 ± 0.1^{ABCD}	2.1 ± 0.1^{ABCD}	2.4 ± 0.2^{A}	1.9 ± 0.1^{AB}	2.2 ± 0.1^{BC}
PS3+PK	2.3 ± 0.1^{C}	2.1 ± 0.2^{ABC}	2.2 ± 0.1^{BCD}	2.5 ± 0.2^{A}	1.8 ± 0.1^{AB}	2.1 ± 0.2^{ABC}
PS3+NPK	2.2 ± 0.1^{BC}	2.7 ± 0.1^{CDE}	2.4 ± 0.1^{CD}	2.5 ± 0.1^{A}	2.1 ± 0.1^{AB}	2.3 ± 0.1^{C}
NPK E1	1.8 ± 0.2^{ABC}	2.7 ± 0.1^{DE}	2.2 ± 0.2^{BCD}	1.7 ± 0.2^{A}	2.4 ± 0.2^{B}	2.1 ± 0.3^{ABC}
NPK E2	1.9 ± 0.1^{ABC}	2.7 ± 0.1^{DE}	2.3 ± 0.2^{BCD}	2.1 ± 0.2^{A}	2.3 ± 0.1^{AB}	2.2 ± 0.1^{BC}
NPK E3	2.3 ± 0.2^{C}	2.8 ± 0.1^{E}	2.5 ± 0.1^{D}	2.6 ± 0.2^{A}	2.2 ± 0.2^{AB}	2.4 ± 0.2^{C}
Mean	1.8 ± 0.1^{a}	2.3 ± 0.1^{b}		2.0 ± 0.1^{a}	1.9 ± 0.1^{a}	

Mean values (±SE) followed by the same letter (a vertical comparison of the effect of fertilizer treatment) are not statistically significantly different.

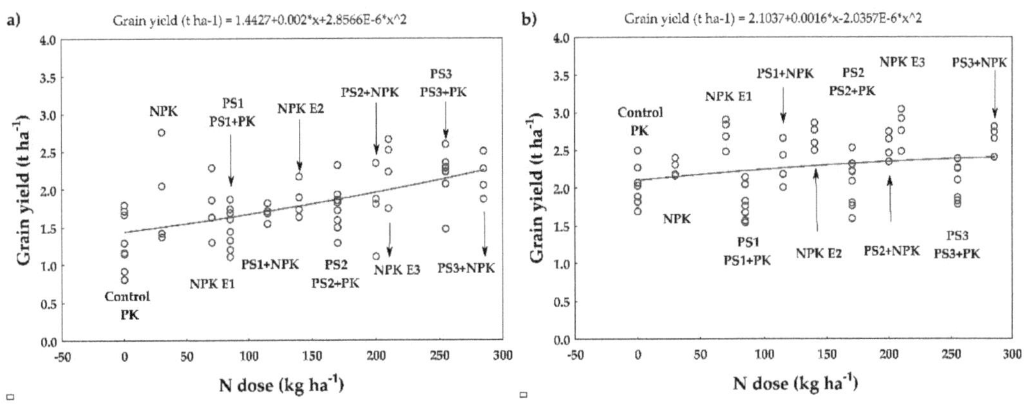

Figure 6. FP grain yield (t ha^{-1}) as affected by N dose in Pernolec in (**a**) 2017 and (**b**) 2020. The average yields (blue points) are interleaved with the quadratic function (red line). The equation of the quadratic model is given above the figure.

Comparing the results from both years (Table 7), we find that the highest average yields were obtained with the NPK E3 treatment (2.5 t ha^{-1}). However, lower, but statistically comparable, yields were obtained with the NPK (30 kg mineral N ha^{-1} with an average yield of 2.1 t ha^{-1}) and PS3 (51 t ha^{-1} with an average yield of 2.1 t ha^{-1}) treatments. This is a very important finding as PS applied in higher doses can completely replace mineral fertilizers and a negative effect of mineral fertilizers on soil pH can be partially avoided.

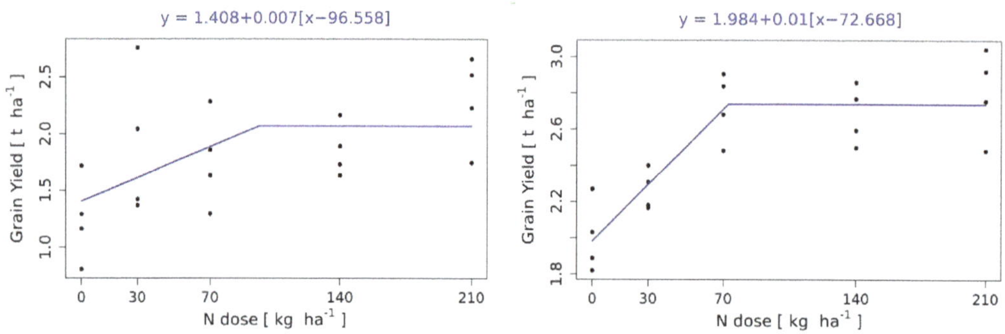

Figure 7. The response of FP yields to different doses of mineral N fertilizers (NPK, NPK E1, NPK E2, NPK E3 treatments) in 2017 (**left**) and 2020 (**right**). Yields (black dots) are interleaved with the linear-plateau model (blue line). The equation of the model is given above the figure.

Straw yields were significantly affected by fertilization in 2020 (d.f. = 14; F = 3.26; $p < 0.05$), with insignificant differences in 2017 (d.f. = 14; F = 3.07; $p < 0.05$, Tukey's test did not confirm ANOVA as multiple comparison methods generally have lower test power than analysis of variance-ANOVA). Straw yield tended to increase with increasing fertilizer rate. The differences between 2017 and 2020 were insignificant. The highest yields obtained were recorded for the PS3+NPK and NPK E3 treatments; however, the PK, PS1 and PS2 treatments also provided statistically comparable yields (Table 7).

4. Discussion

Long-term and regular application of mineral fertilizers, pig slurry, and their combinations significantly affected soil properties and the effect of fertilizers depends on soil conditions (type) of the site. One of the most important soil properties is the value of the pH. Soil pH is considered to be the dominant factor directly influencing other soil properties such as elements' availability [10,71,72] and abundance and representation of plant and microbial communities [73] and their activity [74]. All macronutrients are best available in neutral to alkaline soils, while in acid soils their availability decreases and the availability of elements such as Fe, Mn, B, Zn and Al increases. Changes in pH thus directly affect the soil's ability to supply nutrients to plants. In our case, the lowest pH values were recorded for the NPK treatments (applied alone or in combination with PS, but only in the NPK treatments with the highest N doses, Tables 4 and 5). The same result was recorded worldwide [11,13,21,72,75] and has been known for a long time [76]. The primary driver of downward pH changes is mineral nitrogen, in its ammonium form, because the conversion of the ammonium form to nitrate in soils releases hydrogen, directly affecting its concentration in the soil environment. This can be particularly evident in the case of PK treatments. As mentioned above, Czech conventional crop production is primarily dependent on mineral nitrogen. Add to this the fact that most of the cultivated land is rented and its owners have no idea or do not care about acidification. This leaves room for acidification to run freely. An interesting survey was carried out in the USA, which also shows that acidification is taking place there and that about half of the farmers were not even aware of it [77]. One way to reduce the negative effects of mineral fertilizers on soil pH is to combine mineral fertilizers and organic manures [78]. Co-application of mineral fertilizers and organic manures is often cited as a sustainable method of fertilization, providing high and stable yields and a healthy state of the soil. The unifying element of this approach is organic matter (together with nutrients) [3,5,7,8,13,29,79,80] added to the soil, beneficially affecting soil chemical, physical and microbiological properties. From this point of view, we can support these results only partially as the combined application of PS+NPK provided better pH values than NPK only in Pernolec (Figure 1a), in contrast to Kostelec (lower and comparable to NPK treatment, Figure 1b). This may be due to the overall higher soil organic matter content in Pernolec (Table 1) and the very low organic matter content in the slurry, which seems to be behind the non-significant Cox differences between fertilization treatments in both locations (Tables 4 and 5). The DM of pig slurry usually ranges from 0.7% up to 23% [61] and quality slurry has a dry matter content between 6% and 8% in Czech Republic. In our case, the dry matter content of the available and applied pig slurry was very low, which is probably the reason why the soil organic matter content is slightly higher in the high slurry fertilizer treatments, but not statistically significantly higher compared to the other fertilization treatments.

From the point of view of nutrients, the highest concentrations of macronutrients were always connected with PS+PK, NPK and PS+NPK treatments (Figure 5), while nutrient depletion can be found in Control treatment. PS+PK treatment has a close relationship to nutrient content and a moderate relationship to yields (Figure 5), showing that nitrogen is a limiting element in this treatment and its P and K nutrients are not utilized completely. The combination of mineral fertilizers and organic manures provides high yields while leaving a high micronutrient content in the soil (Figure 5). From the point of view of agriculture in Czech Republic, we can expect that acidification problems will intensify, as mineral nitrogen is important for all agricultural crops and significantly affects yields, which is the most monitored parameter. The application of mineral fertilizers at higher doses (NPK E3, PS2+NPK, PS3+NPK treatments) significantly reduced the soil reaction values at both sites (Kostelec and Pernolec) compared to the Control; a more significant decrease was recorded on the luvisol soil type (Kostelec). Similar findings (decrease in pH in treatments fertilized with mineral fertilizers only) are supported by some other studies [81–84]. The negative effect of acidification on the content of available nutrients (Ca, Mg) in the plough soil horizon is shown in Tables 4 and 5 (in the NPK E3, PS2+NPK, PS3+NPK treatments, low Ca and Mg contents were recorded at both sites). For available nutrients P and K, the acidification effect was predominant in the mineral fertilized treatments (NPK E1-3). This is confirmed by the results of the multicriteria PCA evaluation. These results are in agreement with [85,86], which showed a negative effect of acidification on the regime of available nutrients in the soil. Without the addition of other nutrients (PK treatments), there will be a reduction in the content of these nutrients in the soil (as in the case of Control). The combination of mineral fertilizers and organic manures can partially reduce the negative effect of mineral fertilizers on pH (depending on the location and soil and climate conditions), which is good news, but the lack of organic manures due to reduced livestock production in the country plays against the solution to the current problems.

In terms of pea yields, we can clearly see the dependence of yields on nitrogen, with pea yields increasing with increasing nutrient rates, although the differences are not statistically significant between higher doses of fertilizers. The yields are strongly affected by fertilization and by weather conditions. While nutrient utilization is lower in years with poorer weather conditions, nutrient utilization increases in years with normal conditions. This can be seen in the results of the linear-plateau model, which compared nutrient and yield dependence in 2017 and 2020. Based on this model, we can say that under normal weather conditions the optimum nitrogen rate in Pernolec is around 70 kg ha^{-1}. As the variation from normal conditions increases, the nutrient requirement increases as the optimal dose of N raised to 97 kg ha^{-1} N in 2017. Another important finding is that mineral fertilizers can be completely replaced by PS applied in higher doses (51 t ha^{-1} in our case). PS has a low C:N ratio, and the mineralization of slurries is rapid, providing a huge amount of available nutrients at the beginning of the season before symbiosis with mycorrhizal bacteria fully develops. Replacing mineral fertilizers with PS can provide comparable yields without a negative effect on soil pH value.

Supplementary Materials: The following are available online at https://www.mdpi.com/article/10.3390/land11020187/s1, Table S1: Forms and doses of mineral fertilizers and pig slurry (PS) according to the fertilizer treatments applied in Pernolec. Cumulative doses for the entire three-year crop rotation. Table S2: Forms and doses of mineral fertilizers and pig slurry (PS) according to the fertilizer treatments applied in Kostelec. Cumulative doses for the entire three-year crop rotation. Table S3: The long-term mean precipitation (Mean; 1977–2016 for Pernolec; 1982–2016 for Kostelec; mm) and the sum of precipitation (mm) in individual months in 2017 and 2020 in Pernolec and Kostelec. The comparison between long-term mean and actual (2017, 2020) precipitation was done according to [63]. Table S4: The long-term mean temperature (Mean; 1977–2016 for Pernolec; 1982–2016 for Kostelec; °C) and the average temperature (°C) in individual months in 2017 and 2020 in Pernolec and Kostelec. The comparison between the long-term mean and actual (2017, 2020) temperature was done according to [63]. Table S5: The effect of the year (2017, 2020) and fertilizer treatment on FP grain and straw yield (t ha^{-1}) in Kostelec.

Author Contributions: Conceptualization, L.H., L.M. and E.K.; methodology, E.K.; software, L.H, L.M. and K.K.; formal analysis, L.H., L.M. and K.K.; investigation, L.H., L.M. and K.K.; resources, E.K. and P.Č. writing—original draft preparation, L.H.; writing—review and editing, L.H., L.M. and E.K.; visualization, L.H. and K.K.; supervision, E.K.; project administration, E.K.; funding acquisition, E.K. and P.Č. All authors have read and agreed to the published version of the manuscript.

Funding: This research was funded by the Ministry of Agriculture of the Czech Republic, grant number RO0418, Czech National Agency for Agricultural Research, grant number QK1810010, QK21020155 and QK21010124, and EU project H2020 677407—Soil care for profitable and sustainable crop production in Europe.

Institutional Review Board Statement: Ethical review and approval were waived for this study, due to the non-use of experimental animals or human subjects.

Informed Consent Statement: No human trials were conducted during the study.

Data Availability Statement: Not applicable.

Acknowledgments: We would like to thank the staff at the experimental station in Pernolec, Pavel Beneš and Vera Folejtárová, and the staff of the experiment in Kostelec nad Orlicí, Petr Ivičic, Vlatimil Basař and Jaroslav Málek.

Conflicts of Interest: The authors declare no conflict of interest.

References

1. Yanai, J.; Okada, T.; Yamada, H. Elemental composition of agricultural soils in Japan in relation to soil type, land use and region. *Soil Sci. Plant Nutr.* **2012**, *58*, 1–10. [CrossRef]
2. Domagała-Świątkiewicz, I.; Gąstoł, M. Soil chemical properties under organic and conventional crop management systems in south Poland. *Biol. Agric. Hortic.* **2013**, *29*, 12–28. [CrossRef]
3. Cai, A.; Xu, M.; Wang, B.; Zhang, W.; Liang, G.; Hou, E.; Luo, Y. Manure acts as a better fertilizer for increasing crop yields than synthetic fertilizer does by improving soil fertility. *Soil Tillage Res.* **2019**, *189*, 168–175. [CrossRef]
4. Triberti, L.; Nastri, A.; Baldoni, G. Long-term effects of crop rotation, manure and mineral fertilisation on carbon sequestration and soil fertility. *Eur. J. Agron.* **2016**, *74*, 47–55. [CrossRef]
5. Shang, Q.; Ling, N.; Feng, X.; Yang, X.; Wu, P.; Zou, J.; Shen, Q.; Guo, S. Soil fertility and its significance to crop productivity and sustainability in typical agroecosystem: A summary of long-term fertilizer experiments in China. *Plant Soil* **2014**, *381*, 13–23. [CrossRef]
6. Hemkemeyer, M.; Schwalb, S.A.; Heinze, S.; Joergensen, R.G.; Wichern, F. Functions of elements in soil microorganisms. *Microbiol. Res.* **2021**, *252*, 126832. [CrossRef] [PubMed]
7. Chen, Y.; Camps-Arbestain, M.; Shen, Q.; Singh, B.; Cayuela, M.L. The long-term role of organic amendments in building soil nutrient fertility: A meta-analysis and review. *Nutr. Cycl. Agroecosyst.* **2018**, *111*, 103–125. [CrossRef]
8. Du, Y.; Cui, B.; Zhang, Q.; Wang, Z.; Sun, J.; Niu, W. Effects of manure fertilizer on crop yield and soil properties in China: A meta-analysis. *Catena* **2020**, *193*, 104617. [CrossRef]
9. Shiwakoti, S.; Zheljazkov, V.D.; Gollany, H.T.; Kleber, M.; Xing, B. Macronutrients in soil and wheat as affected by a long-term tillage and nitrogen fertilization in winter wheat–fallow rotation. *Agronomy* **2019**, *9*, 178. [CrossRef]
10. Czarnecki, S.; Düring, R.A. Influence of long-term mineral fertilization on metal contents and properties of soil samples taken from different locations in hesse, Germany. *Soil* **2015**, *1*, 23–33. [CrossRef]
11. Mi, W.; Sun, Y.; Xia, S.; Zhao, H.; Mi, W.; Brookes, P.C.; Liu, Y.; Wu, L. Effect of inorganic fertilizers with organic amendments on soil chemical properties and rice yield in a low-productivity paddy soil. *Geoderma* **2018**, *320*, 23–29. [CrossRef]
12. Maltas, A.; Kebli, H.; Oberholzer, H.R.; Weisskopf, P.; Sinaj, S. The effects of organic and mineral fertilizers on carbon sequestration, soil properties, and crop yields from a long-term field experiment under a Swiss conventional farming system. *L. Degrad. Dev.* **2018**, *29*, 926–938. [CrossRef]
13. Brar, B.S.; Singh, J.; Singh, G.; Kaur, G. Effects of long term application of inorganic and organic fertilizers on soil organic carbon and physical properties in maize-wheat rotation. *Agronomy* **2015**, *5*, 220–238. [CrossRef]

14. Eo, J.; Park, K.C. Long-term effects of imbalanced fertilization on the composition and diversity of soil bacterial community. *Agric. Ecosyst. Environ.* **2016**, *231*, 176–182. [CrossRef]
15. Soman, C.; Li, D.; Wander, M.M.; Kent, A.D. Long-term fertilizer and crop-rotation treatments differentially affect soil bacterial community structure. *Plant Soil* **2017**, *413*, 145–159. [CrossRef]
16. Li, H.; Liu, J.; Li, G.; Shen, J.; Bergström, L.; Zhang, F. Past, present, and future use of phosphorus in Chinese agriculture and its influence on phosphorus losses. *Ambio* **2015**, *44*, 274–285. [CrossRef]
17. Verbeeck, M.; Salaets, P.; Smolders, E. Trace element concentrations in mineral phosphate fertilizers used in Europe: A balanced survey. *Sci. Total Environ.* **2020**, *712*, 136419. [CrossRef]
18. Zwolak, A.; Sarzyńska, M.; Szpyrka, E.; Stawarczyk, K. Sources of Soil Pollution by Heavy Metals and Their Accumulation in Vegetables: A Review. *Water. Air. Soil Pollut.* **2019**, *230*, 1–9. [CrossRef]
19. Kissel, D.E.; Bock, B.R.; Ogles, C.Z. Thoughts on acidification of soils by nitrogen and sulfur fertilizers. *Agrosystems Geosci. Environ.* **2020**, *3*, 1–10. [CrossRef]
20. Kopeć, M.; Gondek, K.; Mierzwa-Hersztek, M.; Jarosz, R. Changes in the soil content of organic carbon nitrogen and sulphur in a long-term fertilisation experiment in czarny potok (Poland). *J. Elem.* **2021**, *26*, 33–46. [CrossRef]
21. Vašák, F.; Černý, J.; Buráňová, Š.; Kulhánek, M.; Balík, J. Soil pH changes in long-term field experiments with different fertilizing systems. *Soil Water Res.* **2015**, *10*, 19–23. [CrossRef]
22. Lassaletta, L.; Billen, G.; Garnier, J.; Bouwman, L.; Velazquez, E.; Mueller, N.D.; Gerber, J.S. Nitrogen use in the global food system: Past trends and future trajectories of agronomic performance, pollution, trade, and dietary demand. *Environ. Res. Lett.* **2016**, *11*, 095007. [CrossRef]
23. Zhu, H.; Chen, C.; Xu, C.; Zhu, Q.; Huang, D. Effects of soil acidification and liming on the phytoavailability of cadmium in paddy soils of central subtropical China. *Environ. Pollut.* **2016**, *219*, 99–106. [CrossRef] [PubMed]
24. Yang, Z.; Ha, L. Analysis and comparison of nutrient contents in different animal manures from Beijing suburbs. *Agric. Sci.* **2013**, *4*, 50–55. [CrossRef]
25. Eghball, B.; Wienhold, B.J.; Gilley, J.E.; Eigenberg, R.A. Biological Systems Engineering: Papers and Publications Mineralization of Manure Nutrients Mineralization of manure nutrients. *J. Soil Water Conserv.* **2002**, *57*, 470–473.
26. Eghball, B. Nitrogen Mineralization from Field-Applied Beef Cattle Feedlot Manure or Compost. *Soil Sci. Soc. Am. J.* **2000**, *64*, 2024–2030. [CrossRef]
27. Hamm, A.C.; Tenuta, M.; Krause, D.O.; Ominski, K.H.; Tkachuk, V.L.; Flaten, D.N. Bacterial communities of an agricultural soil amended with solid pig and dairy manures, and urea fertilizer. *Appl. Soil Ecol.* **2016**, *103*, 61–71. [CrossRef]
28. Šimon, T.; Czakó, A. Influence of long-term application of organic and inorganic fertilizers on soil properties. *Plant Soil Environ.* **2014**, *60*, 314–319. [CrossRef]
29. Suwara, I.; Pawlak-Zaręba, K.; Gozdowski, D.; Perzanowska, A. Physical properties of soil after 54 years of long-term fertilization and crop rotation. *Plant Soil Environ.* **2016**, *62*, 389–394. [CrossRef]
30. De la Torre, A.I.; Jiménez, J.A.; Carballo, M.; Fernandez, C.; Roset, J.; Muoz, M.J. Ecotoxicological evaluation of pig slurry. *Chemosphere* **2000**, *41*, 1629–1635. [CrossRef]
31. Díez, J.A.; Hernaiz, P.; Muñoz, M.J.; de la Torre, A.; Vallejo, A. Impact of pig slurry on soil properties, water salinization, nitrate leaching and crop yield in a four-year experiment in Central Spain. *Soil Use Manag.* **2005**, *20*, 444–450. [CrossRef]
32. Antezana, W.; De Blas, C.; García-Rebollar, P.; Rodríguez, C.; Beccaccia, A.; Ferrer, P.; Cerisuelo, A.; Moset, V.; Estellés, F.; Cambra-López, M.; et al. Composition, potential emissions and agricultural value of pig slurry from Spanish commercial farms. *Nutr. Cycl. Agroecosyst.* **2016**, *104*, 159–173. [CrossRef]
33. Moral, R.; Perez-Murcia, M.D.; Perez-Espinosa, A.; Moreno-Caselles, J.; Paredes, C.; Rufete, B. Salinity, organic content, micronutrients and heavy metals in pig slurries from South-eastern Spain. *Waste Manag.* **2008**, *28*, 367–371. [CrossRef] [PubMed]
34. Antil, R.S.; Singh, M. Effects of organic manures and fertilizers on organic matter and nutrients status of the soil. *Arch. Agron. Soil Sci.* **2007**, *53*, 519–528. [CrossRef]
35. Aula, L.; Omara, P.; Dhillon, J.S.; Fornah, A.; Raun, W.R. Influence of Applied Cattle Manure on Winter Wheat (*Triticum aestivum* L.) Grain Yield, Soil pH and Soil Organic Carbon. *Commun. Soil Sci. Plant Anal.* **2019**, *50*, 2056–2064. [CrossRef]
36. Han, X.; Wang, S.; Veneman, P.L.M.; Xing, B. Change of organic carbon content and its fractions in black soil under long-term application of chemical fertilizers and recycled organic manure. *Commun. Soil Sci. Plant Anal.* **2006**, *37*, 1127–1137. [CrossRef]
37. Ziogas, M.; Ipsilantis, I.; Matsi, T.; Kostopoulou, S. Long-term Fertilization with Liquid Cattle Manure Leaves Legacy Nutrients, but not Organic Carbon and Has No Effect on Soil Microbial and Physical Properties a Year after Last Application. *Commun. Soil Sci. Plant Anal.* **2021**, *52*, 1264–1274. [CrossRef]
38. Furtado e Silva, J.A.M.; do Amaral Sobrinho, N.M.B.; Lima, E.S.A.; García, A.C. Modifications of soil organic matter structure by long-term pig slurry amendment of tropical soil. *Arch. Agron. Soil Sci.* **2020**, *68*, 1–15. [CrossRef]
39. Büyükkılıç Yanardağ, A.; Faz Cano, Á.; Mermut, A.; Yanardağ, İ.H.; Gomez Garrido, M. Organic carbon fluxes using column leaching experiments in soil treated with pig slurry in SE Spain. *Arid L. Res. Manag.* **2020**, *34*, 136–151. [CrossRef]
40. Marszałek, M.; Kowalski, Z.; Makara, A. The Possibility of Contamination of Water-Soil Environment as a Result of the Use of Pig Slurry. *Ecol. Chem. Eng. S* **2019**, *26*, 313–330. [CrossRef]
41. Yuan, Z.; Pan, X.; Chen, T.; Liu, X.; Zhang, Y.; Jiang, S.; Sheng, H.; Zhang, L. Evaluating environmental impacts of pig slurry treatment technologies with a life-cycle perspective. *J. Clean. Prod.* **2018**, *188*, 840–850. [CrossRef]

42. Ghirardini, A.; Grillini, V.; Verlicchi, P. A review of the occurrence of selected micropollutants and microorganisms in different raw and treated manure–Environmental risk due to antibiotics after application to soil. *Sci. Total Environ.* **2020**, *707*, 136118. [CrossRef] [PubMed]
43. Lorencová, E.Š.; Frélichová, J.; Nelson, E.; Vačkář, D. Past and future impacts of land use and climate change on agricultural ecosystem services in the Czech Republic. *Land Use Policy* **2013**, *33*, 183–194. [CrossRef]
44. Lošák, T.; Čermák, P.; Hlušek, J. Changes in fertilisation and liming of soils of the Czech Republic for the past 20 years. *Arch. Agron. Soil Sci.* **2012**, *58*, 17–20. [CrossRef]
45. Kubát, J.; Lipavský, J. Evaluation of organic matter content in arable soils in the Czech Republic. In *Crop Science and Land use for Food and Bioenergy*; Agrobios Publications: Jodhpur, India, 2011; p. 500. ISBN 9788190430982.
46. Wei, W.; Yan, Y.; Cao, J.; Christie, P.; Zhang, F.; Fan, M. Effects of combined application of organic amendments and fertilizers on crop yield and soil organic matter: An integrated analysis of long-term experiments. *Agric. Ecosyst. Environ.* **2016**, *225*, 86–92. [CrossRef]
47. Jaskulska, I.; Lemanowicz, J.; Breza-Boruta, B.; Siwik-Ziomek, A.; Radziemska, M.; Dariusz, J.; Białek, M. Chemical and biological properties of sandy loam soil in response to long-term organic-mineral fertilisation in a warm-summer humid continental climate. *Agronomy* **2020**, *10*, 1610. [CrossRef]
48. Domingo-Olivé, F.; Bosch-Serra, À.D.; Yagüe, M.R.; Poch, R.M.; Boixadera, J. Long term application of dairy cattle manure and pig slurry to winter cereals improves soil quality. *Nutr. Cycl. Agroecosyst.* **2016**, *104*, 39–51. [CrossRef]
49. Zavattaro, L.; Bechini, L.; Grignani, C.; van Evert, F.K.; Mallast, J.; Spiegel, H.; Sandén, T.; Pecio, A.; Giráldez Cervera, J.V.; Guzmán, G.; et al. Agronomic effects of bovine manure: A review of long-term European field experiments. *Eur. J. Agron.* **2017**, *90*, 127–138. [CrossRef]
50. Qaswar, M.; Yiren, L.; Jing, H.; Kaillou, L.; Mudasir, M.; Zhenzhen, L.; Hongqian, H.; Xianjin, L.; Jianhua, J.; Ahmed, W.; et al. Soil nutrients and heavy metal availability under long-term combined application of swine manure and synthetic fertilizers in acidic paddy soil. *J. Soils Sediments* **2020**, *20*, 2093–2106. [CrossRef]
51. Věžník, A.; Král, M.; Svobodová, H. Agriculture of the Czech Republic in the 21st century: From productivism to post-productivism. *Quaest. Geogr.* **2013**, *32*, 7–14. [CrossRef]
52. Czech Statistical Office. *Statistical yearbook of the Czech Republic 2020*; Czech Statistical Office: Prague, Czech Republic, 2021.
53. Af Geijersstam, L.; Mårtensson, A. Nitrogen fixation and residual effects of field pea intercropped with oats. *Acta Agric. Scand. Sect. B Soil Plant Sci.* **2006**, *56*, 186–196. [CrossRef]
54. Buraczyńska, D.; Ceglarek, F.; Gasiorowska, B.; Zaniewicz-Bajkowska, A.; Płaza, A. Cultivation of wheat following pea and triticale/pea mixtures increases yields and nitrogen content. *Acta Agric. Scand. Sect. B Soil Plant Sci.* **2011**, *61*, 622–632. [CrossRef]
55. Škarpa, P.; Školníková, M.; Antošovský, J.; Horký, P.; Smýkalová, I.; Horáček, J.; Dostálová, R.; Kozáková, Z. Response of normal and low-phytate genotypes of pea (*Pisum sativum* L.) on phosphorus foliar fertilization. *Plants* **2021**, *10*, 1680. [CrossRef] [PubMed]
56. Wen, G.; Chen, C.; Neill, K.; Wichman, D.; Jackson, G. Yield response of pea, lentil and chickpea to phosphorus addition in a clay loam soil of central Montana. *Arch. Agron. Soil Sci.* **2008**, *54*, 69–82. [CrossRef]
57. Brkić, S.; Milaković, Z.; Kristek, A.; Antunović, M. Pea yield and its quality depending on inoculation, nitrogen and molybdenum fertilization. *Plant Soil Environ.* **2004**, *50*, 39–45. [CrossRef]
58. Deibert, E.J.; Utter, R.A. Field pea growth and nutrient uptake: Response to tillage systems and nitrogen fertilizer applications. *Commun. Soil Sci. Plant Anal.* **2004**, *35*, 1141–1165. [CrossRef]
59. Beck, H.E.; Zimmermann, N.E.; McVicar, T.R.; Vergopolan, N.; Berg, A.; Wood, E.F. Present and future köppen-geiger climate classification maps at 1-km resolution. *Sci. Data* **2018**, *5*, 1–12. [CrossRef]
60. FAOUN. World reference base for soil resources 2014. In *International Soil Classification System for Naming Soils and Creating Legends for Soil Maps*; FAOUN: Rome, Italy, 2014; ISBN 9789251083697.
61. Risberg, K.; Cederlund, H.; Pell, M.; Arthurson, V.; Schnürer, A. Comparative characterization of digestate versus pig slurry and cow manure–Chemical composition and effects on soil microbial activity. *Waste Manag.* **2017**, *61*, 529–538. [CrossRef]
62. Kowalski, Z.; Makara, A.; Fijorek, K. Changes in the properties of pig manure slurry. *Acta Biochim. Pol.* **2013**, *60*, 845–850. [CrossRef]
63. Kožnarová, V.; Klabzuba, J. Recommendation of World Meteorological Organization to describing meteorological or climatological conditions-Information. *Plant Soil Environ.* **2011**, *48*, 190–192. [CrossRef]
64. Kirk, P.L. Kjeldahl Method for Total Nitrogen. *Anal. Chem.* **1950**, *22*, 354–358. [CrossRef]
65. Mehlich, A. Mehlich 3 soil test extractant: A modification of Mehlich 2 extractant. *Commun. Soil Sci. Plant Anal.* **1984**, *15*, 1409–1416. [CrossRef]
66. Sims, J.R.; Haby, V.A. Simplified colorimetric determination of soil organic matter. *Soil Sci.* **1971**, *112*, 137–141. [CrossRef]
67. Nelson, D.W.; Sommers, L.E. Total Carbon, Organic Carbon, and Organic Matter. In *Total Carbon, Organic Carbon, and Organic Matter: Methods of Soil Analysis Part 3-Chemical Methods*; American Society of Agronomy: Madison, WI, USA, 1996; pp. 961–1010.
68. Arnhold, E. Easynls: Easy Nonlinear Model, R Package Version 5.0. 2017. Available online: https://cran.r-project.org/web/packages/easynls/easynls.pdf (accessed on 11 January 2021).
69. Mangiafico, S. *Functions to Support Extension Education Program Evaluation*; The Comprehensive R Archive Network: Vienna, Austria, 2019; pp. 1–82.

70. Wickham, H.; Averick, M.; Bryan, J.; Chang, W.; McGowan, L.; François, R.; Grolemund, G.; Hayes, A.; Henry, L.; Hester, J.; et al. Welcome to the Tidyverse. *J. Open Source Softw.* **2019**, *4*, 1686. [CrossRef]
71. Xiang, J.; Haden, V.R.; Peng, S.; Bouman, B.A.M.; Visperas, R.M.; Nie, L.; Huang, J.; Cui, K. Improvement in nitrogen availability, nitrogen uptake and growth of aerobic rice following soil acidification. *Soil Sci. Plant Nutr.* **2009**, *55*, 705–714. [CrossRef]
72. Rutkowska, B.; Szulc, W.; Sosulski, T.; Stępień, W. Soil micronutrient availability to crops affected by long-term inorganic and organic fertilizer applications. *Plant Soil Environ.* **2014**, *60*, 198–203. [CrossRef]
73. An, G.; Miyakawa, S.; Kawahara, A.; Osaki, M.; Ezawa, T. Community structure of arbuscular mycorrhizal fungi associated with pioneer grass species Miscanthus sinensis in acid sulfate soils: Habitat segregation along pH gradients. *Soil Sci. Plant Nutr.* **2008**, *54*, 517–528. [CrossRef]
74. Aciego Pietri, J.C.; Brookes, P.C. Relationships between soil pH and microbial properties in a UK arable soil. *Soil Biol. Biochem.* **2008**, *40*, 1856–1861. [CrossRef]
75. Fan, J.; Ding, W.; Chen, Z.; Ziadi, N. Thirty-year amendment of horse manure and chemical fertilizer on the availability of micronutrients at the aggregate scale in black soil. *Environ. Sci. Pollut. Res.* **2012**, *19*, 2745–2754. [CrossRef]
76. Hoyt, P.B.; Hennig, A.M.F. Soil Acidification By Fertilizers and Longevity of Lime Applications in the Peace River Region. *Can. J. Soil Sci.* **1982**, *62*, 155–163. [CrossRef]
77. Mahler, R.L.; Wilson, S.; Shafii, B.; Price, W. Long-Term Trends of Nitrogen and Phosphorus Use and Soil pH Change in Northern Idaho and Eastern Washington. *Commun. Soil Sci. Plant Anal.* **2016**, *47*, 414–424. [CrossRef]
78. Hoffmann, S.; Berecz, K.; Tóth, Z. Soil fertility as affected by long-term fertilization and crop sequence. *Arch. Agron. Soil Sci.* **2010**, *56*, 481–488. [CrossRef]
79. Macholdt, J.; Piepho, H.P.; Honermeier, B. Mineral NPK and manure fertilisation affecting the yield stability of winter wheat: Results from a long-term field experiment. *Eur. J. Agron.* **2019**, *102*, 14–22. [CrossRef]
80. Körschens, M.; Albert, E.; Armbruster, M.; Barkusky, D.; Baumecker, M.; Behle-Schalk, L.; Bischoff, R.; Čergan, Z.; Ellmer, F.; Herbst, F.; et al. Effect of mineral and organic fertilization on crop yield, nitrogen uptake, carbon and nitrogen balances, as well as soil organic carbon content and dynamics: Results from 20 European long-term field experiments of the twenty-first century. *Arch. Agron. Soil Sci.* **2013**, *59*, 1017–1040. [CrossRef]
81. Shi, R.Y.; Liu, Z.D.; Li, Y.; Jiang, T.; Xu, M.; Li, J.Y.; Xu, R. kou Mechanisms for increasing soil resistance to acidification by long-term manure application. *Soil Tillage Res.* **2019**, *185*, 77–84. [CrossRef]
82. Ning, Q.; Chen, L.; Jia, Z.; Zhang, C.; Ma, D.; Li, F.; Zhang, J.; Li, D.; Han, X.; Cai, Z.; et al. Agriculture, Ecosystems and Environment Multiple long-term observations reveal a strategy for soil pH-dependent fertilization and fungal communities in support of agricultural production. *Agric. Ecosyst. Environ.* **2020**, *293*, 106837. [CrossRef]
83. Xiang, X.; Liu, J.; Zhang, J.; Li, D.; Xu, C.; Kuzyakov, Y. Soil & Tillage Research Divergence in fungal abundance and community structure between soils under long-term mineral and organic fertilization. *Soil Tillage Res.* **2020**, *196*, 104491. [CrossRef]
84. Nobile, C.M.; Bravin, M.N.; Becquer, T.; Paillat, J. Chemosphere Phosphorus sorption and availability in an andosol after a decade of organic or mineral fertilizer applications: Importance of pH and organic carbon modi fi cations in soil as compared to phosphorus accumulation. *Chemosphere* **2020**, *239*, 124709. [CrossRef]
85. Barak, P.; Jobe, B.O.; Krueger, A.R.; Peterson, L.A.; Laird, D.A. Effects of long-term soil acidification due to nitrogen fertilizer inputs in Wisconsin. *Plant Soil* **1997**, *197*, 61–69. [CrossRef]
86. Matějková, Š.; Kumhálová, J.; Lipavský, J. Evaluation of crop yield under different nitrogen doses of mineral fertilization. *Plant Soil Environ.* **2010**, *56*, 163–167. [CrossRef]

land

Article

Perceived Causes and Solutions to Soil Degradation in the UK and Norway

Niki Rust [1], Ole Erik Lunder [2], Sara Iversen [3], Steven Vella [4], Elizabeth A. Oughton [1], Tor Arvid Breland [2], Jayne H. Glass [5], Carly M. Maynard [5], Rob McMorran [5] and Mark S. Reed [6,*]

1 Centre for Rural Economy, Newcastle University, Kings Road, Newcastle NE2 1HA, UK; niki_rust@hotmail.co.uk (N.R.); Elizabeth.Oughton@newcastle.ac.uk (E.A.O.)
2 Faculty of Biosciences, Norwegian University of Life Science, 1432 Aas, Norway; Ole@greenhouse.bio (O.E.L.); tor.arvid.breland@nmbu.no (T.A.B.)
3 Department of Agroecology, Aarhus University, Blichers Allé, 8830 Tjele, Denmark; sara.iversen@agro.au.dk
4 School of the Built Environment, Birmingham City University, 15 Bartholomew Row, Birmingham B5 5JU, UK; stevenvella@gmail.com
5 Rural Policy Centre, Department of Rural Economy, Environment & Society, Scotland's Rural College (SRUC), Peter Wilson Building, Kings Buildings, West Mains Road, Edinburgh EH9 3JG, UK; Jayne.Glass@sruc.ac.uk (J.H.G.); Carly.Maynard@sruc.ac.uk (C.M.M.); Rob.McMorran@sruc.ac.uk (R.M.)
6 Thriving Natural Capital Challenge Centre, Department of Rural Economy, Environment & Society, Scotland's Rural College (SRUC), Peter Wilson Building, Kings Buildings, West Mains Road, Edinburgh EH9 3JG, UK
* Correspondence: mark.reed@sruc.ac.uk

Citation: Rust, N.; Lunder, O.E.; Iversen, S.; Vella, S.; Oughton, E.A.; Breland, T.A.; Glass, J.H.; Maynard, C.M.; McMorran, R.; Reed, M.S. Perceived Causes and Solutions to Soil Degradation in the UK and Norway. *Land* **2022**, *11*, 131. https://doi.org/10.3390/land11010131

Academic Editors: Guido Wyseure, Julián Cuevas González, Jean Poesen and Christine Fürst

Received: 24 November 2021
Accepted: 13 January 2022
Published: 14 January 2022

Abstract: Soil quality is declining in many parts of the world, with implications for the productivity, resilience and sustainability of agri-food systems. Research suggests multiple causes of soil degradation with no single solution and a divided stakeholder opinion on how to manage this problem. However, creating socially acceptable and effective policies to halt soil degradation requires engagement with a diverse range of stakeholders who possess different and complementary knowledge, experiences and perspectives. To understand how British and Norwegian agricultural stakeholders perceived the causes of and solutions to soil degradation, we used Q-methodology with 114 respondents, including farmers, scientists and agricultural advisers. For the UK, respondents thought the causes were due to loss of soil structure, soil erosion, compaction and loss of organic matter; the perceived solutions were to develop more collaborative research between researchers and farmers, invest in training, improve trust between farmers and regulatory agencies, and reduce soil compaction. In Norway, respondents thought soils were degrading due to soil erosion, monocultures and loss of soil structure; they believed the solutions were to reduce compaction, increase rotation and invest in agricultural training. There was an overarching theme related to industrialised agriculture being responsible for declining soil quality in both countries. We highlight potential areas for land use policy development in Norway and the UK, including multi-actor approaches that may improve the social acceptance of these policies. This study also illustrates how Q-methodology may be used to co-produce stakeholder-driven policy options to address land degradation.

Keywords: conservation agriculture; deliberative democracy; q-methodology; regenerative agriculture; soil conservation; sustainable land management

1. Introduction

> "Countries can withstand coups d'état, wars and conflict, even leaving the EU, but no country can withstand the loss of its soil and fertility". (Rt Hon Michael Gove, former Secretary of State for the Environment, speaking at the British parliamentary launch of the 'Sustainable Soils Alliance', October 2017).

The ground beneath our feet is not only a substrate upon which we traverse this earth but is also a vital component of our natural capital. Soils are the foundation of terrestrial food production, supporting directly or indirectly 95% of our food production [1]. Along with providing a substrate to grow our food, soils also confer other essential ecosystem services, such as water storage and filtration, nutrient cycling, biodiversity and carbon storage [2]. However, demand for food, increasing human populations and the effects of climate change are placing unprecedented pressures on soil. Over the last 70 years, the supply of global per capita food calories increased by about one-third, with the use of irrigation water roughly doubling and use of inorganic nitrogen fertiliser increasing nearly nine-fold [3]. At the same time, climate change has led to faster rates of warming on land than the global mean and altered precipitation patterns, which have contributed to altered growing seasons and regional crop yield reductions [4]. With rising human populations, coupled with increased individual wealth, it is expected that food demand will grow by as much as 70% by 2050; an estimated 46% of that demand needs to come from increasing food production [5]. This increase in food productivity must be achieved whilst significantly reducing greenhouse gas emissions from agriculture, if warming is to be restricted "well below" 2 °C, as proposed in the Paris Agreement [6]. How this is achieved without negatively impacting soils any further remains a challenge.

Soil quality [1] in many parts of the world is declining due to a combination of physical, chemical and biological degradation coupled with socio-economic drivers, reducing the soil's ability to undertake these important ecosystem functions [7]. Globally, 20–30 gigatons of soil are lost each year due to water erosion [7] and climate change is projected to increase erosion from water and reduce levels of soil organic carbon, especially in drylands [4]. There is thus an urgent need to develop and encourage widespread adoption of effective and profitable sustainable soil management practices [8,9]. This is articulated in Sustainable Development Goal 15, which aims to "protect, restore and promote sustainable use of terrestrial ecosystems, sustainably manage forests, combat desertification, and halt and reverse land degradation and halt biodiversity loss" [10], and its Land Degradation Neutrality target which aims to counterbalance expected losses with measures to achieve equivalent gains within the same type of land [11].

There are many competing methods to deal with agricultural soil degradation at different governance scales: from multilateral policies such as the United Nations Convention to Combat Desertification (UNCCD) and the proposed EU Soils Directive, to national and sub-national policies and measures designed to incentivise and regulate the management of soils. These policies typically seek changes in management at farm and field scales, for example through the adoption of soil-improving cropping systems and other sustainable land management technologies and approaches (e.g., WOCAT [12]). The lack of scalable policy options was cited in the UNCCD's Global Land Outlook [9] as a key barrier to more sustainable land management, but there are no easy solutions given the different social, cultural, economic, environmental and technological contexts in which policies and practices need to operate [8]. Again, the attractiveness and appropriateness of different options for policy and practices differ based on the subjective experience and contrasting knowledge and values of the people the policies are meant to serve.

Policies and practices that can tackle the multiple causes of declining soil quality are urgently needed and stakeholder engagement in the policy formulation process is crucial for this complex issue, given the subjective and value-laden nature of both the causes of and solutions to the challenge. Effectively representing diverse stakeholder perspectives in decision-making processes can lead to better informed, more durable, and flexible outcomes across a wide range of contexts (De Vente et al., 2016). Policies created through deliberative democracy can align better with social and cultural norms, resulting in increased trust and ownership of problems and solutions; together, this can lead to decisions that are more likely to be accepted and implemented, helping to achieve environmental goals more effectively [13,14].

As interest has grown in participatory approaches to policy making and other forms of deliberative democracy, methods have been sought to represent and integrate the range of perspectives, values and beliefs held by citizens in the policy-making process [15]. The application of Q-methodology to the co-production of policy options with stakeholders has been used by Rust [16] and Addams and Proops [17] as a form of deliberative democracy. These studies had the normative goal of representing more diverse perspectives in the policy-making process. They also had a pragmatic goal of improving the quality of decisions or range of policy options based on more comprehensive information inputs and/or improving the acceptability of policies based on deeper insights into the way publics conceptualise environmental issues.

In this study, we used Q-methodology to understand a wide range of stakeholder perspectives that could inform the design of socially acceptable options in agricultural soil management policy and practice. To address the lack of scalable policy options noted above by the UNCCD, this was done in the UK and Norway which are two countries experiencing similar types of soil degradation that are broadly representative of soil quality issues and climatic variation across northern temperate regions of Europe. The countries have contrasting agricultural policy environments, with Norway not being a part of the EU and the UK in the process of leaving the EU when the research was conducted. The countries also have quite different models of social democracy in land governance [18], which provides an opportunity to consider how the different land tenure regimes influence policy formulation and application. Both countries are interested in the co-development of policies to increase food production whilst reducing the environmental impacts of agriculture. By understanding where agricultural stakeholders agreed and disagreed over causes and solutions to declining soil quality in each country, we sought to highlight potential scalable options for land use policy development. This study also illustrates how Q-methodology may be used to co-produce stakeholder-driven policy options to address land degradation.

2. Materials and Methods

2.1. Study Sites

2.1.1. UK

Like much of the rest of Europe, the UK has a long history of unsustainable soil management practices, leading to a loss of soil structure and fertility. About 25% of the total land area of the UK is suitable for arable cropping, with an average farm size of 81 hectares [19]. Currently, soil erosion exceeds the rate of soil formation in many areas in the UK, with around 17% of arable land showing signs of erosion, although as much as 40% may be at risk of further degradation [20]. The cost of soil erosion to the UK has been estimated at £45 million a year, including £9 million in lost production [21]. UK soil is being lost at a rate ten times that which it is created [2], with dramatic economic implications.

A comparison of soil nutrient balances from the year 2000 to 2019 shows a 24% decrease for nitrogen and a 46% decrease for phosphate (in kg per hectare) [19]. Soil erosion, compaction and loss of organic matter are thought to cost arable farmers an average of £5584 per year [22] and English water companies spend £21 million a year on addressing soil erosion [23]. Improving soil management in the UK is therefore not only an environmental but also an economic imperative. Soil quality decline in the UK is more pronounced in arable regions due to the highly intensive practices used, such as monocropping, use of heavy machinery, overuse of chemical inputs and a lack of integration of organic material [21,24].

2.1.2. Norway

Only 3.1% of the total land area in Norway is suitable for arable cropping, with an average farm size of 23.9 hectares in 2016; cereals can only be grown on one third of this area due to limiting natural conditions [25]. Although agricultural policies in Norway advocate multifunctional agriculture [26], regional agricultural specialisation, known as "kanaliseringspolitikken", was introduced in the after-war period, which led to increased agricultural production by incentivising cereal production in lower-lying areas [27]. In the last two decades, the total Norwegian cereal yield has declined due to a reduction in the area used for cropping [28]. Despite this decline, the Norwegian government has set a target of increasing food production by 20% by 2030 from 2010 levels, to meet projected population growth in Norway [29]. Three counties (Akershus, Østfold, and Hedmark) in southeastern Norway produce 60% of the country's cereal; however, soil organic matter (SOM) content has declined in the region, with an average loss of 1% of SOM a year from 1991 to 2001, which is not sustainable [30]. The underpinning governance and institutions (both formal and informal) are strongly communal in character [26]. The long history of collective land management, the regulation of the Norwegian land market and the self-imposed limits to farm scale are in contrast to the generally unregulated land market and existence of larger-scale farms in the UK.

2.2. Research Design

Q-methodology is a mixed-methods approach using interviews to explore participants' subjective understanding of a topic using Q sorts where respondents rate the extent to which they agree with statements, which are then analysed using by-person factor analysis, correlating people with others who hold similar opinions based on their Q-sorts. Q-methodology was chosen due to its capacity to shed light on complex, subjective phenomena where individuals hold differing views and values [30]. It allows for exploration of tensions in knowledge and perspectives between stakeholders that may affect the effectiveness and acceptability of a land use policy. The results can show areas of statistical agreement and disagreement, whilst also revealing distinct narratives emerging from groups of respondents [31,32]. When applied to situations with conflicted stakeholder dynamics, Q-methodology can be useful in identifying common ground among diverse stakeholders in situations where conservation or resource management is contested [16,33]. This makes the method particularly useful for this study due to the above benefits.

2.3. Data Collection

The research undertaken in the UK took place in late 2018 (when the UK was still part of the EU) and in Norway in mid-2019. Q-methodology studies commonly begin by using a qualitative approach, where interviews are undertaken with a range of stakeholders on a study's topic to gather the diversity of opinions on the phenomenon in question. This data collection can be enhanced or replaced with a literature review. This qualitative step is used to develop the "concourse", which is the range of views (listed as statements) held on a topic, followed by a structured, quantitative interview where participants rank the concourse statements, usually based on the extent to which they agree/disagree. During these interviews, qualitative information is gathered from participants on their decision-making processes and preferences. Because the concourse is designed to cover as closely as possible all perspectives on a topic, and participants are chosen to cover the range of views, then random sampling from the wider population is not necessary. Because of non-random sampling and smaller sample sizes, conclusions cannot be generalised but the aim is to understand the range rather than the frequency of the views, and to find points of convergence or divergence of opinion.

The concourse for this study was developed by interviewing 18 European agricultural stakeholders on causes of declining soil quality and corresponding solutions. Interviewees were purposefully chosen to represent researchers, land managers and other stakeholders from ten European countries participating in the wider project, SoilCare, on which this study is based. Ten researchers and eight other stakeholders (representing agricultural unions, farmers and other landowners) were interviewed. An interview guide was used, which was piloted on a subset of the sample population and amended due to feedback. Interview themes and prompts are shown in Table 1. Interviews were undertaken by telephone or Skype and lasted an average of an hour. Free, prior informed consent was obtained from all interviewees and ethical approval was gained from Newcastle University. Interviews were recorded with permission from the participants and later transcribed.

Interviews were conducted in English, apart from one which took place in Italian, which was later translated to English for analysis.

Table 1. Interview guide used to develop the concourse.

Questions	Prompts
What do you see as the main threats to soil quality in Europe?	These may be general or specific to the locality
What roles do you see to changes in cropping practices to overcome these threats or to improve soil quality?	These may be general or specific to the locality or threat
How do you know if these approaches are actually improving the soil?	
Who should have primary responsibility for improving soils in your country or across the EU?	Individual versus collective; Private versus public, etc.
In your experience, why do people promote or adopt soil-improving cropping systems?	Why should people promote or adopt these?
What factors incentivize or prevent soil improvement from farm to landscape scales?	
Name as many reasons as you can why farmers may choose to adopt soil-improving cropping systems or not	

A narrative review was undertaken, based on a broad-based search for relevant material, to provide further evidence to supplement the interviews and further expand the concourse. This review was to ensure that the topic was sufficiently covered by the statements developed from the interview data. Data were then analysed using a thematic analysis focusing on reasons for soil quality decline and solutions for how to fix this. A total of 142 statements was obtained from across all interviews and the literature review, which included statements both for the problem Q-set and the solution Q-set.

Similar statements for each set were merged, whilst trying to retain as far as possible the original wording of the interviews to capture the intent of the source. For both studies, some statements arising from the literature were amended subtly to match the country's context, e.g., changing the statement "EU agricultural policy" to "Norwegian agricultural policy", and adding local problems such as drainage. For the UK study, this resulted in 41 statements for the "problems" Q-set and 34 statements for the "solutions" Q-set, and in Norway, this resulted in 42 problem statements and 36 solution statements (see Tables A1–A4, Appendix A).

A "Q-sort" is the ranking of the Q-set by participants. Data collection for the Q-sort was undertaken via an online survey using Google Forms. The Q-sort survey was first piloted on a subset of the target population and subsequently adapted following feedback to improve question clarity and to include additional statements that were not captured via the interviews or literature review. Participations then ranked the statements on a scale of −2 (strongly disagree) to +2 (strongly agree). The UK survey was distributed via soil-specific newsgroups, British agricultural union members and by sharing on agricultural social media channels. A total of 61 UK respondents undertook the survey: 19 scientists, 19 farmers, 16 agricultural advisers, three water company employees who work in agriculture, two nature conservationists, one agricultural union representative and one civil servant. For the Norwegian study, a link to the survey was distributed in "Plantenytt", a newsletter from the government extension service Norsk Landbruksrådgivning Øst and to a local "soil education group". Forty-two Norwegian farmers took part in the survey, as well as six agricultural advisers and five scientists, totaling 53 respondents. The substantial weighting towards farmers in the Norwegian study was deemed acceptable due to the smaller average farm size in Norway and the historical legacy of communal land management which is embedded in national agricultural institutions [26]. However, the findings of our analysis may need to be interpreted in light of the greater diversity of stakeholders in the UK study.

At the end of the survey, participants were asked what they thought was the leading cause of declining soil quality and the most important solution to solve this problem. Participants could choose a statement from the Q-sort or add a new statement. These open-ended questions were used to find out what, subjectively, respondents thought were the most important drivers for causing declining soil quality and how to fix these. Data from these open-ended questions were analysed via thematic analysis. Quotes in the results section are used to highlight common sentiments as well as responses that stood apart from the rest. Quotes from the Norwegian study were translated into English.

2.4. Analysis

Data from the Q-sorts were analysed using KenQ (https://shawnbanasick.github.io/ken-q-analysis, accessed on 10 January 2022). First, a principal component analysis (PCA) was used to identify the groups of participants who ranked their Q-sorts similarly, also known as a "loaded factor". Flags were automatically added to respondents that significantly loaded onto these factors at $p < 0.05$.

For the UK study, the PCA for the problem Q-sort revealed eight factors with Eigenvalues >1 (which together explained 67% of the variance) but most loaded onto factors 1–4 (which together explained 53% of the variance). Large datasets, such as in this study, run the risk of inflating the Eigenvalues [34]. Because of this, we focused on the first four factors for the problem set as this explained over half the variance. A Varimax rotation was then applied to the four factors, which calculated the highest variability between factors. A z-score was calculated based on the average ranking participants gave to the statement within each factor group. Respondents that significantly loaded onto more than one factor were excluded from subsequent analysis because their inclusion gives little information about the clustering of opinions. Statistical disagreement (and agreement) between participants was set where $p > 0.01$, which meant that the groups of participants did (not) rank the statements differently at the 99% confidence level. The PCA for the solutions Q-sort revealed eight factors with Eigenvalues > 1 (which together explained 79% of the variance) but most loaded onto factors 1–3 (which together explained 65% of the variance). The rest of the solutions analysis followed the same process as with the problem Q-sort.

For the Norwegian study, the analysis followed the same procedure as the UK study. For the problem sort, eight principal components with Eigenvalue above 1 were extracted through the PCA, which explained 69% of the variance. Most of the participants loaded onto the first three problem factors, which together explained 51% of the variation, and these three factors were carried forward for further analysis. For the solution sort, eight factors with Eigenvalues above 1 were extracted, explaining 78% of the variance, though as respondents loaded onto factors 1–3, explaining 63% of the variance, these three factors were used in further analysis.

3. Results

This section describes results from the problems Q-sorts (Table A1, Appendix A: UK; Table A2, Appendix A: Norway) and solutions Q-sorts (Table A3, Appendix A: UK; Table A4, Appendix A: Norway). The number of respondents loading onto each factor (i.e., who ranked statements similarly) is shown in Figure 1 (UK problem Q-sort), Figure 2 (Norway problem Q-sort), Figure 3 (UK solution Q-sort) and Figure 4 (Norway solution Q-sort). Results are grouped under the key defining factors that emerged from each Q-sort, which are summarised in short, narrative phrases based on the main defining traits of each factor. Key areas of consensus and disagreement that emerged across these different groupings are then highlighted.

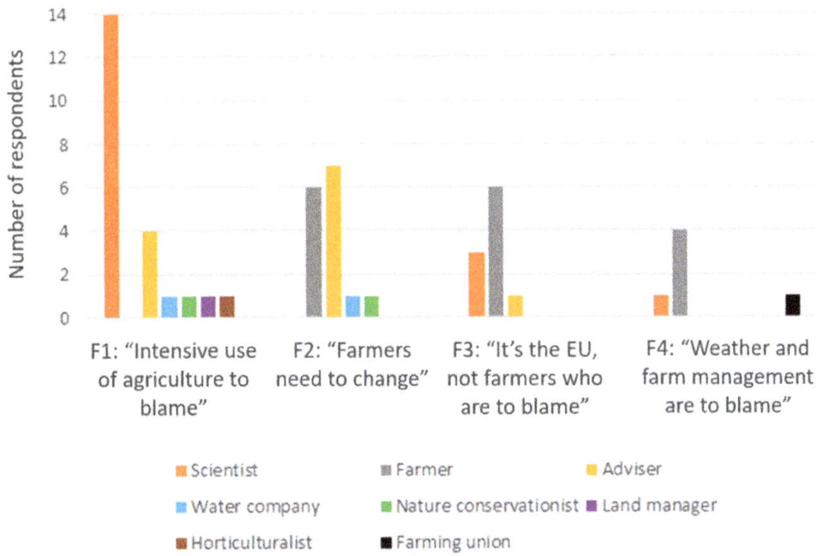

Figure 1. Professions of UK respondents loading onto the four problem factor groups.

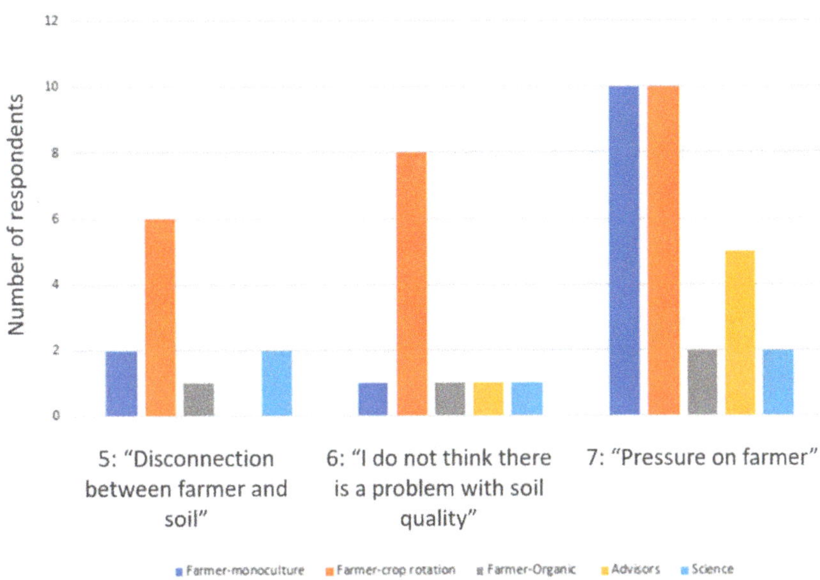

Figure 2. Professions of Norwegian respondents loading onto the three problem factor groups.

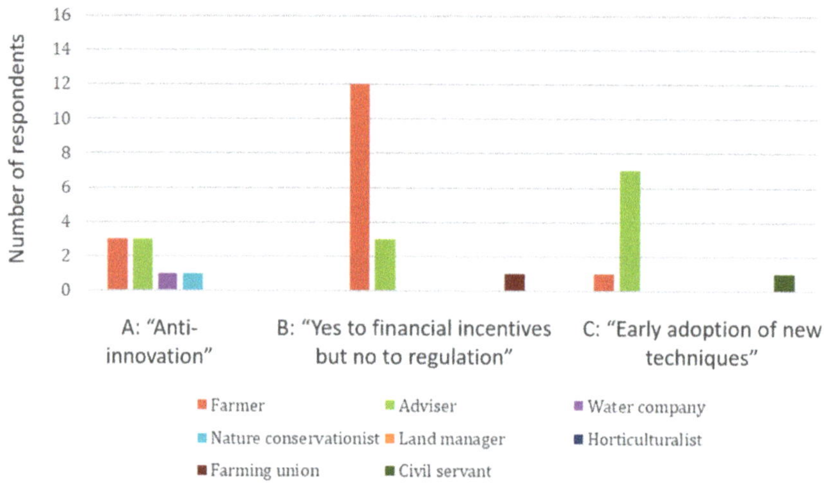

Figure 3. Professions of UK respondents loading onto the three solution factor groups.

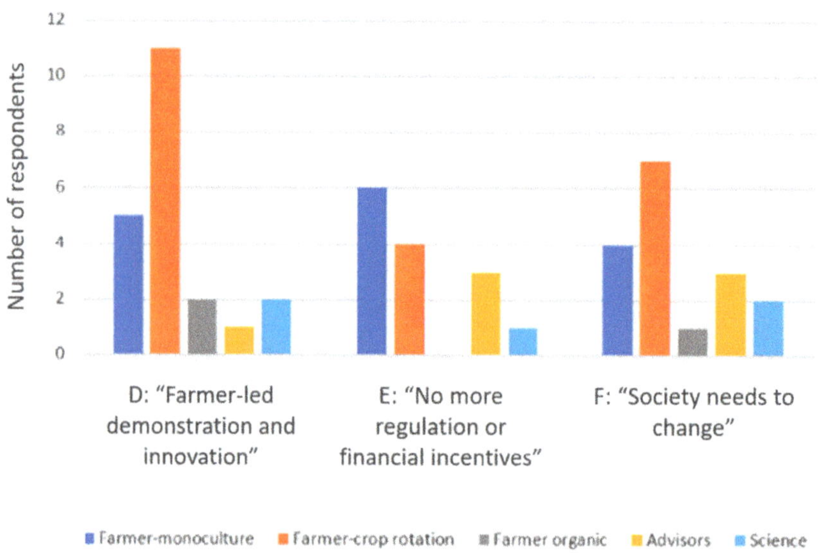

Figure 4. Professions of Norwegian respondents loading onto the three solution factor groups.

3.1. *Perceived Problems Causing Declining Soil Quality*

3.1.1. UK Study

Factor 1: "Intensive Agriculture to Blame"

This factor was defined by respondents who were significantly more likely to think the problems causing declining soil quality were due to "intensive use of soil without time to recover" and "overuse of inputs", more strongly agreeing with these statements than other factors. In contrast, they strongly disagreed that the problem was caused by the fact that "soil has become too saline", ranking this statement more negatively than other factors.

Factor 2: "Farmers Need to Change"

This factor was defined by respondents more strongly agreeing than other factors that "lack of knowledge of soils amongst farmers" and "some traditions of farmers are damaging" were causing problems in soil quality. Conversely, they more strongly disagreed with statements regarding "declining level of nutrient status", "loss of number of wild species" and "I do not believe there is a problem with soil quality" than other respondents.

Factor 3: "It's the EU, Not Farmers That Are to Blame"

Respondents here were defined by more strongly agreeing than other factors with the idea that "EU agricultural policy" was the cause of declining soil quality, along with "lack of knowledge of soils amongst farmers" and "natural local climate constraints". Conversely, they strongly disagreed that the problems were "use of contractors" and "loss of numbers of wild species" compared with other respondents.

Factor 4: "Weather and Farm Management to Blame"

Respondents here more strongly agreed than other factors that "pressure on farmers to produce at a low cost", "choice of cropping system" and "flooding or drought" were causing problems with soil. Conversely, they more strongly disagreed with statements regarding "lack of knowledge of soils amongst farmers", "overuse of inputs" and "distrust of scientists by farmers" were causing problems, when compared with other respondents.

Areas of Agreement and Disagreement

Respondents in all factors strongly agreed that soil quality was declining due to loss of soil structure, and agreed/strongly agreed that compaction, soil erosion, loss of organic matter and insufficient knowledge exchange were other causes (Table 1). In contrast, they did not think that the causes were due to farmers having little control over their land or due to a distrust of scientists by farmers. Conversely, the only area of statistical dissensus between each factor was "lack of knowledge of soils amongst farmers" (Table 1).

Leading Causes of Declining Soil Quality

There were two themes that were frequently mentioned by UK respondents as the leading causes of declining soil quality when answering this open-ended question. The first group blamed market pressures for pushing farmers into intensifying farming, with a sentiment that an ever-increasing drive to produce more food at cheaper costs was a fundamental driver of unsustainable land management, including soil quality decline. This could relate to the Q-set statement 21, "pressure on farmers to produce at low cost", which respondents in Factors 1, 3 and 4 strongly agreed was a cause for declining soil quality. This sentiment is captured by an agricultural adviser (UK2) who said:

> "There is an increasing demand to produce cheaper food for a larger population using the same/declining land area. Pressure is put on producers by supermarkets and the general public to provide food to contracts, often unknowingly, which results in poor management choices".

Conversely, the second group blamed farmers and thought that intensive agricultural practices, such as ploughing and insufficient crop rotation, were the leading causes of soil quality decline. Many respondents felt this was due to a lack of understanding by the farmer of better soil management practices. One nature conservationist (UK4) summarised this theme by saying the problems causing declining soil quality were due to:

> "Traditional' farming practices and cropping, which means too many farmers not being innovative/open to new methods. Time to start re-thinking about how we measure what makes a successful farm-it's not all about productivity".

This sentiment was not reflected in the answers to the problem Q-sort. One of the reasons for this could be that it encapsulates many of the problem statements related to farm management, as it is a multi-faceted and complex problem.

3.1.2. Norway Study

Factor 5: "Disconnection between Farmer and Soil"

Respondents in this factor were more likely to rank "poor management of the soil" as one of the main reasons for the decline in soil quality. This group disagreed more strongly with the statement "I do not think there is a problem with soil quality". Instead, they ranked statements on agricultural practices and farmers' knowledge as leading causes for declining soil quality, such as "farmer has lost the finer touch with his land", "overuse of input like fertiliser and chemicals", and "lack of knowledge of soil amongst farmers".

Factor 6: "There Is No Problem with the Soil Quality"

Respondents in this factor disagreed that agricultural practices are reasons leading to a decline in soil quality such as "intensive agriculture to blame" and "overuse of input like fertilisers and chemicals". They also strongly disagreed with the statements "use of contractors", "too much leased land" or "farmers have lost the finer touch with their land". They agreed more strongly than others with the statements "too little advice on soil-improving practices" and "lack of knowledge-sharing between scientists, advisors, and farmers" as problems for soil quality.

Factor 7: "Industrialised/Intensive Agriculture to Blame"

Respondents in this factor thought the problems were often outside of the farmer's actions and responsibility compared to factor 5, being significantly more likely to agree on "pressure on the farmer to produce at low cost", and "intensive agriculture" than the other factors. This group also more strongly agreed about structural characteristics like "too large farms" and "high share of leased land" as problems of declining soil quality compared with other factor groups.

Areas of Agreement and Disagreement

Respondents in the three factors agreed/strongly agreed that soil quality was declining because of "soil erosion", "repetition of the same crop, year after year", and "loss of soil structure". There were also numerous areas of statistical disagreement between the factor groups (Table A2, Appendix A), such as knowledge/education, environmental conditions and management of the farm.

Leading Causes of Declining Soil Quality

There were two common perceived causes of declining soil quality. The first was an increase in use of large machinery causing soil compaction, captured by the statement (N5):

> "Larger farms stimulate heavier machinery leading to more compaction" and " … modern machinery can drive in unfavourable conditions".

The second aspect was lack of crop rotation, which respondents felt contributed to declining levels of SOM, while some connected monocultures to the regional specialisation policy.

3.2. Perceived Solutions to Address Declining Soil Quality

3.2.1. UK Study

Factor A: "Anti-Innovation"

This factor was defined by respondents more strongly agreeing that there is not much we can do to improve soil quality and that the problems were due to natural climatic constraints. They also disagreed with innovations and increasing early adoption of new techniques to solve the issue, ranking these statements more negatively than other factors.

Factor B: "Yes to Financial Incentives but No to Regulation"

Respondents here more strongly agreed that financial incentives could be a solution but more strongly disagreed that restrictive policies, such as more regulation (including for fertilisers and to reduce water usage) and creating a Soil Directive, would improve soil quality.

Factor C: "Early Adoption of New Techniques"

This factor was defined by respondents more significantly agreeing to increasing early adoption of new techniques as a solution to declining soil quality. They also more strongly disagreed that solutions were maintaining small farms, giving more freedom for farmers to manage their land as they would like, and that farmers have already tried lots of things.

Areas of Agreement and Disagreement

Respondents loading onto the three factors strongly agreed that more research should be done in collaboration with farmers, and all agreed in investing in education and training (Table A3, Appendix A). They also agreed that we should work towards improving trust between farmers and regulatory agencies and initiatives to reduce compaction. Respondents did not think changing the timing of tillage would improve soil quality. There was disagreement on numerous solutions, particularly around maintaining small farms, increasing the early adoption of new techniques and giving more freedom to farmers to manage their land.

The Most Important Perceived Solutions to Addressing Soil Quality Decline

There were two main themes that emerged in the responses to the open-ended question, with the first (and most common) requesting improved knowledge exchange between agricultural stakeholders. This links to the Q-set statements on "more research should be done in collaboration with farmers" and "investing in education and training", to which all factors agreed. This theme can be best encapsulated by a quote from a researcher (UK5) who said the solution lay with:

> "Two-way communication between farmers, researchers and policy makers. Even the best solutions will not work if they can't be shown as favourable or acceptable to the farmer".

The second theme was around suggestions of using soil-improving cropping systems, or derivatives thereof, such as diverse crop rotations, direct drilling and reduced tillage. This related to many of the solution Q-set statements, such as on cover crops, rotation and less use of heavy machinery.

3.2.2. Norway Study

Factor D: "Farmer-Led Demonstration and Innovation"

Respondents in this factor were more likely to rank "setting examples to follow; if one farmer succeeds others will follow", "more innovation" and "more targeted mapping of soil threats" as solutions to declining soil quality than others. This group disagreed more strongly than others on "more small farms" and "reduction of leased land" as solutions to increase soil quality and was the only group that was neutral on the statement "reduce use of heavy machinery".

Factor E: "No More Regulation or Financial Incentives"

Respondents in this group agreed more strongly on "farmers have already tried many measures to improve soil quality" compared with other factors. They disagreed on "more use of cover crops", "financial incentives", "creation of a soil directive", "more regulation of fertiliser use", and "more regulation" as solutions.

Factor F: "Society Needs to Change"

The respondents in this factor distinguished themselves from the others by strongly agreeing on "society needs to change focus on what farmers produce". This group also agreed that the solutions could be to "reduce use of heavy machinery" and "more use of cover crops", though not at the $p < 0.01$ level.

Areas of Agreement and Disagreement

Respondents agreed on "less soil compaction" and "more variation in crop rotation" as ways to improve soil quality (Table A4, Appendix A), as well as on statements related to education, such as "investment in education and training" and "more farmer demonstration days". Respondents did not think "there is not much we can do with the cropping system to improve soil quality" or that the "problems are due to natural, climatic variations". Further, respondents were strongly against "more use of financial penalties" and were neutral or disagreed with "financial incentives" as a solution.

The Most Important Perceived Solution to Declining Soil Quality

More than half of the respondents mentioned "soil organic matter", "cover crops" or "crop rotation" in the open-ended section as solutions to improve soil quality. In addition, more drainage was mentioned by eight respondents as the most critical measure to increase soil quality in the open-answer section, a factor not discussed at all in the UK survey.

4. Discussion

Understanding the range of stakeholder perceptions of the causes of, and solutions to, declining soil quality is useful as it can highlight potential tensions and agreement that might affect the acceptability of land management policies and measures. In our work in the UK and Norway, whilst there were disagreements between respondents on the perceived causes of soil degradation, there was consensus on numerous soil-specific factors, e.g., compaction, soil erosion and loss of organic matter. Both groups agreed that the underlying drivers of declining soil quality were related to wider issues around industrialised agriculture and demand for cheap food, which many farmers felt were out of their control. When it came to solutions, some stakeholders felt that society needs to change in order to address these underlying drivers. Knowledge exchange between agricultural stakeholders was also seen as key. However, many respondents were against further regulation or financial mechanisms including both incentives and penalties.

When focusing on the causes of declining soil quality, studies show that UK soils are threatened by soil erosion, compaction and organic matter decline [21], which reflected the main problems that UK respondents believed were causing declining soil quality. Respondents in Norway also thought soil degradation was due to soil erosion and loss of soil structure, reflecting findings in southeast Norway, where erosion and loss of soil structure have been linked to increased soil compaction [35]. However, Norwegian respondents considered lack of crop rotation as a problem causing soil decline, which was not noted in the UK study, perhaps reflecting the fact that crop rotations were at the time incentivised in the UK via the EU Common Agricultural Policy's three crop rule [36].

Reducing compaction was agreed to be key to improving soil quality for UK and Norwegian respondents. There has been significant interest in the effects of compaction over the last few years in both Norway and the UK, with numerous research projects, training events, innovations and industry-led technology to help address this problem (e.g., [35,37,38]). This suggests compaction is a salient issue for respondents. However, some of the ways for dealing with compaction, such as reducing usage of heavy machinery, were not highly rated by respondents in this survey. More research would be needed to understand why this is.

Industrialisation of the agri-food sector was thought to be a driver of soil degradation. This perception reflects the significant structural changes in southeastern Norway, described by Bjørlo and Rognstad [39] in their report "Barely recognisable" (translated from Norwegian). When analysing the answers to the open-ended question about the main problem causing declining soil quality, both British and Norwegian respondents often highlighted the complex nature of soil degradation, related to external pressures along the food supply chain such as consumers demanding cheap food and agricultural policy and supermarkets dictating farm management. One UK farmer summarised this sentiment succinctly by stating: "give a farmer the right tools and he can put things right but remember he is only a puppet in a political system". Competing demands were thought to be placed on farmers, pulling them in different directions and this was thought to have a negative effect on the soil. To illustrate this tension, one UK adviser stated that "the machinery industry wants to sell big heavy machinery and the agronomy industry wants to sell more chemistry and soil health is the loser". The pressure for farmers to produce more food as cheaply as possible appeared to be part of the symptom of industrialised agriculture where farmers felt trapped and unable to improve their soil quality due to these powerful external market forces.

Whilst respondents in the UK study felt there had been a decline in soil quality, many respondents in factor 6 of the Norwegian study ("I do not think there is a problem with soil quality") did not agree. There can be several reasons to why this may be, such as respondents in factor 6 conceptualising "soil quality" in a different way to others, thereby not considering there to be a decline. For instance, a crop consultant wrote, "I do not think that soil quality has gone down, farmers are harvesting higher and higher yields". This might reflect a more historic definition of the term "soil quality" that focused on productivity rather than wider ecosystem services [40], where continued application of fertiliser and pesticides can mask underlying soil quality issues [41]. In addition, the larger proportion of non-farmers in the UK study group may have resulted in a greater emphasis on declining soil quality, with the scientists in the UK group most commonly identifying industrial agriculture as a causal factor in declining soil quality.

When it came to improving soil quality, many respondents in both studies were neutral towards or disagreed with financial measures, including penalties and incentives. This is a finding also established in other studies [16,42] whereby farmers felt financial incentives in particular were bribes to coerce farmers into doing what others wanted them to do. Whilst there may be some reluctance to agree to using financial incentives to change farmer behaviour, when implemented effectively they can change farmer practices and produce environmental benefits [43,44]. Leaving the EU presents the UK with an opportunity to revolutionise its agricultural policy and there is increasing interest in paying farmers for providing essential ecosystem services, such as soil conservation [45]. However, given that numerous respondents in this study did not think financial incentives could reduce soil degradation, it remains to be seen whether this approach will result in widespread uptake or improved soil quality, especially if fundamental drivers of soil degradation are not also addressed. For instance, if supermarkets and consumers continue to demand and purchase high quantities of cheap food, it is possible that market forces may undermine Government incentive schemes if the profits that farmers make from selling cheap, industrially produced food is more than what the Government can offer. Equally, supermarkets often tie farmers into contracts, with strict requirements on yield and quality of food but limited attention towards how the food is produced. To transform the agri-food system, supermarkets should also start requiring food to be produced in more sustainable ways [46].

Many UK and Norwegian farmers in this study did not believe EU or national intervention could improve soils, such as by creating a Soils Directive, and were also more negative with regards to any form of regulation. In the UK, this may partly be as a result of longstanding political opposition on the issue; in 2012, UK Ministers, together with Germany, France, The Netherlands and Austria, played a key role in blocking an EU Soils Directive [47]. This finding may also reflect the fact that the survey ran during the EU "Brexit" negotiations, where trust in the EU by many UK citizens was at a low point, suggesting a lack of faith in the UK's national application of the Common Agricultural Policy. It remains to be seen whether trust can be rebuilt between British farmers and policymakers as the UK leaves the EU and devises its own agricultural policies.

Investing in education and training were additional solutions that respondents agreed upon. Research has shown that education and training can be effective at spreading awareness and encouraging uptake of more sustainable agricultural practices [48,49]. This may work best with farmers who are more open to learning about new topics and trying novel approaches. However, conservative farmers that are more risk-averse and less willing to change might be less likely to attend training events or try new practices [50] and it could be these farmers that are undertaking the most soil-damaging practices. Targeting these hard-to-reach farmers has continued to prove challenging though one way of addressing this could be to frame knowledge exchange events in ways that attract these farmers by focusing on aspects they are passionate about, and where the event is run by someone whom they respect and relate to. In particular, local peer-to-peer knowledge exchange events have been identified as offering scope for strengthening land manager networks and facilitating behavioural change through exchange of information and experience [51,52]. Opportunities also exist to use online information and integration to influence farmers to change their practice, although more understanding of the effectiveness of this type of approach is needed [53]. Another solution agreed upon by many respondents was to undertake more research in collaboration with farmers. The EU Horizon 2020 funding stream promotes a "multi-actor approach" for agricultural research projects, encouraging a diverse group of stakeholders to work together rather than research solely (or primarily) being conducted by researchers [54]. This approach has many potential benefits as it can help promote greater understanding of different perspectives, building empathy, making research more robust, allowing quicker uptake of results, and grounding research in non-academic stakeholder experiences and knowledge, as well as others [55]. This collaborative approach may help stakeholders understand their epistemological differences and build trust to work together more effectively and respect each other's perspectives. Given that one of the suggested solutions was to build trust between regulators and farmers, future work should encourage participation of regulators in multi-actor projects. In the Norwegian study, "farmer demonstration days" were considered as an agreed solution to improve soil quality, where researchers, extension services and farmers meet to discuss both theoretical and practical aspects of agriculture. These demonstration days could provide a valuable opportunity for knowledge exchange between researchers, farmers and other stakeholders. Similar events are held in the UK and have been met with great success from the farmers attending.

5. Conclusions

To be well-informed, equitable and transparent, public decision-making needs to take account of the views of the diversity of stakeholders they may affect. Taking these perspectives into account in the policy-making process has the potential to deliver more robust decisions that are more beneficial for the environment and more likely to be implemented. Participatory processes can elicit a more inclusive range of perspectives than conventional consultative processes, revealing areas of consensus and disagreement that can inform policy development. They are also able to capture the likely social impacts of proposed policies, which are often neglected in favour of more straightforward environmental and economic appraisals [16,56]. Using Q-methodology to analyse the diversity of stakeholder perspectives, this research has shown that there is a diversity of perspectives on the problems of and solutions to declining soil quality across different professions within the agricultural sector in Norway and the UK.

Respondents in both countries found it easy to agree on the physical processes causing declining soil quality (in both countries, respondents pointed to a loss of soil structure and soil erosion). It was harder to find agreement on social and political drivers from the Q-sorts, other than a lack of knowledge exchange in the UK. However, analysis of qualitative data suggested that respondents primarily blamed industrial agricultural methods, which in turn, they blamed on market drivers, pushing down farm-gate prices (in the UK) and regional specialisation policies (in Norway). Although these drivers of declining soil quality are difficult to address in the short term, and market drivers are outside the control of policymakers, the proposed solutions were pragmatic, focusing primarily on capacity building measures. Respondents in both countries agreed that more investment was needed in training for farmers to use soil-improving cropping systems as an important way of improving soil quality. Respondents in Norway were strongly against the use of financial penalties to encourage the use of these cropping systems, and UK respondents believed that trust needed to be built between farmers and regulators, and that more research needed to be done in collaboration with farmers. It may be possible to engage farmers in action research, including the development of evidence-based training tailored to their needs, drawing on both existing evidence and findings from new collaborative research.

Linked to this, future research might integrate Q methodology with Delphi or other structured elicitation techniques to further triangulate and increase the robustness of findings. Notwithstanding the sample sizes in this research, it is important to note that Q methodology alone should not be used to generalise findings to wider populations, and so these findings should be seen as indicative of the views of some stakeholders in each country, rather than as an authoritative representation of the perspectives of these groups in general.

Although the limited sample makes generalisations inadvisable at national scales, areas of consensus are important for policy makers to understand, as they could indicate areas where policy changes might be more acceptable to a range of stakeholders, addressing the challenge of creating scalable policy options noted by the UNCCD. It is also useful to highlight areas of disagreement among stakeholders, so that further consultation can be carried out to understand the basis of dissensus and its likely impact on policy implementation. For example, in this study we highlighted the diverging view of perspectives on the underlying causes of declining soil quality, which variously blamed farmers (who "have lost touch with their land and are afraid of doing something new"), policy-makers ("it's EU policy, not farmers that are to blame"), the industrial system ("intensive agriculture to blame") and external forces from society ("pressure to produce at low cost").

As policymakers in the UK, Norway and other countries grapple with the challenge of feeding growing populations whilst mitigating climate change, there is a greater need than ever before to develop policies that are acceptable, implementable and sustainable. In this context, policies are needed that address the widest possible range of real and perceived causes of declining soil quality, harnessing the adaptability and ingenuity of farmers and other stakeholders as part of wider attempts to address systemic market and policy failures across the agri-food system. Dealing with soil degradation requires tackling underlying drivers and this study has highlighted numerous solutions for addressing this challenge that are acceptable to a range of agricultural stakeholders.

Author Contributions: Conceptualization: M.S.R., E.A.O.; Data curation: N.R., O.E.L., N.R., S.I., S.V., E.A.O., T.A.B.; Formal analysis: N.R., O.E.L.; Funding acquisition: M.S.R.; Investigation: N.R., O.E.L., S.I., S.V., E.A.O., T.A.B., M.S.R.; Methodology: N.R., O.E.L., S.I., E.A.O., T.A.B., M.S.R.; Supervision: M.S.R., N.R., T.A.B.; Project administration: M.S.R., N.R.; Writing—original draft: N.R., O.E.L., N.R., S.I., S.V., E.A.O., T.A.B., M.S.R.; Writing—review & editing: J.H.G., C.M.M., R.M., M.S.R. All authors have read and agreed to the published version of the manuscript.

Funding: This work was supported by EU Horizon 2020 project 'SoilCare' (Soil Care for profitable and sustainable crop production in Europe, H2020-SFS-2015-2). This funder had no role to play in the design, data collection, analysis or interpretation of results.

Institutional Review Board Statement: The study was conducted according to the guidelines of the Declaration of Helsinki, and approved by the Newcastle University SAGE Faculty Ethics Committee.

Informed Consent Statement: Informed consent was obtained from all subjects involved in the study.

Data Availability Statement: The data that support the findings of this study are available from the corresponding author upon reasonable request and approval from the study sites representative co-authors.

Conflicts of Interest: MR is CEO of Fast Track Impact Ltd. and Research Lead for IUCN UK Peatland Programme. The other authors declare no competing interests.

Appendix A

Table A1. Average Q-sort scores for each of the four factors identified as causing problems by UK respondents. Bold text indicates distinguishing statement at $p < 0.01$, underlined text indicates consensus statements; scale from −2 (strongly disagree) to 2 (strongly agree).

Factors * Problem Statements	1	2	3	4
1. Intensive use of soil without time to recover	**2**	1	0	0
2. Farmers have lost touch with the finer understandings of their land	−2	**1**	**0**	−1
3. Land is being used for other purposes (e.g., grazing, housing, industry)	0	−1	−1	0
4. Farming has become too quantified, where everything is measured	−2	−1	−1	0
5. Some traditions of farmers are damaging	0	**2**	0	1
6. Loss of organic matter	2	2	2	1
7. Loss of numbers of wild species	1	**−2**	**−2**	**−1**
8. Compaction	1	2	1	2
9. Soil has become too saline	−2	−2	−2	−2
10. Declining level of nutrient status	0	**−2**	1	1
11. Overuse of inputs like fertilisers and pesticides	**2**	0	−1	**−2**
12. Repetition of the same crops, year after year	**1**	0	1	**−1**
13. No crop cover over winter	1	1	0	**−1**
14. Loss of soil structure	2	2	2	2
15. Choice of cropping system	0	0	1	**2**
16. Soil tillage practices	**0**	2	2	2
17. Use of contractors	0	1	**−2**	0
18. Poor management	1	1	1	**−1**
19. Farms have become too big	0	−1	−2	0
20. Soil erosion	2	1	2	1
21. Pressure on farmers to produce at low cost	2	**0**	2	2
22. Product demand from national/international markets	**1**	0	0	0
23. Help towards improvements are not given fairly	−1	0	0	1
24. Too many regulations	−2	−2	**0**	**−1**
25. Too much environmental regulation	−2	−2	**0**	**−1**
26. EU agriculture policy	−2	−1	**2**	−1
27. Farmers have little control over their own land	−1	−2	−1	−2
28. Climate change	1	**−1**	**−1**	1
29. Natural local climate constraints	−1	−1	**1**	0
30. Topography of the land	−1	−1	1	1
31. Flooding or drought	0	**−1**	0	2
32. Disconnection between nature-based land use and modern agriculture	1	0	**−1**	0
33. I do not believe that there is a problem with soil quality	−2	**−2**	−2	**0**
34. Distrust of new technology and innovations by farmers	−1	0	−2	−2
35. Fear of doing something new	0	**1**	0	**−1**
36. Distrust of scientists by farmers	−1	−1	−1	**−2**
37. Not enough knowledge being shared	1	1	1	1
38. Distrust between farmers and advisory agencies	−1	**0**	−1	−2
39. Peer pressure by others	−1	0	−2	0
40. Lack of knowledge of soils amongst farmers	**−1**	**2**	**1**	**−2**
41. Modern machinery is too large	**0**	1	**−1**	1

* Factor 1—Intensive use of agriculture to blame; Factor 2—Farmers need to change; Factor 3—It's the EU, not farmers that are to blame; Factor 4—Weather and farm management are to blame.

Table A2. Average Q-sort scores for each of the four factors identified as causing problems by Norwegian respondents. Bold text indicates distinguishing statement at $p < 0.01$, underlined text indicates consensus statements; scale from −2 (strongly disagree) to 2 (strongly agree).

	Factors * Problem Statements	5	6	7
1.	Compaction	2	2	**2**
2.	Soil tillage practices	1	0	1
3.	Lack of use of new technology and innovation	**−2**	0	−1
4.	Use of entrepreneurs/external labor	1	**−1**	1
5.	Intensive agriculture	0	**−1**	1
6.	Not enough knowledge being shared between scientists, advisors and farmers	0	1	**−1**
7.	Loss of organic matter	2	2	2
8.	Soil erosion	2	2	2
9.	Fear of new practices and methods	0	1	**−1**
10.	Local weather and climate	**−1**	1	1
11.	Lack of knowledge on soil amongst farmers	2	1	**0**
12.	Distrust of scientists among farmers	−1	0	**−2**
13.	Soil being used to other types of agriculture (grazing/other plant production)	−2	**−1**	−2
14.	Flooding or drought	**−2**	0	1
15.	Pressure on farmer to produce at a low cost	0	1	**2**
16.	Soil has become too saline	−1	−1	−2
17.	Too little advise on soil-improving practices	**1**	**2**	**−1**
18.	Peer-pressure	−1	−1	**−2**
19.	Choice of crops (cropping system)	1	1	1
20.	Too much environmental regulation	**−2**	**0**	**−1**
21.	Too large farms	**−1**	**−2**	**1**
22.	Topography of the land	**−2**	0	0
23.	Loss of number of wild species	**0**	−2	−2
24.	Farmer has lost touch with the finer understandings of his land	**1**	**−2**	**−1**
25.	Climate change	**−2**	**0**	**1**
26.	Farmer has little control over his own land	−1	−2	−1
27.	Loss of soil structure	2	1	2
28.	Declining level of nutrient status	0	1	0
29.	Distrust between farmers and advisory agencies	−1	−1	**−2**
30.	Poor management of the soil/poor soil management	**2**	1	0
31.	Too many regulations	−2	**0**	−1
32.	Overuse of input like fertilizers and pesticides	**1**	**−2**	**−1**
33.	I do not believe that there is a problem with soil quality	−2	**2**	−2
34.	No cover crop over winter	1	2	1
35.	Disconnection between nature-based agriculture and the modern agriculture	**1**	**−1**	**0**
36.	Product demand from the market	0	−2	0
37.	Repetition of same crop year after year; monoculture	2	1	2
38.	Norwegian agriculture policy	1	**−1**	0
39.	Agriculture has become too quantified, everything is to be measured	−1	−1	**0**
40.	High share of leased land	**0**	**−2**	**1**
41.	Too little drained land	**0**	**2**	2

* Factor 5—Disconnection between farmer and soil; Factor 6—There is no problem with the soil quality; Factor 7—Industrialised agriculture to blame.

Table A3. Average Q-sort scores for each of the four factors identified as solutions to improve soil by UK respondents. Bold text indicates distinguishing statement at $p < 0.01$, underlined text indicates consensus statements; scale from −2 (strongly disagree) to 2 (strongly agree).

Factors * **Solution Statements**	**A**	**B**	**C**
1. Keep updated with new information	<u>0</u>	<u>0</u>	<u>1</u>
2. Farmers have already tried lots of things to improve soil quality	**−2**	**−1**	**−2**
3. More technical advice	0	1	0
4. Setting examples for others to follow	1	0	2
5. More innovations	**−1**	1	1
6. Maintain small farms	**0**	**−1**	**−2**
7. More resting/recuperating of the soil	1	**−1**	0
8. More organic fertilizer	<u>0</u>	<u>0</u>	<u>0</u>
9. More cover crops	2	0	−1
10. More diverse crop rotations	2	0	1
11. Less use of heavy machinery	0	0	0
12. Change the timing of tillage	<u>−1</u>	<u>−1</u>	<u>−1</u>
13. Reduce compaction	<u>1</u>	<u>2</u>	<u>2</u>
14. More targeted mapping of soil threats	<u>0</u>	<u>0</u>	<u>0</u>
15. More financial incentives	0	**1**	0
16. More financial penalties	−2	−2	**0**
17. More freedom for the farmers to manage their land as they would like	**−1**	**0**	**−2**
18. More regulation	**−2**	**−2**	**−1**
19. More regulations for water usage	−1	**−1**	−1
20. More regulations for fertilisers	**−1**	**−2**	**−1**
21. We cannot do much as the problems are down to natural climatic constraints	**−2**	**−1**	**−2**
22. Creation of a 'Soil Directive'	0	**−2**	0
23. More research done in collaboration with farmers	<u>2</u>	<u>2</u>	<u>2</u>
24. More traditional farming practices	**−1**	−1	−1
25. Improve trust between farmers and regulatory agencies	<u>1</u>	<u>1</u>	<u>1</u>
26. Society needs to change focus on what we want to produce	0	0	**−1**
27. Increase early adoption of new techniques	**−1**	**0**	**1**
28. More farmer demonstration days	**0**	1	1
29. More communication and sharing of knowledge between farmers locally	<u>0</u>	<u>2</u>	1
30. More local knowledge and experience	1	0	0
31. More education on environmental impacts	<u>2</u>	1	<u>0</u>
32. Increase knowledge of difference in soil types	1	2	0
33. Invest in education and training	<u>1</u>	<u>1</u>	<u>2</u>
34. There is not much new we can do in terms of soil management	**−2**	**−2**	−1

* Factor A—Anti-innovation; Factor B—Yes to financial incentives but no to regulation; Factor C—Early adoption of new techniques.

Table A4. Average Q-sort scores for each of the four factors identified as solutions to improve soil by Norwegian respondents. Bold text indicates distinguishing statement at $p < 0.01$, underlined text indicates consensus statements; scale from −2 (strongly disagree) to 2 (strongly agree).

	Factor * Solution Statements	D	E	F
1.	Regulation of water usage	−2	−1	−1
2.	More communication and sharing of knowledge between farmers on a local level	0	1	1
3.	Investment in education and training	1	1	2
4.	More use of local knowledge and experience	**0**	2	1
5.	More regulations for fertilizers	**−1**	**−2**	**0**
6.	Less use of heavy machinery	**0**	1	2
7.	More farmer demonstration days	1	1	1
8.	Change the timing of tillage	0	0	−1
9.	More research done in collaboration with farmers	1	0	1
10.	More small farms	**−2**	0	0
11.	Improve trust between farmers and institutions	**−1**	0	0
12.	More resting soil	−1	−2	−2
13.	More innovations	**1**	−1	−1
14.	There is not much we can do with the cropping system to improve soil quality	**−2**	−2	**−2**
15.	Financial incentives (e.g., subsidies)	0	**−1**	0
16.	Setting examples to follow; if a farmer succeed others will follow	**2**	1	0
17.	Creation of a "Soil Directive"	**−1**	**−2**	**−1**
18.	There is not much we can do; problems are due to natural, climatic variations.	**−2**	**−1**	**−2**
19.	Less soil compaction	2	2	2
20.	More diverse crop rotation	2	2	2
21.	More education on environmental impacts	1	**0**	1
22.	More use of organic fertilizer	1	1	0
23.	Society needs to change focus on what farmers produce	0	0	**2**
24.	More advise on use of technology	**0**	0	−1
25.	More targeted mapping of soil threats	**1**	0	0
26.	More cover crops	2	**−1**	2
27.	More traditional agricultural practices	−1	−1	−1
28.	Farmers have already tried many measures to improve soil quality	−1	**1**	−1
29.	More regulation	−1	**−2**	−2
30.	Increase adoption of new techniques	0	−1	−1
31.	More financial penalties	−2	−2	−2
32.	Increase knowledge of soil types	2	2	**1**
33.	Reduce share of leased land	**−2**	1	0
34.	More drainage of agricultural land	2	2	**1**

* Factor D-Farmer-led demonstration and innovation; Factor E—No more regulation or financial incentives; Factor F—Society needs to change.

Note

1 Definitions of "soil quality" vary and have progressed from focusing solely on agricultural production to a broader focus on the complex and diverse functions that soil confers to humans and our environment [7]. Here, we define soil quality as "the capacity of a specific kind of soil to function, within natural or managed ecosystem boundaries, to sustain plant and animal productivity, maintain or enhance water and air quality, and support human health and habitation. In short, the capacity of the soil to function" [57].

References

1. Dominati, E.; Patterson, M.; Mackay, A. A Framework for Classifying and Quantifying the Natural Capital and Ecosystem Services of Soils. *Ecol. Econ.* **2010**, *69*, 1858–1868. [CrossRef]
2. Moncrieff, C.; Draisey, Z. *Saving the Earth: A Sustainable Future for Soils and Water*; WWF: Woking, UK, 2018.
3. Rusinamhodzi, L. *The Role of Ecosystem Services in Sustainable Food Systems*; Elsevier: Amsterdam, The Netherlands, 2020.
4. Intergovernmental Panel on Climate Change. *Climate Change and Land: An IPCC Special Report on Climate Change, Desertification, Land Degradation, Sustainable land Management, Food Security, and Greenhouse Gas Fluxes in Terrestrial Ecosystems*; Intergovernmental Panel on Climate Change: Geneva, Switzerland, 2019.
5. Keating, B.A.; Herrero, M.; Carberry, P.S.; Gardner, J.; Cole, M.B. Food wedges: Framing the global food demand and supply challenge towards 2050. *Glob. Food Secur.* **2014**, *3*, 125–132. [CrossRef]
6. UNFCCC. *Adoption of the Paris Agreement, 21st Conference of the Parties*; United Nations: Paris, France, 2015.
7. FAO and ITPS. *State of the World's Soil Resources*; United Nations Food and Agriculture Orgaisation: Rome, Italy, 2015.
8. Thomas, D.T.; Moore, A.D.; Bell, L.W.; Webb, N.P. Ground Cover, Erosion Risk and Production Implications of Targeted Management Practices in Australian Mixed Farming Systems: Lessons from the Grain and Graze. *Agric. Syst.* **2018**, *162*, 123–135. [CrossRef]
9. UNCCD. *Global Land Outlook*; United Nations: Bonn, Germany, 2017.
10. UN. *The Sustainable Development Goals Report*; United Nations: New York, NY, USA, 2016.
11. Cowie, A.L.; Orr, B.J.; Sanchez, V.M.C.; Chasek, P.; Crossman, N.D.; Erlewein, A.; Louwagie, G.; Maron, M.; Metternicht, G.I.; Minelli, S.; et al. Land in Balance: The Scientific Conceptual Framework for Land Degradation Neutrality. *Environ. Sci. Policy* **2018**, *17*, 25–35. [CrossRef]
12. WOCAT. *Where the Land Is Greener: Case Studies and Analysis of Soil and Water Conservation Initiatives Worldwide*; Liniger, H., Critchley, W., Eds.; World Overview of Conservation Approaches and Technologies: Bern, Switzerland, 2007.
13. De Vente, J.; Reed, M.S.; Stringer, L.C.; Valente, S.; Newig, J. How Does the Context and Design of Participatory Decision Making Processes Affect Their Outcomes? Evidence from Sustainable Land Management in Global Drylands. *Ecol. Soc.* **2016**, *21*, 24. [CrossRef]
14. Jager, N.W.; Newig, J.; Challies, E.; Kochskämper, E. Pathways to Implementation: Evidence on How Participation in Environmental Governance Impacts on Environmental Outcomes. *J. Public Adm. Res. Theory* **2019**, *30*, 383–399. [CrossRef]
15. Dryzek, J.S. Democratization as Deliberative Capacity Building. *Comp. Political Stud.* **2009**, *42*, 1379–1402. [CrossRef]
16. Rust, N.A. Can Stakeholders Agree on How to Reduce Human–Carnivore Conflict on Namibian Livestock Farms? A Novel Q-Methodology and Delphi Exercise. *Oryx* **2017**, *51*, 339–346. [CrossRef]
17. Addams, H.J.; Proops, L.R. *Cheltenham Social Discourse and Environmental Policy: An Application of Q Methodology*; Edward Elgar Publishing: Cheltenham, UK, 2000.
18. Bryden, J.; Riddoch, L.; Brox, O. Conclusions. In *Northern Neighbours: Scotland and Norway since 1800*; Bryden, J., Brox, O., Riddoch, L., Eds.; Edinburgh University Press: Edinburgh, UK, 2015; pp. 282–286.
19. UK National Statistics, Agriculture in the United Kingdom 2020. Report compiled by: Department for Environment, Food and Rural Affairs (England); Department of Agriculture, Environment and Rural Affairs (Northern Ireland); Welsh Government, Knowledge and Analytical Services; The Scottish Government, Rural and Environment Science and Analytical Services. 2020. Available online: https://www.gov.uk/government/collections/agriculture-in-the-united-kingdom (accessed on 23 November 2021).
20. Environment Agency. The State of the Environment: Soil. 2019. Available online: https://assets.publishing.service.gov.uk/government/uploads/system/uploads/attachment_data/file/805926/State_of_the_environment_soil_report.pdf (accessed on 23 November 2021).
21. DEFRA. Safeguarding Our Soils—A Strategy for England. 2009. Available online: https://assets.publishing.service.gov.uk/government/uploads/system/uploads/attachment_data/file/69261/pb13297-soil-strategy-090910.pdf (accessed on 20 December 2021).
22. Tipper, W.A. *Protecting Our Assets: Using Natural Infrastructure Schemes to Support Sustainable Agriculture*; Green Alliance: London, UK, 2017; Available online: https://www.green-alliance.org.uk/resources/protecting_our_assets.pdf (accessed on 23 November 2021).
23. O'Neill, D. *The Total External Environmental Costs and Benefits of Agriculture in the UK*; Environment Agency: Bristol, UK, 2007.
24. Boardman, J. Soil Erosion in Britain: Updating the Record. *Agriculture* **2013**, *3*, 418–422. [CrossRef]
25. Hegrenes, A.; Mittenzwei, K.; Prestegard, S.S. *Norsk Jordbrukspolitikk: Handlingsrom i Endring*; Fagbokforl: Bergen, Norway, 2011.
26. McKee, A.; Vinge, H.; Bjørkhaug, H.; Almås, R. Agricultural Models in Scotland and Norway: A Comparison. In *Land Reform in Scotland: History, Law and Policy*; Combe, M., Glass, J., Tindley, A., Eds.; Edinburgh University Press: Edinburgh, UK, 2020; pp. 388–409.
27. Arnoldussen, A.H.; Forbord, M.; Grønlund, A.; Hillestad, M.E.; Mittenzwei, K.; Pettersen, I.; Tufte, T. *Økt Matproduksjon På Norske Arealressurser*; AgriAnalyse AS: Oslo, Norway, 2014.
28. Vagstad, N.; ' Abrahamsen, U.; Strand, E.; Uhlen, A.K.; Lund, H.J.; Rognlien, A.; Stuve, L.F.; Stabbetorp, E.M.H.; Mangerud, K.; Solberg, H. Økt norsk kornproduksjon. Utfordringer og tiltak. In *Rapport fra Ekspertgruppe til LMD*; Minister of Agriculture and Food of Norway: Oslo, Norway, 2013; 39p.

29. Stortinget, M. *Landbruks-Og Matpolitikken–Velkommen Til Bords*; Landbruks-og Matdepartementet: Oslo, Norway, 2011.
30. Kolberg, D. Regional Specialisation Policies in Norwegian Agriculture should Be Revised. *NMBU Stud. J. Life Sci.* **2016**, *6*, 14–17.
31. Brown, S.R. Q Methodology and Qualitative Research. *Qual. Health Res.* **1996**, *6*, 561–567. [CrossRef]
32. Rastogi, A.; Hickey, G.M.; Badola, R.; Hussain, S.A. Diverging Viewpoints on Tiger Conservation: A Q-Method Study and Survey of Conservation Professionals in India. *Biol. Conserv.* **2013**, *161*, 182–192. [CrossRef]
33. Deary, H.; Warren, C. Trajectories of rewilding: A taxonomy of wildland management. *J. Environ. Plan. Manag.* **2019**, *62*, 466–491. [CrossRef]
34. Coogan, J.; Herrington, N. Q Methodology: An Overview. *Res. Second. Teach. Educ.* **2011**, *1*, 24–28.
35. Seehusen, T.; Riggert, R.; Fleige, H.; Horn, R.; Riley, H. Soil Compaction and Stress Propagation after Different Wheeling Intensities on a Silt Soil in South-East Norway. *Acta Agric. Scand. Sect. B—Soil Plant Sci.* **2019**, *69*, 343–355. [CrossRef]
36. Rural Payments Agency. The Basic Payment Scheme in England. In *Guidance for farmers about the Basic Payment Scheme in 2015*; UK Government: London, UK, 2015. Available online: https://assets.publishing.service.gov.uk/government/uploads/system/uploads/attachment_data/file/649753/BPS_Handbook_-_final_v1.0.pdf (accessed on 23 November 2021).
37. De-Ville, S.; Pattison, I.; Frost, M. COMPACT: A frequency analysis of Ground Penetrating Radar to identify agricultural soil compaction. *AGU Fall Meet. Abstr.* **2018**, *2018*, NS13B-0596.
38. Jordon, M.W. Does mixed vs separate sheep and cattle grazing reduce soil compaction? *Soil Use Manag.* **2021**, *37*, 822–831. [CrossRef]
39. Bjørlo, B.; Rognstad, O. Nesten Ikke Til å Kjenne Igjen. 2019. Available online: https://www.ssb.no/jord-skog-jakt-og-fiskeri/artikler-og-publikasjoner/nesten-ikke-til-a-kjenne-igjen (accessed on 28 November 2021).
40. Erkossa, T.; Itanna, F.; Stahr, K. Indexing Soil Quality: A New Paradigm in Soil Science Research. *Soil Res.* **2007**, *45*, 129–137. [CrossRef]
41. Uhlen, A.K.; Børresen, T.; Deelstra, J.; Waalen, W.; Strand, E.; Bleken, M.A.; Seehusen, T.; Kværnø, S.; Sundgren, T.; Lillemo, M.; et al. Økt kornproduksjon gjennom forbedret agronomisk praksis. En vurdering av agronomiske tiltak som kan bidra til avlingsøkninger i kornproduksjonen. In *NIBIO Rapport*; NIBIO: Oslo, Norway, 2017; Volume 3.
42. Rust, N.A. Media Framing of Financial Mechanisms for Resolving Human–Predator Conflict in Namibia. *Hum. Dimens. Wildl.* **2015**, *20*, 440–453. [CrossRef]
43. Stabbetorp, H. *Landbrukets Utvikling Siden 1950. Utviklingstrekk Av Betydning for Vannkvaliteten*; Vann: Vannforeningen, Norway, 2014.
44. Piñeiro, V.; Arias, J.; Dürr, J.; Elverdin, P.; Ibáñez, A.M.; Kinengyere, A.; Morales Opazo, C.; Owoo, N.; Page, J.R.; Prager, S.D.; et al. A scoping review on incentives for adoption of sustainable agricultural practices and their outcomes. *Nat. Sustain.* **2020**, *3*, 809–820. [CrossRef]
45. DEFRA. *A Green Future: Our 25 Year Plan to Improve the Environment*; HMSO: London, UK, 2018.
46. Rust, N.A.; Ridding, L.; Ward, C.; Clark, B.; Kehoe, L.; Dora, M.; Whittingham, M.J.; McGowan, P.; Chaudhary, A.; Reynolds, C.J.; et al. How to transition to reduced-meat diets that benefit people and the planet. *Sci. Total Environ.* **2020**, *718*, 137208. [CrossRef]
47. Glæsner, N.; Helming, K.; De Vries, W. Do Current European Policies Prevent Soil Threats and Support Soil Functions. *Sustainability* **2014**, *6*, 9538–9563. [CrossRef]
48. Ingram, J. Framing Niche-Regime Linkage as Adaptation: An Analysis of Learning and Innovation Networks for Sustainable Agriculture across Europe. *J. Rural Stud.* **2015**, *40*, 59–75. [CrossRef]
49. Mills, J.; Gaskell, P.; Ingram, J.; Dwyer, J.; Reed, M.S.; Short, C. Engaging farmers in environmental management through a better understanding of behaviour. *Agric. Hum. Values* **2017**, *34*, 283–299. [CrossRef]
50. Kilpatrick, S.; Williamson, J. Farmer Participation in Training. *Rural Soc.* **1996**, *6*, 3–12. [CrossRef]
51. McMorran, R. *Key Drivers for Land Manager Decision Making in Loch Lomond and the Trossachs National Park and Opportunities for Addressing Constraints to Woodland Creation*; SEFARI Fellowship Report; SEFARI: Edinburgh, UK, 2021.
52. O'Brien, L.; Morris, J.; Marzano, M.; Dandy, N. Promoting sustainability behaviours through forestry. *Forestry* **2017**, *90*, 88–98. [CrossRef]
53. Rust, N.A.; Stankovics, P.; Jarvis, R.M.; Morris-Trainor, Z.; de Vries, J.R.; Ingram, J.; Mills, J.; Glikman, J.A.; Parkinson, J.; Toth, Z.; et al. Have farmers had enough of experts? *J. Environ. Manag.* **2021**. [CrossRef] [PubMed]
54. EIP-AGRI. Horizon 2020 Multi-Actor Projects. 2018. Available online: https://ec.europa.eu/eip/agriculture/sites/agri-eip/files/eip-agri_brochure_multi-actor_projects_2017_en_web.pdf (accessed on 23 November 2021).
55. Acquaye-Baddoo, N.A.; Ekong, J.; Mwesige, D.; Neefjes, R.; Nass, L.; Ubels, J.; Visser, P.; Wangdi, K.; Were, T.; Brouwers, J. Multi-actor systems as entry points for capacity development. *Capacity. org J.* **2010**, *41*, 4–7.
56. Kenter, J.O.; O'Brien, L.; Hockley, N.; Ravenscroft, N.; Fazey, I.; Irvine, K.N.; Reed, M.S.; Christie, M.; Brady, E.; Bryce, R.; et al. What are shared and social values of ecosystems? *Ecol. Econ.* **2015**, *111*, 86–99. [CrossRef]
57. Karlen, D.L.; Mausbach, M.J.; Doran, J.W.; Cline, R.G.; Harris, R.F.; Schuman, G.E. Soil quality: A concept, definition, and framework for evaluation. *Soil Sci. Soc. Am. J.* **1997**, *61*, 4–10. [CrossRef]

land MDPI

Article

A Multivariate Approach to Evaluate Reduced Tillage Systems and Cover Crop Sustainability

Felice Sartori, Ilaria Piccoli *, Riccardo Polese and Antonio Berti

Department of Agronomy, Food, Natural Resources, Animals and Environment, University of Padova, Viale dell'Università 16, 35020 Legnaro, Italy; felice.sartori@phd.unipd.it (F.S.); riccardo.polese@unipd.it (R.P.); antonio.berti@unipd.it (A.B.)
* Correspondence: ilaria.piccoli@unipd.it; Tel.: +39-049-827-2841

Abstract: The evaluation of the effects of conservation agriculture during the transition from conventional tillage to no-tillage requires numerous indicators to be considered. For this purpose, we monitored changes in a multi-parameter dataset during a three-year experiment that combined three tillage intensities (conventional tillage—CT; minimum tillage—MT; and no tillage—NT) with three soil covering managements (tillage radish cover crop, winter wheat cover crop and bare soil). Using a multivariate analysis, we developed a Relative Sustainability Index (*RSI*) based on 11 physical (e.g., bulk density and penetration resistance), chemical (e.g., soil organic carbon and pH) and biological soil properties (e.g., earthworm density) to evaluate cropping systems sustainability. The *RSI* was most affected by tillage intensity showing higher *RSI* values (i.e., better performances) in reduced tillage systems. Specifically, the *RSI* under NT was 42% greater than that of CT and 13% greater than that of MT. Soil covering had little impact on the *RSI*. Among the tested parameters, the *RSI* was increased most by saturated hydraulic conductivity (+193%) and earthworm density (+339%) across CT and NT treatments. Our results suggest that conservation agriculture and, particularly, reduced tillage systems, have the potential to increase farm environmental and agronomic sustainability.

Keywords: conservation agriculture; no tillage; minimum tillage; principal component analysis; soil quality index; scoring function

Citation: Sartori, F.; Piccoli, I.; Polese, R.; Berti, A. A Multivariate Approach to Evaluate Reduced Tillage Systems and Cover Crop Sustainability. *Land* **2022**, *11*, 55. https://doi.org/10.3390/land11010055

Academic Editors: Amrakh I. Mamedov, Guido Wyseure, Julián Cuevas González and Jean Poesen

Received: 14 December 2021
Accepted: 30 December 2021
Published: 31 December 2021

Publisher's Note: MDPI stays neutral with regard to jurisdictional claims in published maps and institutional affiliations.

1. Introduction

Conservation agriculture (CA) is defined as the combination of three principles: minimum soil disturbance, permanent soil organic cover and species diversification [1]. In addition to reduced management costs, CA is considered to enhance several ecosystem services (soil physical and chemical properties, soil organic carbon (SOC) and biodiversity) [2,3] and prevent some soil threats, such as soil erosion [4–6]. Although some of these benefits remain less clear, the worldwide adoption of CA grew to 12.5% of 2016 global cropland. At odds with this growth trend, there is Europe, where only 5% of total cropland is managed with CA. One country that has shown particularly limited adoption is Italy—less than 300,000 ha (a mere 2% of agricultural land) [7].

Adoption of CA has suffered slow adoption in Europe primarily due to the long transition time that follows conversion from conventional agriculture to CA before the positive effects are realized. During this crucial period, farmers face reduced crop yield and new equipment expenditures. Conversion to CA also requires a permanent soil covering, yet another cost that would benefit from economic support [8]. A key reason behind the very limited use of CA in Italian agrosystems is the long conversion time (more than five years) required before SOC, fertility and nutrient use efficiency benefits are observed [9,10]. Most studies have considered transition time only as a function of a single parameter, such as soil physical properties [11,12], yield [13–16], net SOC stock [17,18], soil aggregate stability, biodiversity, SOC content [19,20], earthworm density, or CO_2 emission reduction [21]. However, each of these exerts an effect on CA. As such, we suggest that a holistic approach capable of considering multiple parameters may provide a better means by which to evaluate the effects of CA.

The ecosystem services delivered by CA justify the need for environmentally conscious policymakers to consider economic support of the practice through a program such as the "green payments" program already established in the EU Common Agricultural Policy [22]. Alternatively, yield losses and/or other negative effects could be limited or compensated in some fashion. In either case, programs such as these are effective only when the protocol created is adaptable to local area specifics but is assessable by a single, consistent set of criteria. As is often described in the literature, an index represents one way to determine and compare the impact of different management strategies [23,24]. Similarly, the literature has already identified potential soil quality indicators to comprise an index: physical soil property measures (soil hydraulics, penetration resistance and bulk density) [25,26] plus soil aggregate stability [27], soil C and N content and earthworm density [28]. Masto et al. [29] previously adopted a statistical methodology to determine the impact of different management strategies on soil quality and sustainability using a dataset with several soil characteristics, as reported above. The method involved the application of a principal component analysis (PCA) to derive the weight of the different soil parameters in promoting the sustainability. The derived index showed to be a reliable tool to assess the performance and impacts of alternative land uses and soil management options [23].

In this work, three tillage systems (no tillage, minimum tillage and conventional tillage) were combined with three different soil coverings (tillage radish cover crop, winter wheat cover crop and bare soil) to compare the effects of the main factors influencing CA. A multivariate approach was applied to a dataset of soil quality measures taken during a north Italy field experiment. A sustainability index was calculated to compare different treatment combinations as a function of the selected indicator variability. This study aims to determine the short-term effects of reduced tillage and cover crops on the studied parameters. Our starting hypothesis was that a reduced tillage system combined with tillage radish could minimize conversion time side effects and improve soil properties.

2. Materials and Methods

2.1. Experimental Design

The experiment took place at the Lucio Toniolo Experimental Farm, located in Legnaro, PD (NE Italy, 45°21 N; 11°58 E; 6 m a.s.l.). The climate is sub-humid with average temperatures between −1.5 °C in January and 27.2 °C in July. Rainfalls reaches 850 mm annually. The reference evapotranspiration of 945 mm exceeds rainfalls from April to September. The highest rainfalls occur in June (100 m) and October (90 mm), while winter is the driest season, with average rainfalls of 55 mm. The shallow water table ranges in depth from 0.5 to 2 m, with the lowest values recorded in summer.

This three-year study began in spring 2018 and it was designed as a split plot, with two replicates located in a flat area of the Po valley with a maximum slope < 1%. An area of 2 ha was divided into 18 plots, each of 1111 m^2. The soil at the site was Fluvi-Calcaric Cambisol [30] with a silt loam texture (25% clay, 50% silt and 25% sand), pH 7.8, 27.1% total carbonate content, <1% soil organic carbon and <0.1% total nitrogen. The main factor was tillage intensity; conventional tillage (CT) was ploughed to a depth of 30 cm and then harrowed to 15 cm; minimum tillage (MT) was tilled with a harrow to a depth of 15 cm; no tillage (NT) was sod seeded with a zero-tillage seeder that included double disks for furrow openers and press wheels for soil firming. Within each main plot, three winter soil coverings were randomized: tillage radish (TR—*Raphanus sativus* L.), winter wheat (WW—*Triticum aestivum* L.) and bare soil (BS), where only residues from the previous year crop were present. Cover crops were seeded on residues from the main crop (always maize, *Zea mays* L.) in autumn 2018 and 2019.

2.2. Field Surveys

According to what already reported by other authors [25–28], we chose 11 parameters to monitor changes in the condition of the soil: (1) aggregate stability (Agg), (2) bulk density (BD), (3) soil organic carbon (C org), (4) total nitrogen (N tot), (5) gravimetric water content (GWC), (6) penetration resistance (PR), (7) saturated hydraulic conductivity (Ks), (8) earthworm density (EW), (9) mineral nitrogen (N min), (10) pH and (11) cash crop yield (Y) (Table 1). Each parameter was measured at two times. The first measurement was taken immediately after treatment combination adoption (T0) and the second measurement was taken at the end of the three-year period (T1). The method for determination of the measure of each parameter is fully described below.

Table 1. Soil parameters used for building the sustainability index.

Soil Characteristic	Acronym	Used Method
Aggregate stability	Agg	Slakes application
Bulk density	BD	Core method
Soil organic carbon	C org	CNS Elemental analyzer
Total nitrogen	N tot	CNS Elemental analyzer
Gravimetric water content	GWC	Oven-dried at 105 °C
Penetration resistance	PR	Penetrologger
Saturated hydraulic conductivity	*Ks*	Infiltrometer method
Earthworm density	EW	Mustard extraction
Mineral nitrogen	N min	Photometry
pH	pH	1 M KCl solution
Cash crop yield	Y	Oven-dried at 105 °C

A continuous value of Agg was determined. The Slakes application [27,31] was employed to soil aggregates in the 0.2–2 cm fraction sampled from the 0–20 cm soil layer. Three randomly selected aggregates from each sample were analyzed to produce a dimensionless slaking index (*SI*) with a value ≥ 0. A low *SI* (<3) represents high aggregate stability, an *SI* between 3 and 7 indicates moderate stability and an *SI* above 7 indicates that the aggregates have low stability. The *SI* was calculated as the difference between the wet aggregate area (A_t) after 10 min of water saturation and the dry aggregate area (A_{t0}), divided by A_{t0}, as shown in Equation (1).

$$SI = \frac{A_t - A_{t0}}{A_{t0}} \quad (1)$$

The BD was measured in the 0–30 cm soil profile with the core method as described in Grossman and Reinsch [32]. In the studied soil, a BD value of 1.55 g cm^{-3} was considered a limiting condition to the growth of plant roots [33].

The C org and N tot contents were determined from shallow layer (0–30 cm) samples. The soil was air-dried and sieved at 0.5 mm and the inorganic carbon was removed with an acid pre-treatment. Subsequently, SOC and N tot were determined with flash combustion using a CNS Elemental analyzer (Vario Max; Analysensysteme GmbH, Langenselbold, Germany).

Four sampling areas were selected in each plot for GWC and PR measurements. For the PR, the measures were taken from the 0–20 cm layer and an average PR value was calculated. In each sampling area, a disturbed soil core was collected, weighted and oven-dried at 105 °C to determine the GWC. For the PR, the measures were taken in each plot with the Penetrologger (Eijkelkamp, Giesbeek, The Netherland). A PR value above 2.5 MPa was considered a limiting factor to plant root growth [34].

The *Ks* [35] was determined using the double-ring infiltrometer method [36]. An inner ring of 60 cm in diameter was used to measure both the row and inter-row areas in the tillage radish plots. The water within the inner ring was maintained at two levels. As one operator measured the time for the water to reach the lower level from the upper level, another added more water to reach the upper level again. This operation was replicated until the infiltration rate was constant. Meanwhile, the water in the external ring was suspended at an average value between the two levels of the inner ring. Then, the data were analyzed by fitting Philip's equations [37] with the Solver Add-in for Microsoft Excel.

$$i(t) = S \times t^{1/2} + At \tag{2}$$

$$v(t) = \frac{S \times t^{-1/2}}{2} + A \tag{3}$$

where *i(t)* is the water infiltration (m) and *v(t)* is the infiltration rate (m s^{-1}) expressed as a function of time. Parameters S and A were calculated with the Solver add-in by minimizing the square difference between predicted and observed *i(t)* and *v(t)*. The *Ks* was calculated as below and m is a constant equal to 2/3.

$$Ks = \frac{A}{m} \tag{4}$$

The EW was measured with a mustard extraction as described by Valckx et al. [38]. The measure was performed by taking an earthworm extraction from the soil surface using a water-suspended mustard in a 25 × 25 cm^2 frame [38]. First, we used the number of extracted earthworms to score soil quality [39]. A density of <4 was the lowest score or of "poor" soil condition, a density of 4–8 was "moderate" soil condition and the highest density (>8) was "good" soil condition. Then, the earthworm count was compared amongst the different treatment combinations.

We estimated N min based on samples of the 0–20 cm soil layer. Concentrations of ammonium, nitrite and nitrate were measured using a KCl extraction followed by photometry, as described by García-Robledo et al. [40].

Soil pH was determined from air dried, mixed and sieved (0.5 mm) samples taken from the 0–20 cm soil layer. The pH was measured in a 1 M KCl solution (1:2.5 solid–liquid ratio) [41].

At the end of the cropping season, four biomass samples were collected from each subplot to determine maize grain Y at 27% grain moisture. After the harvest, a grain sample was oven dried at 105 °C until it maintained a constant weight to determine the dry mass weight. The Y was expressed in kilogram of dry grain per hectare.

2.3. Data Analyses and Statistics

First, a mixed-effects model was constructed using tillage, covering and their interactions in each monitored year. These effects were treated as fixed effects and the block effects as random. Post hoc pairwise comparisons of least-squares means were performed, using Tukey's method to adjust for multiple comparisons, with a $p < 0.05$.

To calculate a soil quality index, we relied on the method of Masto et al. [23,29]. The procedure requires that indicators be selected once they have been surveyed and normalized with linear or non-linear scoring functions, so that higher scores represent better-performing observations. The indicators and their weights were determined using the multivariate analysis method of Andrews et al. [42,43], which has been adapted and applied to many studies evaluating long-term practices [23,29], combinations of various crop rotations under different residue managements [44,45] and different tillage practices [46].

The sampled data were normalized with a linear scoring function [23] by applying Equations (5)–(7).

$$S = \frac{x_{ij} - x_{i\,min}}{x_{i\,max} - x_{i\,min}} \tag{5}$$

$$S = -\frac{x_{ij} - x_{i\,max}}{x_{i\,max} - x_{i\,min}} \tag{6}$$

$$S = \frac{|x_{ij} - 7|}{|x_i - 7|_{max} - |x_i - 7|_{min}} \tag{7}$$

where $x_{i\ max}$ is the maximum value measured during the *i* parameter survey and $x_{i\ min}$ is the smallest. The *S* value ranges between 0 and 1, which corresponds to the minimum and maximum values, respectively, observed in the *i* parameter. Equation (5) was used as a "more is better" scoring function for C org, GWC, Ks, EW, N min, N tot and Y. Alternatively, the parameters Agg, BD and PR were scored with Equation (6), according to a "less is better" approach. Finally, Equation (7) was used for pH scoring. In this way, treatment combinations that most favorably impacted the parameters scored highest.

The Relative Sustainability Index (*RSI*) was calculated as the sum of the observed parameter score, weighted with principal component analysis weighting factors (*PWs*). These factors were calculated according to Masto et al. [23], by selecting principal components (PCs) explaining at least 10% of the variability. Within each of these PCs, loaded factors (values > |0.2|) were selected and their correlations were measured [43]. In cases in which r > |0.8|, only the factor with the highest load was used for *RSI* calculation, together with all the other uncorrelated highly loaded factors. The percentage of variation explained by each PC provided the *PW*. The *RSI* was calculated with Equation (8).

$$RSI = \sum_{i=1}^{n} PW_i \times S_i \tag{8}$$

To normalize the *RSI*, this was divided by the highest *RSI* value obtained. A total of 36 *RSIs* were calculated, one per treatment combination replication in survey T0 and another in T1.

RSI differences amongst tillage, soil covering and their interaction were tested with mixed models and the model with the smallest Akaike's Information Criterion (AIC) was selected [47]. Post hoc pairwise comparisons of least-squares means were performed using Tukey's method to adjust for multiple comparisons, with a $p < 0.05$. The statistical analyses were performed using Microsoft Excel 2016, ClustVis [48] and SAS (SAS Institute Inc., Cary, NC, USA), version 5.1.

3. Results

Below is a description of the mixed model results comparing changes in the 11 indicators of soil quality under the tested treatments over time (Table 2). Table 3 reports the average 2019 and 2020 values used to calculate RSIs.

Table 2. Comparison of *p*-values among the linear mixed-effect model analysis of observed parameters (Agg—aggregate stability; BD—bulk density; C org—soil organic carbon; N tot—soil total nitrogen; GWC—gravimetric water content; PR—penetration resistance; *Ks*—saturated hydraulic conductivity; EW—earthworm density; N min—mineral nitrogen; Y—yield, CC—cover crop).

	Time	Tillage	CC	Time × Till	Time × CC	Till × CC	Time × Till × CC
Agg	0.001 **	0.111	0.831	0.928	0.507	0.227	0.112
BD	0.155	0.663	0.529	0.043 *	0.469	0.672	0.536
C org	0.715	0.633	0.99	0.35	0.768	0.778	0.882
N tot	0.052	0.188	0.87	0.192	0.545	0.766	0.566
GWC	<0.001 ***	0.255	0.03 *	0.443	0.808	0.677	0.915
PR	<0.001 ***	0.004 **	0.635	0.334	0.815	0.724	0.877
Ks	0.034 *	0.046 *	0.187	0.39	0.564	0.252	0.68
EW	0.389	0.126	0.104	0.006 **	0.199	0.161	0.796
N min	0.451	0.906	0.615	0.169	0.589	0.343	0.501
pH	<0.001 ***	0.159	0.982	0.551	0.612	0.867	0.97
Y	0.84	0.904	0.68	0.76	0.378	0.648	0.589

*, ** and *** mean $p < 0.05$, <0.01 and <0.001, respectively.

Table 3. Descriptive statistics of studied parameters for T0 and T1 surveys (Agg—aggregate stability; BD—bulk density; C org—soil organic carbon; N tot—soil total nitrogen; GWC—gravimetric water content; PR—penetration resistance; *Ks*—saturated hydraulic conductivity; EW—earthworm density; N min—mineral nitrogen; St. Dev.—standard deviation; Var. Coef.—coefficient of variation).

Survey	Parameter	Unit	Min	Max	Mean	St. Dev.	Var. Coef.
T0	Agg	-	2.90	6.10	4.50	0.92	0.20
	BD	g cm^{-3}	1.32	1.54	1.43	0.05	0.04
	C org	%	0.64	1.07	0.83	0.12	0.14
	N tot	‰	0.08	1.09	0.88	0.23	0.26
	GWC	%	20	25	23	1	0.06
	PR	MPa	0.46	1.05	0.70	0.14	0.21
	Ks	m s^{-1}	6.7×10^{-6}	1.7×10^{-4}	3.4×10^{-5}	3.9×10^{-5}	1.15
	EW	n m^{-2}	0.00	16.00	6.17	4.69	0.76
	N min	mg kg^{-1}	12.85	46.90	22.97	8.91	0.39
	pH	-	7.22	7.49	7.36	0.06	0.008
	Yield	Mg ha^{-1}	5.41	12.36	9.96	1.62	0.16
T1	Agg	-	0.30	5.20	3.19	1.25	0.39
	BD	g cm^{-3}	1.36	1.56	1.46	0.06	0.04
	C org	%	0.63	1.01	0.82	0.11	0.13
	N tot	‰	0.74	1.21	1.01	0.13	0.13
	GWC	%	12	22	16	2	0.13
	PR	MPa	0.96	1.96	1.34	0.25	0.19
	Ks	m s^{-1}	8.2×10^{-6}	3.6×10^{-4}	8.7×10^{-5}	1.0×10^{-4}	1.16
	EW	n m^{-2}	0.00	20.00	7.44	6.21	0.83
	N min	mg kg^{-1}	6.49	53.41	26.11	14.37	0.55
	pH	-	6.93	7.22	7.05	0.08	0.01
	Yield	Mg ha^{-1}	9.28	11.09	10.04	0.55	0.05

All values of Agg, GWC and pH were significantly higher at T0 than at T1. Overall, average Agg was higher at T0 (4.50) than at T1 (3.19) and was characterized as of high-to-moderate stability, according to the *SI* range (0.3–6.1). At T0, all aggregate samples, except one, were of moderate stability (>3). The exception sample value, collected from treatment combination MT–BS, was 2.9. During the T1 survey, 44% of the observations were <3 (high aggregate stability) and the lowest values found in the reduced tillage systems (NT and MT). The measures of the GWC were strictly related to the pedoclimatic conditions on the sampling dates, with the GWC ranging from 20% to 25% in 2019 and from 12% to 22% in 2020. Cover crop treatments showed significant effects on GWC, as demonstrated by values of 18.3% in TR and 20.3% in WW, while BS had an intermediate value. Despite the significantly lower pH values at T0 versus T1, the pH values maintained non-critical averages (7.36 in T0 and 7.05 in T1).

Between survey T0 and T1, the N tot, PR and Ks all increased significantly. The N tot rose from 0.88‰ at T0 to 1.01‰ at T1. During each survey, the N tot maintained a modest variability, as indicated by the coefficients of variation at T0 (0.26) and T1 (0.13). The PR test values differed from an average of 0.70 MPa in T0 to an average of 1.34 MPa in T1. In the second survey, the PR was not only significantly higher, but also more variable than it was in T0; all of the PR observations across both surveys registered below the 2.5 MPa threshold. The PR differences occurred among the differing tillage systems. Specifically, CT reported a PR of 0.88 MPa, which proved to be significantly higher than the 1.18 MPa observed under NT. The PR result under MT was intermediate. Last, the *Ks* increased by 158% between T0 (3.4×10^{-5} m s^{-1}) and T1 (8.7×10^{-5} m s^{-1}). This parameter showed it was also significantly impacted by different tillage intensities, as shown by the average *Ks* values of 1.05×10^{-4} m s^{-1} in NT, 3.58×10^{-5} m s^{-1} in CT and an intermediate value in MT.

The parameters BD and EW were affected by the time × tillage interaction. Despite a generally limited effect on the BD across the various treatments, the average BD under CT was lower during the first survey (1.39 g cm^{-3}) versus the second survey (1.45 g cm^{-3}). All measures of BD were less than its 1.55 g cm^{-3} threshold. In the case of EW, variability was higher; it ranged between 0 and 20 (Table 3). Among the treatments, during T1, the EW differences were, on average, significantly higher (13.17) under NT than under CT (3.00).

The C org, Y and N min parameters resulted as unaffected by all factors tested. On average, the C org was 0.83% and displayed only a modest variability within and between the surveys. Similarly, Y (10.00 Mg ha^{-1}, on average) and N min (24.54 mg kg^{-1}, on average) showed no significance among the treatment combinations in the different surveys.

The values presented in Table 3 were normalized. The average of each treatment combination is presented in Figure 1 (biochemical parameters) and Figure 2 (physical parameters). Normalization allows higher values to be associated with parameter improvement and wider areas to represent an overall sustainability increment.

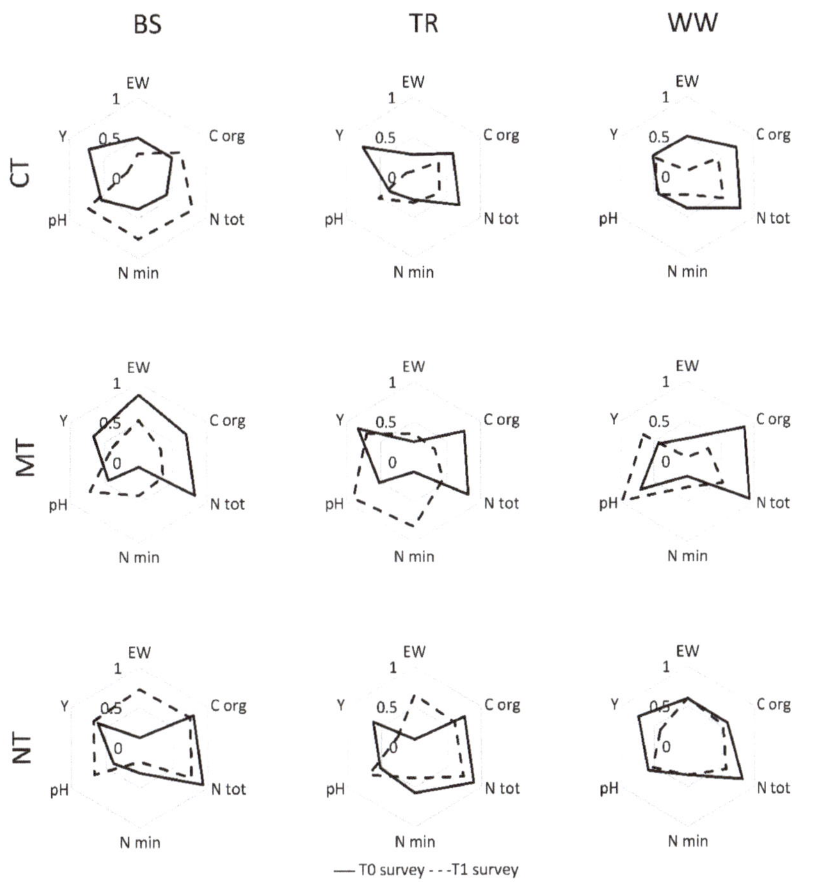

Figure 1. Biochemical parameter scores with average values in treatment combinations at T0 and T1 surveys (C org—soil organic carbon; N tot—soil total nitrogen; EW—earthworm density; N min—mineral nitrogen; Y—yield; BS—bare soil; TR—tillage radish; WW—winter wheat; CT—conventional tillage; MT—minimum tillage; NT—no tillage).

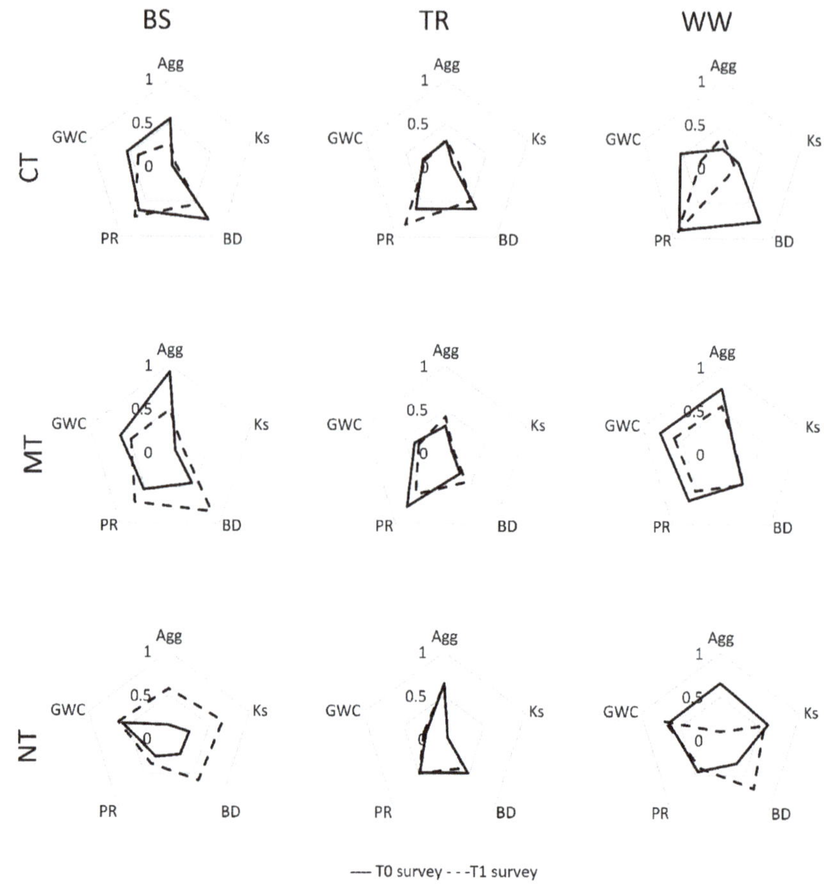

Figure 2. Physical parameter scores with average values in treatment combinations in T0 and T1 surveys (Agg—aggregate stability; BD—bulk density; GWC—gravimetric water content; PR—penetration resistance; *Ks*—saturated hydraulic conductivity; BS—bare soil; TR—tillage radish; WW—winter wheat; CT—conventional tillage; MT—minimum tillage; NT—no tillage).

Figures 1 and 2 show sizeable differences between the treatment combinations and two years. The correlation matrix between each parameter pair is shown in Table 4. As expected, the highest correlation resulted between the C org and N tot (r = 0.924). To identify which of these highly-correlated parameters could best explain treatment variation—and warrant inclusion in the *RSI*—we performed a principal component analysis (PCA).

Table 4. Correlation among the parameters. Boldface indicates highly-correlated values (r > 0.8). (Agg—aggregate stability; BD—bulk density; C org—soil organic carbon; N tot—soil total nitrogen; GWC—gravimetric water content; PR—penetration resistance; *Ks*—saturated hydraulic conductivity; EW—earthworm density; N min—mineral nitrogen).

	Agg	BD	C Org	EW	GWC	*Ks*	N Min	N Tot	pH	PR	Y
Agg	1										
BD	−0.009	1									
C org	−0.131	0.187	1								
EW	−0.274	0.234	0.033	1							
GWC	−0.152	0.299	0.340	0.175	1						
Ks	0.032	0.007	0.256	0.310	0.344	1					
N min	0.385	0.091	−0.134	−0.153	−0.150	−0.037	1				
N tot	−0.165	0.100	**0.924**	0.038	0.328	0.293	−0.022	1			
pH	−0.091	−0.038	0.000	−0.069	0.115	0.100	0.182	−0.086	1		
PR	0.038	0.130	−0.203	−0.271	0.012	−0.469	−0.154	−0.284	−0.039	1	
Y	−0.043	−0.119	0.009	0.104	−0.182	0.144	−0.088	0.024	−0.037	−0.332	1

Table 5 presents the PCA results. Each parameter was weighted according to the treatment variation it explained based on the PC selected.

Table 5. Results of principal component analysis under different treatment combination in different years. Bolded factor loads were considered as high. Bolded and underlined factor loads determined for each variable were those the PC considered in the *RSI* calculation. The weighting factor (*PW*) for each variable was equal to the variation explained by the PC selected (Agg—aggregate stability; BD—bulk density; C org—soil organic carbon; N tot—soil total nitrogen; GWC—gravimetric water content; PR—penetration resistance; *Ks*—saturated hydraulic conductivity; EW—earthworm density; N min—mineral nitrogen).

Principal Components	PC-1	PC-2	PC-3	PC-4
Variation	0.241	0.149	0.135	0.117
Cumulative variation	0.241	0.389	0.525	0.642
Agg	0.196	−0.134	<u>**0.536**</u>	−0.062
BD	0.153	<u>**−0.355**</u>	−0.072	0.420
C org	**0.498**	−0.168	−0.222	−0.334
EW	<u>**0.248**</u>	0.131	0.403	0.441
GWC	<u>**0.355**</u>	−0.326	−0.018	0.302
Ks	<u>**0.378**</u>	0.304	−0.057	0.277
N min	−0.124	0.176	<u>**−0.601**</u>	0.265
N tot	<u>**0.504**</u>	−0.097	−0.244	−0.357
pH	0.010	0.046	−0.169	<u>**0.338**</u>
PR	−0.291	<u>**−0.563**</u>	0.073	−0.056
Y	0.079	<u>**0.502**</u>	0.196	−0.174

The parameters selected in PC-1 were N tot, GWC, Ks and EW. It showed that the N tot should be included in the *RSI* because it had the highest weight and it was highly correlated to the C org. In PC-2, the highly weighted parameters BD, PR and Y were all included in the *RSI* as they showed limited correlation amongst them. In PC-3, the Agg and N min were selected and, in PC-4, the pH was chosen. The *PW* of each parameter equals the variability explained by the PC selected for that specific factor (0.241 for PC-1, 0.149 for PC-2, 0.135 for PC-3 and 0.117 for PC-4). To normalize the RSI, the sum of the weighted parameters was divided by the highest sum of the weighted parameters reported across all observations (1.247). The value was reported under NT–WW (block 1) during the survey T0. The lowest value (0.358) was under CT–TR (block 2) in T1. Then, the resulting *RSI* was expressed by Equation (9).

$$RSI = \frac{0.135Agg + 0.149BD + 0.241EW + 0.241GWC + 0.241Ks + 0.135N\text{min} + 0.241N\ tot + 0.117pH + 0.149PR + 0.149Y}{1.247} \quad (9)$$

Then, mixed models were calculated on *RSI* values, considering the combination of tillage and CC effects. The smallest AIC for the *RSI* linear mixed model was obtained when intercept, tillage and covering were tested as fixed factors and block was a random factor. Table 6 summarizes the *p*-values for the selected mixed model.

Table 6. Linear mixed model analysis of *RSI* output.

	T0			T1		
Effect	**F**	***p***		**F**	***p***	
Intercept	75.81	<0.001	***	81.27	<0.001	***
Tillage	0.20	0.823		5.57	0.019	*
Covering	2.48	0.125		2.88	0.095	

* and *** mean $p < 0.05$ and <0.001, respectively.

Figure 3 displays the average RSIs and corresponding contribution from each parameter to it under each treatment. On average, the GWC (0.09) and N tot (0.13) impacted the *RSI* the most. During T1, their highest scores were in NT (GWC = 0.10 and N tot = 0.13.) Observations of the *Ks* and EW were notable in that they contributed little to the RSI, yet they were high variable across treatments. During T1, the *Ks* averaged 0.08 under NT, which was three-fold the value observed under MT (0.03) or CT (0.02). Similarly, the EW averaged 0.13 in NT, which was double that in MT (0.06) and four-fold the value observed in CT (0.03).

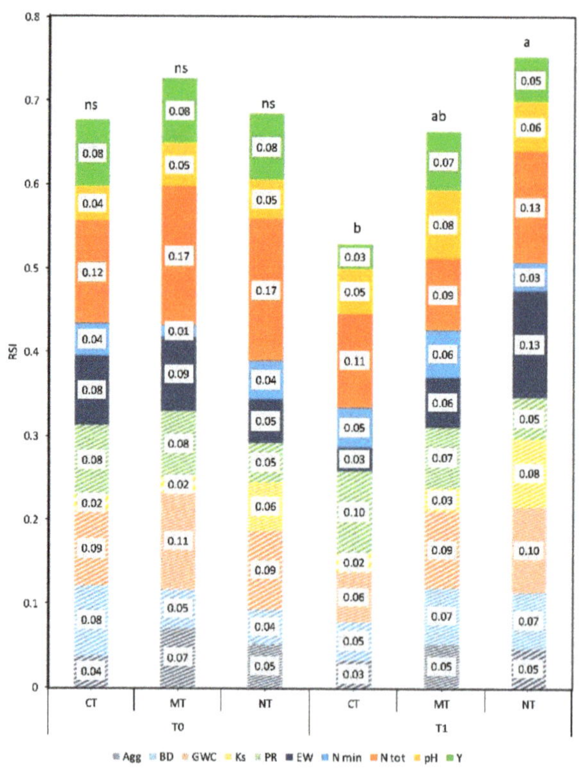

Figure 3. Average *RSIs* and the contribution of each parameter under different tillage systems in different years. Agg—aggregate stability; BD—bulk density; GWC—gravimetric water content; *Ks*—saturated hydraulic conductivity; PR—penetration resistance; EW—earthworm density; N min—mineral nitrogen; N tot—soil total nitrogen; Y—yield; CT—conventional tillage; MT—minimum tillage; NT—no tillage. Different letters represent significant differences of the global treatment *RSI* at $p < 0.05$.

No clear effect was observed for the soil covering treatment in either year and no statistical difference was found. During both years under TR, the minimum *RSI* was always reached (0.604 in 2019 and 0.583 in 2020). Higher values were recorded for coverings WW in 2019 and BS in 2020 (Figure 4).

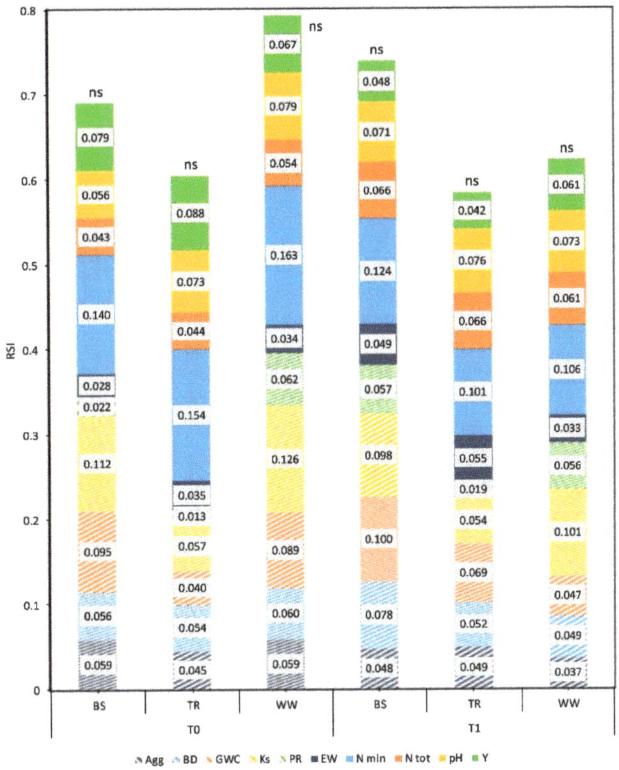

Figure 4. Average *RSI* values for different soil coverings in different years. BS—bare soil; TR—tillage radish; WW—winter wheat. Agg—aggregate stability; BD—bulk density; GWC—gravimetric water content; *Ks*—saturated hydraulic conductivity; PR—penetration resistance; EW—earthworm density; N min—mineral nitrogen; N tot—soil total nitrogen; Y—yield; CT—conventional tillage; MT—minimum tillage; NT—no tillage. Different letters represent significant differences of the global treatment *RSI* at $p < 0.05$.

4. Discussion

In total, 8 of the 11 parameters revealed significant differences between T0 and T1, which suggests that the soil system was changing regardless of the agronomic management applied. One potential cause of these results may be attributed to differences in the environmental conditions at T0 and T1. The GWC was found to be affected by CC. For those who have considered the effects of CC on water cycle, the results have been contradictory. Some have found CC to improve water balance and water availability [49], while others have reported soil water reduction in the subsequent crop after CC termination [50]. Our results, where under WW, the GWC was high and, under TR, the GWC was at its lowest, are also mixed. In both instances, the results can be equally attributed to either better maintenance of soil water content by WW, or higher soil evaporation under TR, due to a lesser soil covering.

The different tillage systems seemed to have a stronger impact, especially on some parameters (BD, PR, *Ks* and EW), if compared to the CC effect. For example, under CT, the BD and PR values aligned with previous evidence that reduced tillage systems can increase soil strength and bulk density, especially in the first years [51]. The reduced BD and PR values were expected tillage effects under CT, given that that they were measured in the 0–20 cm soil layer. This result may also relate to instrument resolution; the PR can be negatively impacted by the high spatial variability in reduced tillage systems [52]. However, in this instance, the BD almost always remained below its threshold (1.55 g cm^{-3}) for which it is known to limit plant root growth in silty loam soils [33]. Similarly, the PR values (0.46–1.96 MPa) fell well below the growth-limiting threshold usually set at 2.5 MPa [34]. Finally, the soil at the experimental site was characterized as having structural inertia in response to management changes [53–55].

Although the BD and PR values worsened slightly (i.e., soil strength increased) under reduced tillage systems, soil function was improved in NT, as evidenced by an increase of 193% in the *Ks* under NT relative to CT during T1. The highest EW value observed in this study may relate to the significant contributions made by earthworm bio-macropores to soil function and, in particular, air and water permeability, even in compacted soils. Earthworms can improve soil structure [56] and hydraulic properties [57] by burrowing and casting. The positive effects of NT on the EW confirm previous studies evidence [9,21,28].

The computation of the *RSI* highlighted the strong effect of the EW as it carried a high relative weight (11%, on average) within the index. It also showed a high variability among the different treatments. Additional parameters that averaged high impact on the *RSI* were the N tot (17%), GWC, (14%) and PR (11%), which, together, accounted for more than 50% of the *RSI*. In addition to the EW, *RSI* variability was driven by the *Ks*, N min and Y. In absolute terms, the *Ks*, N min and Y each had impacts of less than 10% on the *RSI*, but their variation coefficients ranged the highest (from 0.67 for the *Ks* to 0.31 for the Y). These two conditions suggest that this set of measures should be considered as the best to indicate soil quality changes during the conversion from conventional tillage to CA. The *RSI* results also suggest that the *Ks* and EW are two sustainability indicators that were positively affected by NT.

The final *RSI* score evaluates the combination of tillage intensity and soil covering with an holistic approach [58]. It showed the positive effect of NT relative to conventional tillage, even in the short term. Midway between the effects of CT and NT lay the MT system. It mitigated the negative effects on some physical parameters but lessened the improvements of biological parameters (EW). According to Issaka et al. [59], both the minimum and no-tillage systems resulted as sustainable techniques, considering the nutrient cycles. As opposed to other studies [9,10,54], clear negative effects during the transition time were not detected during this three-year experiment.

The limited differences reported for the various soil coverings may be evidence that a CC effect was masked by the strong effects of reduced tillage systems combined with the sampling methods used. It may be that longer conversion times or different sampling methods are required for CC effects to be revealed [60]. Even in the case of BS, a partial and spontaneous covering (weeds) may impact soil properties in a way not unlike that expected with CCs. Indeed, "spontaneous CCs" have provided ecosystem services [61–65]. In the presence of plant residues, microbial diversity [66] could improve to the point where it should even be considered an environmental sustainability indicator [67].

From another perspective, the modest TR effect could relate to sample timing. Most TR-related benefits (improved porosity and pore connectivity) occur only when taproots are degraded. At the same time, reports of short-term tillage radish benefits exist [68,69], although it seems that longer timespans are necessary to exploit the benefits of TR on soil properties [10]. The bio-tillage effect, which was expected from TR, as suggested by Zhang et al. [70], could be masked by earthworm activity in NT treatments, irrespectively from the presence of TR. The high EW values observed under NT could have performed this bio-tillage effect, which, according to the authors, could replace conventional tillage.

Then, even if the WW fibrous root apparatus had a limited impact on soil structure, many Poaceae CC improved overall system sustainability [71,72] and aggregate stability [49,65,73], or nutrient cycles [74–76]. The combination of grass CC and reduced tillage systems proved to positively affect environmental sustainability, fostering biodiversity [77] and soil organic carbon [78].

In conclusion, to correctly evaluate the CA effect, especially on the soil system, a holistic approach should be preferred to consider both the effects on crop production and on soil physics, considering different soil function at different scales.

5. Conclusions

A multivariate analysis of selected sustainability indicators revealed a positive effect of reduced tillage systems management and in particular NT, despite the limited variation in the observed parameters.

Despite the short-term nature of the experiment, this positive result could be the effect of an increase in soil fauna activity, which could have contributed to soil structure improvement. As a consequence, NT seemed to impact soil physics and soil habitability, resulting in a significantly higher *RSI* value. The effect of CC was limited, but WW reported the best results in the short term, with a tendency to have higher RSI values.

Collectively, the combination of NT and WW can be considered the most promising in terms of sustainability improvement. In this study, only the short-term effect of different tillage and soil cover management results were reported. Therefore, longer-term experiments could better evaluate the effects of these management systems on some parameters, such as soil organic carbon, which have a wide impact on sustainability, yet vary little in the short term.

In conclusion, to correctly evaluate the CA effect, especially on the soil system, a holistic approach should be preferred to consider both the effects on crop production and on soil physics, considering different soil functions at different scales.

Author Contributions: Conceptualization, F.S.; formal analysis, F.S. and A.B.; investigation, F.S. and R.P.; data curation, F.S. and R.P.; writing—original draft preparation, F.S.; writing—review and editing, F.S., I.P., R.P. and A.B.; visualization, F.S.; supervision, I.P. and A.B.; project administration, A.B.; funding acquisition, A.B. All authors have read and agreed to the published version of the manuscript.

Funding: The research leading to these results received funding from the European Union HORIZON2020 Programme for Research, Technological Development and Demonstration under Grant Agreement No. 677407 (SOILCARE Project).

Institutional Review Board Statement: Not applicable.

Informed Consent Statement: Not applicable.

Data Availability Statement: The data that support the findings of this study are available from the corresponding author upon reasonable request.

Acknowledgments: We thank Rudi Hessel (project coordinator) and all the work package leaders for coordinating the project activities.

Conflicts of Interest: The authors declare no conflict of interest.

References

1. FAO. Conservation Agriculture. Available online: http://www.fao.org/conservation-agriculture/en/ (accessed on 23 October 2020).
2. Baveye, P.C.; Rangel, D.; Jacobson, A.R.; Laba, M.; Darnault, C.; Otten, W.; Radulovich, R.; Camargo, F.A.O. From Dust Bowl to Dust Bowl: Soils are Still Very Much a Frontier of Science. *Soil Sci. Soc. Am. J.* **2011**, *75*, 2037–2048. [CrossRef]
3. Palm, C.; Blanco-Canqui, H.; DeClerck, F.; Gatere, L.; Grace, P. Conservation agriculture and ecosystem services: An overview. *Agric. Ecosyst. Environ.* **2014**, *187*, 87–105. [CrossRef]
4. Komissarov, M.A.; Klik, A. The Impact of no-Till, conservation, and conventional tillage systems on erosion and soil properties in lower Austria. *Eurasian Soil Sci.* **2020**, *53*, 503–511. [CrossRef]
5. Carretta, L.; Tarolli, P.; Cardinali, A.; Nasta, P.; Romano, N.; Masin, R. Evaluation of runoff and soil erosion under conventional tillage and no-till management: A case study in northeast Italy. *Catena* **2021**, *197*, 104972. [CrossRef]

6. Derpsch, R.; Friedrich, T.; Kassam, A.; Li, H. Current status of adoption of no-till farming in the world and some of its main benefits. *Int. J. Agric. Biol. Eng.* **2010**, *3*, 1–25.
7. Kassam, A.; Friedrich, T.; Derpsch, R. Global spread of Conservation Agriculture. *Int. J. Environ. Stud.* **2019**, *76*, 29–51. [CrossRef]
8. Troccoli, A.; Maddaluno, C.; Mucci, M.; Russo, M.; Rinaldi, M. Is it appropriate to support the farmers for adopting conservation agriculture? Economic and environmental impact assessment. *Ital. J. Agron.* **2015**, *10*, 169–177. [CrossRef]
9. Perego, A.; Rocca, A.; Cattivelli, V.; Tabaglio, V.; Fiorini, A.; Barbieri, S.; Schillaci, C.; Chiodini, M.E.; Brenna, S.; Acutis, M. Agro-environmental aspects of conservation agriculture compared to conventional systems: A 3-year experience on 20 farms in the Po valley (Northern Italy). *Agric. Syst.* **2019**, *168*, 73–87. [CrossRef]
10. Camarotto, C.; Piccoli, I.; Dal Ferro, N.; Polese, R.; Chiarini, F.; Furlan, L.; Morari, F. Have we reached the turning point? Looking for evidence of SOC increase under conservation agriculture and cover crop practices. *Eur. J. Soil Sci.* **2020**, *71*, 1050–1063. [CrossRef]
11. Piccoli, I.; Camarotto, C.; Lazzaro, B.; Furlan, L.; Morari, F. Conservation agriculture had a poor impact on the soil porosity of Veneto low-lying plain silty soils after a 5-year transition period. *Land Degrad. Dev.* **2017**, *28*, 2039–2050. [CrossRef]
12. Pagnani, G.; Galieni, A.; D'Egidio, S.; Visioli, G.; Stagnari, F.; Pisante, M. Effect of soil tillage and crop sequence on grain yield and quality of durum wheat in Mediterranean areas. *Agronomy* **2019**, *9*, 488.
13. Piccoli, I.; Lazzaro, B.; Furlan, L.; Berti, A.; Morari, F. Examining crop root apparatus traits in a maize-soybean-winter wheat rotation under conservation agriculture management. *Eur. J. Agron.* **2021**, *122*, 126171. [CrossRef]
14. Calzarano, F.; Stagnari, F.; D'egidio, S.; Pagnani, G.; Galieni, A.; Di Marco, S.; Metruccio, E.G.; Pisante, M. Durum wheat quality, yield and sanitary status under conservation agriculture. *Agriculture* **2018**, *8*, 140.
15. Vastola, A.; Zdruli, P.; D'Amico, M.; Pappalardo, G.; Viccaro, M.; Di Napoli, F.; Cozzi, M.; Romano, S. A comparative multidimensional evaluation of conservation agriculture systems: A case study from a Mediterranean area of Southern Italy. *Land Use Policy* **2017**, *68*, 326–333. [CrossRef]
16. Ruisi, P.; Giambalvo, D.; Saia, S.; Di Miceli, G.; Frenda, A.S.; Plaia, A.; Amato, G. Conservation tillage in a semiarid Mediterranean environment: Results of 20 years of research. *Ital. J. Agron.* **2014**, *9*, 1–7. [CrossRef]
17. Piccoli, I.; Chiarini, F.; Carletti, P.; Furlan, L.; Lazzaro, B.; Nardi, S.; Berti, A.; Sartori, L.; Dalconi, M.C.C.; Morari, F. Disentangling the effects of conservation agriculture practices on the vertical distribution of soil organic carbon. Evidence of poor carbon sequestration in North-eastern Italy. *Agric. Ecosyst. Environ.* **2016**, *230*, 68–78. [CrossRef]
18. Longo, M.; Piccoli, I.; Minasny, B.; Morari, F. Soil apparent electrical conductivity-directed sampling design for advancing soil characterization in agricultural fields. *Vadose Zone J.* **2020**, *19*, e20060. [CrossRef]
19. Massaccesi, L.; Rondoni, G.; Tosti, G.; Conti, E.; Guiducci, M.; Agnelli, A. Soil functions are affected by transition from conventional to organic mulch-based cropping system. *Appl. Soil Ecol.* **2020**, *153*, 103639. [CrossRef]
20. Piazza, G.; Pellegrino, E.; Moscatelli, M.C.; Ercoli, L. Long-term conservation tillage and nitrogen fertilization effects on soil aggregate distribution, nutrient stocks and enzymatic activities in bulk soil and occluded microaggregates. *Soil Tillage Res.* **2020**, *196*, 104482. [CrossRef]
21. Stagnari, F.; Pagnani, G.; Galieni, A.; D'Egidio, S.; Matteucci, F.; Pisante, M. Effects of conservation agriculture practices on soil quality indicators: A case-study in a wheat-based cropping systems of Mediterranean areas. *Soil Sci. Plant Nutr.* **2020**, *66*, 624–635. [CrossRef]
22. European Commission. *Report from the Commission to the European Parliament and the Council on the Implementation of the European Statistical Programme 2013–2017*; European Commission: Brussels, Belgium, 2017.
23. Masto, R.E.; Chhonkar, P.K.; Purakayastha, T.J.; Patra, A.K.; Singh, D. Soil quality indices for evaluation of long-term land use and soil management practices in semi-arid sub-tropical India. *Land Degrad. Dev.* **2008**, *19*, 516–529. [CrossRef]
24. Xuefang, H.; Juanling, W.; Mingjing, H.; Cong, Z.; Huatao, L. Sustainability of soil-crop systems under different long-term fertilizations in Chestnut Cinnamon soil areas. *Zhongguo Shengtai Nongye Xuebao Chin. J. Eco-Agric.* **2018**, *26*, 1107–1116.
25. Blanco-Canqui, H.; Mikha, M.M.; Presley, D.R.; Claassen, M.M. Addition of Cover Crops Enhances No-Till Potential for Improving Soil Physical Properties. *Soil Sci. Soc. Am. J.* **2011**, *75*, 1471–1482. [CrossRef]
26. Blanco-Canqui, H.; Ruis, S.J. Cover crop impacts on soil physical properties: A review. *Soil Sci. Soc. Am. J.* **2020**, *84*, 1527–1576. [CrossRef]
27. Fajardo, M.; McBratney, A.B.; Field, D.J.; Minasny, B. Soil slaking assessment using image recognition. *Soil Tillage Res.* **2016**, *163*, 119–129. [CrossRef]
28. Crotty, F.V.; Stoate, C. The legacy of cover crops on the soil habitat and ecosystem services in a heavy clay, minimum tillage rotation. *Food Energy Secur.* **2019**, *8*, e00169. [CrossRef]
29. Masto, R.E.; Chhonkar, P.K.; Singh, D.; Patra, A.K. Soil quality response to long-term nutrient and crop management on a semi-arid Inceptisol. *Agric. Ecosyst. Environ.* **2007**, *118*, 130–142. [CrossRef]
30. FAO-UNESCO. *Soil Map of the World. Revised Legend*; Food and Agriculture Organization of the United Nations: Rome, Italy, 2008.
31. Flynn, K.D.; Bagnall, D.K.; Morgan, C.L.S. Evaluation of SLAKES, a smartphone application for quantifying aggregate stability, in high-clay soils. *Soil Sci. Soc. Am. J.* **2020**, *84*, 345–353. [CrossRef]
32. Grossman, R.B.; Reinsch, T.G. 2.1 Bulk density and linear extensibility. In *Methods of Soil Analysis: Part 4 Physical Methods*; Dane, J.H., Topp, C.G., Eds.; Soil Science Society of America: Madison, WI, USA, 2002; pp. 201–228.

33. USDA NRCS. *Soil Quality Information Sheet Soil Quality Resource Concerns: Compaction USDA Natural Resources Conservation Service*; USDA: Washington, DC, USA, 1996.
34. Groenevelt, P.H.; Grant, C.D.; Semetsa, S. A new procedure to determine soil water availability. *Aust. J. Soil Res.* **2001**, *39*, 577–598. [CrossRef]
35. Parr, J.F.; Bertrand, A.R. Water infiltration into soils. *Adv. Agron.* **1960**, *12*, 311–363.
36. Lai, J.; Ren, L. Assessing the Size Dependency of Measured Hydraulic Conductivity Using Double-Ring Infiltrometers and Numerical Simulation. *Soil Sci. Soc. Am. J.* **2007**, *71*, 1667. [CrossRef]
37. Philip, J.R. The theory of infiltration: 4. Sorptivity and algebraic infiltration equations. *Soil Sci.* **1957**, *84*, 257–264. [CrossRef]
38. Valckx, J.; Govers, G.; Hermy, M.; Muys, B.; Valckx, J.; Govers, G.; Hermy, M.; Muys, B. Optimizing Earthworm Sampling in Ecosystems. *Soil Biol.* **2011**, *24*, 19–38.
39. Shepherd, G.; Janssen, H. *Visual Soil Assessment, Field Guide for Cropping*; Landcare Research: Palmerston North, New Zealand, 2000; ISBN 1-877221-92-9.
40. García-Robledo, E.; Corzo, A.; Papaspyrou, S. A fast and direct spectrophotometric method for the sequential determination of nitrate and nitrite at low concentrations in small volumes. *Mar. Chem.* **2014**, *162*, 30–36. [CrossRef]
41. Van Reeuwijk, L.P. Procedures for soil analysis. *Int. Soil Ref. Inf. Cent.* **1986**, *9*, 106.
42. Andrews, S.S.; Mitchell, J.P.; Mancinelli, R.; Karlen, D.L.; Hartz, T.K.; Horwath, W.R.; Pettygrove, G.S.; Scow, K.M.; Munk, D.S. On-Farm Assessment of Soil Quality in California's Central Valley. *Agron. J.* **2002**, *94*, 12–23.
43. Andrews, S.S.; Karlen, D.L.; Mitchell, J.P. A comparison of soil quality indexing methods for vegetable production systems in Northern California. *Agric. Ecosyst. Environ.* **2002**, *90*, 25–45. [CrossRef]
44. Armenise, E.; Redmile-Gordon, M.A.; Stellacci, A.M.; Ciccarese, A.; Rubino, P. Developing a soil quality index to compare soil fitness for agricultural use under different managements in the mediterranean environment. *Soil Tillage Res.* **2013**, *130*, 91–98. [CrossRef]
45. Kumar, M.; Mitra, S.; Mazumdar, S.P.; Majumdar, B.; Saha, A.R.; Singh, S.R.; Pramanick, B.; Gaber, A.; Alsanie, W.F.; Hossain, A. Improvement of Soil Health and System Productivity through Crop Diversification and Residue Incorporation under Jute-Based Different Cropping Systems. *Agronomy* **2021**, *11*, 1622. [CrossRef]
46. Raiesi, F.; Kabiri, V. Identification of soil quality indicators for assessing the effect of different tillage practices through a soil quality index in a semi-arid environment. *Ecol. Indic.* **2016**, *71*, 198–207. [CrossRef]
47. Schabenberger, O.; Pierce, F. *Contemporary Statistical Models for the Plant and Soil Sciences*; CRC Press: Boca Raton, FL, USA, 2001; ISBN 978-1-58488-111-7.
48. Metsalu, T.; Vilo, J. ClustVis: A web tool for visualizing clustering of multivariate data using Principal Component Analysis and heatmap. *Nucleic Acids Res.* **2015**, *43*, W566–W570. [CrossRef] [PubMed]
49. García-González, I.; Hontoria, C.; Gabriel, J.L.; Alonso-Ayuso, M.; Quemada, M. Cover crops to mitigate soil degradation and enhance soil functionality in irrigated land. *Geoderma* **2018**, *322*, 81–88. [CrossRef]
50. Krstić, D.; Vujić, S.; Jaćimović, G.; D'Ottavio, P.; Radanović, Z.; Erić, P.; Ćupina, B. The Effect of Cover Crops on Soil Water Balance in Rain-Fed Conditions. *Atmosphere* **2018**, *9*, 492. [CrossRef]
51. Blanco-Canqui, H.; Ruis, S.J. No-tillage and soil physical environment. *Geoderma* **2018**, *326*, 164–200. [CrossRef]
52. Piccoli, I.; Schjønning, P.; Lamandé, M.; Zanini, F.; Morari, F. Coupling gas transport measurements and X-ray tomography scans for multiscale analysis in silty soils. *Geoderma* **2019**, *338*, 576–584. [CrossRef]
53. Camarotto, C.; Dal Ferro, N.; Piccoli, I.; Polese, R.; Furlan, L.; Chiarini, F.; Morari, F. Conservation agriculture and cover crop practices to regulate water, carbon and nitrogen cycles in the low-lying Venetian plain. *Catena* **2018**, *167*, 236–249. [CrossRef]
54. Piccoli, I.; Furlan, L.; Lazzaro, B.; Morari, F. Examining conservation agriculture soil profiles: Outcomes from northeastern Italian silty soils combining indirect geophysical and direct assessment methods. *Eur. J. Soil Sci.* **2020**, *71*, 1064–1075. [CrossRef]
55. Piccoli, I.; Schjønning, P.; Lamandé, M.; Furlan, L.; Morari, F. Challenges of conservation agriculture practices on silty soils. Effects on soil pore and gas transport characteristics in North-eastern Italy. *Soil Tillage Res.* **2017**, *172*, 12–21. [CrossRef]
56. Bertrand, M.; Barot, S.; Blouin, M.; Whalen, J.; de Oliveira, T.; Roger-Estrade, J. Earthworm services for cropping systems. A review. *Agron. Sustain. Dev.* **2015**, *35*, 553–567. [CrossRef]
57. van Schaik, L.; Palm, J.; Klaus, J.; Zehe, E.; Schröder, B. Linking spatial earthworm distribution to macropore numbers and hydrological effectiveness. *Ecohydrology* **2014**, *7*, 401–408. [CrossRef]
58. Vogel, H.-J.; Balseiro-Romero, M.; Kravchenko, A.; Otten, W.; Pot, V.; Schlüter, S.; Weller, U.; Baveye, P.C. A holistic perspective on soil architecture is needed as a key to soil functions. *Eur. J. Soil Sci.* **2021**, 1–14. [CrossRef]
59. Issaka, F.; Zhang, Z.; Zhao, Z.-Q.; Asenso, E.; Li, J.-H.; Li, Y.-T.; Wang, J.-J. Sustainable conservation tillage improves soil nutrients and reduces nitrogen and Phosphorous losses in maize farmland in southern China. *Sustainability* **2019**, *11*, 2397. [CrossRef]
60. Wagger, M.G.; Denton, H.P. Influence of Cover Crop and Wheel Traffic on Soil Physical Properties in Continuous No-Till Corn. *Soil Sci. Soc. Am. J.* **1989**, *53*, 1206–1210. [CrossRef]
61. Herencia, J.F. Soil quality indicators in response to long-term cover crop management in a Mediterranean organic olive system. *Biol. Agric. Hortic.* **2017**, *34*, 211–231. [CrossRef]
62. de Torres, M.A.R.-R.; Ordóñez-Fernández, R.; Giráldez, J.V.; Márquez-García, J.; Laguna, A.; Carbonell-Bojollo, R. Efficiency of four different seeded plants and native vegetation as cover crops in the control of soil and carbon losses by water erosion in olive orchards. *Land Degrad. Dev.* **2018**, *29*, 2278–2290. [CrossRef]

63. Carpio, A.J.; Castro, J.; Tortosa, F.S. Arthropod biodiversity in olive groves under two soil management systems: Presence versus absence of herbaceous cover crop. *Agric. For. Entomol.* **2019**, *21*, 58–68. [CrossRef]
64. Guzmán, G.; Cabezas, J.M.; Sánchez-Cuesta, R.; Lora, Á.; Bauer, T.; Strauss, P.; Winter, S.; Zaller, J.G.; Gómez, J.A. A field evaluation of the impact of temporary cover crops on soil properties and vegetation communities in southern Spain vineyards. *Agric. Ecosyst. Environ.* **2019**, *272*, 135–145. [CrossRef]
65. Guzmán, G.; Perea-Moreno, A.-J.; Gómez, J.A.; Cabrerizo-Morales, M.Á.; Martínez, G.; Giráldez, J.V. Water Related Properties to Assess Soil Quality in Two Olive Orchards of South Spain under Different Management Strategies. *Water* **2019**, *11*, 367. [CrossRef]
66. Li, Y.; Song, D.; Liang, S.; Dang, P.; Qin, X.; Liao, Y.; Siddique, K.H.M. Effect of no-tillage on soil bacterial and fungal community diversity: A meta-analysis. *Soil Tillage Res.* **2020**, *204*, 104721. [CrossRef]
67. Anyanzwa, H.; Okalebo, J.R.; Othieno, C.O.; Bationo, A.; Waswa, B.S.; Kihara, J. Effects of conservation tillage, crop residue and cropping systems on changes in soil organic matter and maize–legume production: A case study in Teso District. *Nutr. Cycl. Agroecosystems* **2008**, *88*, 39–47. [CrossRef]
68. Toom, M.; Talgre, L.; Mäe, A.; Tamm, S.; Narits, L.; Edesi, L.; Haljak, M.; Lauringson, E. Selecting winter cover crop species for northern climatic conditions. *Biol. Agric. Hortic.* **2019**, *35*, 263–274. [CrossRef]
69. Wittwer, R.A.; van der Heijden, M.G.A. Cover crops as a tool to reduce reliance on intensive tillage and nitrogen fertilization in conventional arable cropping systems. *Field Crops Res.* **2020**, *249*, 107736. [CrossRef]
70. Zhang, Z.; Peng, X. Bio-tillage: A new perspective for sustainable agriculture. *Soil Tillage Res.* **2021**, *206*, 104844. [CrossRef]
71. Diacono, M.; Persiani, A.; Testani, E.; Montemurro, F. Sustainability of agro-ecological practices in organic horticulture: Yield, energy-use and carbon footprint. *Agroecol. Sustain. Food Syst.* **2019**, *44*, 726–746. [CrossRef]
72. Ciaccia, C.; Ceglie, F.G.; Burgio, G.; Madžarić, S.; Testani, E.; Muzzi, E.; Mimiola, G.; Tittarelli, F. Impact of Agroecological Practices on Greenhouse Vegetable Production: Comparison among Organic Production Systems. *Agronomy* **2019**, *9*, 372. [CrossRef]
73. Domagała-Świątkiewicz, I.; Siwek, P.; Bucki, P.; Rabiasz, K. Effect of hairy vetch (*Vicia villosa* Roth.) and vetch-rye (*Secale cereale* L.) biculture cover crops and plastic mulching in high tunnel vegetable production under organic management. *Biol. Agric. Hortic.* **2019**, *35*, 248–262. [CrossRef]
74. Wen, Y.; Zang, H.; Freeman, B.; Ma, Q.; Chadwick, D.R.; Jones, D.L. Rye cover crop incorporation and high watertable mitigate greenhouse gas emissions in cultivated peatland. *Land Degrad. Dev.* **2019**, *30*, 1928–1938. [CrossRef]
75. Fiorini, A.; Maris, S.C.; Abalos, D.; Amaducci, S.; Tabaglio, V. Combining no-till with rye (*Secale cereale* L.) cover crop mitigates nitrous oxide emissions without decreasing yield. *Soil Tillage Res.* **2020**, *196*, 104442. [CrossRef]
76. Norberg, L.; Aronsson, H. Effects of cover crops sown in autumn on N and P leaching. *Soil Use Manag.* **2020**, *36*, 200–211. [CrossRef]
77. Reeves, D.W. The role of soil organic matter in maintaining soil quality in continuous cropping systems. *Soil Tillage Res.* **1997**, *43*, 131–167. [CrossRef]
78. Calegari, A.; Hargrove, W.L.; Rheinheimer, D.D.S.; Ralisch, R.; Tessier, D.; de Tourdonnet, S.; de Fatima Guimarães, M. Impact of long-term no-tillage and cropping system management on soil organic carbon in an oxisol: A model for sustainability. *Agron. J.* **2008**, *100*, 1013–1019. [CrossRef]

land

MDPI

Article

Comparison of Compaction Alleviation Methods on Soil Health and Greenhouse Gas Emissions

Jennifer Bussell [1], Felicity Crotty [2] and Chris Stoate [1,*]

[1] The Game & Wildlife Conservation Trust, Allerton Project, Loddington, Leicester LE7 9XE, UK; jbussell@gwct.org.uk
[2] Centre for Agriculture, Royal Agricultural University, Cirencester GL7 6JS, UK; felicity.crotty@rau.ac.uk
* Correspondence: cstoate@gwct.org.uk; Tel.: +44-1572-717220

Abstract: Soil compaction can occur due to trafficking by heavy equipment and be exacerbated by unfavourable conditions such as wet weather. Compaction can restrict crop growth and increase waterlogging, which can increase the production of the greenhouse gas nitrous oxide. Cultivation can be used to alleviate compaction, but this can have negative impacts on earthworm abundance and increase the production of the greenhouse gas carbon dioxide. In this study, a field was purposefully compacted using trafficking, then in a replicated plot experiment, ploughing, low disturbance subsoiling and the application of a mycorrhizal inoculant were compared as methods of compaction alleviation, over two years of cropping. These methods were compared in terms of bulk density, penetration resistance, crop yield, greenhouse gas emissions and earthworm abundance. Ploughing alleviated topsoil compaction, as measured by bulk density and penetrometer resistance, and increased the crop biomass in one year of the study, although no yield differences were seen. Earthworm abundance was reduced in both years in the cultivated plots, and carbon dioxide flux increased significantly, although this was not significant in summer months. Outside of the summer months, nitrous oxide production increased in the non-cultivated treatments, which was attributed to increased denitrifying activity under compacted conditions.

Keywords: nitrous oxide; N_2O; carbon dioxide; CO_2; greenhouse gas; compaction; earthworms; direct drilling; bulk density

Citation: Bussell, J.; Crotty, F.; Stoate, C. Comparison of Compaction Alleviation Methods on Soil Health and Greenhouse Gas Emissions. *Land* **2021**, *10*, 1397. https://doi.org/10.3390/land10121397

Academic Editors: Guido Wyseure, Julián Cuevas González, Jean Poesen and Amrakh I. Mamedov

Received: 22 November 2021
Accepted: 15 December 2021
Published: 17 December 2021

1. Introduction

Soil compaction is a form of soil degradation, which is an issue worldwide, due to the detrimental effects it has on agricultural productivity, through reduced crop growth, increased soil erosion and nutrient depletion [1]. Within England and Wales, almost 4 million hectares of soil are at risk of compaction [2]. Compaction was identified by DEFRA as one of the three key threats to the agricultural and environmental productivity of soils [3] and one of the ten soil threats identified in Europe [4]. Although soil compaction is not a recent phenomenon, some modern farming techniques can exacerbate the risks, including increasing field size and weight of farm equipment [5,6]. In this study, we specifically looked at the impact of topsoil compaction exerted by trafficking, which can occur when soils are trafficked by heavy equipment, especially in wet conditions [7,8]. The susceptibility of soils to compaction depends on the interaction between soil physical properties and climate; often soils are workable when soil moisture is lower than field capacity, making the window of opportunity for poorly draining soils particularly narrow [5,9]. Heavy clay soils, such as those found at this experimental site, are therefore often prone to compaction when necessary field operations, such as harvest, coincide with wet weather. This may be exacerbated with the impact of climate change making weather patterns more extreme, with warmer wetter winters and increased occurrences of intense storms, potentially reducing machinery working days [10].

Compacted soils have less pore space, and increased bonding between particles, which leads to several problems. They will take more energy to cultivate, and aggregates will be harder to separate [11]. As pore spaces shrink, less space is available for water, and capillary attraction holding water within the soil increases, reducing water availability and plant uptake. Lower pore space and reduced infiltration also reduces soil aeration; this combined with restricted root growth impairs nutrient and water availability, reducing crop growth [9,12]. Compacted soils can also have detrimental effects on soil fauna, most notably, earthworms are often cited as being negatively affected by compaction, due to physical crushing and disruption of their burrow network [13]. Earthworms are also considered one of the key biological engineers needed to improve soil structure after compaction [14].

Due to the poor structure of compacted soils, they can become progressively poorer at absorbing rainfall, becoming more anaerobic over time without ameliorative action [14], which can affect microbial activity, subsequent nutrient cycling and greenhouse gas emissions. Carbon dioxide (CO_2) is produced through many microbial processes [15] and can spike immediately after ploughing due to the flush of CO_2 released from the mixing of the microbial community with decomposable substrates and aerated voids produced through tillage [16]. Pore space and pore connectivity allow for oxygen exchange within the soil, and when these are reduced, oxygen will deplete more rapidly leading to anoxic conditions [15] changing microbial activity. The microbial process of denitrification produces nitrous oxide (N_2O) and is greatest in wet conditions, so less plant available nitrogen can be found in the soil and more nitrogen is lost to the atmosphere as N_2 and N_2O [17]. As N_2O has a global warming potential 298 time higher than that of CO_2 [18], compaction has implications for global warming emissions as well as soil health and productivity.

The efficacy of three methods for mitigating compaction damage was compared with the direct drilled control to see, not only the impact of these methods on crop production, but also their impact on soil health and greenhouse gas emissions. Ploughing was used as the conventional cultivation method for alleviating topsoil compaction. As ploughing aerates the soil profile, it can accelerate the loss of soil organic carbon (SOC) to the atmosphere as CO_2, and destroy soil aggregates, exposing organic carbon for mineralization [19]. The physical process of running a plough through a soil can also have a detrimental effect on earthworm populations [13,20]. Both SOC and earthworm numbers have beneficial impacts on aggregate stability and soil structure [21], improving infiltration and resilience to future compaction. Low disturbance subsoiling (LDS) can be used as an alternative method of compaction alleviation, particularly in the subsoil layer, as the topsoil remains undisturbed. LDS theoretically has lower impact on CO_2 emissions due to the non-inversion nature of the cultivation, reducing the mixing and oxygenation of SOC, and potentially reducing damage to earthworms in the topsoil layers.

Due to poor root exploration in compacted soils and microbial processes occurring in waterlogged soils, there can be lower access to nutrients for plants [7,22]. Mycorrhizal association has been suggested to benefit plants in these conditions, as the excess hyphae network can scavenge nutrients from a larger volume of soil [23,24]. As a final compaction alleviation method, a mycorrhizal inoculant was introduced to help plants overcome the detrimental effects of compaction on nutrient acquisition. The overall aim of the study was to identify the detrimental impacts of topsoil compaction, and to compare methods of alleviating this compaction in terms of their impact on soil compaction, earthworm populations, plant productivity and greenhouse gas emissions.

2. Materials and Methods

The experimental area was set up in October 2017 at the Allerton Project—a 300 hectare mixed arable and livestock research, demonstration and education farm (Game & Wildlife Conservation Trust, Fordingbridge, Hampshire, UK), at Loddington, Leicestershire, UK (N 052°36′53″ W 00°50′31″; 186 m a.s.l). Soils are predominantly a heavy clay loam, UK soil series: Denchworth, texture 47% clay, 31% silt, and 22% sand, soil organic matter 4.2%. To create compaction in the field, a tractor (Massey Ferguson 7720, approximate weight

8 tonnes) was driven across part of the field (100 m × 50 m), so that every area of the plots had been passed over by a tractor wheel twice. The compaction was checked using a cone penetrometer (SC 900, Field Scout, Aurora, IL, USA), taking an average of 10 measurements per plot, and showing an average of 15% higher compaction measured across 45 cm depth, that peaked at an increase of 32% at 7.5 cm depth. Penetration resistance measurements were repeated 4 times across the year.

Plots were arranged across the compacted area in randomized blocks (with tramlines excluded from the experimental treatments), measuring 6 m wide and 40 m long, giving an area of 240 m^2. The effectiveness of cultivation at alleviating the compaction was tested using four treatments: plough, low disturbance subsoiler (LDS), mycorrhizal inoculant (AMF), and a no cultivation direct drilled control. Cultivations took place each year in autumn. Plough plots were ploughed to a depth of 25 cm, then disked to a depth of 10 cm (Väderstad carrier); LDS plots were subsoiled to a depth of 30 cm; AMF plots received a granular application of inoculant SR1:Cereals (Plantworks Ltd., Sittingbourne, UK) drilled with the crop at a rate of 10 kg/ha; while direct drill plots only received a straw rake before drilling. All crops were established using a direct drill (Eco M, Dale Drills, Market Rasen, UK) and standard farm practice was used for the application of manufactured fertiliser and plant protection products, which was consistent across all plots. Following cultivations in October 2017, *Hordeum vulgare* was planted across all plots and harvested in July 2018. The compaction and cultivation treatments were repeated in October 2018 keeping the same plot structure and *Vicia faba* was planted across all plots and harvested in September 2019.

Topsoil bulk density 0–10 cm was measured yearly in spring using a bulk density ring (10 cm depth, 5 cm diameter); three measurements were averaged per plot. Yield was taken from the combine as each plot was harvested. Plant biomass was also taken before combine harvest, by cutting three 0.25 m^2 quadrats per plot, and drying the biomass in an oven at 70 °C until a stable weight was achieved. Earthworm abundance was measured using three replicates of 20 × 20 × 25 cm soil blocks per plot that were removed by spade. Soil was sorted by hand and all worms were counted and weighed.

Greenhouse gas measurements were taken monthly across the two cropping seasons using an FT-IR gas analyser (DX4040, Gasmet, Helsinki, Finland), set to measure CO_2 and N_2O simultaneously, with a 20 cm soil survey chamber attached (Li-cor). Plastic rings (20 cm diameter) were placed in the soil to a depth of 10 cm, allowing a 15 cm lip above the soil, at least 48 h before the first measurement. The chamber formed an airtight seal when placed on top of the rings. Gas flux was measured over 10 min, with the machine set to average measurements over 60 s. The initial 4 min were discarded to allow for gas equilibration in the system, and gas flux was calculated from the increase in gas concentration measured over the remaining 6 min. N_2O was multiplied by 298 to give an equivalent global warming potential to CO_2 to make comparisons between these two greenhouse gasses [18].

Statistical analysis used the Genstat software package [25]. A one-way ANOVA was used for all statistics, with the exception of the penetration resistance analysis. Where multiple measurements were taken, a repeated measures ANOVA was used. For penetration resistance, a principal component analysis (PCA) reduced the dimensionality of data, so comparisons between treatments at all depths could be made. PC1 (containing 75.53% of the overall variation) was used in a repeated measures ANOVA to test between treatments over the multiple measurement times.

3. Results

Penetration resistance showed significantly higher compaction in the uncultivated (AMF and control) plots (p = 0.002), which was mostly due to differences within the 7.5–2 5 cm depth range (Figure 1). There was also a significant impact of measurement time (p = 0.008) due to variation in soil condition over the year.

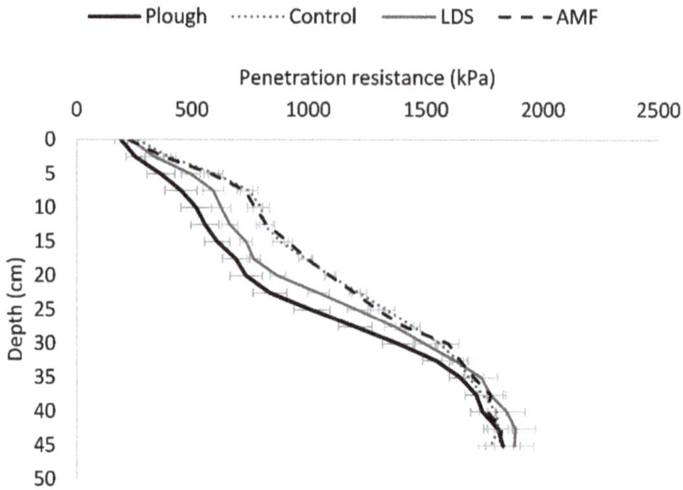

Figure 1. Penetration resistance (kPa) measured 0–45 cm depth through the soil profile. Graph shows average ± SE of all readings taken across the two years of measurements.

Bulk density (0–10 cm) measurements only showed significant results in the first year. Bulk density was lower in the ploughed plots, but surprisingly, significantly higher in the LDS plots ($p < 0.001$) (Figure 2). Bulk density measurements taken in the second year followed the same trend, with plough the lowest and LDS treatment as the highest, but this was not significant.

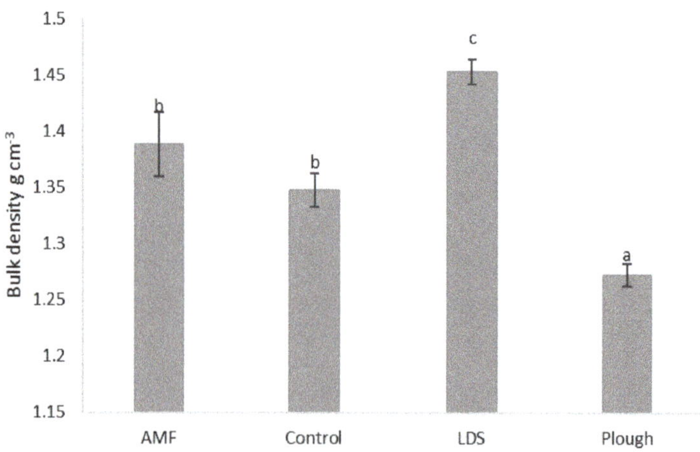

Figure 2. Soil bulk density (g cm^{-3}) measured in the topsoil 0–10 cm in 2018. Bars show mean ± SE. Letters denote significant differences at $p < 0.05$.

Despite the measurable compaction, it was not strong enough to influence yield, with no difference seen in crop yield seen in the two years. For the 2018 barley (*Hordeum vulgare*) crop, overall plant biomass was significantly ($p < 0.001$) higher in the two cultivated plots (Figure 3), but no biomass differences were seen in the subsequent bean crop (*Vicia faba*).

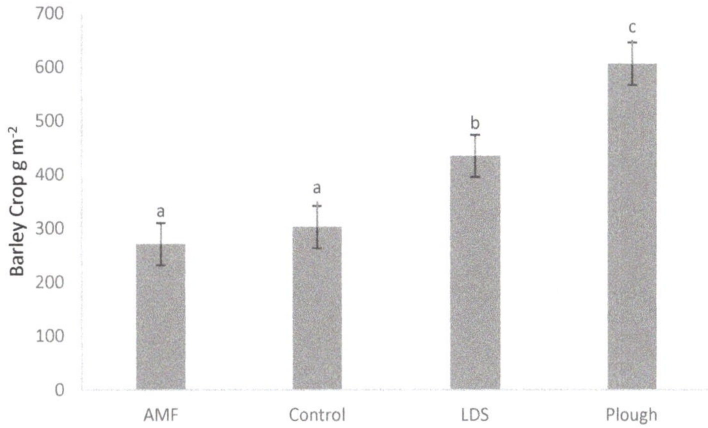

Figure 3. Barley crop (*Hordeum vulgare*) plant biomass measured in May 2018. Bars show mean ± SE. Letters denote significant differences at $p < 0.05$.

Earthworm numbers were higher in the non-cultivated plots (AMF and control) in both years ($p = 0.046$) (Figure 4). There was a highly significant difference between years ($p > 0.001$), with 2019 having less than half the number of worms counted in 2018 (average 411 ± 65 in 2018, 172 ± 28 in 2019).

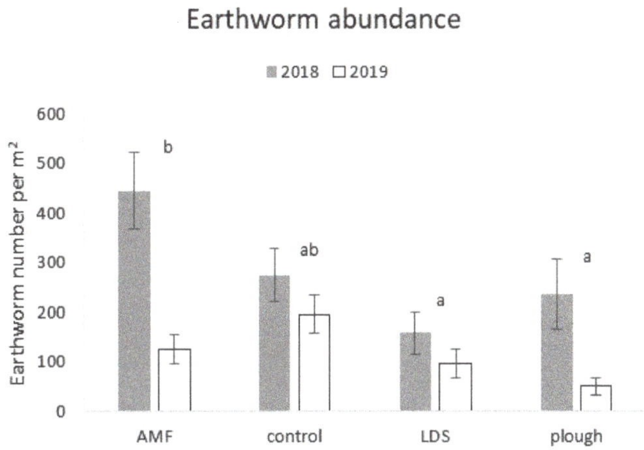

Figure 4. Average earthworm number (per m^2) measured in 2018 and 2019, to a depth of 25 cm. Bars show mean ± SE. Letters denote significant differences between cultivation treatments for both years at $p < 0.05$.

N_2O and CO_2 were measured monthly during cropping. Initially, repeated measures ANOVA showed no significant treatment effects for CO_2 ($p = 0.076$), however, splitting the results by season resulted in significant treatment effects for CO_2 flux in the winter months ($p = 0.034$), with ploughed plots having significantly higher CO_2 emissions (Figure 5). Initial N_2O emissions analysis showed significant treatment differences ($p = 0.046$), with significant differences between sampling times ($p < 0.001$) and a significant interaction between treatment and time ($p = 0.033$). Further investigation showed the interaction was due to much lower N_2O emissions during the warmer drier summer months. Breaking the analysis down into summer months (June, July and August) and all the other months (referred to as winter for simplicity), gave significant treatment effects for winter months ($p = 0.037$), with the AMF and the control plots showing much higher N_2O emissions

(Figure 6), but no significant effects were seen in summer months due to the overall lower emissions. The CO_2 and N_2O results were combined to give total green-house gas emissions for winter. The combined gasses showed no significant treatment differences in total gas fluxes recorded in winter ($p = 0.595$), although the composition of the gas fluxes changes between the plots.

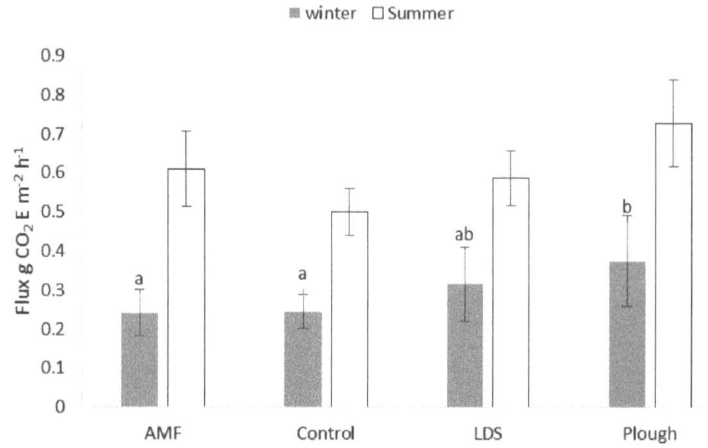

Figure 5. CO_2 flux measured monthly during 2018–2019 cropping and averaged over summer (June, July, August) and winter (all other months). Bars show mean ± SE. Letters denote significant differences between winter treatments at $p < 0.05$.

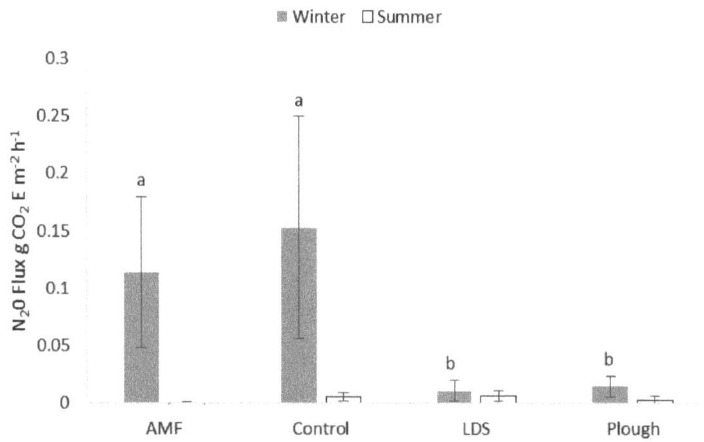

Figure 6. N_2O flux measured monthly during 2018–2019 cropping and averaged over summer (June, July, August) and winter (all other months). N_2O flux is displayed as CO_2 equivalent by 298. Bars show mean ± SE. Letters denote significant differences at $p < 0.05$.

4. Discussion

Direct drilling is established as a management practice that can improve aspects of soil health by leaving the soil undisturbed, which helps build soil organic matter and soil biology, such as earthworm populations. However, it does require soil to be in a fit state for conversion to direct drilling; compaction is a common problem across agricultural land, and can lead to issues including reduced root growth, reduced water and nutrient uptake and overall reduced productivity if not resolved with some form of compaction alleviation. This study compared the effectiveness of ploughing, low disturbance subsoiling and mycorrhizal inoculation as compaction alleviation methods to direct drilling over a compacted area.

Uncultivated plots, AMF and the control plots had significantly higher compaction in the 7.5–20 cm range (Figure 1), when measured using a penetrometer. This is unsurprising, as this is the depth of soil that would have been influenced by the plough and the LDS cultivations. High penetration resistance scores of 1000–2000 kPa are linked with slower root elongation rates [26] and thicker roots, due to the increased pressure needed to penetrate the soil [27], which can be detrimental to plant growth as they require extra energy to explore the soil. A lower proliferation of roots can also result in reduced nutrient and water access for the crop. However, in this experiment compaction only reached above 1000 kPa at a depth of approximately 20 cm (Figure 1), suggesting that compaction in the topsoil where cultivation was used was not large enough to elicit a yield response. In 2018, a plant biomass response was seen in the barley crop, with smaller plants in the uncultivated plots (Figure 3). Previous studies have suggested that monocot crops are more capable of tolerating compaction than dicot crops [28], which could explain the difference between the response in the barley and the bean crop. Bulk density measurements taken in the top 10 cm of soil showed that ploughing resulted in the least densely packed topsoil, while LDS, surprisingly, resulted in the highest compacted topsoil (Figure 2). Subsoilers are designed to alleviate compaction at lower levels, leaving the topsoil relatively undisturbed. However, in some cases, subsoilers have been recorded to increase the compaction at the soil surface [29].

Earthworm numbers have been linked to improved infiltration [30], plant rooting depth [31], aggregate stability [21] and overall plant production [32], making them an excellent indicator of soil biological health [33]. Previous studies have suggested that earthworm populations diminish under cropping compared to pastureland, and under tillage compared to untilled cropped systems [34], which has been attributed to the mechanical damage and destruction of the earthworm habitat [20]. However, there is evidence that soil conditions such as high bulk density and low soil pore space caused by compaction can have adverse effects on earthworm populations, sometimes reducing numbers in uncultivated systems [35]. In the present study, earthworm numbers were significantly reduced under the cultivated treatments LDS and plough, across both years measured (Figure 4), suggesting mechanical damage had reduced the earthworm population, with potential detrimental effect on soil health and plant productivity in these plots. This highlights the trade-off between cultivation to alleviate the damaging effects of compaction, with the disturbance this causes on soil fauna needed for healthy soil processes. A long period of drought in 2018 is likely to have been a factor in the reduction in earthworm abundance between years. As the climate changes and the likelihood of prolonged droughts increases, these deleterious effects on earthworm populations will continue [36].

Direct drilling can reduce CO_2 emissions and lead to an overall accumulation of SOC, due to an increase in aggregate stability and a change in chemical composition of carbon to more recalcitrant forms [37]. N_2O is also a greenhouse gas emitted from soils, but with a far higher global warming potential than CO_2 [18]. The production of N_2O is primarily through denitrification, which increases when water-filled pore spaces within soils are around 65–75% [38]. As direct drilled soils generally have a greater bulk density, particularly if newly converted or previously compacted as in the present experiment, water-filled pore space is often higher favouring denitrification [39]. Bulk density measurements taken in this experiment were significantly lower in ploughed plots, suggesting that water-filled pore spaces would be similarly lower under the ploughed treatment (Figure 3). Higher bulk density was seen in the LDS plots, but this was only measured in the top 10 cm, as the LDS is designed not to disturb the topsoil, but to alleviate compaction at lower depths; there may still have been higher pore space lower down in the soil profile, which the penetration resistance measurement confirms (Figure 1).

Greenhouse gas flux measurements showed overall higher CO_2 emissions in the summer months (June, July and August), when soil activity is at its highest due to warmer temperatures. No significant treatment differences were seen in CO_2 emissions between treatments in these warmer months, but when all other months were analysed together, significantly higher CO_2 emissions were seen from the two cultivated treatments, LDS and plough (Figure 5), due to the mechanical stimulation of organic matter breakdown in the soil [16]. In contrast, N_2O was produced at a much higher rate during the winter months than the summer months. During the winter months, there was significantly higher production of N_2O under the two non-cultivated treatments, AMF and control (Figure 6). These results are similar to those seen by Gregorich et al. [40], who found an increase in N_2O production in compacted soils, which corresponded to precipitation and high soil water content and was not seen in uncompacted treatments under the same conditions. Similarly, additional experiments at the site of our experiment, which used direct drill treatments without prior compaction have not shown this increase in N_2O flux (data not shown). Therefore, the increase in N_2O flux seen within this experiment was attributed to the reduced pore space, and subsequent increased water-filled pores in the compacted soil during the wetter winter months, causing an increase in denitrification activity and N_2O emissions. This has implications for compacted soils exacerbating N_2O emissions under future climate predictions of warmer wetter winters [10].

5. Conclusions

Overall, the efficacy of three methods of mitigating compaction damage to soil health and greenhouse gas emissions were compared with a direct drilled control. Two methods tried to improve soil structure and reduce compaction mechanically in situ, whilst the third, a biological method, attempted to reduce the impact of compaction on plant growth and nutrient acquisition. This study highlighted that compaction alleviation techniques differ in their efficacy as well as differing in their impact on soil health and greenhouse gas emissions. Earthworm abundance, a key indicator of soil health, was significantly reduced in the mechanical alleviation treatments, whilst emissions of CO_2 also increased. However, the link between compaction and increased N_2O emissions during wetter months is concerning, as seen in the AMF treatment and the direct drill control. Considering around 30% of soils in Europe are at risk (or susceptible) to compaction [4] and that winter rainfall is expected to increase due to climate change [18], greenhouse gas emissions may increase, dependent on agricultural (mis)management. This study highlights the importance of understanding how to alleviate compaction if we want to reach our climate emission goals and become net-zero within agriculture by 2040.

Author Contributions: Conceptualization, C.S. and F.C.; methodology, F.C. and J.B.; formal analysis, J.B.; data curation, J.B.; writing—original draft preparation, J.B.; writing—review and editing, C.S. and F.C. All authors have read and agreed to the published version of the manuscript.

Funding: This project has received funding from the European Union HORIZON2020 Programme for research, technological development and demonstration under grant agreement no. 677407.

Institutional Review Board Statement: Not applicable.

Informed Consent Statement: Not applicable.

Data Availability Statement: The data that support the findings of this study are available from the corresponding author upon reasonable request and approval from the study site representatives.

Acknowledgments: Thanks to Plantworks Ltd., Kent, UK, for providing the SR1 Cereals inoculant. Thanks to Gemma Fox for technical support in taking measurements and Phil Jarvis and Oliver Carrick for technical support in creating and maintaining plots.

Conflicts of Interest: The authors declare no conflict of interest. The funders had no role in the design of the study; in the collection, analyses, or interpretation of data; in the writing of the manuscript, or in the decision to publish the results.

References

1. Shah, A.N.; Tanveer, M.; Shahzad, B.; Yang, G.; Fahad, S.; Ali, S.; Bukhari, M.A.; Tung, S.A.; Hafeez, A.; Souliyanonh, B. Soil compaction effects on soil health and crop productivity: An overview. *Environ. Sci. Pollut. Res.* **2017**, *24*, 10056–10067. [CrossRef]
2. Environment Agency; Chief Scientist's Group. *The State of the Environment: Soil*; Environment Agency: London, UK, 2019.
3. DEFRA. *Environmental Standards for Farming—Consultation on Proposed Changes to Standards in Cross Compliance Good Agricultural and Environmental Condition (GAEC) and Related Measures in England*; Department for Environment Food & Rural Affairs: London, UK, 2009.
4. Jones, A.; Panagos, P.; Barcelo, S.; Bouraoui, F.; Bosco, C.; Dewitte, O.; Gardi, C.; Hervás, J.; Hiederer, R.; Jeffery, S.; et al. *The State of Soil in Europe: A contribution of the JRC to the European Environment Agency's Environment State and Outlook Report—SOER 2010*; Publications Office of the European Union: Luxembourg, 2012.
5. Batey, T. Soil compaction and soil management—A review. *Soil Use Manag.* **2009**, *25*, 335–345. [CrossRef]
6. Akker, J.V.D.; Arvidsson, J.; Horn, R. Introduction to the special issue on experiences with the impact and prevention of subsoil compaction in the European Union. *Soil Tillage Res.* **2003**, *73*, 1–8. [CrossRef]
7. Gregory, A.S.; Ritz, K.; McGrath, S.; Quinton, J.; Goulding, K.; Jones, R.J.A.; Harris, J.; Bol, R.; Wallace, P.; Pilgrim, E.S.; et al. A review of the impacts of degradation threats on soil properties in the UK. *Soil Use Manag.* **2015**, *31*, 1–15. [CrossRef]
8. Obour, P.B.; Ugarte, C.M. A meta-analysis of the impact of traffic-induced compaction on soil physical properties and grain yield. *Soil Tillage Res.* **2021**, *211*, 105019. [CrossRef]
9. Chamen, W.T.; Moxey, A.P.; Towers, W.; Balana, B.; Hallett, P. Mitigating arable soil compaction: A review and analysis of available cost and benefit data. *Soil Tillage Res.* **2015**, *146*, 10–25. [CrossRef]
10. Lowe, J.A.; Howard, T.P.; Pardaens, A.; Tinker, J.; Holt, J.; Wakelin, S.; Milne, G.; Leake, J.; Wolf, J.; Horsburgh, K.; et al. *UK Climate Projections Science Report: Marine and Coastal Projections*; Met Office Hadley Centre: Exeter, UK, 2009.
11. Chamen, W.; Cavalli, R. The effect of soil compaction on mole plough draught. *Soil Tillage Res.* **1994**, *32*, 303–311. [CrossRef]
12. Arvidsson, J.; Håkansson, I. Do effects of soil compaction persist after ploughing? Results from 21 long-term field experiments in Sweden. *Soil Tillage Res.* **1996**, *39*, 175–197. [CrossRef]
13. Crotty, F.V. Soil Organisms Within Arable Habitats. In *The Changing Status of Arable Habitats in Europe*; Hurford, C., Wilson, P., Storkey, J., Eds.; Springer: Cambridge, UK, 2020; pp. 123–138.
14. Whalley, R.; Dumitru, E.; Dexter, A. Biological effects of soil compaction. *Soil Tillage Res.* **1995**, *35*, 53–68. [CrossRef]
15. Ball, B.C. Soil structure and greenhouse gas emissions: A synthesis of 20 years of experimentation. *Eur. J. Soil Sci.* **2013**, *64*, 357–373. [CrossRef]
16. Reicosky, D.; Dugas, W.; Torbert, H. Tillage-induced soil carbon dioxide loss from different cropping systems. *Soil Tillage Res.* **1997**, *41*, 105–118. [CrossRef]
17. Højberg, O.; Revsbech, N.P.; Tiedje, J.M. Denitrification in Soil Aggregates Analyzed with Microsensors for Nitrous Oxide and Oxygen. *Soil Sci. Soc. Am. J.* **1994**, *58*, 1691–1698. [CrossRef]
18. IPCC. *Climate Change 2013: The Physical Science Basis. Contribution of Working Group I to the Fifth Assessment Report of the Intergovernmental Panel on Climate Change*; Cambridge University Press: Cambridge, UK; New York, NY, USA, 2013.
19. Liu, X.; Herbert, S.; Hashemi, A.; Zhang, X.; Ding, G. Effects of agricultural management on soil organic matter and carbon transformation—A review. *Plant Soil Environ.* **2011**, *52*, 531–543. [CrossRef]
20. Briones, M.J.; Schmidt, O. Conventional tillage decreases the abundance and biomass of earthworms and alters their community structure in a global meta-analysis. *Glob. Chang. Biol.* **2017**, *23*, 4396–4419. [CrossRef]
21. Six, J.; Bossuyt, H.; Degryze, S.; Denef, K. A history of research on the link between (micro) aggregates, soil biota, and soil organic matter dynamics. *Soil Tillage Res.* **2004**, *79*, 7–31. [CrossRef]
22. Colombi, T.; Keller, T. Developing strategies to recover crop productivity after soil compaction—A plant eco-physiological perspective. *Soil Tillage Res.* **2019**, *191*, 156–161. [CrossRef]
23. Elzenga, J.T.M.; van Veen, H. Waterlogging and plant nutrient uptake. In *Waterlogging Signaling and Tolerance in Plants*; Springer: Berlin/Heidelberg, Germany, 2010; pp. 23–35.
24. Wu, Q.-S.; Zou, Y.-N.; Huang, Y.-M. The arbuscular mycorrhizal fungus Diversispora spurca ameliorates effects of waterlogging on growth, root system architecture and antioxidant enzyme activities of citrus seedlings. *Fungal Ecol.* **2013**, *6*, 37–43. [CrossRef]
25. VSN International. *Genstat for Windows 18th Edition*; VSN International: Hemel Hempstead, UK, 2015; Available online: Genstat.co.uk (accessed on 29 January 2020).
26. Bennie, A.T.P.; Burger, R.D.T. Root characteristics of different crops as affected by mechanical resistance in fine sandy soils. In *Tenth National Congress of the Soil Science Society of Southern Africa*; SSSA: Lusakav, Zambia, 1983; pp. 29–32.
27. Nosalewicz, A.; Lipiec, J. The effect of compacted soil layers on vertical root distribution and water uptake by wheat. *Plant Soil* **2013**, *375*, 229–240. [CrossRef]
28. Sadras, V.O.; O'Leary, G.J.; Roget, D.K. Crop responses to compacted soil: Capture and efficiency in the use of water and radiation. *Field Crop. Res.* **2005**, *91*, 131–148. [CrossRef]
29. Bernier, H.; Bostock, G.; Raghavan, G.S.V.; Broughton, R.S. Subsoiling Effects on Moisture Content and Bulk Density in the Soil Profile. *Appl. Eng. Agric.* **1989**, *5*, 24–28. [CrossRef]
30. Andriuzzi, W.S.; Pulleman, M.M.; Schmidt, O.; Faber, J.H.; Brussaard, L. Anecic earthworms (*Lumbricus terrestris*) alleviate negative effects of extreme rainfall events on soil and plants in field mesocosms. *Plant Soil* **2015**, *397*, 103–113. [CrossRef]

31. Kemper, W.D.; Schneider, N.N.; Sinclair, T.R. No-till can increase earthworm populations and rooting depths. *J. Soil Water Conserv.* **2010**, *66*, 13A–17A. [CrossRef]
32. van Groenigen, J.W.; Lubbers, I.M.; Vos, H.M.; Brown, G.G.; De Deyn, G.B.; van Groenigen, K.J. Earthworms increase plant production: A meta-analysis. *Sci. Rep.* **2014**, *4*, 6365. [CrossRef] [PubMed]
33. Stroud, J.L. Soil health pilot study in England: Outcomes from an on-farm earthworm survey. *PLoS ONE* **2019**, *14*, e0203909. [CrossRef] [PubMed]
34. Chan, K. An overview of some tillage impacts on earthworm population abundance and diversity—Implications for functioning in soils. *Soil Tillage Res.* **2001**, *57*, 179–191. [CrossRef]
35. Söchtig, W.; Larink, O. Effect of soil compaction on activity and biomass of endogeic lumbricids in arable soils. *Soil Biol. Biochem.* **1992**, *24*, 1595–1599. [CrossRef]
36. Singh, J.; Schädler, M.; Demetrio, W.; Brown, G.G.; Eisenhauer, N. Climate change effects on earthworms—A review. *Soil Org.* **2020**, *91*, 114–138.
37. Cooper, H.V.; Sjögersten, S.; Lark, R.M.; Girkin, N.T.; Vane, C.H.; Calonego, J.C.; Rosolem, C.; Mooney, S.J. To till or not to till in a temperate ecosystem? Implications for climate change mitigation. *Environ. Res. Lett.* **2021**, *16*, 054022. [CrossRef]
38. Rabot, E.; Cousin, I.; Hénault, C. A modeling approach of the relationship between nitrous oxide fluxes from soils and the water-filled pore space. *Biodegradation* **2014**, *122*, 395–408. [CrossRef]
39. Palm, C.; Blanco-Canqui, H.; DeClerck, F.; Gatere, L.; Grace, P. Conservation agriculture and ecosystem services: An overview. *Agric. Ecosyst. Environ.* **2014**, *187*, 87–105. [CrossRef]
40. Gregorich, E.; McLaughlin, N.; Lapen, D.; Ma, B.; Rochette, P. Soil Compaction, Both an Environmental and Agronomic Culprit: Increased Nitrous Oxide Emissions and Reduced Plant Nitrogen Uptake. *Soil Sci. Soc. Am. J.* **2014**, *78*, 1913–1923. [CrossRef]

land

MDPI

Article

Soil Water Retention as Affected by Management Induced Changes of Soil Organic Carbon: Analysis of Long-Term Experiments in Europe

Ioanna S. Panagea [1,*], Antonio Berti [2], Pavel Čermak [3], Jan Diels [1], Annemie Elsen [4], Helena Kusá [3], Ilaria Piccoli [2], Jean Poesen [1,5], Chris Stoate [6], Mia Tits [4], Zoltan Toth [7] and Guido Wyseure [1]

1 Department of Earth and Environmental Sciences, KU Leuven, 3001 Leuven, Belgium; jan.diels@kuleuven.be (J.D.); jean.poesen@kuleuven.be (J.P.); guido.wyseure@kuleuven.be (G.W.)
2 DAFNAE Department, University of Padova, 35122 Legnaro, Italy; antonio.berti@unipd.it (A.B.); ilaria.piccoli@unipd.it (I.P.)
3 Crop Research Institute, Prague 6-Ruzyne, 16100 Prague, Czech Republic; pavel.cermak@vurv.cz (P.Č.); kusa@vurv.cz (H.K.)
4 Soil Service of Belgium (BDB), 3001 Heverlee, Belgium; aelsen@bdb.be (A.E.); mtits@bdb.be (M.T.)
5 Institute of Earth and Environmental Sciences, UMCS, 20-035 Lublin, Poland
6 Game & Wildlife Conservation Trust, Allerton Project, Loddington, Leicester LE7 9XE, UK; cstoate@gwct.org.uk
7 Georgikon Campus Keszthely, Institute of Agronomy, Hungarian University of Agriculture and Life Sciences, 2100 Gödöllő, Hungary; toth.zoltan@uni-mate.hu
* Correspondence: ioanna.panagea@kuleuven.be

Citation: Panagea, I.S.; Berti, A.; Čermak, P.; Diels, J.; Elsen, A.; Kusá, H.; Piccoli, I.; Poesen, J.; Stoate, C.; Tits, M.; et al. Soil Water Retention as Affected by Management Induced Changes of Soil Organic Carbon: Analysis of Long-Term Experiments in Europe. *Land* **2021**, *10*, 1362. https://doi.org/10.3390/land10121362

Academic Editor: Manuel López-Vicente

Received: 25 October 2021
Accepted: 8 December 2021
Published: 9 December 2021

Abstract: Soil water retention (SWR) is an important soil property related to soil structure, texture, and organic matter (SOM), among other properties. Agricultural management practices affect some of these properties in an interdependent way. In this study, the impact of management-induced changes of soil organic carbon (SOC) on SWR is evaluated in five long-term experiments in Europe (running from 8 up to 54 years when samples were taken). Topsoil samples (0–15 cm) were collected and analysed to evaluate the effects of three different management categories, i.e., soil tillage, the addition of exogenous organic materials, the incorporation of crop residues affecting SOC and water content under a range of matric potentials. Changes in the total SOC up to 10 g C kg^{-1} soil (1%) observed for the different management practices, do not cause statistically significant differences in the SWR characteristics as expected. The direct impact of the SOC on SWR is consistent but negligible, whereas the indirect impact of SOC in the higher matric potentials, which are mainly affected by soil structure and aggregate composition, prevails. The different water content responses under the various matric potentials to SOC changes for each management group implies that one conservation measure alone has a limited effect on SWR and only a combination of several practices that lead to better soil structure, such as reduced soil disturbances combined with increased SOM inputs can lead to better water holding capacity of the soil.

Keywords: soil organic carbon; soil-water content; no-till; reduced tillage; manure; compost; soil care

1. Introduction

Soil water retention (SWR) is a measure of how much water a particular type of soil can retain. It is an important soil property related to the distribution of pore space and, thus, is highly dependent on soil structure and texture, as well as on other related properties such as soil organic matter (SOM) [1]. SWR is critical for crop growth with a profound influence on crop yield and crop failure and acts as the main source of moisture for the soil's biota, which contributes to land productivity and biological soil health.

The relationship between the volumetric soil water content (θ) and the pressure head (or matric potential head, h) is described by the soil water retention curve (WRC), also known as the soil moisture characteristic curve or pF curve [2]. This curve is characteristic for different soils and is used to predict soil water storage for applications in agronomy, ecology, hydrology and many other soil-related sectors [3–6], as well as in earth systems models [7,8].

For the determination of the WRC, different field and laboratory methods exist [9–11]. Analytical models [12] or regression equations—empirical formulas called pedo-transfer functions (PTF)—are used to predict the WRC values from easily measured or already available soil properties [13–17]. The majority of the PTFs for estimating the WRC use soil texture, bulk density and SOM content as predictors [1,13,17], although the necessity of the latter has been questioned [18] or shown to improve the estimations only for specific soil water potentials [19]. For modelling purposes, analytical functions are fit to a set of observed h-θ values used to represent the continuous WRC. The most common retention functions have been presented by Brooks and Corey [20] and by van Genuchten [21].

The dependence of the SWR on texture and structure has been widely researched and demonstrated [22]. The dependence of the SWR on SOM content has also been proven [23,24], but the results on the quantitative influence of SOM are contradictory and vary with texture, pressure head and soil organic carbon (SOC) content as such [24–26] and therefore need to be further evaluated [27]. Analysing the effect and relationship of SOC content on SWR taking into account different soil textures has shown that the sensitivity of the SWR to SOC changes depending on the soil textural classes and on the SOM content itself [24,26]. For the same SOC increase, soils with coarser textures and low SOC contents present a larger increase in water retention than the finer soils [26], which may also present a decrease [24]. In contrast, for soils with high SOC contents the water retention increases for all textural classes, especially for sandy and silty soils [24]. Nevertheless, as pointed out in a review by Minasny and McBratney [26], a 1% absolute mass increase of SOC (10 g C kg^{-1} soil) has a limited effect on the SWR and can increase the available water capacity by up to 1.16% volumetrically. They also found that the effect is relatively larger for sandy soils. A change in the SOC content also influences the water content at the different pressure points in a different way [26], with field capacity (FC at −33 kPa or pF 2.5) to present higher sensitivity than the wilting point (WP at −1500 kPa or pF 4.2) [24]. Nevertheless, the use of SOM as an auxiliary predictor for the SWR through PTFs has been proven to be redundant when bulk density is also used as a predictor [18].

The different management practices applied in cropping systems affect the soil structure and soil composition, and consequently the SWR and other physical soil properties. Organic and conservation farming (defined as a farming system that promotes practices about maintenance of a permanent soil cover, minimum soil disturbances and crop diversification [28]) can increase the soil water storage through better soil aggregation and improved soil structure [29], but in some cases, the conventional systems yield higher water contents as a result of higher microporosity [30]. The SOC decreases when SOC losses due to erosion or/and mineralization, which can be stimulated also through soil tillage, exceed the organic carbon inputs coming from the addition of exogenous organic inputs (compost or manure) and organic inputs from crop residues (shoots and roots) [31].

Adding more exogenous organic materials such as compost or farmyard manure and the incorporation of crop residues into the soil above the SOC mineralization rate causes an increase in the total SOC in most cases [32–34]. However, the quality and stage of decomposition of exogenous organic materials affect how much of this added carbon remains as stable organic carbon in the soil [31,35]. The degree of maturation of manure and the composition of the compost greatly affects retention rates of organic carbon in the soil [36]. The addition of exogenous organic material increases the volumetric water content at most pressure heads, mainly because of the increase in total porosity [37] and the increase in total SOC. Mulching with or incorporating the crop residues has been proven to significantly impact the SWR in the wet range (pF < 2), but not in the dryer range (pF > 3) [34,38,39].

Reduced or no-tillage is often advocated to increase the SOC in the topsoil, which is important for a good structure, increased infiltration and reduced soil erosion rates when compared to conventional ploughing, but the results are controversial when considering the whole soil profile [31,40,41]. Conservation tillage has been proven to improve chemical soil properties such as SOM or physical soil properties such as aggregate stability of the topsoil, but the effects on soil water content are, in many cases not, significant [42] or controversial [43]. Sometimes the results are not solely dependent on soil tillage type but vary with matric potential [44–46]. The water content tends to be larger in the higher pressures (pF < 1 or wetter part) for conventional tillage when compared with conservation or no-tillage but in the smaller potentials the water content is larger for the conservation or no-tillage practices [44,46]. There are also cases where significant differences in the water content are present only in the more negative (pF > 3 or dryer) matric potentials [47]. López et al. [48] and Kargas and Londra [47] found that reduced and conventional tillage result in similar water content values, whereas no-tillage leads to lower values of water storage. On the other hand, Bescansa et al. [44] found that soil water content was higher in the no-tillage fields when compared to conventional tillage, especially in the drier condition because of the higher available water content caused by increased SOC content and changes in the pore distribution of the untilled soils.

Although previous studies have investigated the effects of management on the soil chemical and physical properties, less attention has been given to the link between combined management practices and SWR. In addition, most studies include a limited number of management practices and intensities and are not replicated in multiple agroecosystems and/or study regions that cover broad environmental gradients (i.e., climatic conditions and soil properties), possibly due to the logistical constraints associated with extensive field work. Finally, a comparison between published data is frequently hindered by methodological discrepancies between studies. To this end, studies that investigate broad management practices and intensities in multiple agroecosystems and regions with distinct environmental conditions are well needed to understand the interactions between soil structure, organic carbon, and water retention.

In our study, we compared seven long-term (8–54 years) experimental setups by sampling the topsoil with identical methods and analysing all samples in the same laboratory. The field experiments have been set up with specific and different objectives, but all together they cover a broad range of tillage practices, fertilization, additions of organic materials and management of crop residues. The objective of this study was to evaluate and quantify comprehensively the effect of different management practices on SOC content and their impact on the water-holding capacity of the soils.

2. Materials and Methods

2.1. Experiments' Descriptions

Topsoil samples were collected from seven different long-term agricultural experiments with different treatments in 5 European countries (the towns, countries, coordinates, start year of the experiment and main soil type of the sites are given in Table 1). In each country, the experiments were setup with different objectives and under different environmental conditions. Although the diversity of the experiments makes it challenging to combine them, they offer a wide range of representative management practices and pedo-climatological conditions. As the original experiments attempted to answer different scientific questions, they include several management treatments. For this research, a subset of treatments was selected from each experiment to include treatments from three main categories. The first category includes different soil tillage treatments (CZ, HU_2, UK), the second category comprises the addition of different types of exogenous organic materials (BE, IT_1c, IT_1p), and the third category deals with the incorporation of crop residues in the topsoil (HU_1, IT_2c, IT_2l). The experiments in Italy are conducted on two different soil types each and, in this study, are analysed as separate experiments: a clay and an initially peaty soil for IT_1 (i.e., IT_1c and IT_1p) and a clay and a loamy soil for experiment IT_2 (i.e., IT_2c and IT_2l) resulting in nine experiments in our study. The selected treatments per experiment are presented in Table 2. At the five study sites, an identical sampling procedure was performed for determining the WRC and SOC.

Table 1. Description of the study sites.

Code	Town, Country	Coordinates (Decimal Degrees)	Agro-Climate Zone [49]	Start of Experiment	Soil Type	Name of Experiment	Reference/Detailed Information
BE	Bierbeek, BE	50.8244 4.79605	Maritime North	1997	Silt Loam	VFG Compost trial	Tits et al. [50]
CZ	Prague-Ruzyně, CZ	50.0880 14.2980	Continental	1995	Silt Loam	Tillage trial	Mühlbachová, Kusá and Růžek, [51]
HU_1	Keszthely, HU	46.7332 17.2295	Pannonian	1983	Silt Loam	Organic & inorganic fertilization trial-IOSDV	Kismányoky and Tóth, [52]
HU_2	Keszthely, HU	46.7346 17.2302	Pannonian	1972	Silt Loam	Soil tillage systems in wheat and maize bi culture	Hoffmann and Kismányoky, [53]
IT_1c	Legnaro, IT	45.3506 11.9497	Maritime South	1964	Silty Clay Loam	Organic & mineral fertilization trial	Giardini, [54]
IT_1p	Legnaro, IT	45.3506 11.9497	Maritime South	1964	Peat* 18% OC initially	Organic & mineral fertilization trial	Giardini, [54]
IT_2c	Legnaro, IT	45.3507 11.9498	Maritime South	1970	Silty Clay Loam	Nitrogen fertilization and crop residue trial	Giardini, [54]
IT_2l	Legnaro, IT	45.3507 11.9498	Maritime South	1970	Silt Loam	Nitrogen fertilization and crop residue trial	Giardini, [54]
UK	Loddington, UK	52.6089 0.83257	Maritime North	2011	Clay loam	Soil Biology and Soil Health	-

Table 2. Details of the soil treatments in the various experiments ‡ Randomized complete block design (RCBD); φ Split plot-randomized complete block design (Split Plot-RCBD).

Code	Name of Experiment/ Experimental Design	Treatments	Replications (#)	Main Crop Types
BE	Vegetable-Fruit-Garden waste (VFG) compost trial ‡	**No organic**: No organic fertilization (control) **45tntriannually**: 45 t/ha compost * applied every three years 15tnannually: 15 t/ha compost * applied yearly **45tnannually**: 45 t/ha compost * applied yearly * C/N ≈ 12	4	Winter wheat, carrots, sugar beet, potatoes
CZ	Tillage trial	**Conventional**: Conventional ploughing (Turning of stubble—furrow opener at 10 cm, Mouldboard plough at 22 cm) (control) **Minimum**: Minimum tillage (Turning of stubble- furrow opener at 10 cm, 30% of crop residues remain on the soil surface) **Zero**: Zero tillage (all residues remain in the soil surface)	4	oil rapeseed, winter wheat, Peas
HU_1	Organic & inorganic fertilization trial-IOSDV φ	**NPK**: Only mineral fertilization/ removal of straw (control) **NPK+FYM**: 35 t/ha 0.5% N, farmyard manure application every 3 years/removal of straw **NPK+STR**: Straw and stalk incorporation completed with 10 kg N*t straw/ha	3	maize, winter wheat, winter barley
HU_2	Soil tillage systems in wheat and maize bi culture φ	**Conventional**: Deep winter ploughing (27–28 cm) + secondary tillage (control) **Minimum**: Disking just before drilling (<15 cm) **Shallow**: Shallow winter disking (<15 cm) +secondary tillage	4	winter wheat, maize

Table 2. *Cont.*

Code	Name of Experiment/ Experimental Design	Treatments	Replications (#)	Main Crop Types
IT_1c	Organic & mineral fertilization trial ‡	**Unfertilized**: No organic or mineral fertilization (control) **Manure L1**: 20 t/ha manure applied annually * **Manure L2**: 40 t/ha manure applied annually *	2	maize, winter wheat, potato, tillage radish (as winter cover crop), ryegrass, silage maize
IT_1p	Organic & mineral fertilization trial ‡	**Unfertilized**: No organic or mineral fertilization (control) **Manure L1**: 20 t/ha manure applied annually * **Manure L2**: 40 t/ha manure applied annually * * Farmyard manure from dairy cows (20% dry matter, 0.5% N, 0.25% P_2O_5, 0.7% K_2O)	2	
IT_2c	Nitrogen fertilization and crop residue trial ‡	**Residue Removal**: Removal of the previous crop residues (control) **Residue incorporation**: Burial of the previous crop residues	3	
IT_2l	Nitrogen fertilization and crop residue trial ‡	**Residue Removal**: Removal of the previous crop residues (control) **Residue incorporation**: Burial of the previous crop residues	3	
UK	Soil Biology and Soil Health ‡	**Conventional**: Ploughing at 25 cm (control) **Direct drilling**: Direct drilling of the seeds into previous crop residues	3	winter wheat, whet, oat

2.2. WRC Points Determination

To estimate the water content at the different points of the WRC, three undisturbed topsoil samples (positioned in the middle of 0–15 cm layer) were collected from each experimental plot (apart from Italy where the plots are too small and only one ring sample per plot could be collected) with the use of a Kopecky ring, of a known volume (100 cm^3). The 177 soil samples taken at different dates do not represent an equal number for each experiment and experimental plot (details about the number of samples per experimental plot are shown in Table 3). The top organic layer was first removed and with the use of suitable equipment (i.e., hammering holders and plastic hammer) to minimize soil disturbances the rings were pushed into the soil to collect the samples, which were stored afterwards at room temperature until analysis.

Table 3. Sampling details per experiment.

Code	Ring Soil Samples per Plot (#)	Sampling Month/Year	Years Applied When Sampling
BE	3	October 2019	22
CZ	3	November 2018	23
HU_1	3	November 2018	35
HU_2	3	November 2018	46
IT_1c	1	November 2018	54
IT_1p	1	November 2018	54
IT_2c	1	November 2018	48
IT_2l	1	November 2018	48
UK	3	April 2019	8

In this paper, we use pF to indicate the soil water potential. The pF is defined as the decimal logarithm of the absolute value of pressure head expressed in cm ($pF = \log_{10} |h|$).

The drainage or drying cycle was used for the determination of the volumetric water content at moisture tensions (suction) from pF 0 to pF 4.2. Sandboxes were used for the determination of the water content at the lower suction values (pF 0.0, pF 1.0, pF 1.8 and pF 2.0) and pressure plates cells (Soilmoisture Equipment Corp.) were used for the determination of the sample water content at pF 2.7, pF3.4 and pF 4.2 [55].

2.3. OC Determination

The largest component and easiest indicator of SOM status to measure is the SOC content and it is used in this report both to present the results and when we refer to content changes in the existing literature. SOC content was determined by dry combustion and mass spectrometry elemental analysis (Carlo-Erba EA 1110, Thermo Scientific, Waltham, MA, USA). Fresh disturbed field topsoil samples (0–15 cm) were taken with a sharp shovel from various spots within each experimental plot, mixed and directly broken to pass a <8 mm sieve. Soil samples were then stored in plastic containers to avoid compaction and disturbance during transportation and then stored in the refrigerator until air-drying could be carried out. All samples were air-dried at 40 °C until a constant mass was achieved and stored in a dark and dry place at room temperature. A subsample of the bulk soil was taken with a soil sample splitter to allow for a random representative sample, crushed manually to a homogenized powder and weighted into an Ag capsule. To determine only the carbon present in organic form carbonates were removed by adding HCl (35%). After drying at 40 °C the soil samples were loaded into the autosampler for combustion with oxygen at 800 °C with the presence of a catalyst. The organic carbon (OC) reacts to carbon dioxide (CO_2) which is quantified by infrared absorption spectroscopy and the mass percentage is determined

2.4. Statistical Analysis and Visualization Tools

The statistical data analysis was performed using R-Studio, R version 3.6.1 [56]. One-way Analysis of Variance (ANOVA) was carried out with the R software [57] to test for differences between treatments. Estimated marginal means (also known as least-squares means) by factors were computed by the least square method using the package "emmeans" [58]. Graphs were produced with the package "ggplot2" [59]. In the present work, statistical significance is assumed at $p < 0.05$. The assumptions of normality and homoscedasticity of the residuals were assessed by visual inspection of Q-Q plots and by plotting the normalized residuals against the fitted values.

3. Results and Discussion

3.1. Soil Organic Carbon

The total SOC as determined in the bulk soil is presented in Figure 1. Generally, SOC is relatively sensitive to the different management treatments and statistically significant differences are present among most of them. In the field experiments sampled, values of total organic carbon vary between 6 and 56 g C kg^{-1} soil (0.6–5.6%). The highest values observed is in the IT_1p experiment, in which an initially peaty soil was treated with different levels of manure. The lowest values were observed in the IT_2l experiment, where removal of residues took place in a loamy soil.

In the organic input experiments, as was expected [33], the higher amount of manure or compost resulted in a higher SOC. Nevertheless, statistically significant SOC differences are observed only between the higher levels of additions and the controls, which in all cases are those with only mineral inputs apart from the IT_1p and IT_1c in which the control is unfertilized treatment. In the IT_1p experiment, no statistically significant differences are observed but despite this, treatments with organic inputs show a trend that follows that same assumption i.e., higher input of organic materials leads to higher SOC. In this experiment, where agricultural management was established in initially peaty soil, there is a reduction in the total SOC over the years because of cultivation, but the decline is lower when manure is added, highlighting the importance of organic fertilizers in maintaining soil fertility in the long term [60].

In the tillage experiments, the treatments with the minimum soil disturbance present statistically significantly higher values of SOC content in the 0–15 cm layer than the treatments where conventional tillage took place, since fresh organic material will be kept concentrated at the topsoil [61,62]. As a result of the no- or reduced- tillage practices, apart from the increased organic inputs from the crop residues which are concentrated in the top layer, and the roots that remain intact in the soil, the carbon outputs by mineralization are reduced in the no-tillage systems. Reduced tillage systems present similar levels of mineralization with the conventional ploughing [61,63] but, according to recent evidence, reduced tillage presents lower CO_2 emissions [64] or higher levels of mineralizable carbon [65] when compared to conventional ploughing, indicating reduced mineralization because of minimal disruption of the stable aggregates. A more stable soil structure and improved aggregation also lead to reduced losses of fertile carbon-rich topsoil because of reduced water erosion. Nevertheless, it should always be kept in mind that only the topsoil 0–15 cm was sampled. Carbon stock changes may be observed in the no- or reduced-tillage experiments within the top layer sampled but if the whole soil profile is considered, conclusions cannot be drawn from our analysis. There is recent evidence to show that when the time since the adoption of the no-tillage system is considered (i.e., at least 6 years application of no-till), the increase of the carbon content in the top layers (0–20 cm) and the no change of the SOC in the deeper layers of the soil profile lead to an overall increase in the carbon stocks under no-tillage [41]. On the other hand, there is also evidence that when the entire soil profile is analysed, soil cultivation methods do not affect the SOC quantity but rather redistribute it in the profile [31,61,66–68].

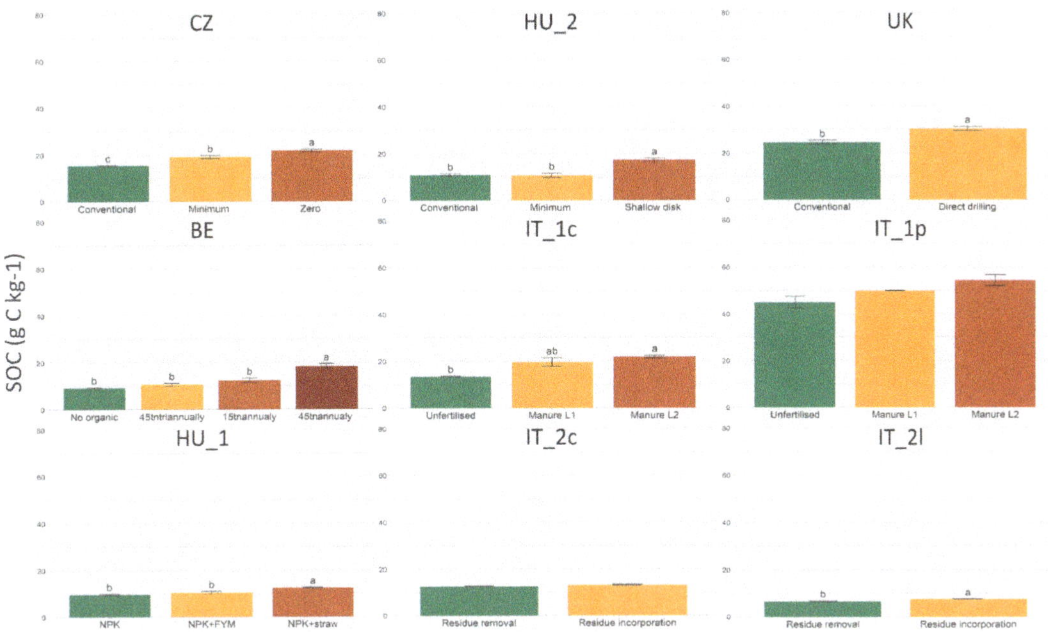

Figure 1. SOC content in the topsoil (0–15 cm) for each study site (see also Table 1 for the codes and Table 2 for a description of the soil improving treatments). Error bars represent the standard error (n -number of treatments replications- is denoted in Table 3 for each experiment).

In the HU_1 experiment, the incorporation of the straw and stalks causes a small but statistically significant difference in the SOC content when compared with the only mineral fertilized treatment and also significantly increased soil aggregate stability [69], supporting this way the physical soil condition. In contrast, in the IT_2c and IT_2l experiments, small or no statistically significant changes are noticed between the treatments. This, on the one hand, confirms the little potential of crop residues for soil improvement [70] and, on the other hand, raises questions if only the incorporation of crop residues without any kind of pre-processing like composting, conversion to biochar or the parallel use of other conservation measures such as reducing tillage, can contribute to the build-up of SOC. Indeed, in the same Italian experiment Dal Ferro et al. [71] recently found that residues incorporation seems to be effective in SOC storage only when coupled with minimum-tillage practice.

The results indicate that the SOC is a sensitive and good indicator to monitor changes caused in the soil quality by management practices in the long-term. The results should always consider the sampling depth, especially when tillage practices with different tillage depths are compared.

3.2. WRC

In Figure 2, the soil water content as a function of the matrix pressures (expressed in pF) is shown for all the study sites as measured in the laboratory conditions which may be different from field conditions. Although all trends looked consistent, whereby more SOC consistently meant a slightly higher water content for a given pF, we detected only a few statistically significant differences (in only 3 out of 9 experiments and in a limited number of pressure points).

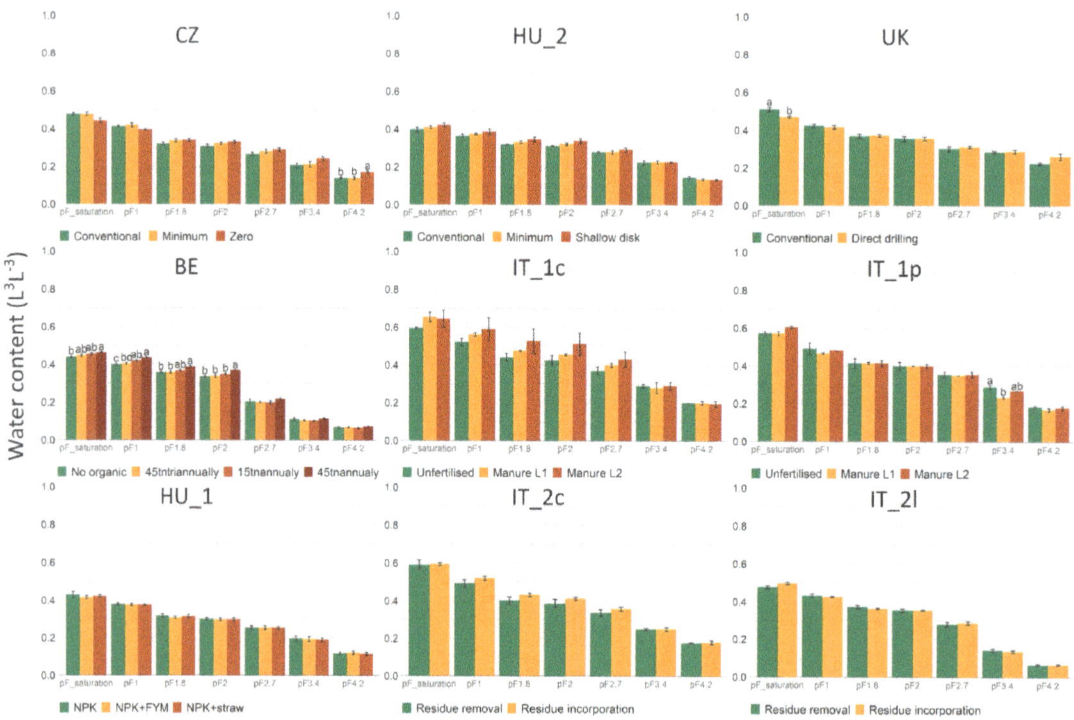

Figure 2. Soil water content at the different pressure points for each study site (see also Table 1 for the codes and Table 2 for a description of the soil improving treatments). Error bars represent the standard error (n -number of treatments replications- is denoted in Table 3 for each experiment).

In the soil tillage experiments, there are no statistically significant changes in the soil water content among the different levels of tillage, but a pattern is observed. In the CZ and UK experiments, we observe that in the higher matric potential range the treatments with the higher levels of disturbances present higher water content, whereas in the lower pressures the pattern is opposite: here, the higher the soil disturbance the lower is the water content. This happens because in the higher pressures (less negative) the macropores and capillary forces play an important role in the total water content, whereas in the lower matric potential the adsorption on the soil particles and the SOM work as water storage pools [72]. These results follow McVay et al. [42], who also observed that tillage methods do not significantly affect the WHC of most of the soils analysed and reported changes only in the higher matric potentials [44,46,47], indicating that tillage mainly influences the volume of larger pore sizes, dominantly influencing water infiltration and aeration of the soil.

Higher organic matter input always led to modestly higher water content at all pressures. Only in the BE experiment, there are statistically significant differences among the treatments in the higher pressures (less negative or wetter part) as a result of an annual addition of 45 tonnes of compost per hectare. It is important to point out that a yearly dose of 45 tonnes of compost is excessive as compared to normal practice. In the IT_1c experiment, in which different levels of farmyard manure are applied in clay soil, the high variability in the water content among the replicates did not allow to detect statistically significant differences, but the same trend is observed i.e., higher organic matter input leads to higher water content at all pressures. In the same experiment with the peaty soil with initially 18% organic carbon, this trend is not noted. The results are consistent with Eusufzai and Fujii [37] who found that organic amended soils present increased water content, especially in the higher pressure points.

In the last experiment group in which the crop residues are incorporated into the soil after the cropping season, our analysis does not reveal consistent changes in the water storage capacity, or at least consistent trends to justify the reported findings in the literature that incorporation of residues impacts the water retention in the higher matric potentials [34,38,39].

An analysis of the effect of different practices in the whole WRC as calculated from the Rosseta version 3 model [17] is presented in Figure S2 of the Supplementary Materials; changes are observed only in the wetter part of the retention curve, whereas the dryer part is not affected by the different applied cropping systems.

3.3. Water Retention as Affected by Carbon Changes and Management Practices

In Figure 3, the percentage differences of water content in relation to the percentage differences of SOC content among the different plots of each experiment are presented as differences with the corresponding control treatment of the relevant block. When all experiments are analysed together, it is observed that the increase of total SOC over the period that each experiment is running, generally causes an increase in water content at all moisture tensions, especially for the higher matric potentials (wetter conditions) at which the regression relationship is also statistically significant (Table 4). This trend is less pronounced at the permanent wilting point (pF 4.2), where the impact is almost negligible. As a result of the negligible increase in the wilting point and the bigger increase in the water content at field capacity, the plant available water increases even with an 1% increase of the total SOC.

Table 4. F-statistic of the linear regression analysis. Significance codes of p-value: ** 0.05, * 0.1.

		Saturation	pF1	pF1.8	pF2	pF2.7	pF3.4	pF 4.2
All experiments	F-statistic	1.32	6.99	3.33	6.17	1.87	3.86	0.42
	p-value	0.27	0.02 **	0.08 *	0.02 **	0.19	0.07	0.53
Exogenous OM	F-statistic	1.32	6.98	3.33	6.17	1.87	3.86	0.42
	p-value	0.27	0.02 **	0.08 *	0.02 **	0.19	0.07 *	0.53
Soil cultivation	F-statistic	0.31	0.98	3.30	2.43	2.25	1.27	0.44
	p-value	0.59	0.34	0.09 *	0.14	0.15	0.28	0.51
Residues	F-statistic	1.49	5.88	1.25	1.69	2.23	0.18	0.18
	p-value	0.25	0.04 **	0.29	0.22	0.17	0.68	0.87

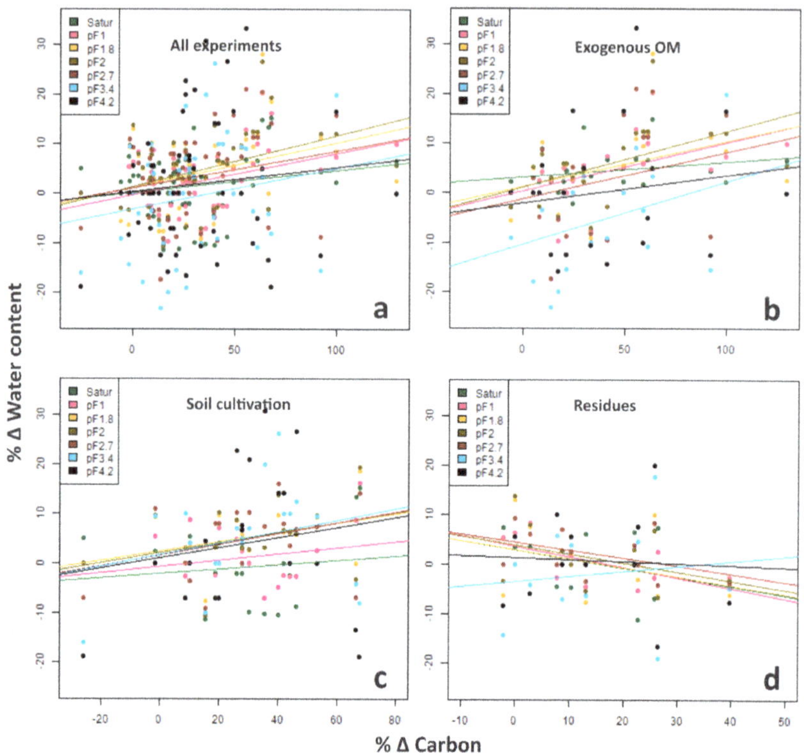

Figure 3. Relationship of the percentage change in soil water content and percentage change in SOC content. The water content and SOC values represent the percentage differences between the different treatments' plots and the control plot of each block of the experiment. The different colours represent the water content change in the different pressure points of the WRC. (**a**): All the experiments plotted together. (**b–d**): Each experimental group plotted separately (**b**): addition of compost or manure, (**c**): tillage experiments, (**d**): residue management).

Comparing the results of the percentage differences of water content in relation with the SOC percentage changes in each experiments group, it is observed that the impact of increased carbon content on the water content does not only depend on the pressure point, texture, and organic carbon content, but also on the applied management practices. Management practices that increase bulk density (no- tillage, reduced OM inputs etc.) decrease the volume fraction of macro-pores, but at the same time increase the volume fraction of both micro- and meso-pores, resulting in an increase of the water content at lower matric potentials and a decrease under wetter conditions [73]. In the soil tillage experiments where the maximum carbon increase observed is about 0.65% as a result of practices that minimize topsoil disturbances, an increase in the soil water content is observed in the lower matric potentials (drier conditions) as a result of increased SOC and surface adsorption [72,74], as also observed from Bescansa et al. [44] and a smaller increase of the water content in the wetter conditions (saturation and pF 1) as a result of changes in the pore distribution and capillary forces [72,74]. These two conditions may lead to a decrease in the plant available water content (AWC) when cultivation practices with less soil disturbance are followed as also mentioned by Hill [75]. Indeed, as shown in Figure S1 in the Supplementary Materials, the AWC in the CZ experiment is statistically significantly lower in the zero-tillage treatment when compared with the minimum tillage, and in the UK experiment it is lower, but not statistically significant in the direct drilling treatment in comparison with the conventional ploughing. In the experiments where exogenous organic material is added, a cca. 1% increase in SOC increases the soil water content under all applied pore water pressures but the increase is lower in the drier conditions and sometimes also a decrease in the soil water content is presented when compared to no addition of organic materials. The addition of organic material leads to increased macro aggregation and therefore increased meso- and macroporosity [37,73] and increased water content in these pores, resulting in less water available for storing in the micropores. A negligible increase in the dryer conditions denotes also that the increase of SOC does not increase the sorbed water or that it is counteracted by the increased macroporosity. Nevertheless, the negligible increase in the water content in the wilting point and the big increase in the water content at field capacity will lead to higher plant available water content (Figure S1—Supplementary Materials), something that impacts positively the plant growth. In the experiments where the residues have been incorporated in the topsoil, the soil water content decreases or remains similar, even if an increase of 0.3% of the total SOC is observed. Despite the long-term application, there were no large SOC changes following the incorporation of the residues. Building up SOC and simultaneously a stable soil structure might be more important than a large increase in SOC content, especially during wetter soil conditions.

There is a strong belief and impression by practitioners and advocates of conservation agriculture and organic farming that an increase in the total SOC increases water retention directly and substantially [76–78]. In this research, the water retention characteristics present the expected but modest trends. However, it is remarkable that even after 54 years of practices that increase SOC, the observed differences in the water content, especially in the lower pressures (drier conditions), are negligible from a practical point of view, and almost all not statistically significant. It was expected that in the lower water pressures (pF > 2.5) where the macropores and capillarity do not have an important impact and surface adsorption and SOC content seems to play the most important role in the soil's water content, the differences in SWR among the different treatments would be noticeable. The statistically significant linear relationship of the carbon change and the water retention mainly in the wetter conditions (Table 4) and not in the drier conditions suggest that the direct impact of the increased SOC on water retention is limited and the indirect impact stronger. The fact that the change of SOC affects the water content under the different matric potential in a different way and is statistically significant only in the higher matric potentials implies that the impact of SOC is indirect and is more linked to the changes in other soil parameters and most probably in soil structure and aggregation status.

4. Conclusions

We analysed different groups of management practices for improving soil quality as applied in long-term experiments in five European countries. We investigated their effect on SOC and the link with the capacity of the soil to retain water at different matric potentials. Our findings suggest that practices that minimize soil disturbances cause an increase in SOC in the topsoil but may lead to decreased plant available water content as a result of the increased water content at wilting point and a less profound increase in water content at field capacity, jeopardizing the crop yield. On the other hand, the different soil-improving management practices that increase the organic materials in the soil (both exogenous and incorporation of residues) contribute to an increase in the soil water availability for the crops, but not because of increased water holding capacity as a result of increased SOC. The addition of organic materials affects the soil structure, and it is more likely that the soil structure—as improved by the SOM—affects the water availability because of more macro and mesopores, rather than because of larger water-holding capacity per soil volume caused by a SOC increase. The better structure formed by higher amounts and more stable SOC and the increase in SWR are important factors leading to increased water infiltration, even under long-term rainy conditions, and promoting several soil functions such as less soil erosion minimised effects of extreme rainfalls and droughts deeper rooting of the crops end enhanced crop productivity. The negligible effect of increased SOC under different management practices during drier conditions, and the increased effect in wetter conditions, implies that the indirect effects of SOC increase in the soil structure are more important and should be considered in future research.

Supplementary Materials: The following are available online at https://www.mdpi.com/article/10.3390/land10121362/s1, Figure S1: Plant Available Water Content as calculated from the difference between water content at field capacity (pF 2) and water content at wilting point (pF 4.2). Figure S2: WRC for each treatment. The lines represent the WRC as calculated from the Rosetta version 3 model with input the average silt clay and sand percentages, the average bulk density of the treatments as measured and the water content at field capacity and wilting point.

Author Contributions: Conceptualization, I.S.P. and G.W.; methodology, I.S.P.; formal analysis, I.S.P.; data curation, I.S.P.; writing—original draft preparation, I.S.P.; writing—review and editing, G.W., J.P., J.D., M.T., A.E., I.P., A.B., C.S., Z.T., H.K. and P.Č.; visualization, I.S.P.; supervision, G.W., J.D. and J.P.; project coordination, G.W.; funding acquisition, G.W. All authors have read and agreed to the published version of the manuscript.

Funding: This research was funded by the project "SoilCare: Soil Care for profitable and sustainable crop production in Europe" from the H2020 Programme under grant agreement n° 677407.

Institutional Review Board Statement: Not applicable.

Informed Consent Statement: Not applicable.

Data Availability Statement: The data that support the findings of this study are available from the corresponding author upon reasonable request and approval from the study sites representative co-authors.

Acknowledgments: We thank all the study sites owners for permitting the use of the experimental fields where this research was done and providing all types of help during the sampling campaign. We thank the laboratory assistant of the Department of Earth and Environmental Sciences of the KU Leuven Lore Fondu for providing valuable guidance and help during the soil samples analysis. ISP thanks Antonios Apostolakis for the extensive discussions and guidance on the statistical analyses and concepts.

Conflicts of Interest: The authors declare no conflict of interest.

References

1. Vereecken, H.; Maes, J.; Feyen, J.; Darius, P. Estimating the soil moisture retention characteristic from texture, bulk density, and carbon content. *Soil Sci.* **1989**, *148*, 389–403. [CrossRef]
2. Stephens, D.B. *Vadose Zone Hydrology*, 1st ed.; CRC Press: Boca Raton, FL, USA, 1996.
3. Grillakis, M.G.; Koutroulis, A.G.; Alexakis, D.D.; Polykretis, C.; Daliakopoulos, I.N. Regionalizing Root-Zone Soil Moisture Estimates From ESA CCI Soil Water Index Using Machine Learning and Information on Soil, Vegetation, and Climate. *Water Resour. Res.* **2021**, *57*, e2020WR029249. [CrossRef]
4. Pan, T.; Hou, S.; Liu, Y.; Tan, Q. Comparison of three models fitting the soil water retention curves in a degraded alpine meadow region. *Sci. Rep.* **2019**, *9*, 18407. [CrossRef] [PubMed]
5. Vereecken, H.; Weynants, M.; Javaux, M.; Pachepsky, Y.; Schaap, M.G.; van Genuchten, M.T. Using Pedotransfer Functions to Estimate the van Genuchten-Mualem Soil Hydraulic Properties: A Review. *Vadose Zone J.* **2010**, *9*, 795–820. [CrossRef]
6. Zhang, L.; Han, J. Improving water retention capacity of an aeolian sandy soil with feldspathic sandstone. *Sci. Rep.* **2019**, *9*, 14719. [CrossRef]
7. Fatichi, S.; Or, D.; Walko, R.; Vereecken, H.; Young, M.H.; Ghezzehei, T.A.; Hengl, T.; Kollet, S.; Agam, N.; Avissar, R. Soil structure is an important omission in Earth System Models. *Nat. Commun.* **2020**, *11*, 522. [CrossRef]
8. Van Looy, K.; Bouma, J.; Herbst, M.; Koestel, J.; Minasny, B.; Mishra, U.; Montzka, C.; Nemes, A.; Pachepsky, Y.A.; Padarian, J.; et al. Pedotransfer Functions in Earth System Science: Challenges and Perspectives. *Rev. Geophys.* **2017**, *55*, 1199–1256. [CrossRef]
9. Lu, D.; Wang, H.; Huang, D.; Li, D.; Sun, Y. Measurement and Estimation of Water Retention Curves Using Electrical Resistivity Data in Porous Media. *J. Hydrol. Eng.* **2020**, *25*, 04020021. [CrossRef]
10. Sun, D.; You, G.; Annan, Z.; Daichao, S. Soil–water retention curves and microstructures of undisturbed and compacted Guilin lateritic clay. *Bull. Eng. Geol. Environ.* **2016**, *75*, 781–791. [CrossRef]
11. Whalley, W.R.; Ober, E.S.; Jenkins, M. Measurement of the matric potential of soil water in the rhizosphere. *J. Exp. Bot.* **2013**, *64*, 3951–3963. [CrossRef]

12. Sun, H.; Lee, J.; Chen, X.; Zhuang, J. Estimating soil water retention for wide ranges of pressure head and bulk density based on a fractional bulk density concept. *Sci. Rep.* **2020**, *10*, 16666. [CrossRef]
13. Cornelis, W.M.; Ronsyn, J.; Van Meirvenne, M.; Hartmann, R. Evaluation of Pedotransfer Functions for Predicting the Soil Moisture Retention Curve. *Soil Sci. Soc. Am. J.* **2001**, *65*, 638–648. [CrossRef]
14. Schaap, M.G.; Leij, F.J.; Van Genuchten, M.T. Rosetta: A computer program for estimating soil hydraulic parameters with hierarchical pedotransfer functions. *J. Hydrol.* **2001**, *251*, 163–176. [CrossRef]
15. Tóth, B.; Weynants, M.; Nemes, A.; Makó, A.; Bilas, G.; Tóth, G. New generation of hydraulic pedotransfer functions for Europe. *Eur. J. Soil Sci.* **2015**, *66*, 226–238. [CrossRef]
16. Wösten, J.H.M.; Pachepsky, Y.A.; Rawls, W.J. Pedotransfer functions: Bridging the gap between available basic soil data and missing soil hydraulic characteristics. *J. Hydrol.* **2001**, *251*, 123–150. [CrossRef]
17. Zhang, Y.; Schaap, M.G. Weighted recalibration of the Rosetta pedotransfer model with improved estimates of hydraulic parameter distributions and summary statistics (Rosetta3). *J. Hydrol.* **2017**, *547*, 39–53. [CrossRef]
18. Zacharias, S.; Wessolek, G. Excluding Organic Matter Content from Pedotransfer Predictors of Soil Water Retention. *Soil Sci. Soc. Am. J.* **2007**, *71*, 43–50. [CrossRef]
19. Contreras, C.P.; Bonilla, C.A. A comprehensive evaluation of pedotransfer functions for predicting soil water content in environmental modeling and ecosystem management. *Sci. Total Environ.* **2018**, *644*, 1580–1590. [CrossRef]
20. Brooks, R.H.; Corey, A.T. *Hydraulic Properties of Porous Media*; Colorado State University: Fort Collins, CO, USA, 1964; Volume 3.
21. Van Genuchten, M.T. A closed-form equation for predicting the hydraulic conductivity of unsaturated soils. *Soil Sci. Soc. Am. J.* **1980**, *44*, 892–898. [CrossRef]
22. Kool, D.; Tong, B.; Tian, Z.; Heitman, J.L.; Sauer, T.J.; Horton, R. Soil water retention and hydraulic conductivity dynamics following tillage. *Soil Tillage Res.* **2019**, *193*, 95–100. [CrossRef]
23. Hudson, B.D. Soil organic matter and available water capacity. *J. Soil Water Conserv.* **1994**, *49*, 189–194.
24. Rawls, W.J.; Pachepsky, Y.A.; Ritchie, J.C.; Sobecki, T.M.; Bloodworth, H. Effect of soil organic carbon on soil water retention. *Geoderma* **2003**, *116*, 61–76. [CrossRef]
25. Manns, H.R.; Parkin, G.W.; Martin, R.C. Evidence of a union between organic carbon and water content in soil. *Can. J. Soil Sci.* **2016**, *96*, 305–316. [CrossRef]
26. Minasny, B.; McBratney, A.B. Limited effect of organic matter on soil available water capacity. *Eur. J. Soil Sci.* **2018**, *69*, 39–47. [CrossRef]
27. Huntington, T. Available Water Capacity and Soil Organic Matter. In *Encyclopedia of Soil Science*; Taylor and Francis: New York, NY, USA, 2007; pp. 139–143.
28. FAO. Conservation Agriculture. 2017. Available online: https://www.fao.org/conservation-agriculture/en/ (accessed on 27 November 2021).
29. Williams, D.M.; Blanco-Canqui, H.; Francis, C.A.; Galusha, T.D. Organic farming and soil physical properties: An assessment after 40 years. *Agron. J.* **2017**, *109*, 600–609. [CrossRef]
30. Głąb, T.; Pużyńska, K.; Pużyński, S.; Palmowska, J.; Kowalik, K. Effect of organic farming on a Stagnic Luvisol soil physical quality. *Geoderma* **2016**, *282*, 16–25. [CrossRef]
31. Chenu, C.; Angers, D.A.; Barré, P.; Derrien, D.; Arrouays, D.; Balesdent, J. Increasing organic stocks in agricultural soils: Knowledge gaps and potential innovations. *Soil Tillage Res.* **2019**, *188*, 41–52. [CrossRef]
32. De Clercq, T.; Merckx, R.; Elsen, A.; Vandendriessche, H. Impact of long-term compost amendments on soil fertility, soil organic matter fractions and nitrogen mineralization. *Acta Hortic.* **2016**, *1146*, 79–86. [CrossRef]
33. Diacono, M.; Montemurro, F. Long-term effects of organic amendments on soil fertility. A review. *Agron. Sustain. Dev.* **2010**, *30*, 401–422. [CrossRef]
34. Karlen, D.L.; Wollenhaupt, N.C.; Erbach, D.C.; Berry, E.C.; Swan, J.B.; Eash, N.S.; Jordahl, J.L. Crop residue effects on soil quality following 10-years of no-till corn. *Soil Tillage Res.* **1994**, *31*, 149–167. [CrossRef]
35. Wang, H.; Boutton, T.W.; Xu, W.; Hu, G.; Jiang, P.; Bai, E. Quality of fresh organic matter affects priming of soil organic matter and substrate utilization patterns of microbes. *Sci. Rep.* **2015**, *5*, 10102. [CrossRef] [PubMed]
36. Bhogal, A.; Nicholson, F.A.; Rollett, A.; Taylor, M.; Litterick, A.; Whittingham, M.J.; Williams, J.R. Improvements in the Quality of Agricultural Soils Following Organic Material Additions Depend on Both the Quantity and Quality of the Materials Applied. *Front. Sustain. Food Syst.* **2018**, *2*, 9. [CrossRef]
37. Eusufzai, M.K.; Fujii, K. Effect of Organic Matter Amendment on Hydraulic and Pore Characteristics of a Clay Loam Soil. *Open J. Soil Sci.* **2012**, *2*, 372–381. [CrossRef]
38. Hiel, M.P.; Chélin, M.; Parvin, N.; Barbieux, S.; Degrune, F.; Aboulkacem, L.; Colinet, G.; Degré, A.; Bodson, B.; Garré, S. Crop residue management in arable cropping systems under a temperate climate. Part 2: Soil physical properties and crop production. A review. *Biotechnol. Agron. Soc. Environ.* **2016**, *20*, 245–256.
39. Mulumba, L.N.; Lal, R. Mulching effects on selected soil physical properties. *Soil Tillage Res.* **2008**, *98*, 106–111. [CrossRef]
40. Wiesmeier, M.; Urbanski, L.; Hobley, E.; Lang, B.; von Lützow, M.; Marin-Spiotta, E.; van Wesemael, B.; Rabot, E.; Ließ, M.; Garcia-Franco, N.; et al. Soil organic carbon storage as a key function of soils—A review of drivers and indicators at various scales. *Geoderma* **2019**, *333*, 149–162. [CrossRef]

41. Cooper, H.V.; Sjögersten, S.; Lark, R.M.; Mooney, S.J. To till or not to till in a temperate ecosystem? Implications for climate change mitigation. *Environ. Res. Lett.* **2021**, *16*, 054022. [CrossRef]
42. McVay, K.A.; Budde, J.A.; Fabrizzi, K.; Mikha, M.M.; Rice, C.W.; Schlegel, A.J.; Peterson, D.E.; Sweeney, D.W.; Thompson, C. Management Effects on Soil Physical Properties in Long-Term Tillage Studies in Kansas. *Soil Sci. Soc. Am. J.* **2006**, *70*, 434–438. [CrossRef]
43. Strudley, M.W.; Green, T.R.; Ascough, J.C. Tillage effects on soil hydraulic properties in space and time: State of the science. *Soil Tillage Res.* **2008**, *99*, 4–48. [CrossRef]
44. Bescansa, P.; Imaz, M.J.; Virto, I.; Enrique, A.; Hoogmoed, W.B. Soil water retention as affected by tillage and residue management in semiarid Spain. *Soil Tillage Res.* **2006**, *87*, 19–27. [CrossRef]
45. Farkas, C.; Birkás, M.; Várallyay, G. Soil tillage systems to reduce the harmful effect of extreme weather and hydrological situations. *Biologia* **2009**, *64*, 624–628. [CrossRef]
46. Karlen, D.L.; Wollenhaupt, N.C.; Erbach, D.C.; Berry, E.C.; Swan, J.B.; Eash, N.S.; Jordahl, J.L. Long-term tillage effects on soil quality. *Soil Tillage Res.* **1994**, *32*, 313–327. [CrossRef]
47. Kargas, G.; Londra, P.A. Effect of tillage practices on the hydraulic properties of a loamy soil. *Desalin. Water Treat.* **2015**, *54*, 2138–2146. [CrossRef]
48. López, M.V.; Arrúe, J.L.; Sánchez-Girón, V. A comparison between seasonal changes in soil water storage and penetration resistance under conventional and conservation tillage systems in Aragon. *Soil Tillage Res.* **1996**, *37*, 251–271. [CrossRef]
49. Ceglar, A.; Zampieri, M.; Toreti, A.; Dentener, F. Observed Northward Migration of Agro-Climate Zones in Europe Will Further Accelerate Under Climate Change. *Earths Future* **2019**, *7*, 1088–1101. [CrossRef]
50. Tits, M.; Elsen, A.; Bries, J.; Vandendriessche, H. Short-term and long-term effects of vegetable, fruit and garden waste compost applications in an arable crop rotation in Flanders. *Plant Soil* **2014**, *376*, 43–59. [CrossRef]
51. Mühlbachová, G.; Kusá, H.; Růžek, P. Soil characteristics and crop yields under differenttillage techniques. *Plant Soil Environ.* **2015**, *61*, 566–572. [CrossRef]
52. Kismányoky, T.; Tóth, Z. Effect of mineral and organic fertilization on soil organic carbon content as well as on grain production of cereals in the IOSDV (ILTE) long-term field experiment, Keszthely, Hungary. *Arch. Agron. Soil Sci.* **2013**, *59*, 1121–1131. [CrossRef]
53. Hoffmann, S.; Kismányoky, T. Soil fertility in a long-term fertilizer trial with different tillage systems. *Arch. Agron. Soil Sci.* **2001**, *46*, 239–250. [CrossRef]
54. Giardini, L. Ongoing Trials. In *Productivity and Sustainability of Different Cropping Systems. 40 Years of Experiments in Veneto Region*; Giardini, L., Ed.; Patron: Bologna, Italy, 2004; pp. 89–97.
55. Klute, A. Water Retention:Laboratory Methods. In *Methods of Soil Analysis; Part 1—Physical and Mineralogical Methods*, 2nd ed.; Klute, A., Ed.; ASA & SSSA: Madison, WI, USA, 1986; pp. 635–662.
56. RStudio Team. *RStudio: Integrated Development Environment for R*; RStudio, Inc.: Boston, MA, USA, 2016.
57. R Core Team. *R: A Language and Environment for Statistical Computing*; R Foundation for Statistical Computing: Vienna, Austria, 2019.
58. Lenth, R. Estimated Marginal Means, aka Least-Squares Means. 2020. Available online: https://cran.r-project.org/package=emmeans (accessed on 8 December 2021).
59. Wickham, H. *ggplot2: Elegant Graphics for Data Analysis*; Springer: New York, NY, USA, 2016; ISBN 78-3-319-24277-4.
60. Morari, F.; Lugato, E.; Berti, A.; Giardini, L. Long-term effects of recommended management practices on soil carbon changes and sequestration in north-eastern Italy. *Soil Use Manag.* **2006**, *22*, 71–81. [CrossRef]
61. Haddaway, N.R.; Hedlund, K.; Jackson, L.E.; Kätterer, T.; Lugato, E.; Thomsen, I.K.; Jørgensen, H.B.; Söderström, B. How does tillage intensity affect soil organic carbon? A systematic review. *Environ. Evid.* **2017**, *6*, 30. [CrossRef]
62. Krauss, M.; Berner, A.; Perrochet, F.; Frei, R.; Niggli, U.; Mäder, P. Enhanced soil quality with reduced tillage and solid manures in organic farming—A synthesis of 15 years. *Sci. Rep.* **2020**, *10*, 4403. [CrossRef]
63. Balesdent, J.; Marriotti, A.; Boisgontier, D. Effect of tillage on soil organic carbon mineralization estimated from 13C abundance in maize fields. *J. Soil Sci.* **1990**, *41*, 587–596. [CrossRef]
64. Ghimire, R.; Norton, U.; Bista, P.; Obour, A.K.; Norton, J.B. Soil organic matter, greenhouse gases and net global warming potential of irrigated conventional, reduced-tillage and organic cropping systems. *Nutr. Cycl. Agroecosyst.* **2017**, *107*, 49–62. [CrossRef]
65. Chen, J.; Zhu, R.; Zhang, Q.; Kong, X.; Sun, D. Reduced-tillage management enhances soil properties and crop yields in a alfalfa-corn rotation: Case study of the Songnen Plain, China. *Sci. Rep.* **2019**, *9*, 17064. [CrossRef]
66. Blanco-Canqui, H.; Lal, R. No-Tillage and Soil-Profile Carbon Sequestration: An On-Farm Assessment. *Soil Sci. Soc. Am. J.* **2008**, *72*, 693–701. [CrossRef]
67. Camarotto, C.; Piccoli, I.; Dal Ferro, N.; Polese, R.; Chiarini, F.; Furlan, L.; Morari, F. Have we reached the turning point? Looking for evidence of SOC increase under conservation agriculture and cover crop practices. *Eur. J. Soil Sci.* **2020**, *71*, 1050–1063. [CrossRef]
68. Piccoli, I.; Chiarini, F.; Carletti, P.; Furlan, L.; Lazzaro, B.; Nardi, S.; Berti, A.; Sartori, L.; Dalconi, M.C.; Morari, F. Disentangling the effects of conservation agriculture practices on the vertical distribution of soil organic carbon. Evidence of poor carbon sequestration in North-Eastern Italy. *Agric. Ecosyst. Environ.* **2016**, *230*, 68–78. [CrossRef]
69. Csitári, G.; Tóth, Z.; Kökény, M. Effects of organic amendments on soil aggregate stability and microbial biomass in a long-term fertilization experiment (Iosdv). *Sustainability* **2021**, *13*, 9769. [CrossRef]

70. Piccoli, I.; Sartori, F.; Polese, R.; Berti, A. Crop yield after 5 decades of contrasting residue management. *Nutr. Cycl. Agroecosyst.* **2020**, *117*, 231–241. [CrossRef]
71. Dal Ferro, N.; Piccoli, I.; Berti, A.; Polese, R.; Morari, F. Organic carbon storage potential in deep agricultural soil layers: Evidence from long-term experiments in northeast Italy. *Agric. Ecosyst. Environ.* **2020**, *300*, 106967. [CrossRef]
72. Hillel, D. *Introduction to Soil Physics*; Academic Press: New York, NY, USA, 1982; ISBN 0-12-348520-7.
73. Assouline, S. Modeling the Relationship between Soil Bulk Density and the Water Retention Curve. *Vadose Zone J.* **2006**, *5*, 554–563. [CrossRef]
74. Lu, N. Generalized Soil Water Retention Equation for Adsorption and Capillarity. *J. Geotech. Geoenviron. Eng.* **2016**, *142*, 04016051. [CrossRef]
75. Hill, R.L. Long-Term Conventional and No-Tillage Effects on Selected Soil Physical Properties. *Soil Sci. Soc. Am. J.* **1990**, *54*, 161–166. [CrossRef]
76. Bot, A.; Benites, J. Creating Drought-Resistant Soil. In *The Importance of Soil Organic Matter Key to Drought-Resistant Soil and Sustained Food Production*; Food and Agriculture Organization of the United Nations (FAO): Rome, Italy, 2005; pp. 35–40.
77. Müller, A.; Bautze, L.; Matthias, M.; Gattinger, A.; Gall, E.; Chatzinikolaou, E.; Meredith, S.; Ukas, T.; Ullmann, L. Organic Farming, Climate Change and Beyond. 2016. Available online: www.ifoam-eu.org (accessed on 8 December 2021).
78. Page, K.L.; Dang, Y.P.; Dalal, R.C. The Ability of Conservation Agriculture to Conserve Soil Organic Carbon and the Subsequent Impact on Soil Physical, Chemical, and Biological Properties and Yield. *Front. Sustain. Food Syst.* **2020**, *4*, 31. [CrossRef]

land

Article

Quantifying Cereal Productivity on Sandy Soil in Response to Some Soil-Improving Cropping Systems

Jerzy Lipiec [1,*] and Bogusław Usowicz [1,2]

1 Institute of Agrophysics, Polish Academy of Sciences, Doświadczalna 4, 20-290 Lublin, Poland
2 Faculty of Civil Engineering and Environmental Sciences, Bialystok University of Technology, Wiejska 45 E, 15-351 Białystok, Poland; b.usowicz@ipan.lublin.pl
* Correspondence: j.lipiec@ipan.lublin.pl

Abstract: Little information is available on the effect of soil-improving cropping systems (SICS) on crop productivity on low fertility sandy soils although they are increasingly being used in agriculture in many regions of the world due to the growing demand for food. The study aimed at quantifying the effect of four soil-improving cropping systems applied on sandy soil on cereal productivity (yield of grain and straw and plant height) in a 4-year field experiment conducted in Poland with spring cereal crops: oat (2017), wheat (2018), wheat (2019), and oat (2020). The experiment included the control (C) and the following SICS: liming (L), leguminous catch crops for green manure (LU), farmyard manure (M), and farmyard manure + liming + leguminous catch crops for green manure together (M + L + LU). To quantify the effect of the SICS, classic statistics and the Bland–Altman method were used. It was shown that all yield trait components significantly increased in the last study year (2020) under SICS with M and M + L + LU. All yield trait components were significantly lower in the dry years (2018–2019) than in the wet years (2017 and 2020). The relatively large rainfall quantity in May during intensive growth at shooting and the scarce precipitation during later growth in the dry year 2019 resulted in a significantly greater straw yield compared to the other dry year 2018. The values of Bland–Altman bias (mean difference between the particular SICS and the control) varied (in kg m^{-2}) from -0.002 for LU in 2019 to 0.128 for M and 0.132 for M + L + LU in 2020. The highest limits of agreement (LoA) were in general noted for all yield trait components (the least even yield) in the most productive SICS including M and M + L + LU in the wet year 2020. The Bland–Altman ratio (BAR) values indicate that quantification of the effects of all soil-improving practices was most uncertain in the dry year 2018 for the grain yield and in the wet year 2020 for the straw yield and much less uncertain for the plant height in all SICS and study years. The results of this study provide helpful information about the effect of the SICS on the different yield trait components depending on the period of their application and weather conditions prevailing during the growing season.

Keywords: soil improving practices; crop response; weather conditions; Podzol soil; Bland–Altman statistics

Citation: Lipiec, J.; Usowicz, B. Quantifying Cereal Productivity on Sandy Soil in Response to Some Soil-Improving Cropping Systems. *Land* **2021**, *10*, 1199. https://doi.org/10.3390/land10111199

Academic Editors: Guido Wyseure, Julián Cuevas González and Jean Poesen

Received: 19 October 2021
Accepted: 4 November 2021
Published: 5 November 2021

1. Introduction

Sandy soils cover globally approximately 900 million ha [1]. They occur in different regions across the world [2–4], particularly in arid or semi-arid regions [5]. In Poland, around 50 percent of soils developed from sands [6,7].

Sandy soils are characterized by low crop productivity. This is mostly attributed to a weakly developed aggregated structure [8], high saturated hydraulic conductivity and permeability and low water-retention capacity due to the high contribution of large pores between sand particles [9–11], low nutrient levels, and poor ability to store and exchange nutrients [1]. Furthermore, after rapid dewatering, the large pores become air-filled first and act as a barrier (discontinuity) to water flow through the smaller pores towards the plant roots in unsaturated soil conditions [12,13].

Another threat limiting crop production on sandy soils is their acidity due to the presence of acid from the post-glacial acidified parent material, leaching exchangeable base cations [14,15], and chemical N fertilization [16]. Soil acidity limits crop productivity by increasing Al3 + toxicity leading to production of short, thick, and shallow plant roots and deficiency of some nutrients in the soil solution [12]. Instead, sandy soils require rather low energy inputs for tillage [17] and warm up rapidly in the spring prior to the growing season to achieve the minimum soil temperature for plant growth [12].

Despite low fertility and quality, sandy soils are increasingly being used for crop production due to the shortage of agricultural land resources [1,18,19] as well as the growing population and demand for food [20,21]. However, arable farming on these soils require large amounts of irrigation and nutrient inputs [22,23] in many areas, which reduces the profitability of agricultural products.

There is a broad agreement that water and nutrient supply for plant growth in sandy soils can be improved by increasing organic matter content [1,24–28]. This is related to the fact that soil organic matter increases plant available water capacity [21,29] by reducing pore diameter [30] and improves the capability of soils to retain and exchange nutrient cations and hold hydrogen ions, thereby neutralizing soil acidity [31]. Furthermore, increase in soil organic carbon (SOC) content in sandy soils is responsible for variation in cation exchange capacity [1].

There are many soil-improving cropping systems to maintain or increase the SOM content. They include application of organic amendments and diversified crop rotation favoring formation of stable soil aggregates, which protect soil organic carbon (SOC) from mineralization [27,32]. Inclusion of legumes fixing nitrogen from the atmosphere in crop rotation reduces the need for mineral nitrogen fertilization, thereby increasing profitability in crop production [33–35], and is one of the ways to meet greening requirements [36]. Furthermore, these practices are important in terms of increasing cereal-based crop rotations that along with conventional treatments (plough) disintegrate soil organic matter by physical disturbance of the soil structure and stability [27]. Increasing the soil organic matter content is part of the global strategy to enhance carbon sequestration stocks, reduce chemical leaching [1,32,37,38], and create drought resilient soils to mitigate global warming effects [21,39].

The aim of the work was to quantify the effect of different soil-improving practices, including application of farmyard manure, liming, and catch crops, on cereal productivity of sandy soil in a 4-year experiment with the use of the statistical Bland–Altman method [40,41]. Plotting the yield differences between a given treatment and the control against their averages and determining the average difference (bias), limits of agreement, and confidence intervals in this method allow quantifying the direct effect of the examined soil-improving cropping systems on crop yields. The Bland–Altman method is widely used in medicine (e.g., [42–44]) and in some satellite research [45,46]. More recent studies showed usefulness of this approach to quantify pure effects of agricultural practices on crop yield and soil physical properties [47,48] and the agreement between methods for determining the Atterberg plastic and liquid limits of soils [49]. This study was inspired by recent literature reviews indicating that, despite their importance, sandy soils have received less research attention compared to other soils [1,5].

2. Materials and Method

2.1. Study Area and Field Experiment

The field experiment (350 × 35 m) was localized in a private farm in Szaniawy, Podlasie region, Poland (51°58′56.5″ N 22°32′22.1″ E) on Podzol soil [50] derived from sandy material of glacial origin. The soil contains 62.9% of sand (2–0.05 mm), 34.8% of silt (0.05–0.002 mm), 2.2% of clay (<0.002 mm), and 0.8% of organic carbon and has pH 4.0 (in H_20) and cation exchange capacity 12.3 cmol kg^{-1}. Such and similar soils predominate in the region and in Poland. A randomized field-experiment was established in autumn 2016 and conducted for four years with the following spring cereal crops: oats (*Avena sativa* L.) (2017), wheat (*Triticum aestivum* L.) (2018), wheat (2019), and oats (2020), which predominate in the crop rotation of the region.

The experiment included the following 5 treatments: (C) control, (L) liming with 5.6 t ha^{-1} CaCO3 (applied once in autumn 2016), (LU) catch crops for green manure including yellow lupin (*Lupinus luteus* L.) with seeding rates in brackets (130 kg ha^{-1}), serradella (*Ornithopus sativus*) (30 kg ha^{-1}), and phacelia (*Phacelia tanacetifolia* Benth.) (3 kg ha^{-1}) sown every year, (M) farmyard manure applied at 30 t ha^{-1} every year in autumn), and (M + L + LU) manure (10 t ha^{-1}) + liming with 5.6 t ha^{-1} $CaCO_3$ + catch crops for green manure including lupines with seeding rates in brackets (130 kg ha^{-1}), serradella (30 kg ha^{-1}), and phacelia (3 kg ha^{-1}) (applied every year), except liming only in autumn 2016. Yellow lupine, serradella, and phacelia are common plants used for green manures in Poland. Each treatment had three replicate plots (35 × 20 m) separated by a 1.0-m margin between the plots. The grain yield, straw yield, and plant height were measured in nine one-square-meter sub-plots in each of the five treatments (three sub-plots × three replicate plots).

Stubble tillage (10 cm) using a cultivator plus tooth harrows was done after harvesting in all treatments (first half of August) and then catch crops were sown in treatments LU and M + L + LU. Next, mouldboard ploughing (20–25 cm) in late autumn and disking (10 cm) and tooth harrowing (6 cm) in spring (2nd half of March) were applied in all treatments (1st half of April) to prepare the seedbed for spring cereals. The autumn ploughing in M and M + L + LU also ploughed down the catch crops for green manure. Weed control and crop protection were carried out by herbicides, insecticides, and fungicides used in the farm where the experiment was conducted in the same manner in all treatments. All management practices were done using light wheel tractors (2.5 to 3.5 Mg mass) to minimize soil compaction effects on crop yield.

2.2. Descriptive Statistics

Descriptive statistics including the mean, standard deviation, coefficient of variation, minimum and maximum values, skewness, and kurtosis were calculated for each yield trait. Pearson correlation coefficients between the yield trait components within the particular years and between the years were determined using STATISTICA 12 PL (StatSoft 2019).

2.3. Bland–Altman Method

The Bland–Altman statistics was adopted to determine the separate effect of the different SICS vs. control plots on the cereal yield trait components. In this method, the differences in the cereal yield trait components (grain, straw, and plant height) between the plots with different SICS and the control plots against the average yield with SICS and the control were graphically presented for each study year. The agreement between the yield in the plots with SICS and the control plots was assessed using bias (average of differences between the yield from the plots with SICS and control plots), the limit of agreement (LoA) defined as bias ±1.96 × standard deviation (SD), confidence intervals (CI) for the bias and LoA defined as ± standard error × the value of t distribution with n–1 degrees of freedom, and the Bland–Altman ratio (BAR) defined as the ratio of half the range of LoA to the mean of the pair including the yield from plots with SICS and control plots, the and regression line from the equation y = ax + b, where y—differences between the plots with different SICS and the control plots, x—average yield from plots with SICS and the control, a—regression coefficient, b—intercept. The BAR values were graded as good, moderate, and insufficient for values (BAR < 0), (0.2 ≤ BAR < 0.4), or (BAR ≥ 0.4), respectively [48].

Root mean square residuals (RMSR) and maximum relative residuals (MRR), which are the differences in the yield between the plots with SICS and the control plots, were determined for all yield trait components and each study year.

3. Results

3.1. Weather Conditions

Figure 1 illustrates the course of monthly mean air temperatures and rainfall sums for 2017, 2018, 2019, and 2020 in the study site. The average temperatures during the growing season (April–September) and the annual temperatures in the successive years were 14.8, 17.1, 15.9, and 15.1 °C and 8.7, 9.3, 10.0, and 9.7 °C, respectively. The respective sums of the growing season and annual precipitation were 424.1, 308.1, 306.2, and 439.8 mm and 670.1, 509.1, 475.9, and 666.2 mm, respectively. The growing season precipitation rates in 2017–2020 were below the long–term average (567 mm).

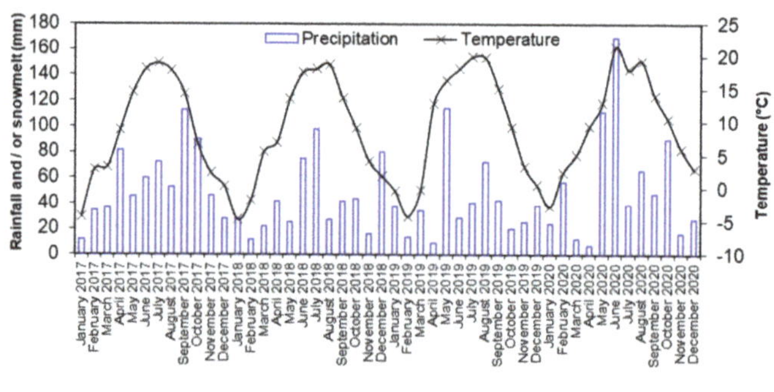

Figure 1. Monthly average air temperatures and sums of precipitation during the period of 2017–2020.

3.2. Descriptive Statistics

Basic statistical parameters of the grain and straw yields and plant height are given in Table 1. The ranges of the mean, minimum, and maximum values of the grain yield (in kg m^{-2}) in the growing seasons 2017–2020 were 0.123–0.361, 0.070–0.240, and 0.180–0.530, respectively. The corresponding ranges for the straw yield (in kg m^{-2}) were 0.205–0.432, 0.140–0.230, and 0.310–0.770 and plant height (in cm): 53.5–90.1, 48–75, and 63–110. The minimum values of all yield trait components were noted in 2018 or 2019 and the maximum values were determined in 2017 or 2020. The largest and similar coefficient variations (CVs) were recorded for the grain yield (17.5–28.1%) and the straw yield (19.7–28.5%), whereas lower values were calculated for the plant height (7.5–10.2%).

According to the classification proposed by Dahiya et al. [51], the CV values for the grain and straw yields were moderate (15–75%) and low (0–15%) for the plant height. The asymmetry (skewness) of the grain and straw yields was positive (0.053–0.930), whereas that of the plant height ranged from positive 0.803 in 2018 to negative −0.052 in 2017. The kurtosis of the grain yield varied from 0.141 in 2018 to negative −0.445 in 2020. The corresponding ranges for the straw yield and the plant height were 0.480 in 2017 to −0.249 in 2019 and −0.349 in 2019 to −1.018 in 2020. In general, the skewness and kurtosis values indicate that the yield trait components were close to the normal distribution, which was slightly flattened in nine cases and slightly slender in three cases.

The response of the cereals to the SICS applied was related to the yield trait components and the study year. The differences in the mean grain yield between the particular treatments and the control in the first three study years (2017–2019) varied from 18.0% to −16.6% (Figure 2a). However, in the last study year (2020), the wheat grain yield increased ($p < 0.05$) in the M and M + L + LU treatments by up to 47.3% and 45.8, respectively, compared to that in the control (0.279 kg m^{-2}). A substantially lower and statistically insignificant increase was observed in the lime (L) and catch crop (LU) treatments (by 10–18%). On average, the cereal grain yield during the experimental period (2017–2020) increased in the L, M, and M + L + LU variants by 2.5, 23.3, and 16.6%, respectively, and slightly decreased in LU (by 0.7%) compared to the control (0.224 kg m^{-2}). Irrespective of the treatment, the grain yields were considerably lower in both dry years 2018 and 2019 (growing season rainfall: 306 and 308 mm) than in the wet years 2017 and 2020 (growing season rainfall: 424 and 440 mm). The grain yields averaged over the treatments were 0.123–0.143 kg m^{-2} in the dry years (2018–2019) and 0.333–0.361 kg m^{-2} in the wet years 2017 and 2020 (Table 1). The inter-annual variations in the grain yields were relatively greater than those between the SICS treatments in all study years.

Table 1. Basic statistics for cereal yield trait components during the period of 2017–2020.

	Year 2017, Oats (kg m^{-2})		**(cm)**	**Year 2018, Spring Wheat (kg m^{-2})**		**(cm)**
Yield	Grain	Straw	Height	Grain	Straw	Height
Number	45	45	45	45	45	45
Mean	0.361	0.350	74.3	0.143	0.205	53.5
SD	0.063	0.069	5.6	0.040	0.040	4.2
CV (%)	17.5	19.7	7.5	28.1	19.3	7.9
Minimum	0.240	0.220	64.0	0.070	0.140	48.0
Maximum	0.530	0.560	85.0	0.250	0.310	63.0
Skewness	0.276	0.596	−0.052	0.883	0.930	0.803
Kurtosis	−0.340	0.480	−1.018	0.141	0.406	−0.514
	Year 2019, Spring Wheat (kg m^{-2})		**(cm)**	**Year 2020, Oats (kg m^{-2})**		**(cm)**
Yield	Grain	Straw	Height	Grain	Straw	Height
Number	45	45	45	45	45	45
Mean	0.123	0.395	71.6	0.333	0.432	90.1
SD	0.026	0.094	5.6	0.072	0.123	9.2
CV (%)	20.9	23.8	7.8	21.5	28.5	10.2
Minimum	0.070	0.200	59.0	0.170	0.230	75.0
Maximum	0.180	0.590	84.0	0.490	0.770	110.0
Skewness	0.249	0.053	−0.229	0.322	0.477	0.300
Kurtosis	−0.289	−0.249	−0.349	−0.445	−0.031	−0.494

The straw yield changes in response to the SICS applied varied in the first three years from 22.3% (in M + L + LU in 2019) to −11.4% (in L in 2018) (Figure 2b). The highest straw yield increment was observed in 2020 in the M and M + L + LU variants, where the straw yield increased by 58.2% and 65.0%, respectively, compared to that in the control (0.340 kg m^{-2}). It is worth noting that this increase in the straw yield in both treatments was relatively greater than that of the grain yield and was reflected in lower grain/straw ratios (Figure 2c). Noteworthy, the similar mean grain yield of spring wheat in the dry years 2018–2019 (0.123–0.143 kg m^{-2}) was accompanied by a considerably (almost twice) higher straw yield (0.395 kg m^{-2}) in the dry year 2019 than in the other dry year 2018 (0.205 kg m^{-2}).

The high straw yield in 2019 was clearly reflected in the considerably lower grain/straw ratios in all treatments (0.287 to 0.331), compared to those in the other years, i.e., 2017 (0.969 to 1.084), 2018 (0.633 to 0.787), and 2020 (0.733 to 0.874) (Figure 2c). The straw yield averaged over the whole experimental period (2017–2020) increased in L, LU, M, and M + L + LU by 1.5, 4.3, 23.8, and 29.0%, compared to the control 0.311 kg m^{-2}, respectively.

The plant height at harvest in the first three years (2017–2019) in the particular SICS was slightly lower (to 5.8%) or higher (to 9.7%) and statistically insignificant compared to the control (Figure 2d). However, in 2020, the plant height was significantly ($p < 0.05$) higher in M (by 16.0%) and in M + L + LU (by 20.7%), compared to the control (86.7 cm). It is worth noting that the plant height response to the particular SICS was relatively lower than that for grain and straw, irrespective of the study year. The plant height averaged over the four study years increased in L, LU, M, and M + L + LU by 3.1, 0.8, 8.4, and 11.6%, respectively, compared to the control (69.4 cm).

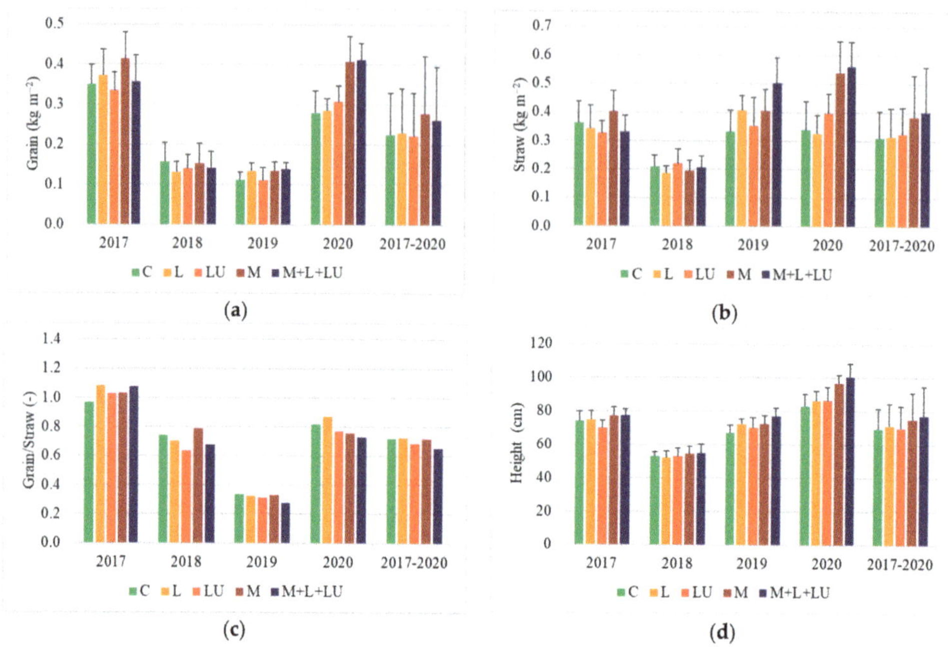

Figure 2. Mean values of the grain yield (**a**), straw yield (**b**), grain to straw yield ratio (**c**), plant height (**d**), and in response to soil-improving cropping systems. C—control, L—liming, LU—leguminous catch crops, M—farmyard manure and M + L + LU—farmyard manure + liming + leguminous catch crops. The vertical bars indicate the standard deviations (n = 9).

3.3. Correlation Coefficients between Yield Trait Components

As can be seen from Table 2, the highest correlation coefficient (r) between straw and grain was determined in the wet and last study year 2020 (0.798), whereas the lowest value was reported in the dry 2018 (0.393); both values were statistically significant ($p < 0.05$). The correlation coefficients between the plant height and the grain yield were more closely correlated in the wet and last study year 2020 (0.776) ($p < 0.05$), compared to the first three study years (0.487–0.596). The lowest correlation coefficients between the plant height and the straw yield were calculated in the dry 2018 (0.189), and the highest values were recorded in the wet and last study year 2020 (0.833) ($p < 0.05$). Noteworthy, there were markedly different r values between the plant height and the straw yield in the two dry years, i.e., 2018 (0.189) and 2019 (0.785), with much higher straw yield (and higher plant height) in the latter at a similar grain yield in both years. In line with this finding, there are significant positive correlations for the grain yield and the plant height between 2018 (0.322) and 2019 (0.333) ($p < 0.05$) in contrast to the insignificant and negative correlation for the straw yield (−0.57). Overall, the highest coefficient correlations between all paired yield trait components were recorded in the last study year.

3.4. Bland–Altman Analysis

Bland–Altman plots including horizontal lines of the bias line (mean difference from the SICS and control plots), limits of agreement (LoA = bias ± 1.96 × SD) along with confidence intervals (CI), regression lines (y = ax + b), and Bland–Altman ratio (BAR, half the range of LoA to the mean differences between the SICS and control plots) describe quantitatively the impact of particular SICS vs. the control on the cereal grain and straw yields and plant height. They are shown in Figures 3–5.

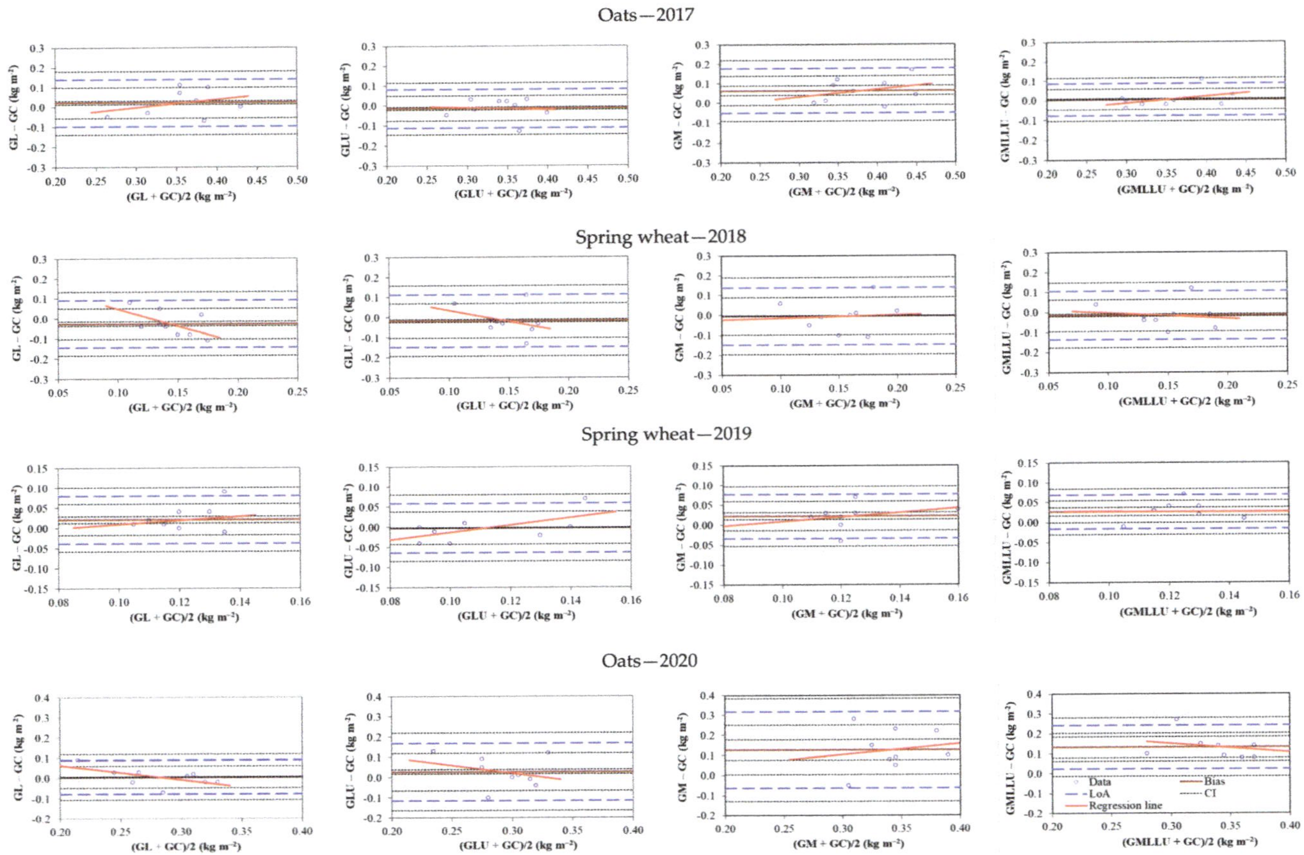

Figure 3. Bland–Altman plots for grain yield (G) (kg m^{-2}) (oats—2017), (spring wheat—2018, 2019), (oats—2020). C—control, L—liming, LU—leguminous catch crops, M—farmyard manure, and MLLU—farmyard manure + liming + leguminous catch crops, LoA—limits of agreement, CI—confidence intervals.

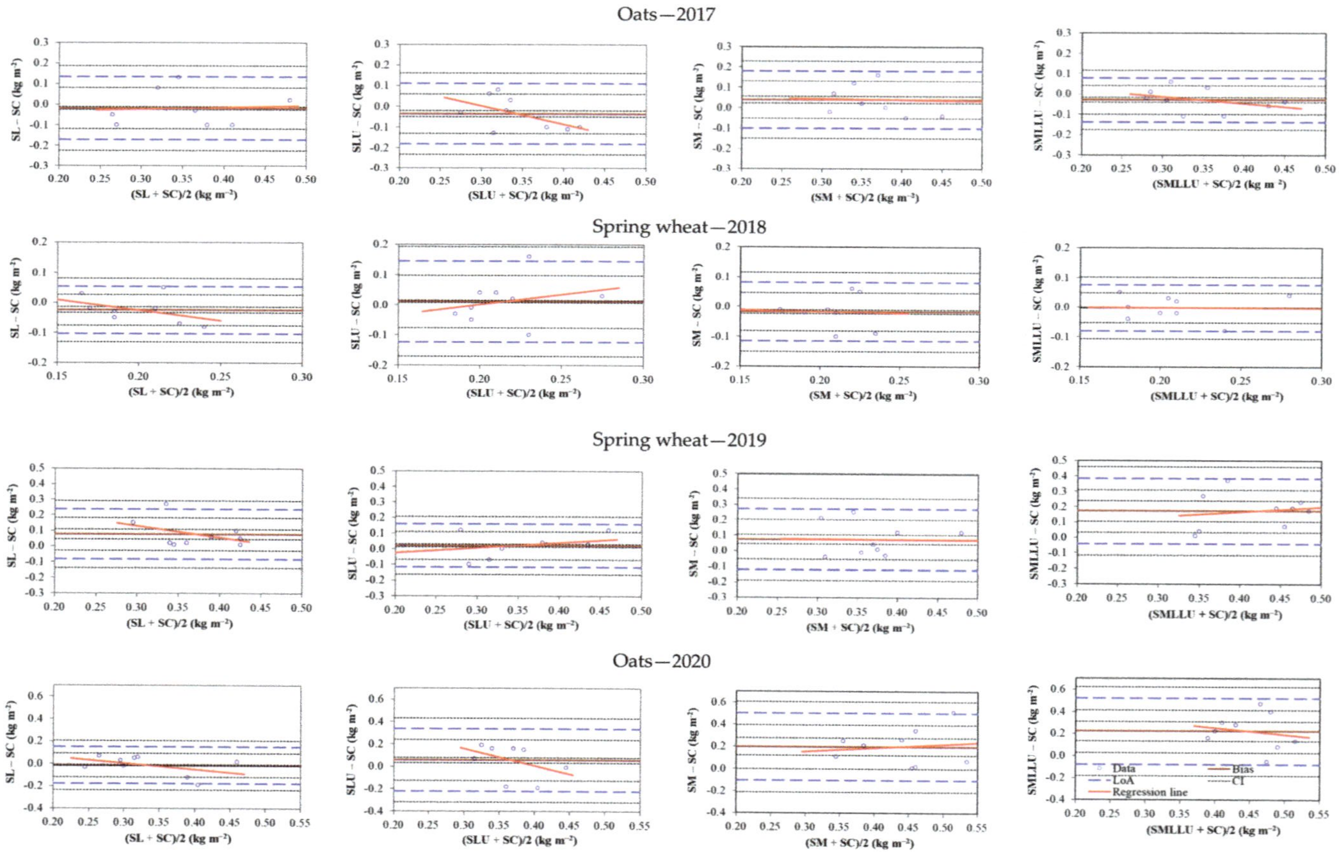

Figure 4. Bland–Altman plots for straw yield (S) (kg m^{-2}) (oats—2017), (spring wheat—2018, 2019), (oats—2020). C—control, L—liming, LU—leguminous catch crops, M—farmyard manure, and MLLU—farmyard manure + liming + leguminous catch crops, LoA—limits of agreement, CI—confidence intervals.

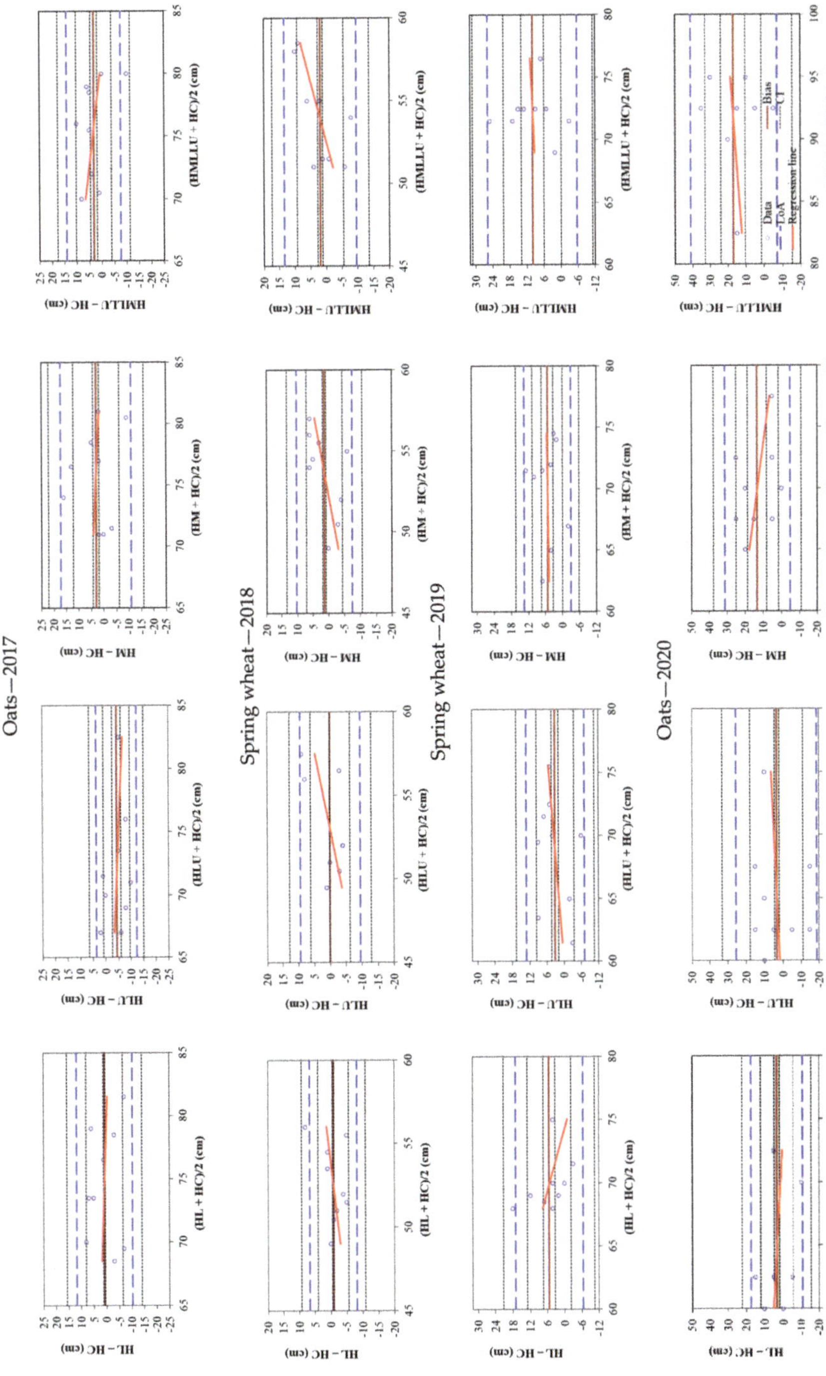

Figure 5. Bland–Altman plots for plant height (H) (cm) (oats—2017), (spring wheat—2018, 2019), (oats—2020). C—control, L—liming, LU—leguminous catch crops, M—farmyard manure, and MLLU—farmyard manure + liming + leguminous catch crops, LoA—limits of agreement, CI—confidence intervals.

Table 2. Correlation coefficients (r) between grain, straw, and height in the study years. The correlation coefficients in bold are significant at $p < 0.05$.

		Oats, 2017			Spring Wheat, 2018			Spring Wheat, 2019			Oats, 2020		
		Grain	**Straw**	**Height**	**Grain**	**Straw**	**Height**	**Grain**	**Straw**	**Height**	**Grain**	**Straw**	**Height**
	Grain	1.000	**0.623**	**0.487**	−0.112	**−0.169**	0.081	0.155	−0.116	0.067	**0.182**	−0.022	0.073
2017	Straw		1.000	**0.562**	**−0.257**	−0.126	−0.014	−0.153	**−0.181**	−0.096	0.116	−0.020	0.057
	Height			1.000	−0.063	−0.001	**0.229**	**0.325**	0.150	**0.230**	**0.306**	0.135	**0.336**
	Grain				1.000	**0.393**	**0.512**	**0.322**	**0.228**	0.109	**0.174**	0.122	−0.007
2018	Straw					1.000	**0.189**	−0.096	−0.057	−0.080	0.136	0.115	0.058
	Height						1.000	**0.329**	**0.346**	**0.333**	**0.217**	**0.198**	0.112
	Grain							1.000	**0.676**	**0.596**	**0.312**	0.121	**0.330**
2019	Straw								1.000	**0.785**	**0.473**	**0.367**	**0.506**
	Height									1.000	**0.375**	**0.326**	**0.483**
	Grain										1.000	**0.798**	**0.776**
2020	Straw											1.000	**0.833**
	Height												1.000

The average differences (biases) indicate that the application of the particular SICS resulted in a lower grain yield (bias < 0) in seven cases and a higher grain yield (bias > 0) in nine cases (Figure 3). The negative biases varied in kg m^{-2} from −0.002 for LU in 2019 to −0.017 for LU in 2018, and positive biased ranged from 0.006 for M + L + LU in 2017 and L in 2020 to 0128 for M and 0.132 for M + L + LU in 2020. It is worth noting that all negative biases occurred in the first three study years and the most positive ones were noted in the last study year. The ranges of LoA for the grain yields were in general similar in 2017 and 2018 in all comparable SICS treatments (except higher in LU in 2018). They decreased considerably in 2019 and then increased in 2020. The increase in 2020 was most pronounced in M and M + L + LU, where the ranges of LoA± in kg m^{-2} (0.318, −0.062 and 0.241, 0.023) were several times greater than those in 2019 (0.077, −0.033 and 0.069, −0.016). The largest ranges of LoA in M and M + L + LU in 2020 correspond with the respective highest maximum values of root mean square residuals (RSMR) (0.16 and 0.143 kg m^{-2}) and maximum relative residuals (MRR) (164.7 and 158.8%) (Figure 6). Irrespective of the treatment, the largest Bland–Altman ratio (BAR) values were noted in 2018 (0.809–0.929) and the lowest were recorded in 2017 (0.231–0.330), which indicates insufficient (BAR $\geq$ 0.4) and moderate (0.2 $\leq$ BAR < 0.4) agreement, respectively, between the grain yield in the SICS and control plots [48].

As to the straw yield, the negative biases occurred in seven of the 16 cases and changed in kg m^{-2} from −0.002 for M + L + LU in 2018 to −0.036 for LU in 2017 (Figure 4). The positive biases varied from 0.011 for LU in 2018 to 0.198 for M and 0.221 for M + L + LU in 2020. The highest positive biases correspond with the highest RMSR (0.251–0.269 kg m^{-2}) and MRR (196.2–204.3%). The ranges of the limits of agreements (LoA = bias ± 1.96 × SD) were in general relatively narrow and similar in all treatments in the first two study years 2017–2018, but increased largely in all SICS treatments in 2019–2020. This increase was most pronounced in 2020 in the M and M + L + LU treatments, where the LoAs ranges approximately doubled compared to those in 2017–2018. The lowest BAR values were recorded (0.311–0.425) in the first study year 2017, and the largest were noted in 2020 (0.490–0.753) in all comparable treatments (Figure 6), which indicates moderate (0.2 $\leq$ BAR < 0.4) and insufficient (BAR $\geq$ 0.4) agreement, respectively, between the straw yield in the SICS and control plots [48].

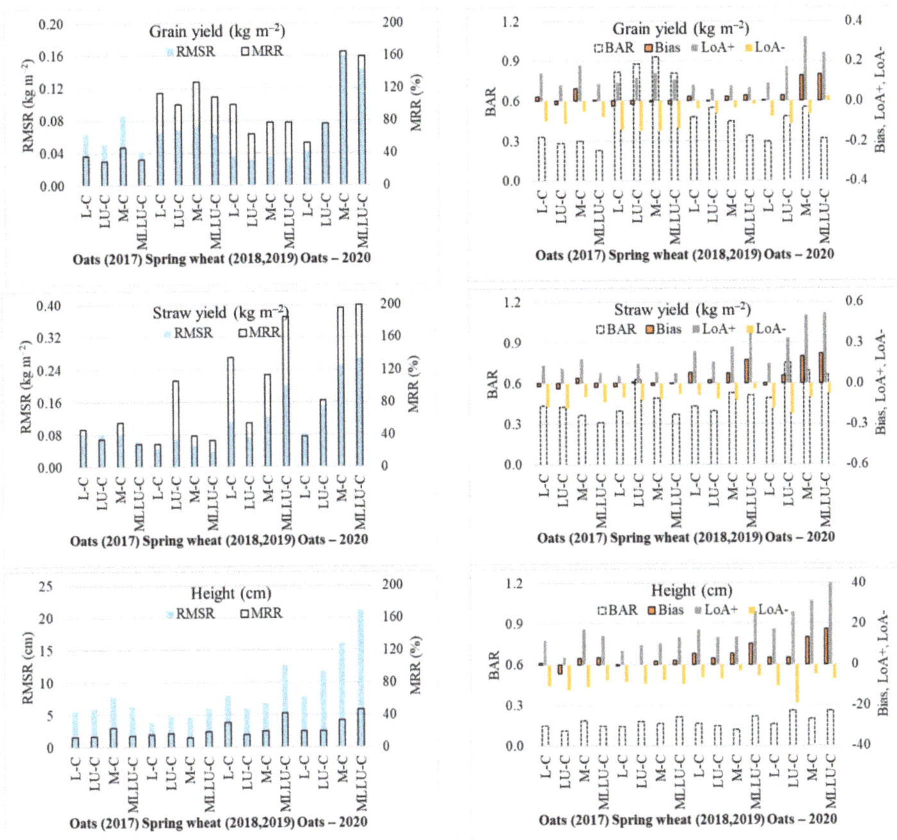

Figure 6. Root mean square residuals (RMSR) and maximum relative residuals (MRR), which are the difference in the yield trait components between the particular SICS and control plots, BAR—Bland–Altman ratio, Bias—mean of the differences, LoA—limit of agreements, SD—standard deviation of the differences.

The bias values for the plant height were mostly positive in 13 of the 16 cases (except two negative values for LU in 2017 and L in 2018 and 0 for LU in 2018). The lowest bias in cm (−4.33) was recorded in 2017 for LU and the highest in 2020 for M (13.33) and M + L + LU (17.22). In comparable treatments, the lowest biases occurred in general in the dry growing season 2018 with the lowest mean plant height among the study years and the highest values in 2020 with the highest plant height (Figure 5). The ranges of LoAs were wider in 2020 than in the other years. The highest BAR values for most of the comparable pairs were recorded in 2020 when the plants were the tallest. In all years except 2017, the highest BAR values were noted for M + L + LU. The BAR values in 8 cases (0.111 to 0.199) and in four cases (0.212–0.262) (Figure 6) indicated good or moderate and moderate agreement [48], respectively, between the grain yield in the SICS and control plots. As can be seen in Figure 6, the BAR values for the plant height were lower than those for the grain and straw yields, irrespective of the treatment and study year. The lower BAR values for the plant height correspond with the higher RMSR and lower MRR values.

The regression lines of the differences between the particular SICS and control plots against the average yield of both indicate that the trends for grain were descending, ascending, or almost unchanged (close to the bias line) depending on the SICS type and study year. Ascending trends were mostly observed for the paired treatment M and Control.

As to straw yield and plant height, the regression lines indicate slightly descending or ascending trends.

Regardless of the SICS type, yield trait component, and study year, the Bland–Altman plots indicate that a bulk of the points are within the limits of agreement (LoA) and outliers—within the confidence intervals (CI) (Figures 3–5).

4. Discussion

4.1. Impact of the Soil-Improving Cropping Systems (SICS) on Yield Trait Components

Our study showed the most pronounced differences in all crop yield trait components between SICS in the fourth and wet study year. A statistically significant and similar increase in the crop yield was found in two SICS, i.e., M consisting of only farmyard manure and M + L + LU providing less farmyard manure and plus lime and cover crops. This significant impact of both treatments may result from the increased nutrient supply from farmyard manure and cover crops although the soil organic matter content increased only slightly (data not shown). Similarly, the high yields in the M and M + L + LU variants imply that organic matter from deficit farmyard manure in the former can be replaced in part by green manure/cover crops, with maintenance of the same productivity. The positive effect of the combined SICS (M + L + LU) on the crop yield in the acid soil can also be enhanced by yield-increasing liming improving the availability of essential nutrients to plants [52]. These results support the recent actions in several countries, including Poland, focused on promotion of incorporating legumes in the intercropping systems and extending agricultural lime application [53–55]. Application of the combined SICS should be considered not only in relation to crop productivity enhancement but also as a sustainable strategy to improve the supply capacity of essential nutrients including fixed atmospheric nitrogen [56], and alleviating the negative effect of soil acidity [52]. It should also be noted that increasing the organic carbon content or even keeping good levels in sandy soils requires a continuous supplying organic materials. This is due to the fact that sandy soils, especially tilled, are well aerated creating conditions conducive to rapid microbial decomposition of organic matter.

4.2. Weather Influences

Our results showed that the cereal yield trait components were largely influenced by both the total rainfall amount and their temporal distribution during the growing seasons. As could be expected, the wheat grain yield was appreciably lower (by more than 50%) in the two dry growing seasons compared to the two wet growing seasons. The analysis of the yield trait components and the weather course further revealed that, in both dry years (2018–2019) with almost the same total amount of rainfalls during the growing season (306–308 mm), the straw yield of spring wheat was by 160–221% higher, depending on treatment, in 2019 than 2018. In turn, the grain yield of spring wheat in 2019, compared to 2018, was not different or slightly lower (by 6.4–24.3%) in the comparable treatments. This opposite response of the yield trait components can be explained by the different distribution of rainfalls during the analyzed growing seasons. The large amounts of rainfalls in May during intensive growth at shooting and the scarce precipitation during later growth in 2019 (Figure 1) may have stimulated top growth. Moreover, the shoot growth in May 2019 may have been favored by the lower temperature (12.5 °C) compared to that in 2018 (18.5 °C) (Figure 1) by changing evaporation rates. The more intensive shoot growth in 2019 vs. 2018 was reflected in the greater straw yield in the former season in all treatments (Figure 4). The results imply that a good water supply at shooting increases allocation of assimilates to shoots while reducing the grain yield. These diverse responses of the yield trait components emphasize the importance of the increasingly frequent episodic (extreme) drought and wet conditions during the growing season associated with climate change [57]. The sensitivity of the yield trait components to weather variation during the growing season in sandy soils can be enhanced by the high permeability and low water holding capacity of these soils, which do not allow storing water for a longer time and efficient use of nutrients [5], and by the relatively shallow root system of spring cereals. Understanding the relations among the yield trait components depending on the weather course during the growing season is important in food and bioethanol production where grain and straw, respectively, are potential feedstocks [58,59].

4.3. Usefulness of the Bland–Altman Method

The use of the Bland–Altman method contributed to improvement of the quantification of the direct (separated) impact of a given soil-improving practice on the cereal yield trait components in reference to the control in different inter-annual weather conditions. For example, the small values of limits of agreement (LoA = bias $\pm$ 1.96 $\times$ SD) for the grain yield in the most yield-producing SICS (M and M + L + LU) in the dry growing season 2019 (with the lowest yield) increased by several times in the wet season 2020 (with the highest yield), indicating that the grain yields were less even in the latter. The reduced evenness of the yield in the wetter growing season 2020 may have resulted in part from the variability in soil water content and the related availability of nutrients from organic matter provided by these two SICS. The variable soil water content in these SICS treatments may have resulted from changes in the soil structure caused by the organic matter amendments and from the natural variability of the soil texture in the study area [19]. This explanation is supported by the fact that water deficit is a dominant crop yield–limiting factor in sandy soils [5,60].

The Bland–Altman plots indicate that the orientations of the regression equation lines for the grain and straw yield in M and M + L + LU, compared to the other SICS, were in most cases close together to the bias lines. This can be indicative of the stabilizing effect of the largest quantity of organic matter provided by both SICS on the yield and uniformity of the yield components. The regression equation lines below or above the bias line indicate a reduction in yield uniformity. It is important to add that if only one treatment, i.e., SICS or the control, in the pair has a wide range of limits of agreement (LoA), the Bland–Altman will always produce wide limits of agreement [61]. This means that poor agreement between the paired SICS and control do not necessarily indicate that the tested SICS has low evenness of crop yields in the replicate sub-plots.

The comparison of the Bland–Altman Ratio of the yield trait components revealed that grain and straw yields, compared to plant height, exhibit appreciably higher uncertainty (at the most comparable paired SICS and control). Even with high uncertainty, analysis of biases and LoA values facilitates assessment of the degree of the causal (positive or negative) effect of particular SICS on the crop yield. This observation along with inter-annual differences in crop yield trait components is important in modelling crop responses to SICS and weather conditions during growing seasons [62].

5. Conclusions

The results of this study indicate the following findings:

(1) Differences in the yield of grain and straw and plant height between all each soil-improving cropping system (SICS) and the control were not significant in the first three study years (2017–2019). In the last study year (2020), however, all yield trait components significantly increased in SICS with the use of farmyard manure (M) and farmyard manure, liming, and catch crops together (M + L + LU) but not in SICS with application of liming (L) and catch crops (LU) alone.

(2) Irrespective of the type of the soil-improving cropping systems, all yield trait components were considerably lower in the dry years (2018–2019) than in the wet years (2017–2020). The inter-annual variations were relatively greater than those between the SICS treatments in all study years. The relatively large amount of rainfalls in May in 2019 during intensive growth at shooting and the scarce precipitation during later growth resulted in a significantly greater straw yield.

(3) The values of Bland–Altman bias (mean difference) varied from (in kg m^{-2}) -0.002 for LU in 2019 to 0.128 for M and 0.132 for M + L + LU in 2020. Irrespective of the yield

trait components, the highest limits of agreement (LoA) were recorded in 2020 in the M and M + L + LU variants, where all yield trait components reached the maximum values.

(4) The highest Bland–Altman ratio (BAR) values suggest that quantification of the effects of all soil-improving cropping practices was most uncertain for the grain yield in the dry year 2018 and for the straw yield in the wet year 2020. The uncertainty for the plant height was much lower than for the grain and straw yield, irrespective of the soil-improving cropping systems and study year.

(5) Overall, the results from the Bland–Altman method well complement classical statistics by providing helpful information for selection of the most yield-producing soil-improving cropping system, depending on weather conditions prevailing during the growing season.

Author Contributions: B.U. and J.L. performed experiments and measurements. B.U. carried out data computation and statistical analysis. All authors have read and agreed to the published version of the manuscript.

Funding: The research leading to these results has received funding from the H2020 under grant agreement number 677407 (SOILCARE).

Institutional Review Board Statement: Not applicable.

Data Availability Statement: Data sharing is not applicable to this article.

Conflicts of Interest: The authors declare no conflict of interest.

References

1. Yost, J.L.; Hartemink, A.E. Chapter four—Soil organic carbon in sandy soils: A review. In *Advances in Agronomy*, 1st ed.; Sparks, D.L., Ed.; Academic Press: Newark, DE, USA, 2019; pp. 217–230.
2. Bronick, C.J.; Lal, R. Manuring and rotation effects on soil organic carbon concentration for different aggregate size fractions on two soils in northeastern Ohio, USA. *Soil Tillage Res.* **2005**, *81*, 239–252. [CrossRef]
3. Thorsen, M.K.; Hopkins, D.W.; Woodward, S.; McKenzie, B.M. Resilience of microorganisms and aggregation of a sandy calcareous soil to amendment with organic and synthetic fertilizer. *Soil Use Manag.* **2010**, *26*, 149–157.
4. Jankowski, M.; Przewoźna, B.; Bednarek, R. Topographical inversion of sandy soils due to local conditions in Northern Poland. *Geomorphology* **2011**, *135*, 277–283. [CrossRef]
5. Huang, J.; Hartemink, A.E. Soil and environmental issues in sandy soils. *Earth Sci. Rev.* **2020**, *208*, 103295. [CrossRef]
6. Białousz, S.; Marcinek, J.; Stuczyński, T.; Turski, R. Soil survey, soil monitoring and soil databases in Poland. In *European Soil Bureau—Research Report*; No. 9 ESB-RR9; European Soil Bureau: Brussels, Belgium, 2000; pp. 263–273. Available online: https://www.academia.edu/34998638/ (accessed on 19 October 2021).
7. Rutkowska, A.; Pikuła, D. Effect of crop rotation and nitrogen fertilization on the quality and quantity of soil organic matter. In *Soil Processes and Current Trends in Quality: Assessment*; Hernandez Soriano, M.C., Ed.; Intech Open: London, UK, 2013; pp. 249–267. Available online: https://www.intechopen.com/chapters/41059 (accessed on 19 October 2021). [CrossRef]
8. Osman, K.T. Management of soil problems. In *Sandy Soils*; Osman, K.T., Ed.; Springer International Publishing: Cham, Switzerland, 2018; pp. 37–65.
9. Peake, L.R.; Reid, G.J.; Tang, X. Quantifying the influence of biochar on the physical and hydrological properties of dissimilar soils. *Geoderma* **2014**, *235–236*, 182–190. [CrossRef]
10. Zhou, H.; Fang, H.; Zhang, Q.; Wang, Q.; Chen, C.; Mooney, S.J.; Peng, X.; Du, Z. Biochar enhances soil hydraulic function but not soil aggregation in a sandy loam. *Eur. J. Soil Sci.* **2019**, *70*, 291–300. [CrossRef]
11. Usowicz, B.; Lipiec, J. Spatial variability of saturated hydraulic conductivity and its links with other soil properties at the regional scale. *Sci. Rep.* **2021**, *11*, 8293. [CrossRef]
12. Gliński, J.; Lipiec, J. *Soil Physical Conditions and Plant Roots*, 1st ed.; CRC Press Taylor & Francis Group: Boca Raton, FL, USA, 1990; p. 260; First Published 1990; ISBN 13: 978-1-351-07670-8.
13. Villagra-Mendoza, K.; Horn, R. Effect of biochar on the unsaturated hydraulic conductivity of two amended soils. *Int. Agrophys.* **2018**, *32*, 373–378. [CrossRef]
14. Krasowicz, S.; Oleszek, W.; Horabik, J.; Dębicki, R.; Jankowiak, J.; Stuczyński, T.; Jadczyszyn, J. Rational management of the soil environment in Poland. *Pol. J. Agron.* **2011**, *7*, 43–58.
15. Behera, S.K.; Shukla, A.K. Spatial distribution of surface soil acidity, electrical conductivity, soil organic carbon content and exchangeable potassium, calcium and magnesium in some cropped acid soils of India. *Land Deg. Dev.* **2015**, *26*, 71–79. [CrossRef]

16. Cai, Z.J.; Wang, B.R.; Xu, M.G.; Zhang, H.M.; He, X.; Zhang, L.; Gao, S. Intensified soil acidification from chemical N fertilization and prevention by manure in an 18-year field experiment in the red soil of southern china. *J. Soils Sediments* **2015**, *15*, 260–270. [CrossRef]
17. Novák, P.; Chyba, J.; Kumhála, F.; Procházka, P. The measurement of stubble cultivator draught force under different soil conditions. *Agron. Res.* **2014**, *12*, 135–142.
18. Schjønning, P.; Heckrath, G.; Christensen, B.T. Threat to soil quality in Denmark. A review of existing knowledge in the context of the EU soil thematic strategy. In *DJF Report Plant Science no. 143*; Aarhus University: Tjele, Denmark, 2009; pp. 11–121. Available online: http://web.agrsci.dk/djfpublikation/index.asp?action=show&id=1073 (accessed on 19 October 2021).
19. Usowicz, B.; Lipiec, J. Spatial variability of soil properties and cereal yield in a cultivated field on sandy soil. *Soil Tillage Res.* **2017**, *174*, 241–250.
20. Reichert, J.M.; Albuquerque, J.A.; Kaiser, D.R.; Reinert, D.J.; Urach, F.L.; Carlesso, R. Estimation of water retention and availability in soils of Rio Grande do Sul. *Rev. Bras. Cienc. Solo* **2020**, *33*, 1547–1560.
21. Lal, R. Soil organic matter and water retention. *Agron. J.* **2020**, *112*, 3265–3277. [CrossRef]
22. Adhikari, K.; Hartemink, A.E. Soil organic carbon increases under intensive agriculture in the Central Sands, Wisconsin, USA. *Geoderma Reg.* **2017**, *10*, 115–125. [CrossRef]
23. Cai, A.; Xu, M.; Wang, B.; Zhang, W.; Liang, G.; Hou, E.; Lua, Y. Manure acts as a better fertilizer for increasing crop yields than synthetic fertilizer does by improving soil fertility. *Soil Tillage Res.* **2019**, *189*, 168–175.
24. Ajayi, A.E.; Holthusen, D.; Horn, R. Changes in microstructural behaviour and hydraulic functions of biochar amended soils. *Soil Tillage Res.* **2016**, *155*, 166–175. [CrossRef]
25. Minasny, B.; McBratney, A.B. Limited effect of organic matter on soil available water capacity. *Eur. J. Soil Sci.* **2017**, *69*, 39–47. [CrossRef]
26. Bhogal, A.; Nicholson, F.A.; Rollett, A.; Taylor, M.; Litterick, A.; Whittingham, M.J.; Williams, J.R. Improvements in the quality of agricultural soils following organic material additions depend on both the quantity and quality of the materials applied. *Front. Sustain. Food Syst.* **2018**, *2*, 9. [CrossRef]
27. EASAC (European Academies' Science Advisory Council). Opportunities for soil sustainability in Europe. In *Policy Report 36*; European Academies Science Advisory Council: Halle, Germany, 2018; pp. 3–38.
28. Villagra-Mendoza, K.; Horn, R. Effect of biochar addition on hydraulic functions of two textural soils. *Geoderma* **2018**, *326*, 88–95.
29. Gluba, Ł.; Rafalska-Przysucha, A.; Szewczak, K.; Łukowski, M.; Szlązak, R.; Vitková, J.; Kobyłecki, R.; Bis, Z.; Wichliński, M.; Zarzycki, R.; et al. Effect of fine size-fractionated sunflower husk biochar on water retention properties of arable sandy soil. *Materials* **2021**, *14*, 1335. [CrossRef]
30. Ibrahim, A.; Horton, R. Biochar and compost amendment impacts on soil water and pore size distribution of a loamy sand soil. *Soil Sci. Soc. Am. J.* **2021**, *85*, 1021–1036. [CrossRef]
31. Bönecke, E.; Meyer, S.; Vogel, S.; Schröter, I.; Gebbers, R.; Kling, C.; Kramer, E.; Lück, K.; Nagel, A.; Philipp, G.; et al. Guidelines for precise lime management based on high-resolution soil pH, texture and SOM maps generated from proximal soil sensing data. *Precis. Agric.* **2021**, *22*, 493–523. [CrossRef]
32. Bolinder, M.A.; Crotty, F.; Elsen, A.; Frąc, M.; Kismanyoky, T.; Lipiec, J.; Tits, M.; Toth, Z.; Kätterer, T. The effect of crop residues, cover crops, manures and nitrogen fertilization on soil organic carbon changes in agroecosystems: A synthesis of reviews. *Mitig. Adapt. Strateg. Glob. Chang.* **2020**, *25*, 929–952. [CrossRef]
33. Kismányoky, T.; Tóth, Z. Effect of mineral and organic fertilization on soil organic carbon content as well as on grain production of cereals in the IOSDV (ILTE) long-term field experiment, Keszthely, Hungary. *Arch. Agron. Soil Sci.* **2013**, *59*, 1121–1131. [CrossRef]
34. Bloch, S.E.; Ryu, M.H.; Ozaydin, B.; Broglie, R. Harnessing atmospheric nitrogen for cereal crop production. *Curr. Opin. Biotechnol.* **2020**, *62*, 181–188. [CrossRef]
35. Popp, M.P.; Purcell, L.C.; Ross, W.J.; Norsworthy, J.S. Profitability of using nitrogen fertilizer or inoculating soybean seed. *Agrosyst. Geosci. Environ.* **2020**, *3*, e20115. [CrossRef]
36. Świtek, S.; Sawinska, Z. Farmer rationality and the adoption of greening practices in Poland. *Sci. Agric.* **2017**, *74*, 275–284. [CrossRef]
37. Zaniewicz-Bajkowska, A.; Rosa, R.; Kosterna, E.; Franczuk, J. Catch crops for green manure biomass yield and macroelement content depending on the sowing date. *Acta Sci. Pol. Agric.* **2013**, *12*, 65–79.
38. Valkama, E.; Lemola, R.; Känkänen, H.; Turtola, E. Meta-analysis of the effects of undersown catch crops on nitrogen leaching loss and grain yields in the Nordic countries. *Agric. Ecosyst. Environ.* **2015**, *203*, 93–101.
39. Soussana, J.-F.; Lutfalla, S.; Ehrhardt, F.; Rosenstock, T.; Lamanna, C.; Havlík, P.; Richards, M.; Wollenberg, E.; Chotte, J.-L.; Torquebiau, E.; et al. Matching policy and science: Rationale for the '4 per 1000—Soils for food security and climate' initiative. *Soil Tillage Res.* **2017**, *188*, 3–15. [CrossRef]
40. Bland, J.M.; Altman, D.G. Statistical method for assessing agreement between two methods of clinical measurement. *Lancet* **1986**, *327*, 307–310.
41. Bland, J.M.; Altman, D.G. Measuring agreement in method comparison studies. *Stat. Methods Med. Res.* **1999**, *8*, 135–160. [CrossRef]
42. Myles, P.S.; Cui, J. Using the Bland–Altman method to measure agreement with repeated measures. *Br. J. Anaesth.* **2007**, *99*, 309–311. [CrossRef] [PubMed]

43. Giavarina, D. Understanding Bland-Altman analysis. *Bio. Chem. Med.* **2015**, *24*, 141–151. [CrossRef] [PubMed]
44. Numao, S.; Uchida, R.; Kurosaki, T.; Nakagaichi, M. Differences in circulating fatty acid-binding protein 4 concentration in the venous and capillary blood immediately after acute exercise. *J. Physiol. Anthropol.* **2021**, *40*, 5. [CrossRef] [PubMed]
45. Usowicz, B.; Marczewski, W.; Usowicz, J.; Łukowski, M.; Lipiec, J. Comparison of surface soil moisture from SMOS satellite and ground measurements. *Int. Agrophys.* **2014**, *28*, 359–369. [CrossRef]
46. Zawadzki, J.; Kędzior, M. Soil moisture variability over Odra watershed: Comparison between SMOS and GLDAS data. *Int. J. Appl. Earth. Obs.* **2016**, *45*, 110–124. [CrossRef]
47. Usowicz, B.; Lipiec, J. Determining the effect of exogenous organic materials on spatial distribution of maize yield. *Sci. Rep.* **2019**, *9*, 19883. [CrossRef]
48. Usowicz, B.; Lipiec, J.; Łukowski, M.; Bis, Z.; Usowicz, J.; Latawiec, A.E. Impact of biochar addition on soil thermal properties: Modelling approach. *Geoderma* **2020**, *376*, 114574. [CrossRef]
49. Rehman, H.U.; Pouladi, N.; Pulido-Moncada, M.; Arthur, E. Repeatability and agreement between methods for determining the Atterberg limits of fine-grained soils. *Soil Sci. Soc. Am. J.* **2020**, *84*, 21–30. [CrossRef]
50. WRB IUSS Working Group. World reference base for soil resources 2014, update 2015. In *International Soil Classification System for Naming Soils and Creating Legends for Soil Maps*; World Soil Resources Reports No. 106; FAO: Rome, Italy, 2015. Available online: https://publications.jrc.ec.europa.eu/repository/handle/JRC91947 (accessed on 19 October 2021).
51. Dahiya, I.S.; Ritcher, J.; Mark, P.S. Soil spatial variability: Review. *Int. J. Trop. Agric.* **1984**, *11*, 1–102.
52. Mkhonza, N.P.; Buthelezi-Dube, N.N.; Muchaonyerwa, P. Effects of lime application on nitrogen and phosphorus availability in humic soils. *Sci. Rep.* **2020**, *10*, 8634. [CrossRef] [PubMed]
53. Tamm, I.; Tamm, Ü.; Ingver, A.; Koppel, R.; Tupits, I.; Bender, A.; Tamm, S.; Narits, L.; Koppel, M. Different leguminous pre-crops increased yield of succeeding cereals in two consecutive years. *Acta Agr. Scand. Sect. B Soil Plant Sci.* **2016**, *66*, 593–601. [CrossRef]
54. Gitari, H.I.; Gachene, C.K.K.; Karanja, N.N.; Kamau, S.; Nyawade, S.; Schulte-Geldermann, E. Potato-legume intercropping on a sloping terrain and its effects on soil physico-chemical properties. *Plant Soil* **2019**, *438*, 447–460. [CrossRef]
55. Jerzak, M.; Mikulski, W. The importance of subsidies for the production of grain legumes for reconstruction of the domestic market of protein raw materials of vegetable origin in Poland. *Probl. Agric. Econ.* **2017**, *2*, 152–163.
56. Wysokinski, A.; Kuziemska, B. The sources of nitrogen for yellow lupine and spring triticale in their intercropping. *Plant Soil Environ.* **2019**, *65*, 145–151. [CrossRef]
57. Kundzewicz, Z.W.; Piniewski, M.; Mezghani, A.; Okruszko, T.; Pińskwar, I.; Kardel, I.; Hov, Ø.; Szcześniak, M.; Szwed, M.; Benestad, R.E.; et al. Assessment of climate change and associated impact on selected sectors in Poland. *Acta Geophys.* **2018**, *66*, 1509–1523. [CrossRef]
58. Dai, J.; Bean, B.; Brown, B.; Bruening, W.; Edwards, J.; Flowers, M.; Karow, R.; Lee, C.; Morgan, G.; Ottman, M.; et al. Harvest index and straw yield of five classes of wheat. *Biomass Bioenerg.* **2016**, *85*, 223–227. [CrossRef]
59. Liu, W.; Hou, P.; Liu, G.; Yang, Y.; Guo, X.; Ming, B.; Xie, R.; Wang, K.; Liu, Y.; Li, S. Contribution of total dry matter and harvest index to maize grain yield—A multisource data analysis. *Food Energy Secur.* **2020**, *9*, e256. [CrossRef]
60. Lipiec, J.; Usowicz, B. Spatial relationships among cereal yields and selected soil physical and chemical properties. *Sci. Total Environ.* **2018**, *633*, 1579–1590. [CrossRef] [PubMed]
61. Van Stralen, K.J.; Dekker, F.W.; Zoccali, C.; Jager, K.J. Measuring agreement, more complicated than it seems. *Nephron Clin. Pract.* **2012**, *120*, c162–c167. [CrossRef]
62. Peng, B.; Guan, K.; Tang, J.; Ainsworth, E.A.; Asseng, S.; Bernacchi, C.J.; Cooper, M.; DeLucia, E.H.; Elliott, J.; Ewert, F.; et al. Towards a multiscale crop modelling framework for climate change adaptation assessment. *Nat. Plants* **2020**, *6*, 338–348. [CrossRef] [PubMed]

land

MDPI

Article

The Impact of Soil-Improving Cropping Practices on Erosion Rates: A Stakeholder-Oriented Field Experiment Assessment

Ioannis K. Tsanis [1,*], Konstantinos D. Seiradakis [1], Sofia Sarchani [1], Ioanna S. Panagea [2], Dimitrios D. Alexakis [3] and Aristeidis G. Koutroulis [1]

1 School of Chemical and Environmental Engineering, Technical University of Crete, Akrotiri, 73100 Chania, Greece; kostas@hydromech.gr (K.D.S.); sofia.sarchani@hydromech.gr (S.S.); aris@hydromech.gr (A.G.K.)

2 Department of Earth and Environmental Sciences, KU Leuven, Celestijnenlaan 200 E, Box 2411, 3001 Leuven, Belgium; ioanna.panagea@kuleuven.be

3 Lab of Geophysical—Satellite Remote Sensing and Archaeo-Environment, Institute for Mediterranean Studies, Foundation for Research and Technology—Hellas, 74100 Rethymno, Greece; dalexakis@ims.forth.gr

* Correspondence: tsanis@hydromech.gr; Tel.: +30-28210-37799

Abstract: The risk of erosion is particularly high in Mediterranean areas, especially in areas that are subject to a not so effective agricultural management–or with some omissions–, land abandonment or wildfires. Soils on Crete are under imminent threat of desertification, characterized by loss of vegetation, water erosion, and subsequently, loss of soil. Several large-scale studies have estimated average soil erosion on the island between 6 and 8 Mg/ha/year, but more localized investigations assess soil losses one order of magnitude higher. An experiment initiated in 2017, under the framework of the SoilCare H2020 EU project, aimed to evaluate the effect of different management practices on the soil erosion. The experiment was set up in control versus treatment experimental design including different sets of treatments, targeting the most important cultivations on Crete (olive orchards, vineyards, fruit orchards). The minimum-to-no tillage practice was adopted as an erosion mitigation practice for the olive orchard study site, while for the vineyard site, the cover crop practice was used. For the fruit orchard field, the crop-type change procedure (orange to avocado) was used. The experiment demonstrated that soil-improving cropping techniques have an important impact on soil erosion, and as a result, on soil water conservation that is of primary importance, especially for the Mediterranean dry regions. The demonstration of the findings is of practical use to most stakeholders, especially those that live and work with the local land.

Keywords: soil erosion; soil-improving crop systems; sustainable land management; sustainable agriculture

Citation: Tsanis, I.K.; Seiradakis, K.D.; Sarchani, S.; Panagea, I.S.; Alexakis, D.D.; Koutroulis, A.G. The Impact of Soil-Improving Cropping Practices on Erosion Rates: A Stakeholder-Oriented Field Experiment Assessment. *Land* **2021**, *10*, 964. https://doi.org/10.3390/land10090964

Academic Editors: Guido Wyseure, Julián Cuevas González and Jean Poesen

Received: 3 August 2021
Accepted: 8 September 2021
Published: 12 September 2021

1. Introduction

Soil erosion is a primary biophysical process involving the detachment of soil particles from a given initial area and their transport and accumulation to a new depositional area [1]. It is considered one of the most severe natural threats worldwide, as it threatens soil fertility, water availability and crop productivity [2]. The risk of erosion is particularly high in Mediterranean areas, especially in areas that are subject to a not so effective agricultural management–or with some omissions–, land abandonment or wildfires [3].

Crete's Mediterranean soils are under imminent threat of desertification, characterized by loss of vegetation, water erosion, and subsequently, loss of soil. In particular, the serious impact of the expected climate change to the southern Mediterranean regions, together with the adoption of crop techniques by many olive groves' farmers that negatively affect the environment, such as intensive tillage, use of chemicals, burning of pruning branches in their fields, may lead to loss of ground's organic matter, putting the fields in possible drought hazard in the upcoming years [4,5]. Several studies have focused on the estimation of average soil erosion in the island. Most of the studies simulate erosion with the use of the revised universal soil loss equation (RUSLE) method. Kazamias et al. [6], for example, estimated average soil erosion rates for the Greek territory at 4.75 Mg ha^{-1} yr^{-1} and indicated that for over 12% of the Greek area higher ratings than 10 Mg ha^{-1} yr^{-1} are observed, mainly located at steep areas. Similar values are estimated for the Cyprus area (11.75 Mg ha^{-1} yr^{-1}) with October and November being the most erosive months [7]. Panagos et al. [8] estimated soil erosion rates between 6 and 8 Mg ha^{-1} yr^{-1} for the region of Crete and Kourgialas et al. [9] suggested ratings of an average 4.85 Mg ha^{-1} yr^{-1} for the western part of the island. Several other investigations assess soil losses as being orders of magnitude higher. Kouli et al. [10] provided an estimate of soil loss of up to 200 Mg ha^{-1} yr^{-1}, and Alexakis et al. [11] suggested that losses of more than 200 Mg ha^{-1} yr^{-1} were recorded for 2018. Polykretis et al. [12] assessed the intra-annual and inter-annual fluctuations, providing similar estimates. Furthermore, changes in rainfall patterns are foreseen to affect soil erosion in the area. Grillakis et al. [13] examined projected changes in erosivity for the island of Crete under three concentration pathways (RCP2.6, RCP4.5 and RCP8.5). Simulations suggest positive changes exceeding 30% for the 2021–2050 period, while for the far future, erosivity decreased with the increase in concentration, ranging from −10% to +30% on average, depending on the scenario and as a result of changes in extremes [14].

Despite the extensive literature devoted in the investigation of soil erosion at the regional scale, few studies focus on the local field scale, and in particular, at the major land use types and associated land management practiced. Olives are the most important crop grown on the island of Crete [15], covering 64% of the arable land and representing 86% of the tree plantations on the island. Despite the problem of phyloxera in the 1980s [16] and the Common Agricultural Policy (CAP) to reduce the area of vineyards, viticulture remains one of the most important production activities of Crete. Olive orchards and vineyards in Crete often suffer from extreme soil erosion by water due to farm slope and recent intensification of tillage practices [17–19]. There is a need to find practices that prevent soil erosion without reducing the profitability of both crops. Less tillage at the olive sites can improve soil health by reducing organic matter decline, keeping soil microbiology intact, and limiting compaction through less machine passes across fields, as well as reducing fuel use and related emissions [20]. In addition, the simplest and most natural way to prevent erosion in vineyards is through planting vegetation. Cover crops keep ground covered over storm events with high rain rates and winds, which can cause erosion [21,22]. Plants establish root systems which stabilize the soil and prevent erosion. Moreover, cover crops can reduce the need for fertilizer and supply organic nitrogen if leguminous [23,24].

Average erosion rates for orange groves on the island are estimated at 1 Mg ha^{-1} yr^{-1}, whereasthe average rates are assessed at 8 Mg ha^{-1} yr^{-1}, which is still above other cultivations. Moreover, in the Chania Prefecture of Crete, orange cultivation is a major crop, but due to severe market competition, producer prices have significantly dropped, leaving little or no profit. Recently, avocado plantations have been proposed as a potentially sustainable alternative over orange groves, with high profitability and increasing demand, but soil erosion for avocados has not been measured [25]. The cultivation is demonstrating to fit in warm seasons of Mediterranean regions [26].

The objectives of the present experimental study are: (i) to compare different tillage practices in olive orchards, as tillage is known to affect soil erosion rates; (ii) to test the use of a vetch cover crop in a vineyard compared to no vetch, with vetch being a nitrogen fixing cover crop; (iii) to compare the erosion rates as well as other soil quality parameters between a field that has remained an orange grove for 45 years and one that was converted to an avocado farm 20 years ago.

2. Materials and Methods

2.1. Study Site

The experiment was conducted on three real farm fields managed by farmers in three different areas of Chania, Crete, Greece. The first field is an olive orchard located at Biolea in Astrikas region, at an altitude of about 260 m and covers an area of about 3000 m^2 with a slope gradient of about 6%. The olive trees were planted in a dense of 90 trees per 1000 m^2. The field is 25 years old and had not been tilled for 7 years before the beginning of the project. The second field is a vineyard located in Alikampos region, in an area of about 3000 m^2 and an altitude of about 254 m. The slope gradient of the field is about 15%. The investment began in 2013. Since then, animal manure is the fertilizer applied every two years in the orchard, while moldboard ploughing at 20 cm depth is a standard farm operation. The field is also drip irrigated during the summer period. The third field is a fruit orchard (orange and avocado) located in Koufos region, in an altitude of about 86 m and covers an area of about 2000 m^2 with a slope gradient of around 10–15%. The orange trees were planted in 1988 and the conversion of part of the orange orchard to avocado orchard occurred in 1998. The Avocado trees' first plantation included 40 trees per 1000 m^2, whereas the orange trees' first plantation included 120 trees per 1000 m^2. The fruit orchards received ammonium sulphate and potassium fertilizers during the past 10 years, applied in the irrigation water during the summer period, whereas solid potassium nitrate is banded on the soil surface in the winter period. Every year, soil mulching with cut branches occurs in the form of wood chips. Manure is also applied every year on the avocado trees. Moldboard ploughing at 20 cm depth occurs every two years and glyphosate is banded on the soil surface every year for weed management. Finally, the field is drip irrigated according to the needs of each summer period. The topsoil of all sites has a clay loam texture according to the USDA classification system.

Each experimental site has a representative meteorological station. The closest meteorological station of the olive orchard field is Kolympari, whereas of the vineyards is Vrysses and of the fruit orchards is Alikianos. However, the time period of temperature, precipitation and evapotranspiration observations is not long. Vrysses started gauging in 2007, Alikianos in 2012, and Kolympari in 2016. Table 1 displays the yearly hydrometeorological records of the 2018–2020 period for the stations of interest.

Table 1. Overview of the yearly temperature, precipitation and ET0 for the experiments.

Station	Period/year	Tmax (°C)	Tmin (°C)	Precip (mm)	ET0 (mm)
Vrysses	2018	24	12	759	1304
Kolympari	2018	23	14	704	1129
Vrysses	2019	23	11	1867	1296
Kolympari	2019	23	14	1332	1137
Vrysses	2020	23	11	1454	1306
Kolympari	2020	23	14	667	1155
Alikianos	2020	23	13	1166	1220

Crete has a typical Mediterranean island environment with about 53% of the annual precipitation occurring in the winter, 23% during autumn and 20% during spring, while there is negligible rainfall during summer [27,28]. The average precipitation for a normal year on the island of Crete is approximately 934 mm, with a markedly non-uniform distribution, a reduction of almost 300 mm from the west to the east part of the island and a strong orographic effect. Noticeable are the high rainfall winters and the dry summers in the Chania Prefecture [14].

Regarding the hydrometeorological conditions during the years of the experiment, on 26 October 2017, as well as on 15 and 24 February 2019, Western Crete suffered excessive rainfall and flooding. The October 2017 event was a high-intensity and short-duration rain event, resulting in flash floods in the low-elevation agricultural and urban areas on the northern part of the Chania Prefecture. Persistent storm events in February 2019 resulted in flooding, extensive riverbank erosion, landslides and rocks throughout the road network of Chania Prefecture, as well as in the collapse of the 111-year-old historical Keritis bridge over Alikianos River. For the entire Chania region, 2018 was a dry year followed by an exceptionally wet 2019, mainly due to the record high precipitation accumulations of February (1202 mm/month for Askifou station, Chania), and a normal 2020.

As for relative mean climate conditions between the study sites during the course of the experiment, safe results cannot be extracted due to the distance of the meteorological stations from the sites, differences in altitude and microclimate. In general, the vineyard site located in Alikampos receives the highest amount of mean annual rainfall (~1400 mm) and has the lowest mean temperature (due to lower minimum temperatures at place). The fruit orchard (orange and avocado) is located in probably the most fertile and intensively cultivated valley of Chania prefecture with an average precipitation of about 1200 mm/year, while the olive orchard site located in Astrikas receives less precipitation and has higher mean annual temperature, despite the higher altitude.

2.2. Experimental Setup

The beginning of the experiment was in 2017, involving a control versus treatment (soil-improving cropping system, SICS) experimental design. At the olive orchard field, two treatments about soil cultivation were tested. A normally tilled area, which served as the control plot and tilled twice within SoilCare project (November 2017 and May 2019), was compared with the no-tilled one, which was the SICS plot. The tillage method was moldboard ploughing at 20 cm depth. Two olive varieties were located in the experimental organic farm of 0.29 ha, *Olea europaea* and *Koroneiki*. At the vineyard site, the experiment compared a vetch (*Vicia faba*) cover crop plot with one without a vetch cover crop. The plot with no vetch served as the control area and the other plot, which was tilled and seeded with vetch, served as the SICS area. The grape variety was *Vitis vinifera* and the plots were located on a corporate organic farm of 0.46 ha. All the farms were fully operated and managed by the farmers, and because of practical management issues, no replicate plots could be designed. At the fruit orchard field, the experiment compared an orange orchard area, served as the control plot, with a rotation crop area of avocado trees, served as the treatment (SICS). The orange orchard variety was *Citrus* × *sinensis*, whereas the crop switch variety was *Persea Americana*, and the plots were located on a family conventional farm of 0.5 ha.

Soil loss rate assessments of both the olive orchards and vineyards' fields were undertaken through cross sections' measurements. The total soil loss is estimated by the erosion/deposition (ER) equation:

$$ER = \frac{VOL \times BD}{TA} \quad (1)$$

where VOL is the volume (m^3), BD is the bulk density (kg/m^3), and TA is the total effective area (m^2).

In the olive orchard site, the soil loss rate monitoring occurred from November 2018 to June 2021 (32 months) with two cross sections, one per plot, having lengths of 5 and 5.8 m for the SICS and control plot, respectively. In the vineyard site, soil loss rate was monitored from January 2019 to June 2021 (2.5 years) through six cross sections, three per plot, of lengths ranging from 1.64 to 2.2 m. At the fruit orchards, soil loss rate assessments were undertaken through soil pins' measurements. The soil loss rate monitoring occurred from May 2018 to June 2021 (37 months), with three to four soil pins per plot, placed per 0.5 m. Figure 1 displays the three farm fields in which experiments were conducted.

Figure 1. (**a**) Tilled plot (up) and non-tilled plot (down) at the olive orchard site, (**b**) positions of the cross sections (CS) in which soil erosion measurements were performed at the vineyard site, and (**c**) oranges and avocado trees at the fruit orchard field.

Biophysical measurements were also performed both in the control and SICS plots regarding soil texture [29], saturated hydraulic conductivity [30,31], water stable aggregates [32], bulk density [33], mineral nitrogen [34,35], available phosphorous [36], exchangeable potassium, sodium and magnesium [37], soil organic carbon [38], soil pH [39], soil electrical conductivity [40] and earthworm count [41] in the three experimental fields at the end of October for the years 2019 and 2020.

Soil texture was measured with the Bouyoucos hydrometer method [42]. Saturated hydraulic conductivity was measured with the Beerkan method [43] but not in fully dry conditions as shown in Figure 2a. Soil aggregate stability was counted by sampling about 100 g of three to four soil aggregates from the topsoil per plot, which was air-dried for 20 days and thereafter immersed in water on a mesh of 0.4 cm diameter. The aggregates were observed for a few minutes for slaking [44]. The steps followed to perform the measurement are indicated in Figure 3. Bulk density was measured in a laboratory as an indicator of soil compaction. For its assessment, three soil samples from topsoil (10–20 cm) and three soil samples from subsoil (40–50 cm) per experimental plot (control and SICS) were taken with a metal ring with known volume of 246.42 cm^3. Figure 2b displays the collection of a soil sample for the bulk density measurement. The following procedure concerned, first of all, the weighting of an ovenproof container in which each one soil sample per time was placed on; the soil was dried for one night in a conventional oven at 105 °C and then weighted. The difference between the two weight measurements divided by the soil volume gave the calculation of the bulk density [44]. A mixed soil sample was collected from each experimental plot, with soil from under ten trees using a Z-shape sampling methodology in order to estimate mineral nitrogen, available phosphorous, exchangeable potassium, sodium and magnesium, soil organic carbon (SOC), soil pH, and soil electrical conductivity. All samples collected were air-dried, grounded to pass a 2-mm sieve, and analyzed for selected chemical properties. Concerning the available forms of the nutrients, NO_3^--N was extracted with 1M KCl and determined with spectrophotometry at wavelengths 210 and 270 nm. Olsen P was extracted by 0.5M $NaHCO_3$ with pH 8.5 and was quantitatively determined with molybdenum blue-ascorbic acid method [36] by using Vis-UV spectrophotometry. Exchangeable cations K^+, Na^+, and Mg^{2+} were extracted with 1M ammonium acetate, having pH 7 [45], and were analyzed by the inductively coupled plasma method ICP-OES. pH was measured in a soil/water suspension at a 1:2 ratio; SOC was determined with the wet oxidation method [38], whereas the electrical conductivity was measured in the saturation paste extract [46]. Earthworm density was evaluated as an indicator of the biological health and condition of the soil per experimental plot. The procedure followed was the mixing of 2 L of water with 20 g of mustard seed, the pouring of the half mixed on a 25 cm × 25 cm sample plot where vegetation and leaves were removed, and the observation of worms that came to surface over a period of 5 min, and then the pouring of the remaining mix and the waiting of another 5 min to gather worms that came to the surface. Figure 2c,d indicate the procedure followed for earthworm counting in the olive and vineyard field respectively.

Figure 2. (**a**) Infiltration rate experiment, (**b**) soil sampling for bulk density measurement, (**c**) pouring mix of water and mustard for earthworm test in Astrikas, and (**d**) worms coming out to surface after the pouring of the mix in Alikambos.

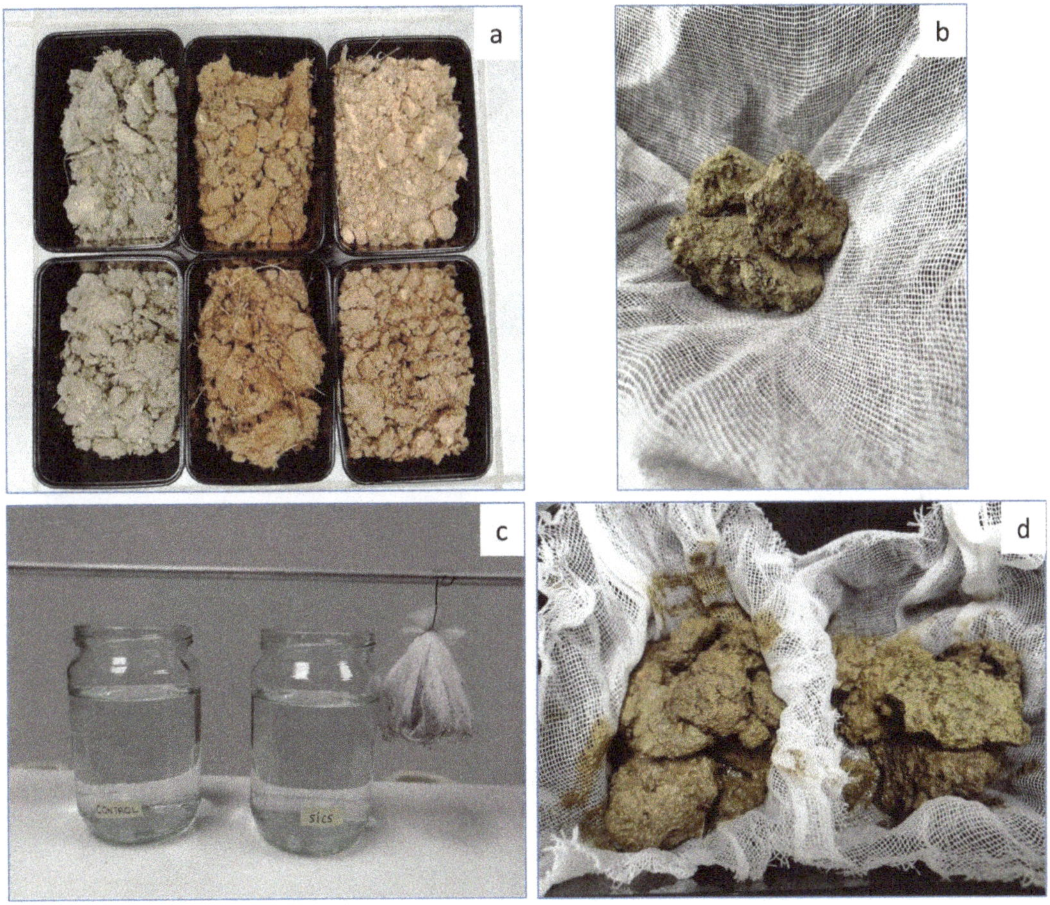

Figure 3. (**a**) Soil aggregates from both plots (control and SICS) of the three study sites: air-drying before slaking test, (**b**) 100 g of three aggregates of the Astrikas field to be used for the slaking test, (**c**) soil aggregates into the mesh before being immersed to water, and (**d**) soil aggregates after the immerse to water.

2.3. Stakeholder Engagement

Throughout the SoilCare project, two stakeholders' workshops and two stakeholders' meetings were held, either with physical presence or virtually. The first workshop occurred on 21 March 2017 at the Technical Chamber of Crete (West Crete Chamber), where 12 persons (4 female and 8 male) participated. The main participants were farmers, agronomists and researchers. The stakeholders introduced themselves and justified their interest in the SoilCare project objectives. They were asked to place themselves on a stakeholder matrix that determined the scale of motivation and perceived influence, graded from low to high. The stakeholders also participated and contributed their experience and knowledge on drivers, barriers and solutions for the soil erosion threat in their area. During the workshop, commonly accepted and applied practices to combat soil erosion, along with their benefits and drawbacks, were discussed for further evaluation and potential outspread to all farmers. The stakeholders were also asked to rank, in a scale from 1 to 6, suggested ways to receive information about SoilCare during the lifetime of the project, as well as the display ways they would receive information regarding new SoilCare practices. The questionnaire they had to fill also concerned their preference on dissemination manners of information and advice of farming practices being used in SoilCare, their three main questions that they would like to be answered when evaluating whether to apply a new practice, as well as the way they would normally find out about new farming practices, beyond the project.

A meeting was held in March 2018 on the premises of Technical University of Crete (TUC), which was attended by 6 people (2 women and 4 men). All the stakeholders were updated about the progress of the field experiments related to soil loss monitoring in the three agricultural fields and the installation of six sediment fences/traps for collecting deposited soil at all the study sites. The stakeholders were asked to evaluate the results of monitoring and soil loss collection thus far. Through a constructive discussion between stakeholders and researchers, it was agreed that the research should focus more on monitoring of soil erosion/deposition implementing additional approaches. In particular, among the following actions of the researchers, there would be the installation of triangular or square grid for monitoring sheet erosion, monitoring of soil roughness with means of images (stereopairs), multi-temporal monitoring and recording of rills in the study areas, as well as correlation of soil organic matter and spectral data (field spectroradiometric measurements) in terms of soil erodibility at a farm and watershed scales.

A research activity was afterward held within the period from April to June 2019 in Chania, which concerned the individual interviewing of 4 stakeholders/farmers (4 men) since it was not possible for them to be gathered in a group. Two of them were involved in the olive orchard experiment, one in the vineyard experiment, and one in the avocado/orange experiment. Toward the specific research activity, the stakeholders had to describe the expected benefits and impacts of the SICS being tested in their field. They also had to identify and describe the key barriers and enablers to SICS adoption. The factors evaluated were economic (farm and market) conditions, biophysical conditions (climate and geomorphology), technical barriers, knowledge and information barrier, and sociocultural factors, as well as institutional or policy regulations. Moreover, the stakeholders were asked to identify and assess feasibility of actions to promote SICS adoption at national and/or (sub)regional level by ranking the enabler and barrier factors from not so important to very important. In addition, they were requested to point out actions that would remove barriers and support enablers as well.

The final stakeholder workshop occurred via online meetings of small groups during various dates in February and March 2021 due to the COVID-19 pandemic restrictions in place. A total of 18 persons (6 female and 12 male) participated and discussed project findings. The main groups were farmers, researchers and agronomists. After the end of the online presentations, the participants, both men and women, and especially the farmers, raised useful questions. Indicatively, the olive groves' farmers wanted clarifications regarding tillage avoidance, especially in the dry season. The vineyard farmers showed particular interest in the application of the experiments. The orange cultivators were interested in understanding the way of further improving the biological health and condition of soil on avocado trees. An important question raised was whether an avocado market will exist for the trees currently planted which will be placed into production in five years. Afterward, the attendees validated whether the project findings were plausible and/or consistent with their understanding. They also identified the benefits that they gained from SoilCare already, as well as the ones that they found important for the future from the project findings. The stakeholders were requested as well to have an active role and state the way they can disseminate the project findings to more people who can benefit from them. The engagement of the stakeholders continued as they were asked to report the way that they would like to be supported in using or implementing project/research findings. Toward the end of the workshop, the attendees mentioned what impressed them most and how to implement what they learned from the workshop.

3. Results

3.1. Impacts on Soil Erosion

3.1.1. Olive Orchard Site

Concerning the sediment fences, although part of the study area was tilled, minimum difference was evident in the collected deposited soil between tilled and no-tilled area. This was considered as a common phenomenon as only few-deposited soil may be collected during winter when it rains more often, due to the presence of naturally grown winter cover crops retaining rainwater. Nevertheless, intensified tillage, which occurred twice in 18 months, contributed to increased soil erosion, as visually observed by the exposed rooting system. Apart from tillage, irrigation also increased soil erosion with the irrigated trees showing shallow roots accompanied with topsoil erosion. Regarding the cross-sectional soil loss measurements, results showed that the no-till treatment had a considerable impact on soil erosion rates. Soil loss rate monitoring revealed that the application of no-till treatment reduced mean soil erosion by over 14%, roughly from 3.3 to 2.9 Mg/ha during the 2.5 years experiment (November 2018 to June 2021).

3.1.2. Vineyard Site

Extreme storm events occurred on 15 February 2019 and 24 February 2019. The nearby rain station recorded the exceptional accumulation of 726.2 mm during this period. These events created rills in the examined field. In the vetch plot, the rills were shorter compared to the no vetch plot when compared visually. The application of the vetch treatment had a direct impact on soil erosion over the 2.5-year monitoring period (January 2019 to June 2021). Soil loss rate monitoring revealed that the vetch coverage reduced mean soil erosion by over 12% (roughly from 3.4 Mg/ha in the no vetch plot to 3 Mg/ha in the vetch plot) during the 2.5 years experiment.

3.1.3. Fruit Orchard Field

An extreme rainfall event occurred on 26 October 2017, leading to more than 2 kg of soil trapped in the sediment fences of 3 m^2 area, corresponding to about 7 Mg/ha. In the rest of the monitored period, the sediment traps did not collect considerable amounts of soil after events of light rain. Further extreme precipitation events, which caused severe flooding in the wider area, occurred in February 2019, triggering further erosion in the field. Field measurements showed that the crop switch to avocado trees significantly reduced the mean soil erosion compared to the orange orchards (control) over 3 years of monitoring (May 2018 to June 2021). Soil loss rate monitoring revealed that the avocado conversion caused over 34% reduce in mean soil erosion, roughly from 4.6 to 3 Mg/ha, during the 3-year experiment.

A one-way ANOVA was performed to compare the effect of the different treatments on the erosion rates, with 90% CI. The analysis revealed that there was a statistically significant difference in the erosion rates between the avocado and orange treatments ($p = 0.096$) (Figure 4iii) but not for the other treatments ($p = 0.745$ for the olive orchard, $p = 0.561$ for the vineyard). Figure 4i,ii indicate the mean soil erosion at the control and SICS plot of the olive orchard and vineyard site respectively after 2.5-year of monitoring through cross sections.

3.2. Impacts on Soil Properties

3.2.1. Olive Orchard Site

Topsoil bulk density was slightly higher in the no-till plot. Bottom soil bulk density was found at the same levels in both plots. Exchangeable magnesium had an increasing trend in both plots from 2018 to 2020. Mineral nitrogen and available phosphorus concentrations were lower in the no-till plots, both in 2019 and 2020. The soil organic carbon rate had an increasing trend in both plots from 2018 to 2020, and was slightly higher at the last year, which was probably due to the animal manure application. The crop yield was the same at both plots (till and no-till) and was increased in 2020 compared to the years

2018 and 2019. Earthworms' density per m^2, which can be used as a sensitive indicator to management changes were substantially higher in the non-tilled plot compared to the tilled one in the 2020 measurement, indicating better soil health and condition. Weed infestation was slightly higher (10%) in the non-tilled plot compared to the tilled one, which cannot be assumed as a considerable high hazard.

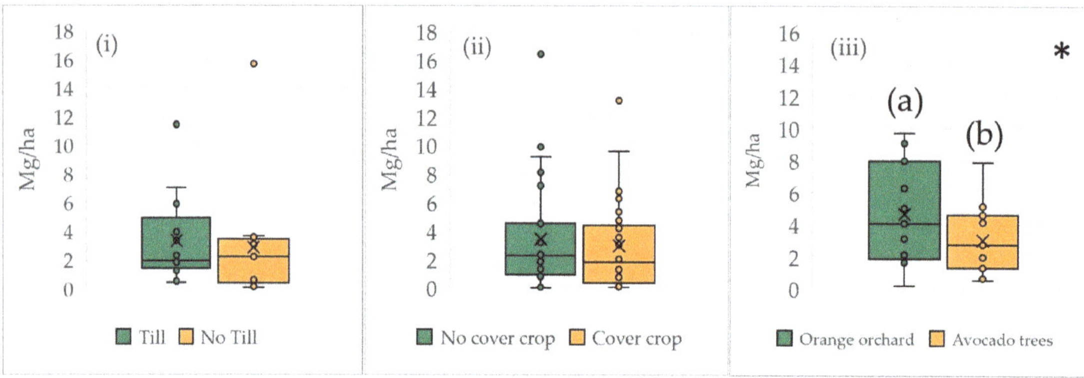

Figure 4. Mean soil erosion (Mg/ha) in **(i)** tilled and non-tilled plot at the olive orchard site during the 2.5-year of monitoring, **(ii)** no vetch and vetch plot at the vineyard site during the 2.5-year of monitoring, and **(iii)** orange and avocado plot at the fruit orchard field during the 3-year monitoring (* denotes significant differences at a 90% CI level).

3.2.2. Vineyard Site

By the end of 2020, top- and bottom soil bulk densities of the vetch plot were lower compared to the no vetch plot, indicating good soil functioning, improved water and solute movement as well as soil aeration. A soil aggregate stability test resulted in good soil stability and resistance to erosion for both plots; however, for the vetch applied plot, slaking effect was slightly less observed, indicating better structure maintenance. The soil organic carbon did not follow a specific trend; it was relatively satisfactory around 4% in both plots (control and SICS) from 2018 to 2020. The crop yield was the same at both plots (control and SICS), having a slightly decreasing trend during the 3-year monitoring. Earthworms per m^2, which is a soil health indicator, were considerably higher in the vineyard with the cover crop applied. The percentage of weed infestation was 20% less in the vineyard with the cover crop.

3.2.3. Fruit Orchard Field

The soil organic carbon rate was higher in the avocado trees compared to the orange orchards. The saturated hydraulic conductivity was considerably higher in the avocado trees plot compared to the orange trees plot, in the 2020 measurement. The exchangeable magnesium was also higher in the avocado trees compared to the orange orchards during the 3-year monitoring. The level of weed infestation was 10% less in the avocado field compared to the orange trees field. Electric conductivity values indicated high salinity levels in both plots, while higher values were observed for avocado trees.

Due to the lack of replicates, there was no efficient way to place error bars on the graphical values of soil properties. Nevertheless, the repetition of measurements every year demonstrated an important variation in properties' values, as shown in Figure 5 regarding the 2.5-year of monitoring of the olive orchard site, Figure 6 concerning the 2.5-year of monitoring of the vineyard site, and Figure 7, which concerned the soil properties after 3-year monitoring of the fruit orchard field.

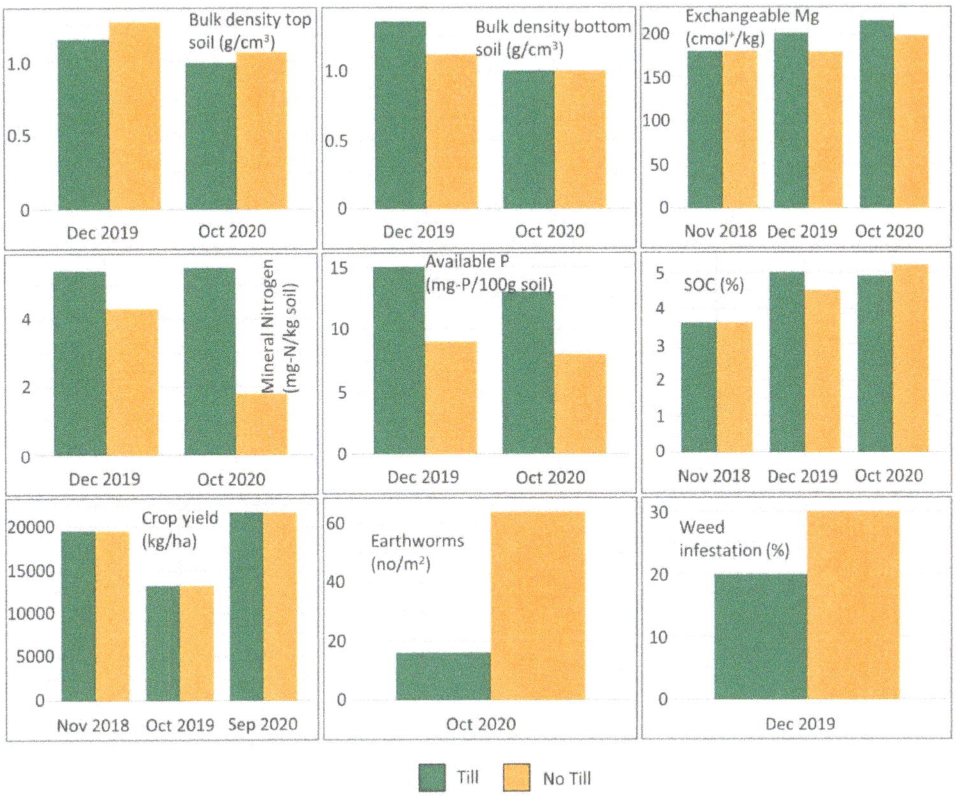

Figure 5. Soil properties in tilled and non-tilled plot at the olive orchard site during the 2.5-year monitoring.

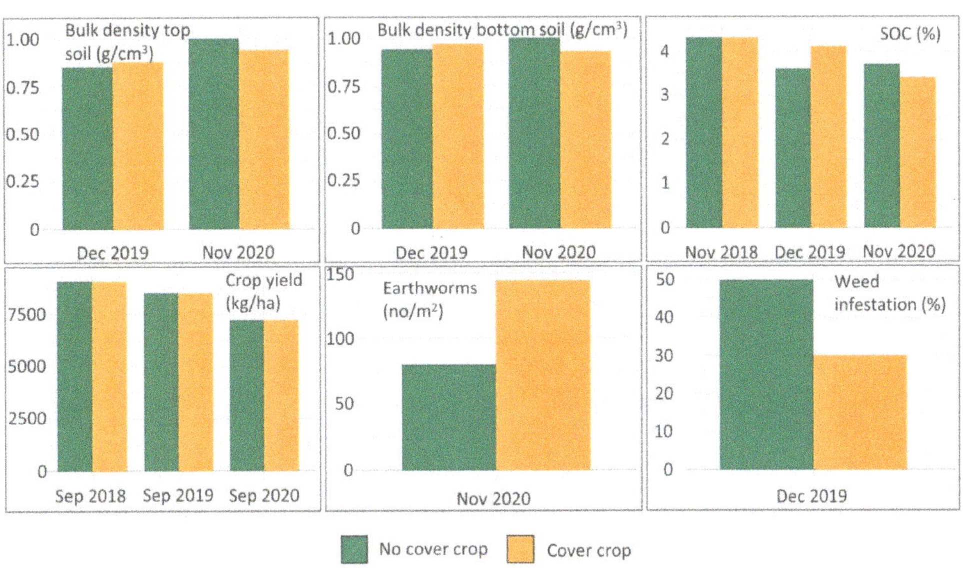

Figure 6. Soil properties in no vetch and vetch plot at the vineyard site during the 2.5-year monitoring.

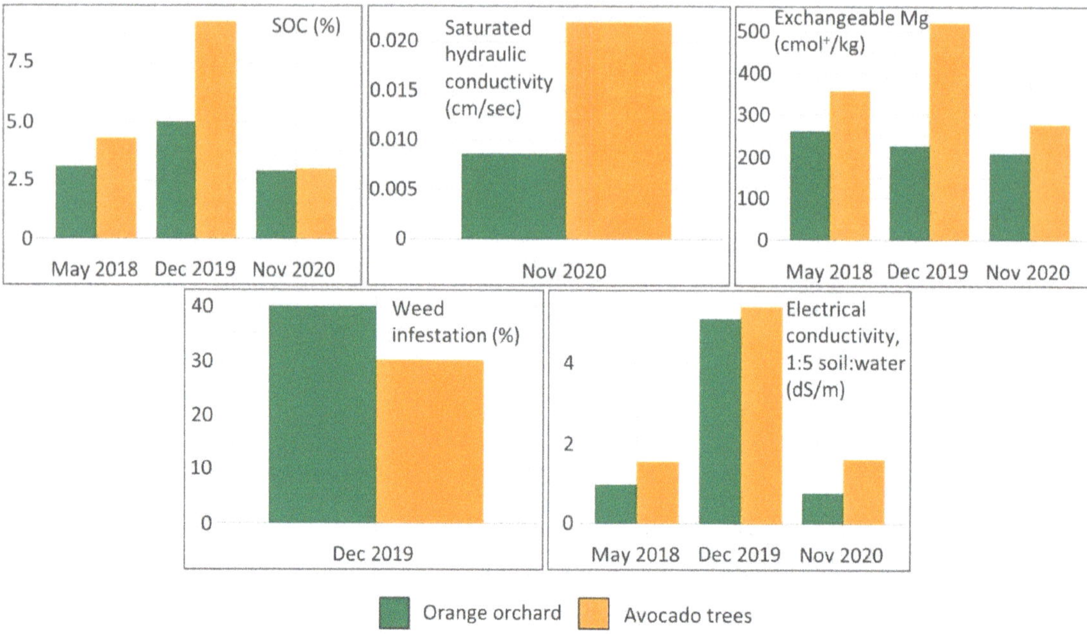

Figure 7. Soil properties in orange and avocado plot at the fruit orchard field during the 3-year monitoring.

3.3. Assessment of Stakeholder Engagement

3.3.1. First Stakeholder Workshop (21 March 2017)

Local stakeholders underlined that soil erosion mainly depends on geomorphology (slope), soil type, vegetation cover, climate, socio-economic and policy drivers, including human activities (land management, soil conservation techniques). They punctuated that soil erosion causes soil infertility, resulting in reduced production and thus income. Land users and agronomists showed the effects and the indicators of increased soil erosion on their cultivations, such as exposed tree roots, rills, reduced soil organic matter, soil pillars. They also seemed aware of the functions and services offered by the soil, as well as the impact of losing them. A few of the stakeholders perceived the soil functions and services as were presented in the meeting, but with high uncertainty on whether the knowledge will be consolidated. However, the meeting presentation was essential to discuss limits and effects of soil erosion before beginning a meaningful conversation at the local scale. The discussion with the stakeholders highlighted knowledge gaps of the interested parts regarding the extent that the erosion affects the performance and quality of production, the extent that different cultivation practices within the same crop or a total crop change affect the rate of soil erosion, and the most promising erosion mitigation approaches and technologies as well. Major barriers in adopting land management practices include the local administration lack of partaking in decision making, yet can implement and control what is already decided; agronomists perceive lack of interest of farmers (typically those with lower education level), including in regular testing of soil quality of their fields, being reactive rather than proactive; farmers' cooperatives often realize that competitive farmers are noncommitted to the cooperatives' objectives, whereas outsiders perceive cooperatives as less sustainable and profit oriented, or simply disagree with how resources are allocated; there is lack of financial motives (or motive awareness) for stakeholders; educational institutions lack the legislative freedom and means to interact with stakeholders for pilot/prototype practices for soil and land management. The stakeholders considered that government, local municipalities and individual farmers were responsible for providing solutions.

3.3.2. Research Activity (April to June 2019)

Based on the evaluation of the questionnaires, the olive farmers stated that the money saved from the no tillage practice is counterbalanced by the weed spraying costs or the time of cutting them off. The climate of Crete acts as an enabler for olives, whereas steep slopes, stones and rocks may significantly affect the crop. Drought may further raise the rooting system of the crop higher, affecting mostly the tilled plots. The vineyard farmer expected the yield to be increased through the cover crop. However, because of no directly visible benefits, the farmer, although willing to adopt the practice, stated the need for long-term experiments such that the benefits are proven and quantified. Yet, the presence of different crop within the vineyard competes with the soil structure in dry years. One farmer also noted that there is sufficient access to knowledge and information, but there is difficulty in convincing to test new alternative cultivation methods; it is preferable to recognize the benefits from other farmers and then adopt the new techniques. The avocado yield is expected to be profitable after the fifth year of the crop change, since the investment costs of the crop change process are high. However, there is a high interest for this investment since the climate conditions on Crete seem identical for the avocado crop. In addition, the avocado demand from European markets is expected to increase. The farmer asked for additional knowledge and advice from an experienced agronomist and identically visiting of demonstration sites.

The farmers claimed the need for guidance and advisory services of great expertise on soil data, soil analysis, fertilizers′ use, as well as extra skills to the specific SICS practices to remove barriers. They also indicated the necessity for financial support and incentives to adopt SICS practices. In addition, they suggested that the organization of workshops could support enablers with successful studies and practical applications that would emphasize the pros and cons of the proposed SICS. All farmers agreed that other farmers with experience in the specific SICS practices can provide important information to them. Social networks and videos can also help farmers adopt new techniques. New cultivation practices according to EU regulations can also be promoted from the State through seminars and programs.

3.3.3. Final Stakeholder Workshop (February and March 2021)

After the end of the online presentations, the soil specialists/consultants and the researchers had well understood the project results in the three fields of its application. The farmers generally found the presentations helpful in understanding the conceptualization of the problems faced in the three field studies, as well as the results obtained. They found the vetch cover crop easy to be applied, and the no-tillage practice feasible. Several of the orange cultivators realized that switching crops to avocados would bring them a great financial profit in long-term, while at the same time soil erosion would be reduced in their fields.

Regarding the benefits gained from SoilCare thus far, all the farmers stated that they gained better knowledge of their fields and the soil properties and functions as well as of the soil erosion's negative impacts and the way these can be avoided. The consultants noted the necessity to inform farmers of soil improving techniques, as well as the requirement for proper training in applying these techniques correctly. Concerning the benefits of the project findings for the future, farmers were willing to apply the proposed SICS practices to new sites or to the rest of the parts of their fields already tested. Several were interested in examining the tests which concerned the soil properties. The consultants were motivated to use the results and present them in workshops and other organized events aimed at farmers. Certain researchers were interested in monitoring the study fields for another 2–3 years, with the agreement of the farm owners, to examine if soil erosion continued to decrease in the SICS plot and at what rate.

About the dissemination of the project findings to more people who can benefit, the farmers suggested that they may share the results with other farmers either through the cooperative or through discussion with nearby growers. Another suggestion was the co-organization of events with local organizations, municipalities, or farmer cooperatives at the local level. Others proposed the local media. The consultants offered to organize training events for farmers in order to strengthen their skills on innovative soil improving mechanisms. Among the suggestions of the researchers for dissemination were informative brochures and workshops about the findings, in situ exhibitions of SoilCare case studies in Crete, video demonstration of SICS solutions, as well as guidance documents about new soil practices addressed both to farmers and agronomists.

As regards to the way the stakeholders would like to be supported in using or implementing project findings, the farmers seek subsidies for new machinery, seeds, including to avoid loans in the case of avocado investment. Several look for policy opportunities, guidelines for crop change, further development of the agricultural associations, and consulting services as well. The consultants seek additional seminars organized by government agencies on the way that they should train the farmers about the benefits of SICS. The researchers look for project funding for new SICS mechanisms and involvement of both farmers and stakeholders.

4. Discussion and Conclusions

Soil-improving cropping systems (SICS) application seems to play an alleviating role in soil loss processes, therefore it is recommended that farmers be properly informed about the tested practices within their fields.

No tillage practice is substantially beneficial for controlling soil erosion (over 14%), improving soil health and keeping good soil structure. Olive farmers should consider reducing tillage practices in olive orchards, control the tillage depth, and at the same time, limit its application especially during severe drought periods. In addition, the biological health and condition of the no-tilled plots were clearly better compared to the tilled ones. Water and solute movement as well as soil aeration were appropriate, including in the case of no-tillage. Weed management is a deterrent factor for this practice.

Vetch application is an inexpensive solution and is recommended to control soil erosion. The correct application of cover crop is a determinant in improving soil quality. Specifically, the biological health and condition of the vetch cover plots were clearly better compared to the no vetch. Furthermore, water and solute movement as well as soil aeration were slightly improved in the case of cover crop application.

Avocado farms, besides having significantly higher financial benefits, can also maintain a comparably overall good soil quality. However, the earthworm density experiment displayed that the biological health, and condition of the avocado plot were inferior to the orange tree plots. Conversely, water and solute movement as well as soil aeration were in good status for both cultivations, as identified by the top and bottom soil bulk density experiments. Moreover, the application of regular manuring resulted in higher values of SOC in the avocado field. The improvement of soil quality and structure through the increase of SOC as well as the control of soil erosion was additionally achieved by the organic material that accumulates from the intense foliage of avocados trees, resulting in a thick layer of organic material on the ground.

Different trends were found for erosion/deposition for the various cultivations and different treatment methods; however, the only statistically significant results were obtained for the orange to avocado treatment. In the olive tree cultivation, average erosion/deposition for the SICS plot with no tillage was 2.86 Mg/ha, ranging from 0.07 to 15.66 Mg/ha, depending on the observation periods. The control applied in this site was tillage twice in the period of study (November 2017 and May 2019). Tillage resulted to increased spread and mean erosion/deposition values of 3.33 Mg/ha (ranging from 0.43 to 11.42 Mg/ha). The application of crop cover treatment had a direct impact on mean and spread of soil erosion/deposition in the vineyard cultivation. The application of vetch treatment reduced mean erosion by 13% (reduced from 3.41 Mg/ha in the control to 2.98 Mg/ha in the SICS plot). A similar trend was found for the crop change experiment. Mean soil erosion/deposition reduced by over 34% (reduced from 4.66 Mg/ha in the orange to 3.04 Mg/ha in the avocado plot) by changing orange to avocado trees according to the measurements. The spread of the magnitude of these processes was also reduced. Comparing the three different experiments, lower values of soil erosion/deposition were obtained for olive orchards, and this may be due to the lower slope of the plot (~6%). Regarding vineyard plot, although the study site is located in an area with a slope of ~15%, the small sediment losses can be mainly attributed to the higher concentration of silt and clay and low sand content of the soil. The sandy soil (55.3% sand) of the orange and avocado plot may be the main reason of higher erosion/deposition values observed. An additional reason for not monitoring significant differences in the two experimental fields may also be the short-term application of the SICS treatments, whereas the conversion of orange to avocado happened already in 1998 and thus can be considered as a long-term experiment.

This experiment demonstrates that soil improving cropping techniques have a significant impact on soil erosion, and as a result, on soil water conservation, which is of primary importance, especially for the Mediterranean dry regions. As reported in other studies, tillage erosion is considered to be one of the most important processes of land degradation in cultivated areas. The effect of tillage in soil erosion was also recorded during the SoilCare experiment, including for the minimum tillage practice that was applied as a soil-improving cropping method. Results of the study also show that crop cover treatment (vetch) and crop type change have a substantial impact on soil erosion/deposition (12% to 34% lower, respectively). The proposed sustainable soil improving practices are already being applied in many parts of the region. In particular, the change in procedure from orange to avocado trees has been adopted by many farmers as a response to the reduced orange prices and the high income from avocado cultivation. These results highlight the crucial role of soil-improving cropping systems for sustainable land management.

Author Contributions: Conceptualization and methodology, I.K.T., K.D.S., S.S., I.S.P., D.D.A. and A.G.K.; validation, S.S. and A.G.K.; formal analysis, S.S. and A.G.K.; data curation, K.D.S.; writing—original draft preparation, S.S. and A.G.K.; writing—review and editing, I.K.T., K.D.S., S.S., I.S.P., D.D.A. and A.G.K.; visualization, S.S. and A.G.K.; supervision, I.K.T.; funding acquisition, I.K.T. All authors have read and agreed to the published version of the manuscript.

Funding: The research leading to these results has received funding from the H2020 under grant agreement n° 677407 (SOILCARE).

Institutional Review Board Statement: Not applicable.

Informed Consent Statement: Not applicable.

Data Availability Statement: Data sharing is not applicable to this article.

Acknowledgments: The authors would like to thank Ioannis Daliakopoulos and Anthi-Eirini Vozinaki for their support.

Conflicts of Interest: The authors declare no conflict of interest.

References

1. Gelagay, H.S.; Minale, A.S. Soil loss estimation using GIS and Remote sensing techniques: A case of Koga watershed, Northwestern Ethiopia. *Int. Soil Water Conserv. Res.* **2016**, *4*, 126–136. [CrossRef]
2. Rahman, M.R.; Shi, Z.H.; Chongfa, C. Soil erosion hazard evaluation—An integrated use of remote sensing, GIS and statistical approaches with biophysical parameters towards management strategies. *Ecol. Modell.* **2009**, *220*, 1724–1734. [CrossRef]
3. García-Ruiz, J.M.; Nadal-Romero, E.; Lana-Renault, N.; Beguería, S. Erosion in Mediterranean landscapes: Changes and future challenges. *Geomorphology* **2013**, *198*, 20–36. [CrossRef]
4. Michalopoulos, G.; Kasapi, K.A.; Koubouris, G.; Psarras, G.; Arampatzis, G.; Hatzigiannakis, E.; Kavvadias, V.; Xiloyannis, C.; Montanaro, G.; Malliaraki, S.; et al. Adaptation of Mediterranean Olive Groves to Climate Change through Sustainable Cultivation Practices. *Climate* **2020**, *8*, 54. [CrossRef]

5. Kavvadias, V.; Papadopoulou, M.; Vavoulidou, E.; Theocharopoulos, S.; Koubouris, G.; Psarras, G.; Manolaraki, C.; Giakoumaki, G.; Vasiliadis, A. Effect of sustainable management of olive tree residues on soil fertility in irrigated and rain-fed olive orchards. *J. Water Clim. Chang.* **2018**, *9*, 764–774. [CrossRef]
6. Kazamias, A.P.; Sapountzis, M. Spatial and temporal assessment of potential soil erosion over Greece. *Eur. Water* **2017**, *59*, 315–321.
7. Karydas, C.G.; Panagos, P. Modelling monthly soil losses and sediment yields in Cyprus. *Int. J. Dig. Earth* **2016**, *9*, 766–787. [CrossRef]
8. Panagos, P.; Christos, K.; Cristiano, B.; Ioannis, G. Seasonal monitoring of soil erosion at regional scale: An application of the G2 model in Crete focusing on agricultural land uses. *Int. J. Appl. Earth Obs. Geoinf.* **2014**, *27*, 147–155. [CrossRef]
9. Kourgialas, N.N.; Koubouris, G.C.; Karatzas, G.P.; Metzidakis, I. Assessing water erosion in Mediterranean tree crops using GIS techniques and field measurements: The effect of climate change. *Nat. Hazards* **2016**, *83*, 65–81. [CrossRef]
10. Kouli, M.; Soupios, P.; Vallianatos, F. Soil erosion prediction using the Revised Universal Soil Loss Equation (RUSLE) in a GIS framework, Chania, Northwestern Crete, Greece. *Environ. Geol.* **2008**, *57*, 483–497. [CrossRef]
11. Alexakis, D.D.; Tapoglou, E.; Vozinaki, A.-E.K.; Tsanis, I.K. Integrated Use of Satellite Remote Sensing, Artificial Neural Networks, Field Spectroscopy, and GIS in Estimating Crucial Soil Parameters in Terms of Soil Erosion. *Remote Sens.* **2019**, *11*, 1106. [CrossRef]
12. Polykretis, C.; Alexakis, D.D.; Grillakis, M.G.; Manoudakis, S. Assessment of Intra-Annual and Inter-Annual Variabilities of Soil Erosion in Crete Island (Greece) by Incorporating the Dynamic "Nature" of R and C-Factors in RUSLE Modeling. *Remote Sens.* **2020**, *12*, 2439. [CrossRef]
13. Grillakis, M.G.; Polykretis, C.; Alexakis, D.D. Past and projected climate change impacts on rainfall erosivity: Advancing our knowledge for the eastern Mediterranean island of Crete. *CATENA* **2020**, *193*, 104625. [CrossRef]
14. Tsanis, I.K.; Koutroulis, A.G.; Daliakopoulos, I.N.; Jacob, D. Severe climate-induced water shortage and extremes in Crete. *Clim. Chang.* **2011**, *106*, 667–677. [CrossRef]
15. Chartzoulakis, K.S.; Paranychianakis, N.V.; Angelakis, A.N. Water resources management in the Island of Crete, Greece, with emphasis on the agricultural use. *Water Policy* **2001**, *3*, 193–205. [CrossRef]
16. Tzortzakakis, E.A. On the occurrence of Xiphinema index Thorne et Allen in grapevine areas of the Heraklion province, Crete, Greece. *Nematol. Mediterr.* **2012**, *40*, 67–68.
17. Karydas, C.G.; Sekuloska, T.; Silleos, G.N. Quantification and site-specification of the support practice factor when mapping soil erosion risk associated with olive plantations in the Mediterranean island of Crete. *Environ. Monit. Assess.* **2009**, *149*, 19–28. [CrossRef] [PubMed]
18. Arnaez, J.; Lasanta, T.; Ruiz-Flaño, P.; Ortigosa, L. Factors affecting runoff and erosion under simulated rainfall in Mediterranean vineyards. *Soil Tillage Res.* **2007**, *93*, 324–334. [CrossRef]
19. Rodrigo-Comino, J.; Davis, J.; Keesstra, S.D.; Cerdà, A. Updated Measurements in Vineyards Improves Accuracy of Soil Erosion Rates. *Agron. J.* **2018**, *110*, 411–417. [CrossRef]
20. Kairis, O.; Karavitis, C.; Kounalaki, A.; Salvati, L.; Kosmas, C. The effect of land management practices on soil erosion and land desertification in an olive grove. *Soil Use Manag.* **2013**, *29*, 597–606. [CrossRef]
21. Gómez, J.A.; Llewellyn, C.; Basch, G.; Sutton, P.B.; Dyson, J.S.; Jones, C.A. The effects of cover crops and conventional tillage on soil and runoff loss in vineyards and olive groves in several Mediterranean countries. *Soil Use Manag.* **2011**, *27*, 502–514. [CrossRef]
22. Schütte, R.; Plaas, E.; Gómez, J.A.; Guzmán, G. Profitability of erosion control with cover crops in European vineyards under consideration of environmental costs. *Environ. Dev.* **2020**, *35*, 100521. [CrossRef]
23. López-Vicente, M.; Calvo-Seas, E.; Álvarez, S.; Cerdà, A. Effectiveness of Cover Crops to Reduce Loss of Soil Organic Matter in a Rainfed Vineyard. *Land* **2020**, *9*, 230. [CrossRef]
24. Novara, A.; Gristina, L.; Guaitoli, F.; Santoro, A.; Cerdà, A. Managing soil nitrate with cover crops and buffer strips in Sicilian vineyards. *Solid Earth* **2013**, *4*, 255–262. [CrossRef]
25. Ragkos, A.; Papoutsi, G.; Bardounioti, M. Why invest on innovative production? A qualitative evaluation of the emerging avocado sector in Crete, Greece. In Proccedings of the X International Agriculture Symposium, Agrosym 2019, Jahorina, Bosnia and Herzegovina, 3–6 October 2019; pp. 1686–1691.
26. Kourgialas, N.N.; Dokou, Z. Water management and salinity adaptation approaches of Avocado trees: A review for hot-summer Mediterranean climate. *Agric. Water Manag.* **2021**, *252*, 106923. [CrossRef]
27. Koutroulis, A.G.; Tsanis, I.K.; Daliakopoulos, I.N.; Jacob, D. Impact of climate change on water resources status: A case study for Crete Island, Greece. *J. Hydrol.* **2013**, *479*, 146–158. [CrossRef]
28. Koutroulis, A.G.; Vrohidou, A.-E.K.; Tsanis, I.K. Spatiotemporal characteristics of meteorological drought for the Island of Crete. *J. Hydrometeorol.* **2011**, *12*, 206–226. [CrossRef]
29. Staff, S.S.D. *Soil Survey Manual. Agriculture Handbook No. 18*; United States Department of Agriculture: Washington, DC, USA, 2017.
30. Bagarello, V.; Di Prima, S.; Iovino, M.; Provenzano, G. Estimating field-saturated soil hydraulic conductivity by a simplified Beerkan infiltration experiment. *Hydrol. Process.* **2014**, *28*, 1095–1103. [CrossRef]

31. Braud, I.; De Condappa, D.; Soria, J.M.; Haverkamp, R.; Angulo-Jaramillo, R.; Galle, S.; Vauclin, M. Use of scaled forms of the infiltration equation for the estimation of unsaturated soil hydraulic properties (the Beerkan method). *Eur. J. Soil Sci.* **2005**, *56*, 361–374. [CrossRef]
32. Emerson, W.W. A classification of soil aggregates based on their coherence in water. *Soil Res.* **1967**, *5*, 47–57. [CrossRef]
33. Burt, R. *Soil Survey Staff. Soil Survey Laboratory Methods Manual. Soil Survey Investigations Report 42, Version 5.0*; United States Department of Agriculture—Natural Resources Conservation Service: Washington, DC, USA, 2014.
34. Janssen, H.-I.; Koopmann, R. *Determination of Ammonium and Nitrate in Soil, Biowaste and Sewage Sludge.* European Standard. 2005. Available online: https://horizontal.ecn.nl/docs/society/horizontal/STD6162_NH4-N.pdf (accessed on 23 July 2021).
35. Bremner, J.M.; Mulvaney, C.S. Total Nitrogen Methods of Soil Analysis. Part 2. Chemical and Microbiological Properties. *Ed. CA Black. Amer. Soc. Agron. Inc. Pub. Agron. Ser.* **1982**, *9*, 595–624.
36. Olsen, S.R.; Sommers, L.E. *Phosphorus Methods of Soil Analysis. Part. 2. Chemical and Microbiological Properties*; American Society of Agronomy Inc.: Madison, WI, USA, 1982; pp. 403–430.
37. Sager, M. A simplified extraction schema to for the analytical characterization of apple orchard soils. *J. Soils Sedim.* **2016**, *16*, 1193–1202. [CrossRef]
38. Wakley, H.; Black, I.A. An examination of the method for determining soil organic matter and a proposed modification of the chromic acid method. *Soil Sci.* **1934**, *37*, 29–38. [CrossRef]
39. Reeuwijk, L. *Procedures for Soil Analysis*, 5th ed.; ISRIC: Wageningen, The Netherlands, 1995; ISBN 9789066720527.
40. He, Y.; DeSutter, T.; Hopkins, D.; Jia, X.; Wysocki, D.A. Predicting ECe of the saturated paste extract from value of EC1: 5. *Can. J. Soil Sci.* **2013**, *93*, 585–594. [CrossRef]
41. Valckx, J.; Govers, G.; Hermy, M.; Muys, B. Optimizing earthworm sampling in ecosystems. In *Biology of Earthworms*; Springer: Berlin/Heidelberg, Germany, 2011; pp. 19–38.
42. Bouyoucos, G.J. Improved hydrometer method for making particle size analysis. *Agron. J.* **1962**, *54*, 464–465. [CrossRef]
43. Lassabatère, L.; Angulo-Jaramillo, R.; Ugalde, J.M.S.; Cuenca, R.; Braud, I.; Haverkamp, R. Beerkan Estimation of Soil Transfer Parameters through Infiltration Experiments—BEST. *Soil Sci. Soc. Am. J.* **2006**, *70*, 521–532. [CrossRef]
44. Alaoui, A.; Schwilch, G.; Bachmann, F.; Panagea, I.; Wyseure, G.; Hessel, R. *Monitoring Plan for Study Sites*; Scientific Report 10, Deliverable D4.2, Work Package 4 of the EU-project SoilCare; University of Bern: Bern, Switzerland, 2018.
45. Thomas, G.W. *Exchangeable Cations, Methods of Soil Analysis: Part 2 Chemical and Microbiological Properties*; American Society of Agronomy: Madison, WI, USA, 1982.
46. Rhoades, J.D. Salinity: Electrical conductivity and total dissolved solids. *Methods Soil Anal. Part 3 Chem. Methods* **1996**, *5*, 417–435.

land

MDPI

Article

Soil Quality Assessment after 25 Years of Sewage Sludge vs. Mineral Fertilization in a Calcareous Soil

Ana Simoes-Mota [1], Rosa Maria Poch [2], Alberto Enrique [1], Luis Orcaray [3] and Iñigo Virto [1,*]

[1] Departamento Ciencias, IS-FOOD, Universidad Pública de Navarra, 31006 Pamplona, Spain; anaclaudia.simoes@unavarra.es (A.S.-M.); alberto.enrique@unavarra.es (A.E.)
[2] Departament de Medi Ambient i Ciències del Sòl, ETSEA, Universitat de Lleida, 25198 Lleida, Spain; rosa.poch@udl.cat
[3] Área de Innovación, Sección de Sistemas Sostenibles, Instituto Navarro de Tecnologías e Infraestructuras Agroalimentarias, 31610 Villava, Spain; lorcaray@intiasa.es
* Correspondence: inigo.virto@unavarra.es

Abstract: The aim of this work was to identify the most sensitive soil quality indicators and assess soil quality after long-term application of sewage sludge (SS) and conventional mineral fertilization for rainfed cereal production in a sub-humid Mediterranean calcareous soil. The treatments included six combinations of SS at different doses (40 t ha^{-1} and 80 ha^{-1}) and frequencies (every 1, 2 and 4 years), plus a control with mineral fertilization, and a baseline control without fertilization. Twenty-five years after the onset of the experiment, 37 pre-selected physical, chemical and biological soil parameters were measured, and a minimum data set was determined. Among these indicators, those significantly affected by treatment and depth were selected as sensitive. A principal component analysis (PCA) was then performed for each studied depth. At 0–15 cm, PCA identified three factors (F1, F2 and F3), and at 15–30 cm, two factors (F4 and F5) that explained 71.5% and 67.4% of the variation, respectively, in the soil parameters. The most sensitive indicators (those with the highest correlation within each factor) were related to nutrients (P and N), organic matter, and trace metals (F1 and F4), microporosity (F2), earthworm activity (F3), and exchangeable cations (F5). Only F3 correlated significantly (and negatively) with yield. From these results, we concluded that soil quality can be affected in opposite directions by SS application, and that a holistic approach is needed to better assess soil functioning under SS fertilization in this type of agrosystem.

Keywords: soil quality assessment; sewage sludge; long-term effect; Mediterranean soils

Citation: Simoes-Mota, A.; Poch, R.M.; Enrique, A.; Orcaray, L.; Virto, I. Soil Quality Assessment after 25 Years of Sewage Sludge vs. Mineral Fertilization in a Calcareous Soil. *Land* **2021**, *10*, 727. https://doi.org/10.3390/land10070727

Academic Editors: Julián Cuevas González, Guido Wyseure and Jean Poesen

Received: 15 June 2021
Accepted: 8 July 2021
Published: 10 July 2021

1. Introduction

In the framework of circular economy and the European Green Deal goals, land application of sewage sludge (SS) is suggested as one of the most economical and ecological sludge disposal methods [1,2]. When properly managed, it is seen as way to prevent environmental pollution [3], recycle nutrients, and decrease the need for commercial fertilizers [4,5]. Sewage sludge, in general, has a high content of organic carbon, nutrients (particularly N and P) [6], and trace elements (S, Mg, Ca), and can promote the proliferation and activity of soil micro and mesofauna [2]. As a consequence, amending soils with SS can improve some soil properties, such as organic matter and nutrient content, soil porosity, bulk density, aggregate stability, or available water holding capacity [2,3,7–9].

However, SS can also contain trace metals and persistent organic pollutants, which present a harmful risk to the environment and can be transferred to crops [10]. Indeed, larger studies on the effect of different organic amendments on soil quality [11] observed that the overall effect can be positive, although some aspects, such as soil contamination or grain quality, may be compromised, depending on the type of amendment used. In particular, the consequences of SS application on soil chemical properties and the accumulation of contaminants have been extensively studied [5,12,13] and different strategies to minimize the risk associated with SS application have been developed [14].

In this framework, the European Commission implemented the EU Directive 86/278/EEC with regulatory guidelines on SS application and the concentration of toxic elements [15]. Other regulations exist at the national and regional level worldwide. Nonetheless, the consequences of the continuous application of SS on other soil quality indicators, particularly the interaction between indicators and their relation to soil functions, have received less attention [11]. These consequences are dependent not only on the composition and frequency of SS application, but also on the pedoclimatic and agronomic characteristics of each site [2,16,17]. The importance of considering all dimensions and properties of soil (in terms of its physical, chemical, biological, and organic matter properties) is intrinsic to any such study [17–19].

In the last decades, the literature on the study of soil quality has grown substantially, with an increasing emphasis on the inclusion of this concept associated with agriculture and other land uses, which should consider the many relationships between soil functions and the ecosystem services they provide [20,21]. Within the diversity of the approaches developed, there is an agreement on the existence of a series of basic steps in the evaluation of soil quality, among which the selection of appropriate indicators stands out for its special relevance [22]. Global reviews of these indicators [22–24] point out the most frequently proposed ones as those related to the organic fraction and soil reaction, together with those referring to the status of some nutrients, porosity (density), and water retention.

In any case, the selection of soil quality indicators needs to be made by simultaneously considering the soil functions and/or the services associated with them that are to be evaluated, and the local conditions imposed by the soil-climatic characteristics at the site under consideration. Some examples of the use of this approach in Navarre, Spain [25,26] for the evaluation of soil quality in agrosystems managed according to conservation agriculture criteria showed that the most appropriate indicators can vary in a relatively short lapse of time with a change of context, such as the transformation from rainfed to irrigated. The identification of these indicators, therefore, must also consider the management context.

In the particular context of farmlands in semiarid and sub-humid Mediterranean regions, which are usually depleted in organic matter [27], and therefore especially sensitive to soil degradation, SS addition to croplands has been seen as a promising practice, due to its high content in organic matter and nutrients [28]. Still, physical degradation of the soil may occur depending on the quality of SS, the doses and frequency of application, and the pedo-environmental conditions [29]. An adequate assessment of soil quality in this context needs, therefore, to be holistic (comprise chemical, physical and biological soil indicators of relevance in this type of agroecosystem [30]). In addition, adequate soil quality indicators have ben conceptually defined not only as sensitive to changes in soil condition, but also as precocious in their reaction as possible, easy to measure, and, if possible, available in common soil datasets [17,31]. Ideally, soil quality indicators should also comprise information measured at the field level (in addition to laboratory analysis), and be easily understandable by famers and policy-makers [32].

Regional studies after long-term applications, and with extreme rates of application, seem useful to better understand the actual effect of SS application on soil quality in these conditions [23]. In this context, this study aimed to identify the most sensitive soil quality indicators and, by studying their correlations, to understand the effect of the long-term application of SS on the overall soil quality of a cultivated calcareous soil after 25 years of SS application at different rates and doses by comparing it with conventional mineral fertilization in a controlled experimental field in Mediterranean sub-humid conditions. We hypothesized that the amount and frequency of SS used might induce differences in the chemical, physical and biological condition of this soil that might be interrelated and explained by the selected soil quality indicators.

2. Materials and Methods

2.1. Site and Experimental Design

The long-term experimental field site in Arazuri, Navarra, NE Spain (42°48′ N, 1°43′ W, 396 m a.s.l.) was established in 1992 to assess the effect of the continuous application of SS on agricultural soil quality and productivity. The climate in the area is temperate Mediterranean, with a humid water regime, according to Papadakis [33]. Mean annual precipitation is 750 mm year^{-1}, and mean annual Thornthwaite's evapotranspiration, 687 mm year^{-1} [34]. The soil in this field is calcareous (approx. 20% of calcium carbonate in the tilled layer) with a clay-loam texture in the topsoil (31% clay, 30% silt, 39% sand) [35], is well-drained and has no salinity problems. It has been classified as a Calcaric Cambisol [36]. The soil's main physical–chemical characteristics in the tilled layer (0–30 cm) at the control plots are summarized in Table 1.

Table 1. Physical and chemical properties of the soil tilled layer (0–30 cm) for the control plots. Values are given as the mean ± standard deviation (n = 3).

Soil Physical and Chemical Properties	
pH (water 1:2:5)	8.67 ± 0.03
Electrical Conductivity (µs cm^{-3} at 25 °C) (soil:water extract 1:2.5)	169 ± 10
Bulk density (g cm^{-3})	1.59 ± 0.08
Carbonates (%)	16.0 ± 2.1
Clay (%)	27.72 ± 1.03
Organic Carbon (%) (Walkley–Black)	1.35 ± 0.02

The experimental design consists of a random factorial block design with eight treatments with three replicates (n = 3), each plot with an area of 35 m^2 (10 m × 3.5 m). The treatments included six combinations of SS at different doses (40 t ha^{-1} and 80 ha^{-1}) and frequencies (every 1, 2 and 4 years), plus a control with the usual mineral fertilization in the area (46% urea and ammonium sulphate), and a baseline control without SS or mineral fertilization. Sewage sludge treatments were denoted after the dose and frequency (40-1, 40-2, 40-4, 80-1, 80-2, and 80-4). Mineral-fertilized and baseline controls were noted as MF and C, respectively. Both doses and frequency of SS were chosen according to the common practices in the area, and to get the highest possible rates in the plots with high doses and frequencies.

The crops used corresponded to the most frequent rainfed rotation of 3 years in the area (cereal–cereal–no cereal), managed with annual tillage with a 30 cm deep moldboard plow, and application of phytosanitary products according to the crops' needs each year. The most common cereal crops used were wheat (*Triticum aestivum* L.) and barley (*Hordeum vulgare* L.), with sunflower (*Helianthus annuus* L.) as the non-cereal crop in the rotation. Sewage sludge was produced in the urban wastewater treatment plant from the city of Pamplona (population 330,000), with primary and secondary treatments, stabilized through anaerobic digestion and mechanical dewatering. Sewage sludge characteristics, as described by [10], are summarized in Table 2. The SS was applied each campaign in September, 3 to 4 weeks before sowing, using a 3.5 m wide spreader trailer, followed by moldboard plowing down to 30 cm. Mineral fertilization with a commercial fertilizer purchased from a local provider was carried out before sowing. Wheat sampling was carried out in June in the year of study, at harvest. Each field replicate was harvested with a plot-scale combine, and grain yields were recorded. Grain weights were taken directly from the combine, and grain samples were collected to analyze their water content, to obtain yield data on a dry-mass basis.

Table 2. Physical and chemical properties of the sewage sludge. Values are given as the mean ± standard deviation (n = 3).

Sewage Sludge Physical and Chemical Properties	
pH	8.16 ± 0.03
Electric Conductivity (µs cm^{-3})	1795 ± 28
Dry material (%)	18.1 ± 0.4
Volatile matter (% of dry substance)	62.8 ± 1.9
C/N	5.35 ± 0.08
Total N (%)	5.85 ± 0.13
Ammonium-N (%)	0.75 ± 0.02
Phosphorus (P_2O_5) (%)	5.59 ± 0.22
Potassium (K_2O) (%)	0.62 ± 0.05
Iron (Fe) (%)	1.68 ± 0.04
Calcium (CaO) (%)	7.98 ± 0.29

2.2. *Soil Sampling and Analysis*

Soil sampling was carried out 25 years after the onset of the experiment, in September, at each treatment and replicate at two depths (0–15 cm and 15–30 cm) after the crop cycle was completed, and at the furthest moment in time from previous soil alterations, except for samples for the physical properties, which were sampled in June before harvesting to avoid the possible effect of harvesting machinery. Disturbed and undisturbed samples were collected for the various analyses. Disturbed soil samples were collected for the 0–15 cm and 15–30 cm depths using an Edelman-type auger (Ø = 5 cm) or a shovel. Three subsamples were collected per plot for each depth increment and combined to obtain a composite sample. Immediately after sampling, a portion of the composite soil was stored at 4 °C for further biological analyses. Part of the sample was gently pushed through a 6 mm sieve. These aggregates were air dried and used for aggregate stability determinations. The remainder of the soil was air-dried and ground to pass through a 2 mm sieve. Undisturbed core samples were collected in triplicate using bevel-edged steel rings (Ø = 5 cm, total volume = 100 cm^3) for the 0–15 cm and 15–30 cm depth increments to determine soil bulk density (ρb), permeability, and water retention characteristics. Undisturbed soil samples were also collected using Kubiena boxes for thin section analysis. For earthworm population assessment, two 20 × 20 × 30 cm soil blocks were extracted from each treatment in all replicates.

2.2.1. Soil Physical Properties

The soil's physical condition was assessed using properties related to compaction and porosity, aggregation, and water flow and storage. Bulk density and penetration resistance (PR) were measured to assess compaction and porosity. As explained above, the core method was used to determine ρb [37]. Penetration resistance was measured at 9 points per field replicate to a depth of 60 cm using a field penetrometer (Rimik CP20, Agridy Rimik Pty Ltd., Toowoomba, Qld, Australia). Measurements were made after a rainy period to avoid differences in water content between treatments. Measurements were recorded every 15 mm, and PR for 0–15 and 15–30 cm were calculated as weighted depth averages.

Dry aggregate stability was determined by placing 100 g of dry aggregates (<6 mm) in the top of a column of sieves of 4, 2, 1, 0.5, and 0.25 mm openings, and shaking in a rotary movement at 60 strokes/min for 60 s in a Retsch VS 100 device (Retsch GmbH & Co., Haan, Germany). For wet aggregate stability, a constant shower-like flux (6 L/min) of distilled water was applied from the top of the same set of sieves while sieving (60 strokes/min, 60 s). We used a mechanical sample divisor (Retsch GmbH & Co., Haan, Germany) to ensure that the initial distribution of aggregates was similar among replicates. Aggregate size distribution and stability were expressed as the mean weight diameter (MWD) after dry and wet sieving [38]. The stability of the aggregates was also evaluated using the mass proportion of water-stable aggregates (WSA) > 0.25 mm [39]. Soil saturated permeability (Ks) was measured on undisturbed soil cores after saturation with deionized water under a vacuum using a laboratory permeameter (Eijkelkamp Soil & Water, Giesbeek, The Netherlands).

Soil water retention at −33 kPa, −50 kPa, and −90 kPa was determined on intact soil cores, and sieved (<2 mm) soil samples were used for water retention assessment at −1500 kPa. Samples were placed on pressure plate extractors (Soil Moisture Equipment Corp., Santa Barbara, CA, USA). Volumetric water was calculated using ρb. Available water-holding capacity (AWHC) was calculated as the difference between volumetric water content at field capacity (−33 kPa) and wilting point (−1500 kPa). From these data, as described in [40,41], the model proposed by [42] was used to estimate the equivalent pore diameter corresponding to each of the water potentials. According to this model, the equivalent pore diameter was 9 μm for −33 kPa, and 0.2 μm for −1500 kPa. This allowed us to obtain the equivalent size ranges of micropores in each sample, expressed as the proportion of each pore range (<0.2 μm, 0.2–9 μm, and >9 μm), as well as the proportion of pores able to retain water (0.2–9 μm) over those able to store water available for plants (>0.2 μm). These were denoted as PØ < 0.2, PØ 0.2–9, PØ > 9 and PØ 0.2–9 (>0.2), respectively.

Soil thin sections were prepared from undisturbed soil samples as described in [43]. Image analysis was used in these sections to determine parameters related to macroporosity. For this, a scanned image was obtained per thin section under two light conditions: parallel polarizers and crossed polarizers. They were processed using Image J [44] to obtain digital binary images. From each binarized thin section, five random images (10 × 10 mm) were selected using an adaptation of the method used by [45], where a grid of 27 squares (1 cm^2 each) was placed in each scanned section from which the eligible squares were chosen using a random number generator. From these, pore-size distribution analysis was carried out based on an open mathematical algorithm: the Quantim4 library [46]. The area occupied by pores was divided into five intervals according to the pore's apparent diameter: 100–400 μm; 400–1000 μm; 1000–2000 μm; >2000 μm. The proportion of the area (equivalent to volume proportion over total soil volume) occupied by pores with diameters between 400–1000 μm was selected for this study because of their special relevance when describing structure (size of planar voids or fissures), and also because these pores can result from the activity of mesofauna [47].

2.2.2. Soil Chemical Properties

All chemical analyses were performed on air-dried sieved (<2 mm) samples. Total N was analyzed using the Kjeldahl digestion method. Available P was determined as described by [48]. Exchangeable K and Na were quantified using atomic absorbance after extraction with NH4OAc 1N [49]. The soil electrical conductivity (EC) and soil pH were measured in distilled water (1:2.5). Soil pH was determined with a Crison GLP22 pH meter (Crison Instruments, S.A., Barcelona, Spain). Conductivity was read with a Crison GLP32 conductivity meter (Crison Instruments, S.A., Barcelona, Spain).

Carbonate concentration was measured in a modified Bernard's calcimeter [50] by quantifying the CO_2 produced after treating a soil sample with HCl. Available trace metals (Cu, Mn, Ni, Zn, Cd and Pb) at the 0–15 cm depth were analyzed as DTPA(C14H23N3O10)-extractable concentrations from air-dried soil samples, using the extraction procedure described in the international standard ISO 14870:2001 [51], as described in [10]. In short, an extraction solution was prepared by mixing, first, 0.735 g of $CaCl_22H_2O$, 0.984 g of DTPA and 7.46 g of triethanolamine ($C_6H_{15}NO_3$), diluted with 800 mL of deionized water, and the pH was adjusted to 7.3 with HCl. Subsequently, in a 100 mL wide-mouth polypropylene container, 20 g of soil and 40 mL of the solution were mixed and stirred for 2 h at 20 °C on a reciprocating shaker at 30 rpm. Then, a fraction of the extract was decanted and centrifuged for 10 min at 6000 rpm. The supernatant was filtered with a membrane filter with a pore size of 0.45 μm and collected for analysis. The extracts were analyzed earlier than 48 h from their preparation, by ICP-MS in a 7700x analyzer (Agilent Technologies, Santa Clara, CA, USA), following the UNE-EN 17053 standard [52].

2.2.3. Soil Organic Matter and Biological Properties

Soil organic C (SOC) was determined by wet oxidation on air-dried sieved (<2 mm) samples [53]. The fraction of soil organic matter defined as particulate organic matter (POM) based on its size (>53 μm) [54] was isolated by dispersion and sieving of 10 g of air-dried soil [55]. Organic C in the form of POM (POM-C) was determined by wet oxidation.

Earthworms were collected crumbling the 20 × 20 × 30 cm soil blocks by hand, placing the worms in a glass jar, and weighing to obtain a fresh weight for each field replicate [56]. This allowed us to determine the total biomass (g per m^{-2}), the abundance (number of individuals per m^{-2}), and the average size (g per individual).

Microbial biomass carbon (MBC) was measured by comparing extractable C from non-fumigated and chloroform ($CHCl_3$)-fumigated soil [57]. Carbon concentration in the extract (chromic acid dissolution) was analyzed by sulfuric digestion and subsequent spectrophotometry. The functional diversity of the soil microbial population was studied through the analysis of the community-level physiological profiles (CCPLs) in fresh samples by studying the C source utilization patterns observed using a Biolog Ecoplates™ microplating system (Biolog, Hayward, CA, USA), as described in [58]. Ecoplates™ were designed for determining CLPPs of terrestrial communities and comprise 31 C substrates that are major ecologically relevant compounds. One g equivalent dry weight of soil was mixed with 9 mL of autoclaved Mili-Q ultra-pure water and shaken in an orbital shaker at 125 rev min^{-1} for 1 h. After shaking, samples were left to settle and then a 1:100 dilution was inoculated onto Biolog Ecoplates™. The plates were incubated at 30 °C and color development was read twice a day at 595 nm using a microplate reader (Thermo Scientific Multiskan® EX Waltham, Massachusetts, USA). Average well color development (AWCD) was determined by calculating the mean of every well's absorbance value at each reading time. The number of substrates used by the soil microbial community (NSU), equivalent to species richness [59], was quantified as the number of wells showing corrected absorbance values >0.25 at the onset of the exponential microbial growth in the Biolog Ecoplate™ microplates (59 h).

2.3. *Statistical Analysis*

The selection of the most sensitive soil quality indicators, and the assessment of the effect of SS application on soil quality, was conducted in a two-step procedure, as in [26,60]. First, a multivariate analysis of variance (MANOVA) was performed to test whether there was a significant effect of the categorical independent variables (fertilization and depth) on at least one of the physical, chemical, or biological variables studied. For this study, the eight fertilization treatments (SS application, MF and C) were considered as a factor, because, despite some treatments receiving equivalent cumulative amounts of added SS, for some soil properties, the effect of SS or mineral fertilization might be different depending on the frequency, whereas some others might be affected by the accumulation of SS applications with time.

Then, a univariate analysis of variance (ANOVA) for the different soil variables was performed to examine for significant influences of fertilization treatment and depth. Only the variables for which the F statistic for SS application or fertilization treatments was significant ($p < 0.05$) were retained for further analysis. In a second step, factor analysis was used to group the retained variables into statistical factors based on their correlation structure. Principal component analysis (PCA) was used as the method for factor extraction. To eliminate the effect of different units of variables, factor analysis was performed using the correlation matrix on the standardized values of the measured soil properties; each variable had mean = 0 and variance = 1 (total variance = number of variables [61]). We used the determinant of the correlation matrix as an indicator to identify the existence of correlations among variables. As in Imaz et al. (2010) [25] and Apesteguía et al. (2017) [26], using the correlation matrix, principal components (factors) with eigenvalues > 1.5 were retained and subjected to varimax rotation with Kaiser to estimate the proportion of the variance of each attribute explained by each selected factor (loadings), and by all factors (communalities).

A high communality for a soil attribute indicates that a high proportion of its variance is explained by the factors. In contrast, a low communality for a soil attribute indicates that much of that attribute's variance remains unexplained. Less importance should be ascribed to soil attributes with low communalities when interpreting the factors [62].

To evaluate the effects of the studied SS application or mineral fertilization treatments on the extracted factors, factor scores for each sample point were calculated and ANOVA was performed on the new score variables. Homogeneous groups among treatments were detected using Duncan's test ($p < 0.05$, unless otherwise indicated). Only factors that differed among treatments were retained for further consideration. Soil attributes were then assigned to the factor for which their loading was the highest [61]. For each retained factor, highly weighted attributes were selected as possible soil quality indicators. We considered as highly weighted those within 10% of the highest factor loading, as in [63].

Finally, a correlation analysis was conducted for wheat yields against the scores of the extracted factors with PCA. For all analysis, the significance level was set at $p < 0.05$ unless otherwise indicated. All statistical treatments were performed using IBM SPSS Statistics 27.0 [64].

3. Results

3.1. Identification of Indicators

The different treatments and sampling depth significantly affected the physical, chemical, and biological properties evaluated. The analysis of variance for each individual parameter studied showed some significant differences for the different treatments and depths, and for some parameters, a significant interaction of both (Table 3). As the goal of this study was to identify the most sensitive soil quality indicators and to assess overall soil quality from PCA, and not to assess each individual parameter, only the significance of the analysis is provided for each parameter. As depth had a significant effect in some of the parameters studied, the factor analysis performed to select soil quality indicators was performed separately for the two sampling depths. Soil parameters that were not significantly affected by treatment per studied depth were not considered for the following analysis.

At the 0–15 cm depth, these were K_s, water retention at −33 kPa and −90 kPa, AWHC, PØ < 0.2, PØ 400–1000 in equivalent diameter, carbonates, available Mn, available Pb, the POM-C/SOC ratio, MBC, diversity indexes (AWCD and NSU), and earthworm abundance (individuals m^{-2}). At the 15–30 cm, among those studied at that depth, the parameters excluded for PCA were all physical parameters, except for PØ < 0.2, the POM-C/SOC ratio, and the microbial functional diversity indexes (AWCD and NSU).

3.1.1. 0–15 cm Depth

The correlation matrix for the 21 selected indicators (determinant < 0.0001) showed several significant correlations on 101 pairs out of 210 (Table S1). The highest positive and significant correlations were found between available P vs. available Zn, available Cu, and available Ni ($p < 0.001$), and also between SOC vs. available Zn and available Ni ($p < 0.001$).

The PCA identified three factors (F1, F2 and F3) with eigenvalues > 1.5 for the 0–15 cm depth, which together explained 70.2% of the variance of the 21 selected indicators (Table 4).

The soil properties with high loadings for these factors were considered potential good soil quality indicators (Table 5). F1 showed high loadings (within 10% of the one with the highest loading, or close) for available P, EC, available Zn, available Cu, available Ni, available Cd, available Pb, and SOC. F1 can therefore be associated to organic matter and chemical parameters. F2 showed high loadings for water retention at −50, the total proportion of pores 0.2–0.9 μm (PØ 0.2–0.9), and the proportion of pores 0.2–0.9 μm over total pores > 0.2 μm (PØ $0.2–9_{(>0.2)}$). F2 can therefore be associated with water retention in the soil. Finally, F3 grouped earthworms' biomass (g m^{-2}) and earthworm average size (g i^{-1}) as the properties with the highest loading. F3 would represent the behavior of earthworm populations as affected by treatments in this field.

Table 3. Results of the analysis of variance (ANOVA) for all soil properties.

Soil Quality Indicators	Depths Studied	R^2	Treatment (T)	Depth (D)	T × D
Physical			ANOVA (*p*-value)		
Bulk density	2	0.517	0.058	0.001	0.709
PR	2	0.948	0.144	0.000	0.992
K_s	2	0.310	0.543	0.150	0.538
Water −33	2	0.376	0.124	0.815	0.479
Water −50	2	0.730	0.000	0.001	0.000
Water −90	2	0.261	0.741	0.163	0.660
AWHC	2	0.338	0.230	0.317	0.622
PØ < 0.2	2	0.613	0.018	0.000	0.397
PØ > 0.2–9	2	0.335	0.444	0.682	0.297
PØ > 9	2	0.687	0.010	0.000	0.029
PØ $0.2–9_{(>0.2)}$	2	0.429	0.407	0.015	0.235
PØ 400–1000	1	0.223	0.704	NA	NA
MWD dry	1	0.708	0.002	NA	NA
MWD wet	1	0.567	0.033	NA	NA
WSA	1	0.648	0.008	NA	NA
Chemical					
Available P	2	0.910	0.000	0.149	1.000
Total N	1	0.701	0.000	NA	NA
Electrical conductivity	2	0.827	0.000	0.543	0.208
pH	2	0.831	0.000	0.000	0.973
Exchangeable K	2	0.529	0.102	0.001	0.433
Exchangeable Na	2	0.720	0.000	0.361	0.573
Carbonates ($CaCO_3$)	1	0.032	0.999	NA	NA
Available Mn	1	0.531	0.055	NA	NA
Available Zn	1	0.908	0.000	NA	NA
Available Cu	1	0.870	0.000	NA	NA
Available Ni	1	0.888	0.000	NA	NA
Available Cd	1	0.696	0.003	NA	NA
Available Pb	1	0.703	0.003	NA	NA
Organic matter and biological					
SOC	2	0.878	0.000	0.000	0.290
POM-C	2	0.702	0.001	0.000	0.123
POM-C/SOC	2	0.510	0.361	0.001	0.105
AWCD	2	0.449	0.909	0.000	0.199
NSU	2	0.571	0.335	0.000	0.173
MBC	2	0.612	0.046	0.000	0.507
Earthworms' biomass (g m^{-2})	1	0.587	0.024	NA	NA
Earthworms' abundance (ind/m^{-2})	1	0.465	0.121	NA	NA
Earthworms' average size (g/ind)	1	0.725	0.001	NA	NA

Table 4. Eigenvalue, percentage, and cumulative variance explained by factor analysis using the correlation matrix of the standardized data of soil parameters at 0–15 cm (F1, F2 and F3) and at 15–30 cm (F4 and F5) depths.

Depth	Factors	Eigenvalue [1]	Percentage (%)	Cumulative (%)
0–15 cm	F1	9.562	43.463	43.463
	F2	4.533	20.603	64.067
	F3	1.654	6.089	70.156
15–30 cm	F4	5.005	50.046	50.046
	F5	1.848	18.477	68.523

[1] Only factors with eigenvalues > 1.5 are shown.

3.1.2. 15–30 cm Depth

Using the nine selected indicators for this depth, a correlation matrix was developed for the 15–30 cm depth (determinant < 0.0001), which showed significant correlations on 17 pairs out of 36 (Table S2). The most significant correlations ($p < 0.001$) were observed between available P vs. SOC and POM-C, EC vs. SOC and POM-C, and finally SOC vs. POM-C.

Table 5. Proportion of variance explained using varimax rotation for each of the retained factors and communalities for the selected soil properties for the 0–15 cm depth.

Soil Indicators	F1	F2	F3	Communalities
PR	0.954	0.175	−0.125	0.612
Water −50	0.852	−0.014	−0.207	0.922
PØ 0.2–9	0.800	0.163	−0.099	0.700
PØ > 9	−0.765	0.212	−0.108	0.862
PØ 0.2–$9_{(>0.2)}$	0.111	0.000	−0.097	0.955
WSA	0.980	0.053	−0.116	0.595
MWD dry	0.962	0.005	−0.133	0.500
MWD wet	0.949	0.104	−0.112	0.603
Av P	0.837	−0.166	−0.170	0.963
Total N	0.059	0.949	−0.022	0.768
EC	0.079	0.797	0.138	0.831
pH	0.010	−0.910	0.152	0.647
Ext Na	0.047	0.975	−0.053	0.900
Av Zn	−0.535	0.362	0.306	0.982
Av Cu	0.041	0.359	−0.092	0.972
Av Ni	0.087	−0.629	0.183	0.947
Av Cd	−0.659	−0.319	0.256	0.912
SOC	0.933	−0.004	−0.097	0.948
POM-C	0.553	−0.125	−0.272	0.783
Earthworms g/m^2	−0.304	−0.004	0.851	0.846
Earthworms g/i	−0.196	−0.214	0.866	0.863

The PCA extracted two factors (F4 and F5) with eigenvalues > 1.5, explaining 68.5% of the variance between indicators at this depth (Table 4). For F4, the highest loadings corresponded to EC, available P, and SOC. Like F1 at 0–15 cm, F4 can be associated to organic matter and chemical parameters. Regarding F5, exchangeable K and Na were the properties with the highest loadings. F5 would therefore represent exchangeable cations (Table 6).

Table 6. Proportion of variance explained using varimax rotation for each of the retained factors and communalities for the selected soil properties for the 15–30 cm depth.

Soil Indicators	F4	F5	Communalities
PØ > 0.2	0.955	0.032	0.925
Av P	0.891	0.082	0.836
Total N	0.954	−0.008	0.934
EC	−0.865	−0.360	0.881
pH	0.173	0.825	0.728
Ext K	−0.114	0.859	0.753
Ext Na	0.031	−0.008	0.888
SOC	0.898	0.033	0.836
POM-C	0.684	−0.005	0.468
MBC	0.318	0.522	0.708

3.2. Sensitivity of PCA Factors to Treatment

All factor scores were sensitive to treatments (Table 7). The scores for F1 were significantly different in C and MF than in all treatments with different doses of SS. The scores for F2 differed significantly in C, MF, 40-4, 80-1, and 80-2 from 40-1, 40-2, and 80-4. The scores for F3 were significantly different in C than in MF, 80-2, and 40-2, with the rest of treatments showing intermediate values.

Table 7. Effect of treatment on factor scores from PCA ($p < 0.05$).

	Mean Scores				
Treatment	**F1**	**F2**	**F3**	**F4**	**F5**
40-1	0.310 b	−1.321 a	0.828 bc	−0.139 b	1.485 c
40-2	−0.244 b	−1.247 a	−0.939 a	−0.637 ab	0.564 bc
40-4	0.257 b	0.818 b	−0.205 ab	0.377 c	0.757 bc
80-1	1.896 c	0.546 b	0.064 ab	1.963 d	−0.169 ab
80-2	0.367 b	0.842 b	−0.490 a	0.792 c	−0.781 a
80-4	−0.215 b	−0.996 a	−0.011 ab	−0.323 b	−0.795 a
MF	−1.162 a	0.794 b	−0.907 a	−0.909 a	−1.075 a
C	−1.121 a	0.564 b	1.660 c	−1.123 a	0.015 ab
Treatment (p-value)	< 0.001	< 0.001	0.002	0.000	0.002

In each column, different letters denote different Duncan's homogeneous groups.

Factor scores for both F4 and F5 were sensitive to treatment (Table 7). For F4, scores were significantly different for C and MF than for all treatments receiving SS. Among them, 80-1 had the highest load, and all other treatments with SS displayed intermediate values. Finally, F5 scores were different for 40-1 than for 80-2, 80-4, and MF, with the other treatments showing intermediate values.

3.3. Yield

Average wheat yield was statistically different among treatments ($p < 0.001$, Table 8). MF treatment had the highest values, followed by 40-1, 40-2 and 40-4. The treatment with the highest rate of sludge, 80-1, showed the lowest yield apart from the baseline control treatment (C), for which yield was less than half compared with the MF and 40-1 treatments. No significant correlations were observed for the factor scores and yield, except for F3 (Pearson's correlation coefficient = −0.638, $p < 0.05$).

Table 8. Crop yield results (kg ha^{-1}) treatments with the same letters are not statistically different ($p < 0.001$). Values are given as the mean ± standard deviation ($n = 3$).

Treatment	Yield (kg ha^{-1})
40-1	8408 ± 921 c
40-2	8752 ± 473 c
40-4	8722 ± 460 c
80-1	6470 ± 1265 b
80-2	7558 ± 480 bc
80-4	7783 ± 782 bc
MF	8877 ± 462 c
C	3505 ± 824 a

Different letters denote different Duncan's homogeneous groups.

The interaction between yield and F1, F3, and F4 scores, as well as the interaction between F1 and F4 grouped by treatment are represented through scatter plot graphics in Figure 1.

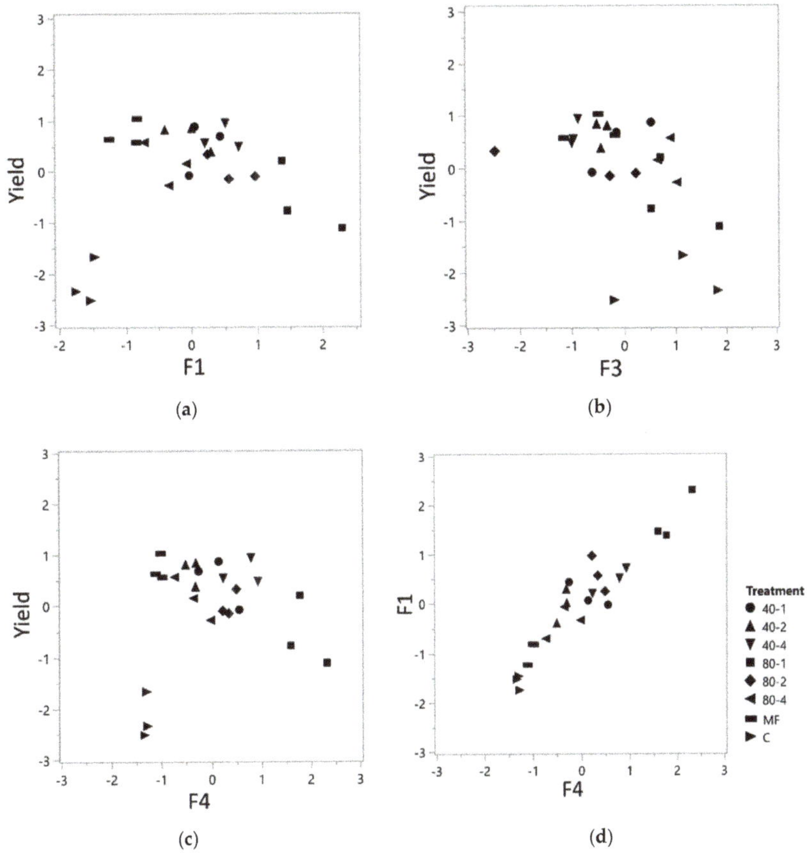

Figure 1. Relationship between soil quality assessment factors selected through PCA and yield. Treatments (n = 3). (**a**) Yield × F1; (**b**) yield × F3; (**c**) yield × F4; (**d**) F1 × F4.

4. Discussion

4.1. Selection of Soil Quality Indicators

4.1.1. Sensitivity to Management

The results of the preliminary ANOVA (Table 3) indicated that some of the pre-selected soil properties were indeed sensitive to SS application and fertilization management in the experimental field used in this study. This response was also different for the two studied depths. At 015 cm, most of the physical and chemical indicators originally considered were shown to be sensitive to treatments, as well as organic matter indicators and earthworms, whereas indicators related to the soil microbial community seemed to be less sensitive. The sensitivity of those preselected was, however, lower in the 15–30 cm depth, despite all treatments receiving annual inversion tillage at 0–30 cm, which suggests a depth stratification in the response to treatments. The relevance of the stratification of the response of soil properties to management [39] and the relatively low sensitivity of SOC stratification have been largely discussed [65,66].

Regarding the selection of the most sensitive indicators, at 0–15 cm, sensitive indicators included water retention at −50, microporosity (PØ > 0.2–9, PØ > 9 and PØ > 0.2–9$_{(>0.2)}$), PR, aggregate stability (WSA, MWD$_d$, MWD$_w$), available P, total N, EC, pH, extractable Na, available trace metals except for Mn and Pb, SOC, POM-C, earthworms' biomass, and average size. At 15–30 cm, they were PØ < 0.2, available P, total N, EC, pH, exchangeable

Na, and K, SOC, POM-C, and MBC. The most sensitive indicators in the physical indicator group were therefore those related to water retention, microporosity, or aggregate stability. These soil properties are well known to be sensitive to changes in soil when SS or other organic amendments are used [3,67], as they represent the changes induced in soil structure as a response to increased inputs of organic matter [68].

On the contrary, non-sensitive indicators included ρ_b, K_s, AWHC, and PØ 400–1000. This can be explained, at least partially, by the fact that all treatments receive the same mechanical management, comprising annual moldboard tillage, seedbed conditioning, and seeding, as well as mechanical harvesting. In terms of ρ_b and K_s, the effect of these operations can counteract that of the addition of organic matter with SS, or from crop residues. Other studies have also shown that ρ_b can be less sensitive to changes in the soil's physical condition associated to changes in organic matter than other physical indicators, such as those related to aggregate stability. For instance, [69] found no differences in ρ_b between a conventional tillage treatment and long-term no-tillage inducing gains in SOC on a silt loam soil, while differences were found among tillage treatments concerning WSA and MWD.

These results can also be understood as some changes in the soil's physical condition being more noticeable than total porosity or AWHC. In this sense, there is evidence that some soil pore-size ranges can be more affected by changes in SOC gains than others. Indeed, Kirchmann et al. (2002) [70] reported microporosity (1–5 μm) as more sensitive to management than macropores in an Inceptisol amended with different exogenous organic matter, which they related to changes in SOC concentration. This coincides with our observation of a greater sensitivity of micro porosity indicators, especially in the 0–15 cm depth, than PØ 400–1000 in size, which may indicate that this porosity interval is more related to management (planar voids, fissures due to tillage), equal to all treatments, than to the activity of mesofauna.

Among chemical indicators, those related to nutrients (N and P), EC, and pH, as well as trace metals, proved to be the most sensitive ones at 0–15 cm. The response of these indicators to SS addition and/or mineral fertilization was expected, and has been reported in many previous studies on the use of SS in agriculture [12,71,72]. In particular, the accumulation of trace metals with SS addition has been already reported and studied in this soil [10,73]. Changes in pH and EC have also been systematically reported in soils amended with SS, in contrast with MF or non-amended soils [12,74,75] and related to the content in soluble salts in SS (Table 1).

Carbonate content was included in the original collection of indicators because it has been observed that the repeated addition of SS and other sources of organic matter can result in changes in the amount and typology of soil carbonates [76,77]. The content in carbonates of the studied soil (Table 1) seems, however, elevated enough not to be sensitive to these changes, although the observed sensitivity of pH (Table 3) suggests that some changes could be expected in the future if the repeated addition of SS results in some sort of acidification in this soil. In fact, changes in pH after SS application can occur due to proton release due to nitrification process, as observed by Tamir et al. (2013) [78] and Huang & Chen (2009) [79] after application of animal manure or sewage sludge compost, respectively, which causes $CaCO_3$ dissolution. For instance, Eid et al. (2021) [80] have recently reported a significant decrease in soil pH (from 8.5 to 7.7) in a short-term pot experiment when the soil was incubated with SS at doses > 30 g kg^{-1} soil.

At 15–30 cm (no data on trace metals were available), exchangeable K and Na were also observed to be sensitive. A positive correlation was found between Ex Na and pH (Table S2). Other studies confirm this correlation [9,81] between exchangeable cations and pH, related to the increase in the amount of exchangeable cations, Na and K in this case, resulting from the leaching process, contributing to the pH acidification. As in most of the studies conducted on SS application [8,12,82], SOC and POM-C were seen to be highly sensitive to treatments at both depths [83]. In our study, however, an important observation was that while both SOC and POM-C revealed to be sensitive to the treatment, the proportion of POM-C over SOC was not. This suggests that the differences in SOC and POM-C between treatments would correspond to the amount of both, and not to differences in the quality of organic matter, or at least, to the proportion of labile C. Other studies using SS have also observed that the long-term addition of SS results in changes mostly in the stock of SOC, rather than in differences in its composition [84], which they attributed to crop residues being the most relevant source of SOC compared to SS.

Finally, in relation to biological indicators, several studies have reported changes in the soil microbial biomass or diversity with SS application [75,85,86], since the microbial community is triggered by the increase of labile C [87], which was indeed sensitive to treatments in our experimental field (Table 3). However, other studies, such as Urra et al. (2019) [73], conducted in the same experimental field, or Picariello et al. (2020) [88] in an incubation experiment, did not find a significant response of soil microbial biomass to the treatments studied, as was the case in our study at 0–15 cm. This might be due to the changes observed within soil chemical parameters. For instance, Lloret et al. (2016) [89] showed that changes in EC can hinder microbial activity in a calcareous soil. Similar to our results, Roig et al. (2012) [7] found no correlation between basal respiration, a known indicator of soil biological activity, and the use of SS in the same experimental field 10 years before our collection of samples. In this sense, it has been reported that the soil microbial community can be more sensitive to tillage practices than to soil organic matter management [90,91]. Earthworm indicators were, however, clearly sensitive to treatments in our study. Their response to soil and organic management has been widely studied, and reported in the sub-humid and semi-arid areas of the region [55,92,93]. In contrast, at the 15–30 cm depth, MBC was found to be significantly sensitive to treatments (Table 3). This suggests some stratification of microbial biomass, as observed for other indicators such as ρ_b and PR, also showing significant differences with depth. As stated above, microbial biomass can respond better to tillage and changes in the soil physical–chemical condition than to changes in organic matter, as can be seen by the lack of significant correlations between MBC and the organic matter parameters at this depth (Table S2). In fact, as in the study by [89], significant correlations were found between MBC, EC, and pH at this depth (Table S2).

In summary, those indicators showing the highest sensitivity to management included some of the originally selected ones, but not all. Among those with the highest sensitivity, physical, chemical, biological, and organic-matter related indicators were included, which supports the idea of a holistic approach being needed to understand changes in soil when SS is tested as an agricultural amendment.

4.1.2. Grouping and Selection of Indicators

The most significant correlations were observed for P and SOC with trace metal availability at 0–15 cm, and for P and EC with SOC in the 15–30 cm depth. These correlations suggest that the addition of SS, which overall implied the addition of different accumulated doses of organic compounds, also implied an enrichment in P, trace metals, and, very likely, soluble compounds. Zoghlami et al. (2020) [12] reported a concomitant change in SOC, P, and exchangeable Na with higher doses of SS application. A correlation between organic matter accumulation and an increase in EC has been also observed in similar studies [12,75]. These observations again put in evidence the relevance of paying attention to all consequences of the addition of SS when assessing their effects in soil, as well as the interaction among soil properties. The correlations observed at both depths were reflected in the results of the PCA. At the depth of 0–15 cm, the selected indicators had different loadings in the three factors retained (F1, F2 and F3), so that F1 received high loadings from organic matter and chemical parameters, while F2 was mostly associated to water retention, and F3 to earthworm populations (Table 5). These results suggest that that the responses of soil water retention and earthworms were not directly correlated to that observed for organic matter, P, and trace metals. This can be explained by different means.

First, although the soil's physical condition and porosity are known to interact with soil organic matter in most soils, the observed discrimination of water retention indicators (F2) from those related to SOC and nutrients (F1) can be related to the particular mineral composition of this soil, which contained 16% carbonates in the studied depth (Table 1). Carbonates are known to interfere with the soil physical stability [94], and can be a factor of stabilization of soil structure in calcareous soils, making it less dependent on SOC than in other soil types [95,96]. In addition, the observed correlation between SOC and EC at both depths and exchangeable Na (at 15–30 cm) suggests that those treatments displaying SOC gains would also result in an increase in soluble salts, which are a known factor of soil structure destabilization [68]. Second, the lack of correlation between earthworm indicators (F3) and those related to SOC and nutrients (F1) and water retention (F2) indicated that their presence and abundance did not directly respond in this soil to the amount of organic C stored in each treatment, nor to physical indicators related to water retention. This suggests that their activity in the studied soil would be dependent on other factors such as toxicity or compaction.

In this framework, following the criterion for selecting the soil attributes with the highest sum of correlation coefficients (Table 5) as the most appropriate soil quality indicators [25,63], our results showed that SOC, available P and trace metals, microporosity and water retention at low water potential (−50 kPa), and earthworms would be those selected at the 0–15 cm depth. Available P would also be selected at 15–30 cm, together with SOC and EC, and exchangeable monovalent cations (Na and K). It must be noted that EC was also a secondary driving factor at 0–15 cm for F1. The relevance of SOC as an indicator of changes in soil resulting from exogenous organic inputs is logical and has been demonstrated in many cases [3,12,84,97]. Indeed, in a recent study conducted at a regional level, the addition of exogenous C has been proven to be the most efficient strategy to increase SOC stocks in the region of study [98]. SOC is also known to correlate well with other fertility indicators, such as the cation exchange capacity, in soils where clay mineralogy is rather stable, like the one used in this study. The linear relation between applying SS to the soil and the enhance of P, trace metals, and soluble salts, as stated above, is also well documented [12,81,99]. The calcareous nature of this soil can explain, at least partially, the accumulation of both P and trace metals [100], resulting in these indicators displaying a high correlation with the addition of SS. Their value as indicators for this type of soils seems relevant, as both are related to environmental risks. At the same time, the selection of earthworms as the most sensitive biological indicator supports their increasingly recognized role as universal soil biological indicators [22,101,102].

Finally, it can be noted in relation to the selection of indicators that, although the pre-selection of indicators was performed based on expertise, the approach used in this study was statistical. This approach has been seen to sometimes result in unexpected or contradictory selection of indicators [11]. Nevertheless, in our study, the most sensitive indicators were in harmony with most soil quality assessments [22], and seemed adequate if the aim of soil quality studies is to provide practical information, with low cost analysis and with influence on the ecosystem services provided by the soil in the particular conditions of calcareous soils under SS application.

4.2. Soil Quality Assessment

A soil quality assessment can be performed based on the scores of the factors selected, and on the link between these factors and the soil functions under study, and the ecosystem services provided by these functions [19,22]. In this case, the goal being to test the effect of SS and MF on an agricultural soil, the main function to be assessed would be biomass production (yield).

Our results showed that amending the studied soil with SS can result in similar yields as with mineral fertilizers, as reported by Jaber et al. (2005) [103] and Obriot et al. (2016) [11] on the use of municipal solid waste (MSW). When compared to the control, yields were high and like MF in the SS treatments with intermediate doses (40-1; 40-2; 40-4; 80-2; 80-4), but the treatment with the highest dose (80-1) implied a decrease in yield (Table 7). The same was previously documented by Mantovi et al. (2005) [72], who reported lower yields on the highest doses on a on a winter wheat–maize–sugar beet rotation fertilized, due to excess of N and wheat lodging. Differently, Cherif et al. (2009) [104] observed that a high dose of municipal solid waste compost (80 t ha^{-1}) enhanced wheat yield by 239%.

In relation to the relationship between factors issued from PCA and yield, for F1, PCA discriminated the treatments with intermediate doses (40-1; 40-2; 40-4; 80-2; 80-4) from treatment 80-1 with the highest accumulated SS dose (Table 7). In addition, both MF and C were differentiated from the SS treatments for this factor. In the 15–30 cm depth analysis, F4, which, as F1, was associated with nutrient dynamics, trace metals, and organic matter, also displayed a clear discrimination of MF and C from the treatments receiving SS, and treatment 80-1 from the other treatments with SS (Table 8). A significant correlation was indeed found between F1 and F4 (Figure 1), indicating that the effects of the treatments tested in this group of soil properties were similar in the two studied depths. This suggests that the continuous application of SS as an organic amendment to the agricultural soil of this study had a different impact on the soil chemical indicators and organic matter than mineral fertilization, or even no fertilization, and that, within those treatments with SS, the dose would also have different effects in this sense.

However, although F1 and F4 correlated with nutrient dynamics, organic matter, and trace metals, no significant correlation was found between these factors and yield. This can be explained because the few indicators selected by our analysis of F1 and F4 (available P, total N, EC, trace metals, and SOC) are known to have implications on yield, which act in opposite directions [105]. Indeed, as can be seen in Figure 1a, the only treatment not following a correlation between F1 scores and yields was the one without any type of fertilization (treatment C). Among the other treatments, the correlation was negative, suggesting that the increasing use of SS (cumulative dose) would result in an overall negative effect on the scores in F1 and F4 [72]. From another point of view, and concerning trace metals, their high loadings in F1 indicate that the assessment of their bioavailability can be of use to assess changes in soil (which may affect yield) in the conditions of the study. The studies [10,73] have recently explained the link between SS application and trace metal accumulation in the soil and crops of this experimental field.

In relation to soil physical indicators, under the site environmental conditions, F2 appeared as a relevant factor to assess soil functions related to structure and water storage. However, the scores of F2 did not show a clear trend among treatments or different fertilization management practices and were not either correlated to yield. Still, the high loadings for microporosity and water storage parameters suggest that they might be of relevance when assessing soil quality in the field under study.

Finally, on the biological condition of this soil, factor F3 (obtaining the highest loadings from earthworm indicators) was the only factor that significantly correlated with yield. The analysis of this factor separated the C treatment from those fertilized (MF and all treatments with some amount of SS), supporting its potential as an indicator for changes induced in the soil by mineral or organic fertilization. The correlation with yield was negative, suggesting that the changes induced by mineral and organic fertilization in this soil, which would result in increased yield, might be detrimental for earthworm populations. Indeed, reports on the effect this type of fertilization has on the earthworm population are contradictory [106]. A better understanding on the relationship between earthworms and soil management in calcareous soils like the one studied here is needed. In the region, it has been observed that earthworms can be positively affected by the reduction of tillage, and the concomitant gains in SOC in long-term trials [55,92,107]. Our results suggest that, when conventional tillage is used, fertilization may become a major driver of earthworm abundance in this type of soil.

Overall, the soil quality assessment on a crop field with 25 years of SS application revealed several implications regarding this type of fertilization. Sewage sludge had a direct effect on nutrient and organic matter input, as well as on trace metals. Yield results indicated that the soil amended with SS was capable of accomplish similar yields as with MF.

5. Conclusions

The goal of this study was to identify the most sensitive soil indicators to assess changes in soil quality after long-term application of SS and MF on a cultivated calcareous soil, and to understand these changes in a controlled experimental field in Mediterranean sub-humid conditions.

A selection of physical, chemical, and biological indicators, as conducted using PCA in this study, was possible, resulting in SOC, available P, total N, EC, trace metals, and earthworms as the most sensitive indicators of changes in the calcareous soil of study. These showed up, therefore, as the most reliable indicators in the long-term monitoring of the effect of SS application in the conditions of the study. These indicators have been frequently identified in other studies on the response of agricultural soils to management and are, in general, commonly reported and easy to monitor.

The study also showed that the overall response of soil quality to the managements tested (SS application and MF) was not linear or straightforward. As hypothesized, the amount and frequency of SS used induced differences in the soil's chemical, physical, and biological condition. However, the overall effect of SS application was more evident on organic matter, nutrients, and trace metals than on the soil's physical condition or earthworms. However, physical indicators and earthworms were highly sensitive to management, and therefore seem useful for assessing changes in soil, not necessarily related to yield. Indeed, the response of the factors issuing from PCA to the treatments tested, and their correlation with yield (which was not always positive or significant) showed that soil quality can be affected in opposite directions by the type of fertilization (mineral vs. organic), or even by the use or not of fertilizers. This supports the idea that a holistic approach, including soil chemical, physical, and biological indicators, is needed to assess soil functioning in this type of agrosystems, while using yield as the only indicator of soil performance may lead to incomplete diagnosis.

Supplementary Materials: The following are available online at https://www.mdpi.com/article/10.3390/land10070727/s1. Table S1: Correlation among measured soil attributes considered for FA in the 0–15 cm depth across all management treatments.; Table S2: Correlation among measured soil attributes considered for FA in the 15–30 cm depth across all management treatments. Earthworm total biomass (EW g m^{-2}), abundance (EW I m^{-2}), and average size (EW g i^{-1}).

Author Contributions: Conceptualization, A.S.-M., R.M.P., A.E., L.O. and I.V.; methodology, A.S.-M., R.M.P., A.E., L.O. and I.V.; Validation, A.S.-M., R.M.P. and I.V.; formal analysis, A.S.-M., R.M.P. and I.V.; investigation, A.S.-M., R.M.P. and I.V.; resources, A.S.-M., R.M.P., A.E., L.O. and I.V. data curation, A.S.-M., R.M.P. and I.V.; writing—original draft preparation, A.S.-M., R.M.P. and I.V.; writing—review and editing, A.S.-M., R.M.P., A.E. and I.V.; project administration L.O.; funding acquisition, I.V. and L.O. All authors have read and agreed to the published version of the manuscript.

Funding: This project has received funding from the National Institute for Agricultural and Food Research and Technology (INIA) through the RTA2017-00088-C03-01 project and from the European Union's H2020 research and innovation programme under Marie Sklodowska-Curie grant agreement No. 801586.

Institutional Review Board Statement: Ethical review and approval were waived for this study, as no humans or animals were used in the study.

Informed Consent Statement: Not applicable.

Data Availability Statement: The data presented in this study are available on request from the corresponding author. The data are not publicly available due to their belonging to a multi stakeholder project.

Acknowledgments: We are grateful for the services and disposition of the Commonwealth of Pamplona, which contributed the experimental plot and the sludge from the Arazuri treatment plant. We also thank C. Gonzaléz from the Soil Science Laboratory of the Public University of Navarra, and A. Zaragüeta and R. Antón for their assistance in sampling and some analysis.

Conflicts of Interest: The authors declare no conflict of interest.

References

1. Metcalf, E. *Wastewater Engineering—Treatment, Disposal and Reuse*, 3rd ed.; McGraw Hill: New York, NY, USA, 2003.
2. Singh, R.P.; Agrawal, M. Potential benefits and risks of land application of sewage sludge. *Waste Manag.* **2008**, *28*, 347–358. [CrossRef]
3. Annabi, M.; Le Bissonnais, Y.; Le Villio-Poitrenaud, M.; Houot, S. Improvement of soil aggregate stability by repeated applications of organic amendments to a cultivated silty loam soil. *Agric. Ecosyst. Environ.* **2011**, *144*, 382–389. [CrossRef]
4. Sommers, L.E. Chemical composition of sewage sludges and analysis of their potential use as fertilizers. *J. Environ. Qual.* **1977**, 225–232. [CrossRef]
5. Siebielec, G.; Siebielec, S.; Lipski, D. Long-term impact of sewage sludge, digestate and mineral fertilizers on plant yield and soil biological activity. *J. Clean. Prod.* **2018**, *187*, 372–379. [CrossRef]
6. Fytili, D.; Zabaniotou, A. Utilization of sewage sludge in EU application of old and new methods-A review. *Renew. Sustain. Energy Rev.* **2008**, *12*, 116–140. [CrossRef]
7. Roig, N.; Sierra, J.; Martí, E.; Nadal, M.; Schuhmacher, M.; Domingo, J.L. Long-term amendment of Spanish soils with sewage sludge: Effects on soil functioning. *Agric. Ecosyst. Environ.* **2012**, *158*, 41–48. [CrossRef]
8. Soria, R.; Ortega, R.; Bastida, F.; Miralles, I. Role of organic amendment application on soil quality, functionality and greenhouse emission in a limestone quarry from semiarid ecosystems. *Appl. Soil Ecol.* **2021**, *164*, 103925. [CrossRef]
9. Skowrońska, M.; Bielińska, E.J.; Szymański, K.; Futa, B.; Antonkiewicz, J.; Kołodziej, B. An integrated assessment of the long-term impact of municipal sewage sludge on the chemical and biological properties of soil. *Catena* **2020**, *189*, 104484. [CrossRef]
10. Zaragüeta, A.; Enrique, A.; Virto, I.; Antón, R.; Urmeneta, H.; Orcaray, L. Effect of the long-term application of sewage sludge to a calcareous soil on its total and bioavailable content in trace elements, and their transfer to the crop. *Minerals* **2021**, *11*, 356. [CrossRef]
11. Obriot, F.; Stauffer, M.; Goubard, Y.; Cheviron, N.; Peres, G.; Eden, M.; Revallier, A.; Vieublé-Gonod, L.; Houot, S. Multi-criteria indices to evaluate the effects of repeated organic amendment applications on soil and crop quality. *Agric. Ecosyst. Environ.* **2016**, *232*, 165–178. [CrossRef]
12. Zoghlami, R.I.; Hamdi, H.; Mokni-Tlili, S.; Hechmi, S.; Khelil, M.N.; Ben Aissa, N.; Moussa, M.; Bousnina, H.; Benzarti, S.; Jedidi, N. Monitoring the variation of soil quality with sewage sludge application rates in absence of rhizosphere effect. *Int. Soil Water Conserv. Res.* **2020**, *8*, 245–252. [CrossRef]
13. Abreu-Junior, C.H.; de Lima Brossi, M.J.; Monteiro, R.T.; Cardoso, P.H.S.; da Silva Mandu, T.; Nogueira, T.A.R.; Ganga, A.; Filzmoser, P.; de Oliveira, F.C.; Firme, L.P.; et al. Effects of sewage sludge application on unfertile tropical soils evaluated by multiple approaches: A field experiment in a commercial Eucalyptus plantation. *Sci. Total Environ.* **2019**, *655*, 1457–1467. [CrossRef] [PubMed]
14. Collivignarelli, M.C.; Abbà, A.; Frattarola, A.; Miino, M.C.; Padovani, S.; Katsoyiannis, I.; Torretta, V. Legislation for the reuse of biosolids on agricultural land in Europe: Overview. *Sustainability* **2019**, *11*, 6015. [CrossRef]
15. Directive Council. European Comission Council Directive of 1 2 June 1986 on the Protection of the Environment, and in Particular of the Soil, when Sewage Sludge is Used in Agriculture. *Off. J. Eur. Communities* **1986**, *181*, 6–12.
16. Guo, Z.; Zhang, Z.; Zhou, H.; Wang, D.; Peng, X. The effect of 34-year continuous fertilization on the SOC physical fractions and its chemical composition in a Vertisol. *Sci. Rep.* **2019**, *9*, 1–10. [CrossRef] [PubMed]
17. Doran, J.W.; Parkin, T.B. Defining and Assessing Soil Quality. In *Defining Soil Quality for a Sustainable Environment*; Soil Science Society of America, Inc.: Madison, WI, USA, 1994.
18. Andrews, S.S.; Karlen, D.L.; Cambardella, C.A. The Soil Management Assessment Framework. *Soil Sci. Soc. Am. J.* **2004**, *68*, 1945–1962. [CrossRef]
19. Adhikari, K.; Hartemink, A.E. Linking soils to ecosystem services—A global review. *Geoderma* **2016**, *262*, 101–111. [CrossRef]
20. Dominati, E.; Patterson, M.; Mackay, A. A framework for classifying and quantifying the natural capital and ecosystem services of soils. *Ecol. Econ.* **2010**, *69*, 1858–1868. [CrossRef]
21. Drobnik, T.; Greiner, L.; Keller, A.; Grêt-Regamey, A. Soil quality indicators—From soil functions to ecosystem services. *Ecol. Indic.* **2018**, *94*, 151–169. [CrossRef]
22. Bünemann, E.K.; Bongiorno, G.; Bai, Z.; Creamer, R.E.; de Deyn, G.; de Goede, R.; Fleskens, L.; Geissen, V.; Kuyper, T.W.; Mäder, P.; et al. Soil quality—A critical review. *Soil Biol. Biochem.* **2018**, *120*, 105–125. [CrossRef]
23. Bai, Z.; Caspari, T.; Gonzalez, M.R.; Batjes, N.H.; Mäder, P.; Bünemann, E.K.; de Goede, R.; Brussaard, L.; Xu, M.; Ferreira, C.S.S.; et al. Effects of agricultural management practices on soil quality: A review of long-term experiments for Europe and China. *Agric. Ecosyst. Environ.* **2018**, *265*, 1–7. [CrossRef]
24. Karlen, D.L.; Obrycki, J.F. Measuring rotation and manure effects in an iowa farm soil health assessment. *Agron. J.* **2019**, *111*, 63–73. [CrossRef]

25. Imaz, M.J.; Virto, I.; Bescansa, P.; Enrique, A.; Fernandez-Ugalde, O.; Karlen, D.L. Soil quality indicator response to tillage and residue management on semi-arid Mediterranean cropland. *Soil Tillage Res.* **2010**, *107*, 17–25. [CrossRef]
26. Apesteguía, M.; Virto, I.; Orcaray, L.; Bescans, P.; Enrique, A.; Imaz, M.J.; Karlen, D.L. Tillage effects on soil quality after three years of irrigation in Northern Spain. *Sustainability* **2017**, *9*, 1476. [CrossRef]
27. Lesschen, J.P.; Stoorvogel, J.J.; Smaling, E.M.A.; Heuvelink, G.B.M.; Veldkamp, A. A spatially explicit methodology to quantify soil nutrient balances and their uncertainties at the national level. *Nutr. Cycl. Agroecosystems* **2007**, *78*, 111–131. [CrossRef]
28. Marschner, P.; Kandeler, E.; Marschner, B. Structure and function of the soil microbial community in a long-term fertilizer experiment. *Soil Biol. Biochem.* **2003**, *35*, 453–461. [CrossRef]
29. Hamdi, H.; Benzarti, S.; Manusadžianas, L.; Aoyama, I.; Jedidi, N. Solid-phase bioassays and soil microbial activities to evaluate PAH-spiked soil ecotoxicity after a long-term bioremediation process simulating landfarming. *Chemosphere* **2007**, *70*, 135–143. [CrossRef]
30. Salomé, C.; Coll, P.; Lardo, E.; Metay, A.; Villenave, C.; Marsden, C.; Blanchart, E.; Hinsinger, P.; Le Cadre, E. The soil quality concept as a framework to assess management practices in vulnerable agroecosystems: A case study in Mediterranean vineyards. *Ecol. Indic.* **2016**, *61*, 456–465. [CrossRef]
31. Andrews, S.S.; Carroll, C.R. Designing a Soil Quality Assessment Tool for Sustainable. *Ecol. Soc. Am.* **2001**, *11*, 1573–1585.
32. Karlen, D.L.; Ditzler, C.A.; Andrews, S.S. Soil quality: Why and how? *Geoderma* **2003**, *114*, 145–156. [CrossRef]
33. Papadakis, J. Climatic Tables for the World. *Soil Sci.* **1961**, *93*, 76. [CrossRef]
34. Gobierno de Navarra Meteorología y Climatología de Navarra. Available online: http://meteo.navarra.es/ (accessed on 25 May 2021).
35. Gee, G.W.; Bauder, J.W. Particle-size Analysis. In *Methods of Soil Analysis: Part 1 Physical and Mineralogical Methods, 5.1*, 2nd ed.; Klute, A., Ed.; American Society of Agronomy, Inc.: Madison, WI, USA; Soil Science Society of America, Inc.: Madison, WI, USA, 1986.
36. FAO. *International Soil Classification System for Naming Soils and Creating Legends for Soil Maps*; FAO: Rome, Italy, 2014.
37. Carter, M.R. *Soil Sampling and Methods of Analysis*; Lewis Publishers: Boca Raton, FL, USA, 1993.
38. Bosch-Serra, D.; Yagüe, M.R.; Poch, R.M.; Molner, M.; Junyent, B.; Boixadera, J. Aggregate strength in calcareous soil fertilized with pig slurries. *Eur. J. Soil Sci.* **2017**, *68*, 449–461. [CrossRef]
39. Franzluebbers, A.J. Water infiltration and soil structure related to organic matter and its stratification with depth. *Soil Tillage Res.* **2002**, *66*, 197–205. [CrossRef]
40. Bescansa, P.; Imaz, M.J.; Virto, I.; Enrique, A.; Hoogmoed, W.B. Soil water retention as affected by tillage and residue management in semiarid Spain. *Soil Tillage Res.* **2006**, *87*, 19–27. [CrossRef]
41. Fernández-Ugalde, O.; Virto, I.; Bescansa, P.; Imaz, M.J.; Enrique, A.; Karlen, D.L. No-tillage improvement of soil physical quality in calcareous, degradation-prone, semiarid soils. *Soil Tillage Res.* **2009**, *106*, 29–35. [CrossRef]
42. Rose, C.W. *Agricultural Physics*; Pergamon: New York, NY, USA, 1966.
43. Benyarku, C.A.; Stoops, G. *Guidelines for Preparation of Rock and Soil Thin Sections and Polished Sections*; Departament de Medi Ambient i Ciències del Soòl, Universitat de Lleida: Lleida, Spain, 2005.
44. Rasband, W. *ImageJ, 1.40*; Toronto Western Research Institute: Toronto, ON, Canada, 2015.
45. Virto, I.; Fernández-Ugalde, O.; Barré, P.; Imaz, M.J.; Enrique, A.; Bescansa, P.; Poch, R.M. Análise micromorfológica da influencia da composição mineral do solo na agregação a curto prazo em solos semiáridos de clima mediterrânico. *Spanish J. Soil Sci.* **2013**, *3*, 116–129. [CrossRef]
46. Vogel, H.J. Quantim4 C/C++ Library for Scientific Image Processing. Available online: https://www.ufz.eu/index.php?en=39198 (accessed on 25 May 2021).
47. FAO. *State of Knowledge of Soil Biodiversity–Status, Challenges and Potentialities*; FAO: Rome, Italy, 2020; ISBN 9789251335826.
48. Olsen, S.R.; Sommers, L.E. Phosphorus. In *Methods of Soil Analysis. Part 2*; American Society of Agronomy: Madison, WI, USA, 1982; pp. 403–430.
49. Knudsen, D.; Peterson, G.A.; Pratt, P.F. Lithium, sodium, potassium. In *Methods of Soil Analysis*; Page, A.L., Miller, R.H., Keeney, D.R., Eds.; Soil Science Society of America: Madison, WI, USA, 1982; pp. 225–246.
50. Sherrod, L.A.; Dunn, G.; Peterson, G.A.; Kolberg, R.L. Inorganic Carbon Analysis by Modified Pressure-Calcimeter Method. *Soil Sci. Soc. Am. J.* **2002**, *66*, 299–305. [CrossRef]
51. ISO. *Soil Quality—Extraction ofTrace Elements by Buffered DTPA Solution*; ISO: Geneva, Switzerland, 2001.
52. AENOR. *Alimentos para Animales. Métodos de Muestreo y Análisis. Determinación de Elementos Traza, Metales Pesados y Otros Elementos en los Alimentos Para Animales por ICP-MS (UNE-EN 17053)*; AENOR: Madrid, Spain, 2008.
53. Nelson, D.W.; Sommers, L.E. Total carbon, organic carbon and organic matter. In *Methods of Soil Analysis*; Page, A.L., Miller, R.H., Keeney, D.R., Eds.; Soil Science Society of America and Agronomy Society of America: Madison, WI, USA, 1982; pp. 539–579.
54. Cambardella, C.A.; Elliott, E.T. Particulate Soil Organic-Matter Changes across a Grassland Cultivation Sequence. *Soil Sci. Soc. Am. J.* **1992**, *56*, 777–783. [CrossRef]
55. Virto, I.; Imaz, M.J.; Enrique, A.; Hoogmoed, W.; Bescansa, P. Burning crop residues under no-till in semi-arid land, Northern Spain—Effects on soil organic matter, aggregation, and earthworm populations. *Aust. J. Soil Res.* **2007**, *45*, 414–421. [CrossRef]
56. Baker, G.H.; Lee, K.E. Earthworms. In *Soil Sampling and Methods of Analysis*; Carter, M.R., Ed.; Canadian Society of Soil Science and Lewis Publishers: Boca Raton, FL, USA, 1993; pp. 359–371.

57. Vance, E.D.; Brookes, P.C.; Jenkinson, D.S. An extraction method for measuring soil microbial biomass C. *Soil Eiol. Biochem.* **1987**, *19*, 703–707. [CrossRef]
58. Mijangos, I.; Becerril, J.M.; Albizu, I.; Epelde, L.; Garbisu, C. Effects of glyphosate on rhizosphere soil microbial communities under two different plant compositions by cultivation-dependent and -independent methodologies. *Soil Biol. Biochem.* **2009**, *41*, 505–513. [CrossRef]
59. Zak, J.C.; Willig, M.R.; Moorhead, D.L.; Wildman, H.G. Functional diversity of microbial communities: A quantitative approach. *Soil Biol. Biochem.* **1994**, *26*, 1101–1108. [CrossRef]
60. Govaerts, B.; Sayre, K.D.; Deckers, J. A minimum data set for soil quality assessment of wheat and maize cropping in the highlands of Mexico. *Soil Tillage Res.* **2006**, *87*, 163–174. [CrossRef]
61. Shukla, M.K.; Lal, R.; Ebinger, M. Determining soil quality indicators by factor analysis. *Soil Tillage Res.* **2006**, *87*, 194–204. [CrossRef]
62. Brejda, J.J.; Karlen, D.L.; Smith, J.L.; Allan, D.L. Identification of Regional Soil Quality Factors and Indicators. *Soil Sci. Soc. Am. J.* **2000**, *64*, 2125. [CrossRef]
63. Andrews, S.S.; Karlen, D.L.; Mitchell, J.P. A comparison of soil quality indexing methods for vegetable production systems in Northern California. *Agric. Ecosyst. Environ.* **2002**, *90*, 25–45. [CrossRef]
64. SPSS Inc. *Statistical Software, SPSS 27.0.*; SPSS: Chicago, IL, USA, 2021.
65. Stockfisch, N.; Forstreuter, T.; Ehlers, W. Ploughing effects on soil organic matter after twenty years of conservation tillage in Lower Saxony, Germany. *Soil Tillage Res.* **1999**, *52*, 91–101. [CrossRef]
66. Dimassi, B.; Cohan, J.P.; Labreuche, J.; Mary, B. Changes in soil carbon and nitrogen following tillage conversion in a long-term experiment in Northern France. *Agric. Ecosyst. Environ.* **2013**, *169*, 12–20. [CrossRef]
67. Zuo, W.; Gu, C.; Zhang, W.; Xu, K.; Wang, Y.; Bai, Y.; Shan, Y.; Dai, Q. Sewage sludge amendment improved soil properties and sweet sorghum yield and quality in a newly reclaimed mudflat land. *Sci. Total Environ.* **2019**, *654*, 541–549. [CrossRef]
68. Rabot, E.; Wiesmeier, M.; Schlüter, S.; Vogel, H.J. Soil structure as an indicator of soil functions: A review. *Geoderma* **2018**, *314*, 122–137. [CrossRef]
69. Sekaran, U.; Sagar, K.L.; Kumar, S. Soil aggregates, aggregate-associated carbon and nitrogen, and water retention as influenced by short and long-term no-till systems. *Soil Tillage Res.* **2021**, *208*, 104885. [CrossRef]
70. Kirchmann, H.; Gerzabek, M.H. Pore size changes in a long-term field experiment with organic amendments. *Dev. Soil Sci.* **2002**, *28*, 419–423. [CrossRef]
71. Alvarenga, P.; Farto, M.; Mourinha, C.; Palma, P. Beneficial Use of Dewatered and Composted Sewage Sludge as Soil Amendments: Behaviour of Metals in Soils and Their Uptake by Plants. *Waste Biomass Valorization* **2016**, *7*, 1189–1201. [CrossRef]
72. Mantovi, P.; Baldoni, G.; Toderi, G. Reuse of liquid, dewatered, and composted sewage sludge on agricultural land: Effects of long-term application on soil and crop. *Water Res.* **2005**, *39*, 289–296. [CrossRef] [PubMed]
73. Urra, J.; Alkorta, I.; Mijangos, I.; Epelde, L.; Garbisu, C. Application of sewage sludge to agricultural soil increases the abundance of antibiotic resistance genes without altering the composition of prokaryotic communities. *Sci. Total Environ.* **2019**, *647*, 1410–1420. [CrossRef]
74. Paz-Ferreiro, J.; Gascó, G.; Gutiérrez, B.; Méndez, A. Soil biochemical activities and the geometric mean of enzyme activities after application of sewage sludge and sewage sludge biochar to soil. *Biol. Fertil. Soils* **2012**, *48*, 511–517. [CrossRef]
75. Vafa, H.J.; Raiesi, F.; Hosseinpur, A. Sewage sludge application strongly modifies earthworm impact on microbial and biochemical attributes in a semi-arid calcareous soil from Iran. *Appl. Soil Ecol.* **2016**, *100*, 45–56. [CrossRef]
76. Sainju, U.M.; Caesar-TonThat, T.; Lenssen, A.W.; Evans, R.G.; Kolberg, R. Long-Term Tillage and Cropping Sequence Effects on Dryland Residue and Soil Carbon Fractions. *Soil Sci. Soc. Am. J.* **2007**, *71*, 1730–1739. [CrossRef]
77. Bughio, M.A.; Wang, P.; Meng, F.; Qing, C.; Kuzyakov, Y.; Wang, X.; Junejo, S.A. Neoformation of pedogenic carbonates by irrigation and fertilization and their contribution to carbon sequestration in soil. *Geoderma* **2016**, *262*, 12–19. [CrossRef]
78. Tamir, G.; Shenker, M.; Heller, H.; Bloom, P.R.; Fine, P.; Bar-Tal, A. Organic N mineralization and transformations in soils treated with animal waste in relation to carbonate dissolution and precipitation. *Geoderma* **2013**, *209–210*, 50–56. [CrossRef]
79. Huang, C.; Chen, Z.-S. Carbon and nitrogen mineralization of sewage sludge compost in soils with a different initial pH. *Soil Sci. Plant Nutr.* **2009**, *55*, 715–724. [CrossRef]
80. Eid, E.M.; Shaltout, K.H.; Alamri, S.A.M.; Alrumman, S.A.; Hussain, A.A.; Sewelam, N.; Ragab, G.A. Monitored Sewage Sludge Application Improves Soil Quality, Enhances Plant Growth, and Provides Evidence for Metal Remediation by Sorghum bicolor L. *J. Soil Sci. Plant Nutr.* **2021**. [CrossRef]
81. Hamdi, H.; Hechmi, S.; Khelil, M.N.; Zoghlami, I.R.; Benzarti, S.; Mokni-Tlili, S.; Hassen, A.; Jedidi, N. Repetitive land application of urban sewage sludge: Effect of amendment rates and soil texture on fertility and degradation parameters. *Catena* **2019**, *172*, 11–20. [CrossRef]
82. Soriano-Disla, J.M.; Navarro-Pedreño, J.; Gómez, I. Contribution of a sewage sludge application to the short-term carbon sequestration across a wide range of agricultural soils. *Environ. Earth Sci.* **2010**, *61*, 1613–1619. [CrossRef]
83. Lorenz, K.; Lal, R.; Ehlers, K. Soil organic carbon stock as an indicator for monitoring land and soil degradation in relation to United Nations' Sustainable Development Goals. *L. Degrad. Dev.* **2019**, *30*, 824–838. [CrossRef]
84. Paetsch, L.; Mueller, C.W.; Rumpel, C.; Houot, S.; Kögel-Knabner, I. Urban waste composts enhance OC and N stocks after long-term amendment but do not alter organic matter composition. *Agric. Ecosyst. Environ.* **2016**, *223*, 211–222. [CrossRef]

85. Antolín, M.C.; Pascual, I.; García, C.; Polo, A.; Sánchez-Díaz, M. Growth, yield and solute content of barley in soils treated with sewage sludge under semiarid Mediterranean conditions. *F. Crop. Res.* **2005**, *94*, 224–237. [CrossRef]
86. Xue, D.; Huang, X. The impact of sewage sludge compost on tree peony growth and soil microbiological, and biochemical properties. *Chemosphere* **2013**, *93*, 583–589. [CrossRef]
87. Odlare, M.; Arthurson, V.; Pell, M.; Svensson, K.; Nehrenheim, E.; Abubaker, J. Land application of organic waste—Effects on the soil ecosystem. *Appl. Energy* **2011**, *88*, 2210–2218. [CrossRef]
88. Picariello, E.; Pucci, L.; Carotenuto, M.; Libralato, G.; Lofrano, G.; Baldantoni, D. Compost and sewage sludge for the improvement of soil chemical and biological quality of mediterranean agroecosystems. *Sustainability* **2021**, *13*, 26. [CrossRef]
89. Lloret, E.; Pascual, J.A.; Brodie, E.L.; Bouskill, N.J.; Insam, H.; Juárez, M.F.D.; Goberna, M. Sewage sludge addition modifies soil microbial communities and plant performance depending on the sludge stabilization process. *Appl. Soil Ecol.* **2016**, *101*, 37–46. [CrossRef]
90. Carter, M.R. The influence of tillage on the proportion of organic carbon and nitrogen in the microbial biomass of medium-textured soils in a humid climate. *Biol. Fertil. Soils* **1991**, *11*, 135–139. [CrossRef]
91. Kabiri, V.; Raiesi, F.; Ghazavi, M.A. Tillage effects on soil microbial biomass, SOM mineralization and enzyme activity in a semi-arid Calcixerepts. *Agric. Ecosyst. Environ.* **2016**, *232*, 73–84. [CrossRef]
92. Bescansa, P.; Virto, I.; Fernández-Ugalde, O.; Imaz, M.J.; Enrique, A. Casting Activity of Scherotheca gigas in No-Till Mediterranean Soils: Role in Organic Matter Incorporation and Influence of Aridity. *Appl. Environ. Soil Sci.* **2010**, *2010*, 1–6. [CrossRef]
93. Valdez, A.S.A.S.; Bosch-Serra, À.D.À.D.; Yagüe, M.R.; Poch, R.M.R.M.; Puigpinós, E. Earthworm community and soil microstructure changes with long-term organic fertilization. *Arch. Agron. Soil Sci.* **2020**, *66*, 957–970. [CrossRef]
94. Rasmussen, C.; Heckman, K.; Wieder, W.R.; Keiluweit, M.; Lawrence, C.R.; Berhe, A.A.; Blankinship, J.C.; Crow, S.E.; Druhan, J.L.; Hicks Pries, C.E.; et al. Beyond clay: Towards an improved set of variables for predicting soil organic matter content. *Biogeochemistry* **2018**, *137*, 297–306. [CrossRef]
95. Fernández-Ugalde, O.; Virto, I.; Barré, P.; Gartzia-Bengoetxea, N.; Enrique, A.; Imaz, M.J.; Bescansa, P. Effect of carbonates on the hierarchical model of aggregation in calcareous semi-arid Mediterranean soils. *Geoderma* **2011**, *164*, 203–214. [CrossRef]
96. Rowley, M.C.; Grand, S.; Verrecchia, É.P. Calcium-mediated stabilisation of soil organic carbon. *Biogeochemistry* **2018**, *137*, 27–49. [CrossRef]
97. Domingo-Olivé, F.; Bosch-Serra, À.D.; Yagüe, M.R.; Poch, R.M.; Boixadera, J. Long term application of dairy cattle manure and pig slurry to winter cereals improves soil quality. *Nutr. Cycl. Agroecosyst.* **2016**, *104*, 39–51. [CrossRef]
98. Antón, R.; Arricibita, F.J.; Ruiz-Sagaseta, A.; Enrique, A.; De Soto, I.; Orcaray, L.; Zaragüeta, A.; Virto, I. Soil organic carbon monitoring to assess agricultural climate change adaptation practices in Navarre, Spain. *Reg. Environ. Chang.* **2021**, *21*, 1–15. [CrossRef]
99. Nicolás, C.; Kennedy, J.N.; Hernández, T.; García, C.; Six, J. Soil aggregation in a semiarid soil amended with composted and non-composted sewage sludge-A field experiment. *Geoderma* **2014**, *219*, 24–31. [CrossRef]
100. Jordán, M.M.; Almendro-Candel, M.B.; Navarro-Pedreño, J.; Pardo, F.; García-Sánchez, E.; Bech, J. Bioavailability, mobility and leaching of phosphorus in a Mediterranean agricultural soil (ne Spain) amended with different doses of biosolids. *Environ. Geochem. Health* **2020**, *3*. [CrossRef] [PubMed]
101. Phillips, H.R.P.; Guerra, C.A.; Bartz, M.L.C.; Briones, M.J.I.; Brown, G.; Crowther, T.W.; Ferlian, O.; Gongalsky, K.B.; van den Hoogen, J.; Krebs, J.; et al. Global distribution of earthworm diversity. *Science* **2019**, *366*, 480–485. [CrossRef]
102. Van Groenigen, J.W.; Lubbers, I.M.; Vos, H.M.J.; Brown, G.G.; de Deyn, G.B.; van Groenigen, K.J. Earthworms increase plant production: A meta-analysis. *Sci. Rep.* **2014**, *4*, 1–7. [CrossRef] [PubMed]
103. Jaber, F.H.; Shukla, S.; Stoffella, P.J.; Obreza, T.A.; Hanlon, E.A. Impact of organic amendments on groundwater nitrogen concentrations for sandy and calcareous soils. *Compost Sci. Util.* **2005**, *13*, 194–202. [CrossRef]
104. Cherif, H.; Ayari, F.; Ouzari, H.; Marzorati, M.; Brusetti, L.; Jedidi, N.; Hassen, A.; Daffonchio, D. Effects of municipal solid waste compost, farmyard manure and chemical fertilizers on wheat growth, soil composition and soil bacterial characteristics under Tunisian arid climate. *Eur. J. Soil Biol.* **2009**, *45*, 138–145. [CrossRef]
105. Rorat, A.; Kacprzak, M.; Vandenbulcke, F.; Plytycz, B. Soil amendment with municipal sewage sludge affects the immune system of earthworms Dendrobaena veneta. *Appl. Soil Ecol.* **2013**, *64*, 237–244. [CrossRef]
106. Barrera, I.; Andrés, P.; Alcañiz, J.M. Sewage sludge application on soil: Effects on two earthworm species. *Water. Air Soil Pollut.* **2001**, *129*, 319–332. [CrossRef]
107. Postma-Blaauw, M.B.; de Goede, R.G.M.; Bloem, J.; Faber, J.H.; Brussaard, L. Soil biota community structure and abundance under agricultural intensification and extensification. *Ecology* **2010**, *91*, 460–473. [CrossRef]

land

Review

Soil Compaction Prevention, Amelioration and Alleviation Measures Are Effective in Mechanized and Smallholder Agriculture: A Meta-Analysis

Peipei Yang [1], Wenxu Dong [2], Marius Heinen [3], Wei Qin [4] and Oene Oenema [1,4,*]

[1] Department of Soil Quality, Wageningen University, P.O. Box 47, 6700 AA Wageningen, The Netherlands; peipei.yang@wur.nl
[2] Hebei Key Laboratory of Soil Ecology, Center for Agricultural Resources Research, Institute of Genetics and Developmental Biology, Chinese Academy of Sciences, 286 Huaizhong Road, Shijiazhuang 050021, China; dongwx@sjziam.ac.cn
[3] Team Soil, Water and Land Use, Wageningen Environmental Research, P.O. Box 47, 6700 AA Wageningen, The Netherlands; marius.heinen@wur.nl
[4] College of Resources and Environmental Sciences, China Agricultural University, Beijing 100193, China; wei.qin@cau.edu.cn
* Correspondence: oene.oenema@wur.nl

Abstract: Background: The compaction of subsoils in agriculture is a threat to soil functioning. Measures aimed at the prevention, amelioration, and/or impact alleviation of compacted subsoils have been studied for more than a century, but less in smallholder agriculture. Methods: A meta-analysis was conducted to quantitatively examine the effects of the prevention, amelioration, and impact alleviation measures in mechanized and small-holder agriculture countries, using studies published during 2000~2019/2020. Results: Mean effect sizes of crop yields were large for controlled traffic (+34%) and irrigation (+51%), modest for subsoiling, deep ploughing, and residue return (+10%), and negative for no-tillage (−6%). Mean effect sizes of soil bulk density were small (<10%), suggesting bulk density is not a sensitive 'state' indicator. Mean effect sizes of penetration resistance were relatively large, with large variations. Controlled traffic had a larger effect in small-holder farming than mechanized agriculture. Conclusion: We found no fundamental differences between mechanized and smallholder agriculture in the mean effect sizes of the prevention, amelioration, and impact alleviation measures. Measures that prevent soil compaction are commonly preferred, but amelioration and alleviation are often equally needed and effective, depending on site-specific conditions. A toolbox of soil compaction prevention, amelioration, and alleviation measures is needed, for both mechanized and smallholder agriculture.

Keywords: compacted subsoils; crop yield; mechanized agriculture; smallholder agriculture; soil bulk density; soil penetration resistance; tillage

Citation: Yang, P.; Dong, W.; Heinen, M.; Qin, W.; Oenema, O. Soil Compaction Prevention, Amelioration and Alleviation Measures Are Effective in Mechanized and Smallholder Agriculture: A Meta-Analysis. *Land* **2022**, *11*, 645. https://doi.org/10.3390/land11050645

Academic Editors: Guido Wyseure, Julián Cuevas González and Jean Poesen

Received: 6 April 2022
Accepted: 24 April 2022
Published: 27 April 2022

1. Introduction

Soil compaction is defined as the 'densification of soil and the distortion of soil structure', which cause the deterioration or loss of one or more soil functions [1,2]. Compacted soils have a relatively high soil bulk density and soil strength, a low number of macro pores, and a relatively high tortuosity, and thereby a low hydraulic conductivity and water infiltration rate [3,4]. These phenomena increase the risks of temporal water logging, runoff, and erosion [5]. Compacted soils impede root elongation and development, and thereby limit soil nutrient uptake and crop development, which in turn causes yield loss [6,7]. The altered soil aeration and wetness and the decreased root growth and crop production also affect soil biodiversity and biological activity, and thereby nutrient transformations and greenhouse gas emissions [4]. Decreased aeration and increased wetness may also predispose compacted soils to infection of root rot diseases [8]. Compacted soils are widespread and have been recognized as a global threat for modern agriculture [9,10]. Greatest concerns relate to subsoil compaction, because of the difficulty to ameliorate subsoil compaction [11,12].

Compacted soils are not easily recognized. This relates especially to compacted subsoils. There are various measures to assess subsoil compaction, e.g., [3], but there is little routine monitoring of soil compaction in practice. Yet, the concerns for soil compaction in the scientific literature is steadily increasing (Figure S1). This increased attention is especially related to the impacts of the increasing mechanization and wheel loads of machines in agriculture [13]. It was noted that a significant fraction of arable farmers in Germany are aware of the risk of intensive field traffic and high axle loads for subsoil compaction, but that this awareness had not yet led to adequate changes in practice [14]. Indeed, the impacts of human-induced (sub)soil compaction seem to increase over time [10,15,16].

Next to human induced soil compaction, through trafficking and ploughing (forming traffic and plough pans in the subsoil), soils may become compacted through natural processes, e.g., during peri-glacial conditions, or as a result of the illuviation of soil colloids, cracking and swelling processes (combined with topsoil tumbling down to the subsoil when cracks are open), heavy rains, and soil trampling by animals. Soils may have a compacted subsoil also because of an abrupt textural or mineralogical change with depth, due to a different geo-genetic origin [3]. The susceptibility of soils to compaction differs greatly. Most susceptible are soils with low soil organic matter content and a high content of silt (particles with a size of 20 to 50 μm). These soils often have a low structural stability and may be characterized as 'sealing, crusting, and hardsetting' [8,17].

Measures to ameliorate compacted subsoils and/or to alleviate their impacts have been explored almost as long as the problem has been realized [18,19]. Hence, many studies have examined the effectiveness of amelioration and alleviation measures, including deep tillage, subsoiling, reduced tillage, crop rotation, reduced trafficking, and using soil amendments. Results of these studies have been discussed and summarized in some excellent reviews. For example, Ungar and Kaspar [6] reviewed studies examining root growth in compacted soils and suggested that tillage and growing deep-rooted crops in rotations will help avoid subsoil compaction and alleviate negative impacts. Soane and Van Ouwerkerk [20] summarized the early studies related to the nature and alleviation of soil compaction. While reviewing the literature since the early 1990s, Hamza and Anderson [21] identified eight practices to avoid, delay, or prevent soil compaction, and suggested that specific combinations of measures are most effective. The review of Batey [3] largely confirmed the suggestions of Hamza and Anderson [21] and emphasized the need for the monitoring of soil compaction in practice. Nawaz et al. [4] reviewed models simulating soil compaction and the effects of soil compaction, while Chamen et al. [22] reviewed studies examining the costs and benefits of measures aimed at ameliorating soil compaction. Schneider et al. [23] quantitatively examined the effects of deep tillage on crop yield, using a meta-analysis of data mainly from Europe and North America, and observed that deep tillage effects were highly site-specific. Shaheb et al. [7] reviewed how soil compaction affected different crop types and listed twelve management strategies to alleviate soil compaction. Most studies focused on mechanized agriculture and paid little attention to smallholder agriculture. Of a different nature, Kodikara et al. [24] reviewed how soil compaction can be improved in civil engineering and transport.

Evidently, soil compaction is a complex and persistent phenomenon affecting the sustainability of crop production in modern agriculture in large areas of the world. The threat of subsoil compaction for crop production is thought to be most severe in mechanized agriculture with high axle loads on wet soils [2,12,25,26]. However, there are also reports on subsoil compaction in smallholder agriculture in China, for example, as a result of long-term soil cultivation practices, irrigation, and natural conditions [27]. It is unclear whether the effects of amelioration and alleviation measures are different between mechanized and smallholder agriculture. Machine weight is much less and ploughing depth is also less in smallholder agriculture than in mechanized agriculture. We hypothesized that amelioration and alleviation measures are more effective in smallholder agriculture than in highly mechanized agriculture, because compacted soil layers are likely more shallow in smallholder agriculture, and thus easier to remediate.

We conducted a systematic review of the quantitative effects of measures aimed at preventing and ameliorating compacted subsoils or at alleviating the impacts of soil compaction on crop yield and soil physical properties, using a meta-analysis of published studies conducted in areas with smallholder farms (mainly China), and in mechanized agriculture in Europe, America, and Australia. We categorized measures in three groups (Table S1), largely following Hamza and Anderson [21] and Chamen et al. [22]: (i) measures aimed at avoiding and preventing subsoil compaction, including minimized and controlled trafficking, zero and minimum tillage (rotary tillage and shallow harrowing); (ii) measures aimed at remediating compacted subsoils, including subsoiling, deep ploughing, and crop rotation; and (iii) measures aimed at alleviating the effects of compacted subsoils, including residue return, controlled irrigation, and manure application. This categorization of measures also fits in the DPSIR framework [1] [2].

The objectives of our study were (1) to quantitatively examine the effects of measures aimed at avoiding and ameliorating soil compaction and at alleviating the impacts of compacted subsoils on crop yield, soil bulk density, and soil penetration resistance, using results of published studies; and (2) to examine the effectiveness of measures in smallholder and mechanized agriculture. We focused on the period 2000–2019/2020, because of the existence of some excellent reviews covering the earlier period, and because studies on smallholder agriculture conducted before 2000 are relatively scarce.

2. Materials and Methods

2.1. Data Collection and Screening

We searched for peer-reviewed publications investigating the effectiveness of measures to address compacted (sub)soils, using Web of Science and China Knowledge Resource Integrated Database (CNKI, for Chinese studies not published in English language). Search terms were ("soil compaction" OR "compacted soil" OR "compacted subsoil" OR "subsoil compaction") AND ("yield" OR "biomass") AND ("density" OR "penetration" OR "soil cone index") in titles, keywords, and abstracts. In Web of Science, conference proceedings and non-English publications were excluded. This search gave 719 publications published between 2000 and 2019 (until 1 August 2019). The search in the China Knowledge Resource Integrated Database yielded 74 additional publications (from 2000 to August 2019).

The search process was followed by a screening procedure that was based on the following criteria: (1) field studies must include side by side comparisons of soil compaction prevention, remediation and/or alleviation treatments, and control (or reference) treatments; (2) for each paired comparison, treatments and reference treatments have the same location, cropping system, cropping management, and year; (3) grain yields and/or biomass yields were reported; (4) soil bulk density and/or soil penetration index data were reported; (5) the test crops were cereals, including wheat, maize, barley, oat, and sorghum; (6) location(s), year(s), and basic soil information of the experiment(s) were stated. Only studies with cereal crops as test crops were included. One reason for this is the importance of cereal crops in global food supply [28], and the other reason is that the results are likely more robust when using crops with similar root morphology and physiology [7]. Grain yield and/or biomass yield were used as crop response indicators.

Following the aforementioned screening procedure, we obtained 400 comparisons (paired observations) of crop yields from 54 studies in 28 countries from Web of Science, and 157 comparisons of crop yields from 23 studies from CNKI. Treatment measures were recorded and grouped. THe results of crop yield and soil bulk density/penetration resistance were extracted from each study, as well as characteristics related to location, experimental year(s), and soil clay content (Table S1). In cases where crop yield and/or soil bulk density and/or penetration results were presented in figures only, values were extracted using the GetData Graph Digitizer (https://apps.automeris.io/wpd/ (accessed on 1 January 2020)).

2.2. Categorization of the Measures

The paired observations were allocated to a category of measures, i.e., prevention, remediation, or alleviation measures. There is some degree of arbitrariness in the allocation of measures. For example, the choice of crop type and crop rotation was categorized as remediation measure but could have been categorized as prevention or alleviation measures equally well. Further, alleviation measures were thought to alleviate the effects of soil compaction, but may contribute also to remediation or prevention, depending on the environmental and management conditions. Thus, irrigation, fertilization, manure application, and straw return were thought to alleviate the impacts of compacted subsoils on root growth (their limited ability to take up water and nutrients from compacted subsoils).

Conventional (random) traffic was chosen as reference treatment for controlled traffic. In this case, a comparison was made between random (deliberate) trafficking and minimal or controlled trafficking, to infer the effects of controlled trafficking indirectly. Thus, random trafficking was used as reference treatment (worst-case), while minimal trafficking or controlled trafficking as the remediation treatment. The reference treatment of manure application was no manure application, while residue return was compared to no residue return. Crop rotation effects were compared to effects of mono-cropping.

Soil bulk density and soil penetration resistance results were grouped into three depth intervals: 0–20 cm (topsoil), 20–40 cm (upper subsoil), and 40–60 cm (lower subsoil). This grouping was seen as a compromise for comparing smallholder and mechanized agriculture. The depth of soil cultivation in smallholder agriculture is commonly less than 20 cm but in mechanized agriculture often a bit deeper, depending also on tillage system. Moreover, about 80% of the roots of most cereal crops are in the upper 40 cm and more than 95% of the roots are in the upper 60 cm of the soil [29,30].

Smallholder farms are mostly found in east and south Asia, Africa, and some countries of Latin America [31], and mechanized agriculture with relatively high axle loads in North America, Oceania, Europe, and west Asia. Therefore, studies conducted in south and east Asia and Africa were considered to be small-holder farming, while studies conducted in America, Europe, Australia, and west Asia were considered to be in mechanized agriculture. For more detailed information of the database composition, see Tables S1 and S2.

2.3. Data Analysis

Our meta-analysis basically followed the same approach as the one described by Qin et al. [32]. We used the natural logarithm of the ratio of the response variable of two treatments as the effect size [33]: $\ln(R) = \ln(x_t/x_c)$, where R is the ratio, x is the response variable, and subscripts t and c refer to the specific treatment and control treatment. The response variable was either crop yield (x = Y), dry bulk density (x = BD), or penetration resistance (x = PR).

For the calculation of a grouped effect size, a linear mixed-effect model was used for which we used the R-package 'nlme' [34]. Mixed-effect models are preferred to fixed-effect models for statistical testing in ecological data synthesis because their assumption of variance heterogeneity is more likely to be satisfied [33]. In our study, results of treatments addressing soil compaction were set as fixed effects and study numbers were set as random effects, to allow accounting for variances among studies. We used the equal weighting method (e.g., [35]) when comparing studies with different number of replicates. The ln(R) of the individual pairwise comparison was used as the dependent variable. The mean effect size and the 95% confidence intervals (CIs) of each categorical group were estimated. The significance of the effects was statistically assessed at the 0.05 confidence level. In the graphs (forest plots), the effect-size of each treatment was transformed back and converted to a percentage change in crop yield, dry bulk density, or penetration resistance relative to the control or reference treatment, i.e., data were presented as $(R-1)*100\%$. In case the value zero in such a forest plot falls outside the 95% CI, the given average value (effect size) is assumed to be significantly different from zero.

3. Results

3.1. Overview of the Dataset

Our dataset consisted of 557 yield comparisons, 620 soil bulk density comparisons, and 592 soil penetration resistance comparisons. About half of the number of bulk density comparisons dealt with the topsoil (346), and half with the subsoil (274). More yield comparisons were from countries with predominantly small-holder farming (S-farming) (323) than from countries with predominantly mechanized agriculture (M-agriculture) (234). More yield observations were related to prevention (221) and remediation measures (205) than alleviation measures (131, Figure 1a). Yield observations of prevention measures were found more in M-agriculture countries than in S-farming countries. The number of yield observations related to remediation and alleviation measures was two times larger with S-farming than M-agriculture (Figure 1b,c).

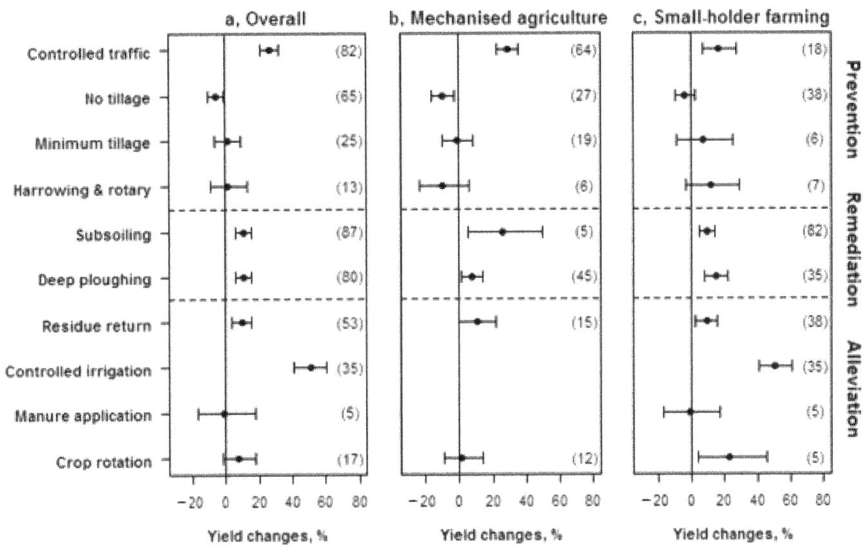

Figure 1. Relative changes in crop yield (%) in response to soil compaction prevention, remediation and alleviation measures; means of all results (**a**); means of results from countries with mechanized agriculture (**b**); means of results from countries with small-holder farming (**c**). Dots show means of treatments, error bars indicate 95% confidence intervals. Numbers in the parentheses indicate number of comparisons.

3.2. Effects of Measures on Crop Yields

Five out of ten measures examined had positive effects on crop yields, including prevention, remediation, and alleviation measures ($p < 0.05$, Figure 1a). Relatively large mean effect sizes were noted for controlled traffic (+26%) and irrigation (+51%). Mean effect sizes were also significantly positive for subsoiling, deep ploughing, residue return, and crop rotation (+8% to +11%). Minimum tillage and manure application did not display significant effects, while no tillage had a negative mean effect on crop yield (−6%).

Differences between S-farming and M-agriculture in the mean effect sizes of prevention, remediation, and alleviation measures on crop yields were relatively small (Figure 1b,c). The mean effect size of controlled traffic on crop yield was two time higher in M-agriculture (+38%) than in S-farming (+16%). However, the number of comparisons was much larger in M-agriculture (88) than in S-farming (21). Subsoiling was more studied in S-farming than in M-agriculture during the last 20 years and the mean effect on crop yield in S-farming was positive (+8%). Controlled irrigation and manure application were examined in S-farming but not in M-agriculture as possible measures to alleviate the effects of compacted subsoils.

Evidently, controlled irrigation had a large effect size, but it is not realistic to ascribe this effect merely to the alleviation of soil compaction. Likely, crop yields in the reference treatments were limited by drought and not only by compacted subsoils.

3.3. *Effects of Measures on Soil Bulk Density*

The measures had a relatively small effect on the soil bulk density of the top soil and subsoil (Figure 2a,d), compared to their effects on crop yields (Figure 1). Relative mean changes in bulk density were in the range of 0–9%. For the subsoil, which is most critical, controlled traffic, deep ploughing, subsoiling, residue return, and crop rotation decreased soil bulk density by on average 2–9% ($p < 0.05$; Figure 2d). Controlled irrigation increased bulk density in the topsoil and subsoil, while minimum tillage increased subsoil bulk density by 3% ($p < 0.05$; Figure 2d).

Essentially all comparisons related to the effects of subsoiling and deep ploughing on subsoil bulk density originated from S-farming. As a consequence, no proper comparison can be made between S-farming and M-agriculture on the effects of subsoiling and deep ploughing. This holds for alleviation measures as well. Controlled trafficking decreased soil bulk density in both topsoil and subsoil, and S-farming and M-agriculture.

3.4. *Effects of Measures on Soil Penetration Resistance*

Soil penetration resistance responded to the measures in a similar way as bulk density, but the relative changes were larger (Figure 3a,d). Controlled traffic treatments had on average 33% lower penetration resistance in topsoils and 26% lower resistance in subsoils than the reference treatments. Subsoiling and deep ploughing decreased penetration resistance by 13% to 20% ($p < 0.05$, Figure 3d). No tillage increased penetration resistance in the topsoil but not in the subsoil.

Observations on subsoiling and deep ploughing originated mainly from S-farming countries, where these measures decreased penetration resistance. Residue return decreased penetration resistance in both topsoil and subsoil in S-farming. The number of comparisons for residue return was too low in M-agriculture to make firm statements. Irrigation slightly decreased penetration resistance in the topsoil but not in the subsoil in S-farming.

3.5. *Effects of Experimental Duration*

More than 80% of the comparisons dealt with short-term experiments (1~3 years; Table S1). Tillage treatments (deep ploughing, subsoiling, no tillage, minimum tillage) accounted for almost half (47%) of the long-term experiments (≥4 years), followed by controlled traffic (23%). For controlled traffic, the relative effect size for crop yield and for subsoil bulk density tended to increase over time (Figure 4a). For crop yield, the effect size was 33% in short-term and 37% in long-term experiments, while subsoil bulk density was 4% lower in short-term and 6% lower in long-term experiments compared to the reference treatments ($p < 0.05$; Figure 4b,c). For deep ploughing, the relative effect size for crop yield and bulk density decreased over time. In short-term (1~3 yrs) experiments, mean effect sizes were statistically significant on crop yields and bulk density ($p < 0.05$), but not in long-term (≥4 yrs) experiments. Similar results were found for no tillage (Figure 4).

3.6. *Effects of Soil Texture*

Soil texture (silt and clay contents) and soil organic matter content affect the susceptibility of soils to compaction and also likely influence the effect sizes of measures. A clay content of 17.5% is commonly used as a threshold value in soil compaction evaluation. Soils with <17.5% clay are considered to be more susceptible to compaction than soils with ≥17.5% clay [36]. Thus, we compared the effect sizes of measures for soils with <17.5% clay with soils having ≥17.5% clay. Yield effects were on average similar for the two textural classes (Figure 5). However, light-textured soils (<17.5% clay) showed greater responses to prevention and amelioration measures than heavy-textured soils (≥17.5% clay). This was most notable for controlled traffic. Effect sizes for yield differed by more than a factor two

(+49% vs. +19%; $p < 0.05$), for subsoiling (+12% vs. 3%), and deep ploughing (13% vs. 8%; $p < 0.05$).

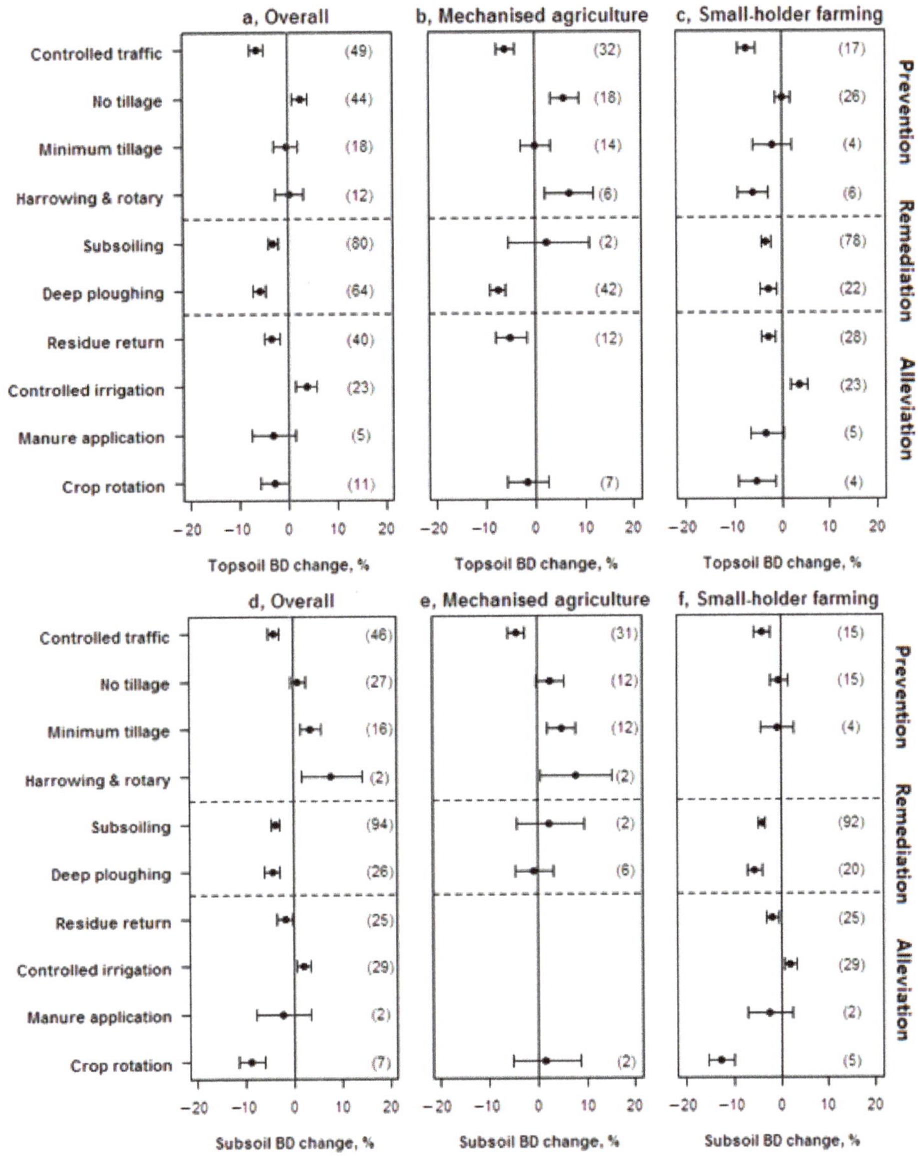

Figure 2. Relative changes in soil bulk density (BD) in response to soil compaction prevention, remediation and alleviation measures for the topsoil (**a**–**c**) and for the subsoil (**d**–**f**); means of all results (**a**,**d**); means of results from M-agriculture (**b**,**e**); means of results from S-farming (**c**,**f**). Dots show means of treatments, error bars indicate 95% confidence intervals. Numbers in the parentheses indicate number of comparisons.

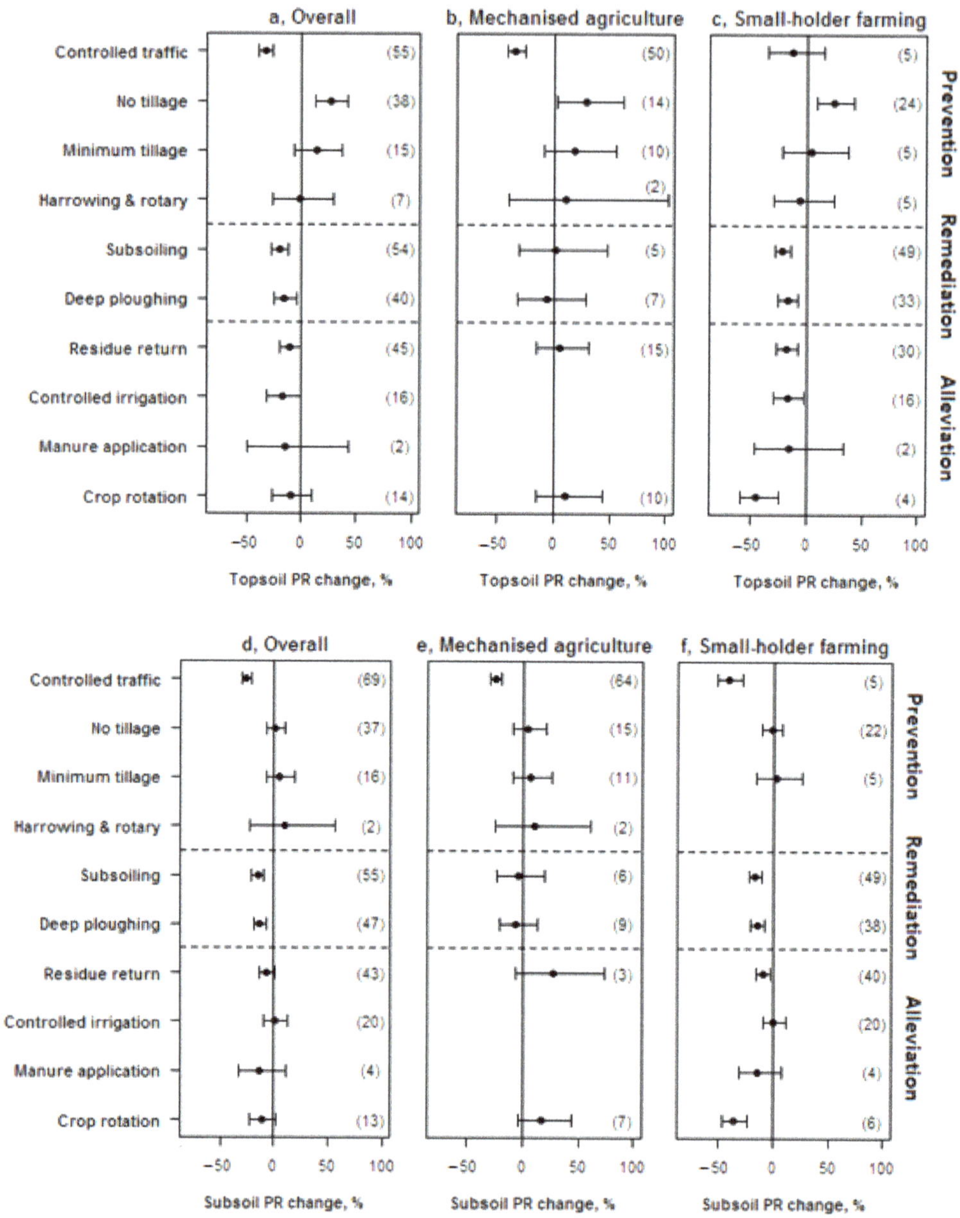

Figure 3. Relative changes in soil penetration resistance (PR) in response to soil compaction prevention, remediation and alleviation measures for the topsoil (**a**–**c**) and for the subsoil (**d**–**f**); means of all results (**a**,**d**); means of results from M-agriculture (**b**,**e**); means of results from S-farming (**c**,**f**). Dots show means of treatments, error bars indicate 95% confidence intervals. Numbers in the parentheses indicate number of comparisons.

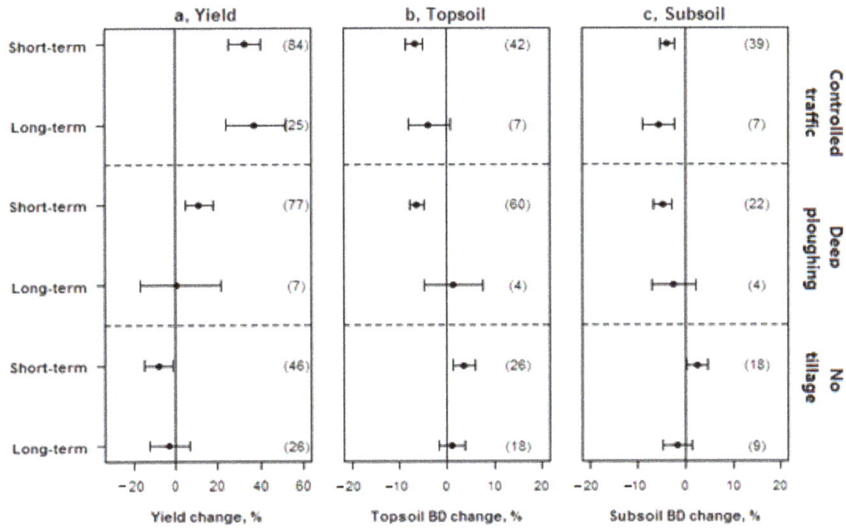

Figure 4. Relative changes in crop yield (**a**) and soil bulk density (BD; for top soil, (**b**); and subsoil, (**c**)) in response to various soil compaction prevention, remediation and alleviation measures; means and standard deviations of results from short-term (<4 years), and long-term (≥4 years) field experiments.

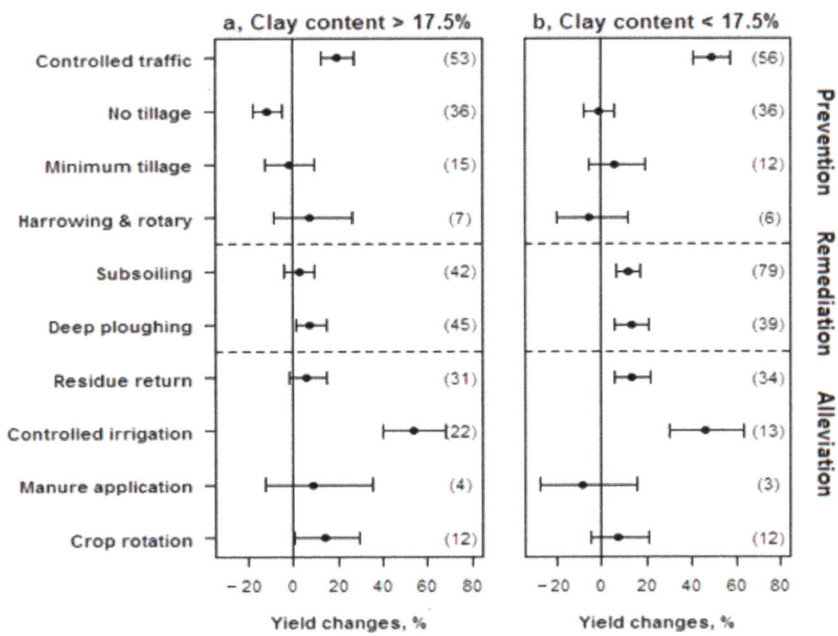

Figure 5. Relative changes in crop yield (%) in response to soil compaction prevention, remediation and alleviation measures; means of results from clay soil (clay content ≥17.5%) (**a**); means of results from sandy soil (clay content < 17.5%) (**b**). Dots show means of treatments, error bars indicate 95% confidence intervals. Numbers in the parentheses indicate number of comparisons.

4. Discussion

4.1. Understanding the Cause-Effect Relationships

The cause–effect relationships of soil compaction and its mitigation measures can be analyzed and understood through the 'driving forces, pressures, state, impact, responses' (DPSIR) framework [2]. In agriculture, the driving forces often stem from the economic incentives to produce more and to lower costs, especially in affluent countries [11,13]. This leads to more intensive soil cultivation and the use of larger and heavier machines, which exerts literally pressure on the soil. This pressure may lead to a densification of the (sub)soil, i.e., compacted (sub)soils, with impacts on water infiltration, root and crop growth, microbiological processes, and gaseous emissions, e.g., [3]. The response of farmers and land managers may be directed towards avoiding or preventing soil compaction, i.e., addressing the driving forces and pressures, or they may focus on the amelioration of compacted soils, i.e., addressing the state, or at alleviating the impacts of compacted soils, or both (Figure 6). Thus, the three categories of measures distinguished in our meta-analysis (Table S2; Figure 1) address different aspects of the cause–effect chain of soil compaction.

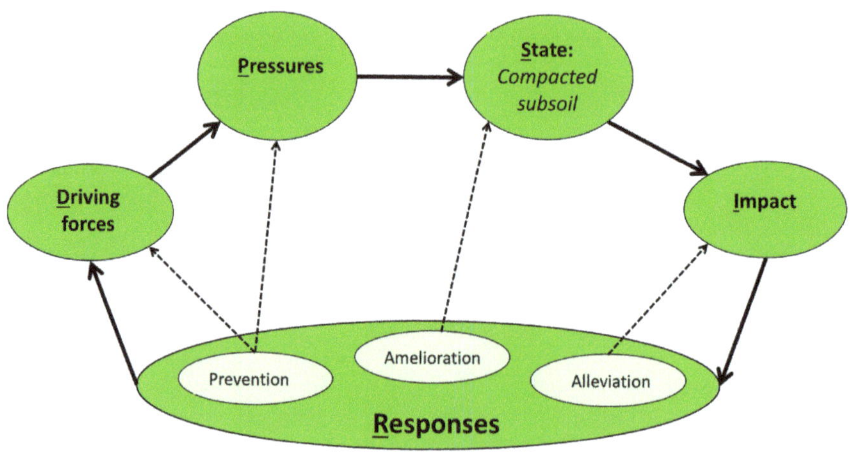

Figure 6. The Driver-Pressure-State-Impact-Response (DPSIR) concept with focus on soil compaction. The response measures indicate which part of the DPSIR chain is being addressed by the measures.

Avoiding, preventing, and precautionary strategies are preferred above amelioration and alleviation strategies, also because of the complexities and imperfections of the latter [2,37]. However, large areas in the world have naturally compacted subsoils (e.g., [8,17]), or have been compacted by human activities in the past [15], and thus will need amelioration and alleviation strategies. Moreover, the susceptibility of soils to densification and the farming and environmental conditions greatly differ across the world, suggesting that region- and farm-specific strategies will be needed, and thus a toolbox of options and strategies. Our meta-analysis contributes to this toolbox by examining quantitatively the effects of both prevention, amelioration, and alleviation measures.

Depending on the strategy, different indicators may be used for evaluating the effectiveness of the strategy. Lebert et al. [37] discussed indicators for precautions against soil compaction (pressure indicators) and for the impairment of subsoil structure through compaction (state indicators). For the first, they proposed the 'pre-compression stress' and 'loading ratio', which can be calculated for different soils, but need soil type specific calibration [37]. For assessing the impairment of subsoil structure, they proposed three indicators, i.e., air capacity (>5% air filled porosity at a water suction of pF 1.8), saturated water conductivity (<10 cm day^{-1}), and a visual classification of the soil morphology (combination of a 'spade diagnosis' and measurements of the effective bulk density and packing density). The second suggested indicator (saturated water conductivity) is basically an impact indicator (and not a state indicator). Soil bulk density was not recommended as an indicator for identification of 'harmful' soil compaction, because 'there is no critical threshold and classification scheme' according to the authors [37]. However, for the related 'packing density' (bulk density corrected for clay content) indicator, there are criteria [38]. Håkansson and Lipiec [39,40] reviewed the usefulness of the relative soil bulk density, or the degree of compactness, which was defined as the dry bulk density in percent of a reference dry bulk density of the same soil obtained by a standardized, long-term, uni-axial compression test at a stress of 200 kPa. Evidently, the measurements of the state of soil compaction are labor-intensive, and thus costly, especially when considering spatial within-field variations [41,42]. As a result, routine monitoring of the state of soil compaction in farmers' fields is not common practice. Indeed, it appears costly and there is debate about appropriate indicators and their interpretation. We observed that soil bulk density and penetration resistance are most commonly used as indicators for assessing the state of soil compaction in field experiments to test measures aimed at preventing, ameliorating, and/or alleviating soil compaction. However, bulk density is not a sensitive indicator (e.g., relative changes in soil bulk density following the implementation of measures are relatively small; Figure 2), while penetration resistance is very sensitive to variations and changes in soil moisture content. Based on uni-axial tests, Panayiotopoulos et al. [43] showed that for a compression stress up to 300 kPa the dry bulk density changed up to 5~15%. This suggests that extreme changes in dry bulk density are not likely to occur. Further, measurements of penetration resistance should be performed at pressure heads of about −100 cm. It is, however, unlikely that this was the case in all studies. This may explain why a large variability in penetration resistance was found in the reviewed studies.

Impact indicators relate to the changes in soil ecosystem functioning following a change in the densification of the soil and associated changes in pore size distributions, tortuosity, and soil structure. Possible impact indicators are crop yield, hydraulic conductivity, run-off and ponding, and emissions of CO_2, CH_4, and N_2O [3,44]. There are no critical thresholds and classification schemes for assessing changes in soil functions, perhaps apart from hydraulic conductivity [37]. Yet, comparisons can be made between situations without and with compacted (sub)soils as in our meta-analysis. Crop yield is probably the most powerful indicator in farmers' practice, because of its influence on farm income, although part of a yield penalty may be nullified through alleviation measures, including irrigation and fertilization.

In conclusion, the DPSIR framework is useful for analyzing and understanding the cause–effect relationship of soil compaction, but further work is needed to derive a proper set of indicators and threshold values.

4.2. Impacts of Measures in Small-Holder Farming and Mechanized Agriculture

The mean effect of controlled traffic on crop yield was 38% (range 32–45%) in mechanized agriculture (M-agriculture) and 16% (range 6–27%) in small-holder farming (S-farming). The wide range of yield effects is roughly in the same range as reported by Antille et al. [16] in a review of 20 studies for various crops. The yield of crops was 0–98% higher when grown in the absence of field traffic compared to the yield of crops grown under typical traffic intensities. Controlled traffic was introduced in commercial-scale farming in the 1990s, initially in Australia and subsequently in Europe and northern America [45,46]. The net economic benefit of controlled traffic increases with farm area. Conversely, the yield effect of controlled traffic needs to be relatively large to make controlled traffic economically attractive in small farms [16,22]. It is therefore no surprise that the number of experimental studies was much larger in M-agriculture than S-farming (Figure 1b,c). Interestingly, the mean yield effect of controlled traffic was on average a factor of two smaller in S-farming than in M-agriculture, which may indeed reflect differences in axel loads between S-farming and M-agriculture.

Zero-tillage minimizes the traffic of soil-cultivating tractors and was therefore considered to be a preventive measure for soil compaction, but it does not necessarily control the traffic of other (e.g., harvesting) machines in the field. There is a lot of interest in zero-tillage and minimum tillage (e.g., [47], as it saves labor and fuel cost, minimizes erosion (especially when combined with surface mulching), and contributes to enhanced soil carbon sequestration. However, it increases N_2O emissions and decreases crop yield. The latter is in agreement with our findings (Figure 1). Further, it tends to increase the soil bulk density and penetration resistance of the topsoil (Figures 2 and 3). The no-till (or reduced-till) compacted topsoils limit root penetration and plant growth [48], while crop residues remaining on the soil surface may increase the incidence of viruses and plant pathogens [49], and lower the soil temperature [50,51]. Our study indicates that current zero-tillage and minimum tillage practices are much less effective as a preventive measure for soil compaction than controlled traffic. However, there is a need for more soil physical and soil structural measurements (including bulk density) of the subsoil in no-till systems to confirm our findings.

Deep ploughing and subsoiling increased crop yields by on average 10% and 9%, respectively, though with relatively large uncertainty bars (Figure 1a). These mean effects were derived mainly from studies conducted in S-farming and reported between 2000 and 2019/2020. Schneider et al. [23] reported rather similar mean positive effects of deep tillage on crop yield (6%), based on a meta-analysis of 45 studies (67 field experiments) that were mainly conducted in Europe and North America between 1918 and 2014 (only three studies were reported after 2000, namely one from North America, one from Argentina, and one from China). They noted that the popularity of deep tillage decreased from the 1970s. Peralta et al. [52] also found positive mean effects of subsoiling on the yield of maize (+6%) and soybean (+26%) in no-till systems in Argentina, using a meta-analysis of 32 field studies. Our study indicates that positive effects of deep tillage on crop yields also hold for smallholder farming, notably China, for both deep tillage and subsoiling. Schneider et al. [23] found that the mean effect size of deep tillage on crop yield depended on the silt content of the topsoil, the density of the subsoil, and drought, but not on the deep tillage method (subsoiling vs. deep ploughing and deep mixing) and tillage depths. The strong interference by drought agrees with our observation that irrigation alleviates the effects of compacted subsoils and greatly increases crop yield (Figure 1). The effect of deep ploughing on crop yield decreased over time (Figure 4). A similar trend was observed in the meta-analysis studies of Schneider et al. [23] and Peralta et al. [52]. The decreasing effect of deep tillage over time is likely the result of re-compaction [22,53]. Our analyses indicate that deep tillage decreased soil bulk density (Figure 2) and penetration resistance (Figure 3) of the topsoil and subsoil. Similar decreases were noted for the topsoil by Peralta et al. [52], but neither Peralta et al. [52] nor Schneider et al. [23] reported changes in soil bulk density and/or penetration resistance for the subsoil in response to deep tillage.

Alleviation measures mainly aim to lessen the negative impacts of compacted subsoils on root and crop growth. Roots elongate less in compacted and dry soils due to a combination of mechanical impedance and water stress [54], and thereby have less access to soil moisture and nutrients. Irrigation thus greatly alleviates the negative impacts of compacted subsoils on crop yield. The mean effect size of irrigation on crop yield was 50% (Figure 1). However, irrigation increased soil bulk density in the topsoil and subsoil (Figure 2). These results are based on observations in S-farming countries only, i.e., mainly China. Crop residue return or surface mulching also had a positive on crop yield, likely because of its effect on soil water preservation [32]. Crop residue return decreased soil bulk density (Figure 2), possibly as a result of enhanced soil carbon sequestration [47]. Only a few studies explicitly examined the effects of manure application on alleviating impacts of compacted subsoils on crop yield. No significant effects on crop yields were found, but manure application in S-farming tended to decrease soil bulk density, possiblY through enhancing soil organic carbon contents [55,56]. In summary, alleviation measures 'treat the symptoms but not the root cause', yet some of these measures can be highly effective, also in cases where amelioration measures were not much effective.

4.3. Managing Soil Compaction

A common opinion is that 'the best way to manage soil compaction is to prevent it from happening'. The popularity of controlled traffic and reduced or no till practices reflects this opinion. The increasing wheel loads and weight of agricultural machinery in practice in especially Europe and North America during the last 60 years do not reflect this opinion. The increase in machinery weight has resulted in an increase in subsoil compaction, which may have contributed to crop yield stagnation and to an increase in the incidence of flooding in Europe [13]. The cascade of possible impacts from soil compaction beyond field and farm scales (e.g., increased risk of flooding, runoff, and erosion) could be seen as driver for actions by policy [57,58]. However, soil compaction is not subject to a coherent set of rules in, for example, the European Union (EU), and is also not mentioned in the recent EU soil strategy for 2030 [59]. Thus, farmers depend on the insights and guidelines of their own and their advisors when it comes to handling soil compaction, while there are essentially no monitoring data concerning farmers' fields.

There is less risk of soil compaction by machines in small-holder farming in China, for example, than in the mechanized agriculture of Europe, North America, and Oceania. There is also no governmental policy aimed at preventing soil compaction in China. However, the intensive cultivation practices and irrigation, and the silty texture of the dominant loess soils in north China are conducive to soil compaction, and there is therefore a continuous search for soil conservation practices that decrease the risk of soil compaction and improve soil structure [60,61]. A combination of tillage practices in sequence appears to be the best strategy [62–64]. This holds for no-till as well. However, it has to be combined with subsoiling once in a few years, as also discussed for the no-till agriculture in Argentina by Peralta et al. [44]. The need for combining tillage practices in China also follows indirectly from the increasing interest in subsoiling during the last two decades (e.g., Figure 1 [24]).

The FAO voluntary guidelines for sustainable soil management do provide technical and policy recommendations to prevent and mitigate soil compaction [65]. Though qualitative and without threshold values, these guidelines are interesting because they address not only the machines and vehicles in the field, but also the importance of crop type and crop rotation, soil organic matter content, soil macrofauna, and microbial and fungal activities. Amelioration measures are not explicitly mentioned, apart from the recommendation to also grow crops with strong tap roots able to penetrate and break up compacted soils. Next to soil compaction, the FAO guidelines also present recommendations to prevent and mitigate nine other soil threats [65]. The need for a more coherent and integrated soil management concept was also recently emphasized by Rietra et al. [47]. They presented a roadmap for developing high-yielding, soil-improving, and environmentally sound cropping systems. This roadmap involves an iterative selection and optimization of site and farm specific crop husbandry and soil management practices, including the selection of machines that minimize soil compaction.

Evidently, preventing soil compaction from happening is too simple a strategy to address soil compaction. Rather, a toolbox of strategies and management practices is needed, which can be used to develop and implement site-specific management measures. Our study provides evidence that both prevention, amelioration, and alleviation measures have value, depending on the site-specific conditions. These measures provide net economic benefits for farms in most cases, through increases in crop yields and resource use efficiency [22,66]. The selection of the most appropriate measures will likely improve, and the effectiveness of these measures will likely increase, when more data become available at the farm level, related to the state and impact of soil compaction, through routine monitoring.

4.4. Limitations of Our Study

We focused on the recent literature (2000–2019/2020), because there are some excellent papers that reviewed and analyzed the older literature, e.g., [23,67], and not many studies have been conducted in small-holder agriculture before 2000. We examined literature from both mechanized agriculture and small-holder farming to make comparisons between these two types of agricultural systems, based on the literature from 2000–2019/20. We note that the literature from S-farming countries from before 2000 has not been analyzed in a systematic manner yet, apart from the studies by Hoogmoed et al. [68], and the reviews by Laker and Nortjé [8], and Peralta et al. [52].

Further, we note that the machine weight is rapidly increasing over time [69], not only in M-agriculture countries, but also in some S-farming countries. Hence, the rough categorization in S-farming and M-agriculture countries may not be the best way to examine differences between mechanized and smallholder agriculture, although this comparison provided new insights, e.g., related to the type of measures applied in the two types of agriculture.

Crop types may respond differently to compacted soils and thereby also to prevention, amelioration, and alleviation measures, because of differences in root morphology and physiology [54,70]. We selected cereals as test crops because these were mostly used and have a more or less uniform response. Thereby, we excluded 183 studies with non-cereal test crops out of the 719 available studies (25%).

Further, we excluded studies that combined various measures, e.g., controlled traffic combined with no tillage, controlled traffic combined with deep tillage, tillage combined with residue management levels, and irrigation combined with subsoiling. The exclusion of these studies does not mean that these studies are less relevant. Instead, it requires another study to infer useful conclusions from these combined-measures studies.

5. Conclusions

Our meta-analysis included 77 studies from 28 countries (32 studies from 16 countries for mechanized agriculture (M-agriculture), and 45 studies from 12 countries for smallholder farming (S-farming)) all related to the effectiveness of soil compaction prevention, amelioration, and alleviation measures. These studies were published between 2000 and 2019/2020 and thus are relatively recent. Prevention measures were mostly studied in M-agriculture, while remediation and alleviation measures were mostly studied in S-farming.

Soil compaction prevention, through controlled traffic, had a positive effect on crop yield in both M-agriculture (+38%) and S-farming (+16%) countries, and led to a lower soil bulk density in topsoil and subsoil (−4% to −6%), and to a lower soil penetration resistance (−26% to −33%). These results confirm earlier estimates for M-agriculture countries but now show that controlled traffic also holds promise for S-farming. However, it is not clear whether controlled traffic is economically profitable in S-farming. Soil compaction prevention through no-till had negative effect on crop yield, while bulk density was increased, in both M-agriculture and S-farming.

Soil compaction amelioration through deep tillage (including subsoiling) had positive effects on crop yields (+9% to +10%), while soil bulk density was decreased by about 3%. These results confirm earlier observations for M-agriculture, but we show that these observations are also valid for S-farming. The relatively large number of studies related to deep tillage in S-farming suggest that subsoil compaction is increasingly seen as a constraint to crop production in the countries with S-farming.

Irrigation was an effective alleviation measure for subsoil compaction, though only reported for S-farming. The large mean effect size for crop yield (+51%) reflects that compacted soils impede root elongation and thereby enhance the impacts of drought, though the effect of irrigation likely relates not only to alleviation of drought related to compacted subsoils. Crop residue mulching and manure application had a small effect on alleviating compacted subsoils.

Soil penetration resistance and bulk density were mostly used as state indicators. Effect sizes of measures on soil bulk density were small (<10%), indicating that bulk density is not a sensitive indicator for assessing the effects of measures. Effect sizes of crop yield as an impact indicator were relatively large, but variable because of interfering factors (climate, soil texture).

A toolbox of soil compaction prevention, amelioration, and alleviation measures is also needed because the cause of soil compaction and the responses of measures are site-specific. Our meta-analysis indicates that such a toolbox is needed for M-agriculture and S-farming.

Supplementary Materials: The following supporting information can be downloaded at: https://www.mdpi.com/article/10.3390/land11050645/s1, Figure S1: Changes in the number of papers published per year, studying the relationships between soil compaction and crop yield.; Table S1: Data and information used for this meta-analysis; Table S2: Summary overview of effects sizes of soil compaction prevention, remediation and alleviation measures on crop yields, soil bulk density (BD), and soil penetration resistance (PR), shown for three categories of studies.

Author Contributions: Conceptualization and writing, P.Y. and O.O.; data collection and data analysis, P.Y. and W.Q. Review and editing, P.Y., W.D., M.H. and O.O. All authors have read and agreed to the published version of the manuscript.

Funding: The present study was carried out in the EU project SoilCare ("Soil care for profitable and sustainable crop production in Europe"), EU grant agreement 677407; https://www.soilcare-project.eu (accessed on 1 February 2022), and National Key Research and Development Program of China (2021YFD190100202). First author (P.Y.) received funding from China Scholarship Council (File No. 201408130093).

Institutional Review Board Statement: Not applicable.

Informed Consent Statement: Not applicable.

Data Availability Statement: Not applicable.

Acknowledgments: This study builds on SoilCare report 07 (https://www.soilcare-project.eu/resources/deliverables; accessed on 1 February 2022); we thank the co-authors of that report for their contributions.

Conflicts of Interest: The authors declare no conflict of interest.

Notes

1 The DPSIR framework stands for Driving forces, Pressure, State, Impact and Responses. It allows for analyzing and understanding the cause-effect chain of soil comapction in a systematic manner, as further discussed in the Discussion section.

References

1. Schjønning, P.; van den Akker, J.H.; Keller, T.; Greve, M.H.; Lamandé, M.; Simojoki, A.; Stettler, M.; Arvidsson, J.; Breuning-Madsen, H. Soil compaction. In *Soil Threats in Europe—Status, Methods, Drivers and Effects on Ecosystem Services*; EU Joint Research Centre: Petten, The Netherlands, 2016; Chapter 6. [CrossRef]
2. Schjønning, P.; van den Akker, J.J.H.; Keller, T.; Greve, M.H.; Lamandé, M.; Simojoki, A.; Stettler, M.; Arvidsson, J.; Breuning-Madsen, H. Driver-Pressure-State-Impact-Response (DPSIR) Analysis and Risk Assessment for Soil Compaction-A European Perspective. *Adv. Agron.* **2015**, *133*, 183–237. [CrossRef]
3. Batey, T. Soil compaction and soil management—A review. *Soil Use Manag.* **2009**, *25*, 335–345. [CrossRef]
4. Nawaz, M.F.; Bourrie, G.; Trolard, F. Soil compaction impact and modelling. A review. *Agron. Sustain. Dev.* **2013**, *33*, 291–309. [CrossRef]
5. Alam, M.K.; Salahin, N.; Islam, S.; Begum, R.A.; Hasanuzzaman, M.; Islam, M.S.; Rahman, M.M. Patterns of change in soil organic matter, physical properties and crop productivity under tillage practices and cropping systems in Bangladesh. *J. Agric. Sci.* **2017**, *155*, 216–238. [CrossRef]
6. Unger, P.W.; Kaspar, T.C. Soil Compaction and Root-Growth—A Review. *Agron. J.* **1994**, *86*, 759–766. [CrossRef]
7. Shaheb, M.R.; Venkatesh, R.; Shearer, S.A. A Review on the Effect of Soil Compaction and its Management for Sustainable Crop Production. *J. Biosyst. Eng.* **2021**, *46*, 417–439. [CrossRef]
8. Laker, M.; Nortjé, G. Review of existing knowledge on subsurface soil compaction in South Africa. *Adv. Agron.* **2020**, *162*, 143–197. [CrossRef]
9. Caon, L.; Vargas, R. Threats to Soils: Global trends and Perspectives. In *A Contribution from the Intergovernmental Technical Panel on Soils, Global Soil Partnership*; Global Land Outlook Working Paper; Pierzynski, G., Ed.; Food and Agriculture: Rome, Italy, 2017.
10. Sonderegger, T.; Pfister, S. Global Assessment of Agricultural Productivity Losses from Soil Compaction and Water Erosion. *Environ. Sci. Technol.* **2021**, *55*, 12162–12171. [CrossRef]

11. Alakukku, L.; Weisskopf, P.; Chamen, W.C.T.; Tijink, F.G.J.; van der Linden, J.P.; Pires, S.; Sommer, C.; Spoor, G. Prevention strategies for field traffic-induced subsoil compaction: A review Part 1. Machine/soil interactions. *Soil Tillage Res.* **2003**, *73*, 145–160. [CrossRef]
12. Techen, A.K.; Helming, K.; Brueggemann, N.; Veldkamp, E.; Reinhold-Hurek, B.; Lorenz, M.; Bartke, S.; Heinrich, U.; Amelung, W.; Augustin, K.; et al. Soil research challenges in response to emerging agricultural soil management practices. *Adv. Agron.* **2020**, *161*, 179–240. [CrossRef]
13. Keller, T.; Sandin, M.; Colombi, T.; Horn, R.; Or, D. Historical increase in agricultural machinery weights enhanced soil stress levels and adversely affected soil functioning. *Soil Tillage Res.* **2019**, *194*, 104293. [CrossRef]
14. Ledermüller, S.; Fick, J.; Jacobs, A. Perception of the Relevance of Soil Compaction and Application of Measures to Prevent It among German Farmers. *Agronomy* **2021**, *11*, 969. [CrossRef]
15. Soane, B.; Van Ouwerkerk, C. Implications of soil compaction in crop production for the quality of the environment. *Soil Tillage Res.* **1995**, *35*, 5–22. [CrossRef]
16. Antille, D.L.; Peets, S.; Galambošová, J.; Botta, G.F.; Rataj, V.; Macak, M.; Tullberg, J.N.; Chamen, W.C.T.; White, D.R.; Misiewicz, P.A.; et al. Soil compaction and controlled traffic farming in arable and grass cropping systems. *Agron. Res.* **2019**, *17*, 653–682. [CrossRef]
17. Daniells, I.G. Hardsetting soils: A review. *Soil Res.* **2012**, *50*, 349–359. [CrossRef]
18. Passioura, J.; Leeper, G. Soil compaction and manganese deficiency. *Nature* **1963**, *200*, 29. [CrossRef]
19. Wilkins, S.M.; Wilkins, H.; Wain, R. Chemical treatment of soil alleviates effects of soil compaction on pea seedling growth. *Nature* **1976**, *259*, 392–394. [CrossRef]
20. Soane, B.; Van Ouwerkerk, C. Soil compaction problems in world agriculture. In *Developments in Agricultural Engineering*; Elsevier: Amsterdam, The Netherlands, 1994; pp. 1–21. [CrossRef]
21. Hamza, M.A.; Anderson, W.K. Soil compaction in cropping systems—A review of the nature, causes and possible solutions. *Soil Tillage Res.* **2005**, *82*, 121–145. [CrossRef]
22. Chamen, W.C.T.; Chamen, W.T.; Moxey, A.P.; Towers, W.; Balana, B.; Hallett, P.D. Mitigating arable soil compaction: A review and analysis of available cost and benefit data. *Soil Tillage Res.* **2015**, *146*, 10–25. [CrossRef]
23. Schneider, F.; Don, A.; Hennings, I.; Schmittmann, O.; Seidel, S.J. The effect of deep tillage on crop yield–What do we really know? *Soil Tillage Res.* **2017**, *174*, 193–204. [CrossRef]
24. Kodikara, J.; Islam, T.; Sounthararajah, A. Review of soil compaction: History and recent developments. *Transp. Geotech.* **2018**, *17*, 24–34. [CrossRef]
25. Obour, P.B.; Ugarte, C.M. A meta-analysis of the impact of traffic-induced compaction on soil physical properties and grain yield. *Soil Tillage Res.* **2021**, *211*, 105019. [CrossRef]
26. Hargreaves, P.R.; Baker, K.L.; Graceson, A.; Bonnett, S.; Ball, B.C.; Cloy, J.M. Soil compaction effects on grassland silage yields and soil structure under different levels of compaction over three years. *Eur. J. Agron.* **2019**, *109*, 125916. [CrossRef]
27. Chen, S.; Yang, P.; Zhang, Y.; Dong, W.; Hu, C.; and Oenema, O. Responses of Cereal Yields and Soil Carbon Sequestration to Four Long-Term Tillage Practices in the North China Plain. *Agronomy* **2022**, *12*, 176. [CrossRef]
28. FAO. Crop Prospects and Food Situation #2. In *Quarterly Global Report*; FAO: Rome, Italy, 2021. [CrossRef]
29. Drew, M.; Saker, L. Assessment of a rapid method, using soil cores, for estimating the amount and distribution of crop roots in the field. *Plant Soil* **1980**, *55*, 297–305. [CrossRef]
30. Liu, S.; Song, F.; Zhu, X.; Xu, H. Dynamics of root growth and distribution in maize from the black soil region of NE China. *J. Agric. Sci.* **2012**, *4*, 21–30. [CrossRef]
31. Lowder, S.K.; Skoet, J.; Raney, T. The number, size, and distribution of farms, smallholder farms, and family farms worldwide. *World Dev.* **2016**, *87*, 16–29. [CrossRef]
32. Qin, W.; Hu, C.; Oenema, O. Soil mulching significantly enhances yields and water and nitrogen use efficiencies of maize and wheat: A meta-analysis. *Sci. Rep.* **2015**, *5*, 16210. [CrossRef]
33. Hedges, L.V.; Gurevitch, J.; Curtis, P.S. The meta-analysis of response ratios in experimental ecology. *Ecology* **1999**, *80*, 1150–1156. [CrossRef]
34. Pinheiro, J.; Bates, D.; DebRoy, S.; Sarkar, D.; Team, R.C. R Core Team (2017) nlme: Linear and Nonlinear Mixed Effects Models. R Package Version 3.1-131. Computer Software. 2017. Available online: https://CRAN.R-project.org/package=nlme (accessed on 1 April 2022).
35. Van Groenigen, J.W.; Lubbers, I.M.; Vos, H.M.; Brown, G.G.; De Deyn, G.B.; Van Groenigen, K.J. Earthworms increase plant production: A meta-analysis. *Sci. Rep.* **2014**, *4*, 6365. [CrossRef]
36. Brus, D.J.; Van Den Akker, J.J.H. How serious a problem is subsoil compaction in the Netherlands? A survey based on probability sampling. *Soil* **2018**, *4*, 37–45. [CrossRef]
37. Lebert, M.; Böken, H.; Glante, F. Soil compaction—indicators for the assessment of harmful changes to the soil in the context of the German Federal Soil Protection Act. *J. Environ. Manag.* **2007**, *82*, 388–397. [CrossRef] [PubMed]
38. Jones, R.J.; Spoor, G.; Thomasson, A. Vulnerability of subsoils in Europe to compaction: A preliminary analysis. *Soil Tillage Res.* **2003**, *73*, 131–143. [CrossRef]
39. Hakansson, I.; Lipiec, J. A review of the usefulness of relative bulk density values in studies of soil structure and compaction. *Soil Tillage Res.* **2000**, *53*, 71–85. [CrossRef]

40. Rabot, E.; Wiesmeier, M.; Schlüter, S.; Vogel, H.J. Soil structure as an indicator of soil functions: A review. *Geoderma* **2018**, *314*, 122–137. [CrossRef]
41. Lipiec, J.; Usowicz, B. Spatial relationships among cereal yields and selected soil physical and chemical properties. *Sci. Total Environ.* **2018**, *633*, 1579–1590. [CrossRef]
42. Yang, P.; Reijneveld, A.; Lerink, P.; Qin, W.; Oenema, O. Within-field spatial variations in subsoil bulk density related to crop yield and potential CO_2 and N_2O emissions. *Catena* **2022**, *213*, 106156. [CrossRef]
43. Panayiotopoulos, K.; Salonikiou, E.; Siaga, K.; Germanopoulou, V.; Skaperda, S. Effect of uniaxial compression on water retention, hydraulic conductivity and the penetration resistance of six Greek soils. *Int. Agrophysics* **2003**, *17*, 191–197.
44. Tullberg, J.; Antille, D.L.; Bluett, C.; Eberhard, J.; Scheer, C. Controlled traffic farming effects on soil emissions of nitrous oxide and methane. *Soil Tillage Res.* **2018**, *176*, 18–25. [CrossRef]
45. Soane, B.D.; Ball, B.C.; Arvidsson, J.; Basch, G.; Moreno, F.; Roger-Estrade, J. No-till in northern, western and south-western Europe: A review of problems and opportunities for crop production and the environment. *Soil Tillage Res.* **2012**, *118*, 66–87. [CrossRef]
46. Tullberg, J.; Yule, D.; McGarry, D. Controlled traffic farming—From research to adoption in Australia. *Soil Tillage Res.* **2007**, *97*, 272–281. [CrossRef]
47. Rietra, R.; Heinen, M.; Oenema, O. A Review of Crop Husbandry and Soil Management Practices Using Meta-Analysis Studies: Towards Soil-Improving Cropping Systems. *Land* **2022**, *11*, 255. [CrossRef]
48. Troccoli, A.; Maddaluno, C.; Mucci, M.; Russo, M.; Rinaldi, M. Is it appropriate to support the farmers for adopting conservation agriculture? Economic and environmental impact assessment. *Ital. J. Agron.* **2015**, *10*, 169–177. [CrossRef]
49. Pittelkow, C.M.; Liang, X.; Linquist, B.A.; Van Groenigen, K.J.; Lee, J.; Lundy, M.E.; Van Gestel, N.; Six, J.; Venterea, R.T.; Van Kessel, C. Productivity limits and potentials of the principles of conservation agriculture. *Nature* **2015**, *517*, 365–368. [CrossRef] [PubMed]
50. Graven, L.; Carter, P. Seed quality effect on corn performance under conventional and no-tillage systems. *J. Prod. Agric.* **1991**, *4*, 366–372. [CrossRef]
51. Ali Omar, N.; Whittenton, J.B.; Williams, J.J.; Watts, F.; Henry, W.B. Sub-Optimal Temperature Effects on Hybrid Corn Seed and Seedling Performance. *Seed Technol.* **2018**, *39*, 129–142.
52. Peralta, G.; Alvarez, C.R.; Taboada, M.A. Soil compaction alleviation by deep non-inversion tillage and crop yield responses in no tilled soils of the Pampas region of Argentina. A meta-analysis. *Soil Tillage Res.* **2021**, *211*, 105022. [CrossRef]
53. Spoor, G. Alleviation of soil compaction: Requirements, equipment and techniques. *Soil Use Manag.* **2006**, *22*, 113–122. [CrossRef]
54. Bengough, A.G.; McKenzie, B.M.; Hallett, P.D.; Valentine, T.A. Root elongation, water stress, and mechanical impedance: A review of limiting stresses and beneficial root tip traits. *J. Exp. Bot.* **2011**, *62*, 59–68. [CrossRef]
55. Mujdeci, M.; Isildar, A.A.; Uygur, V.; Alaboz, P.; Unlu, H.; Senol, H. Cooperative effects of field traffic and organic matter treatments on some compaction-related soil properties. *Solid Earth* **2017**, *8*, 189–198. [CrossRef]
56. Ranaivoson, L.; Naudin, K.; Ripoche, A.; Affholder, F.; Rabeharisoa, L.; Corbeels, M. Agro-ecological functions of crop residues under conservation agriculture. A review. *Agron. Sustain. Dev.* **2017**, *37*, 26. [CrossRef]
57. Horn, R.; Domżżał, H.; Słowińska-Jurkiewicz, A.; Van Ouwerkerk, C. Soil Compaction Processes and Their Effects on the Structure of Arable Soils and the Environment. *Soil Tillage Res.* **1995**, *35*, 23–36. [CrossRef]
58. Thorsøe, M.H.; Noe, E.B.; Lamandé, M.; Frelih-Larsen, A.; Kjeldsen, C.; Zandersen, M.; Schjønning, P. Sustainable soil management-Farmers' perspectives on subsoil compaction and the opportunities and barriers for intervention. *Land Use Policy* **2019**, *86*, 427–437. [CrossRef]
59. European Commission. *699 Final: EU Soil Strategy for 2030 Reaping the Benefits of Healthy Soils for People, Food, Nature and Climate*; European Commission: Brussels, Belgium, 2021.
60. Chen, K.; Shaw, B. Public communication of soil conservation practices: A large-scale content analysis of Wisconsin's agricultural trade publications. *J. Soil Water Conserv.* **2022**, *77*, 184–197. [CrossRef]
61. Zhou, L.; Monreal, C.M.; Xu, S.; McLaughlin, N.B.; Zhang, H.; Hao, G.; Liu, J. Effect of bentonite-humic acid application on the improvement of soil structure and maize yield in a sandy soil of a semi-arid region. *Geoderma* **2019**, *338*, 269–280. [CrossRef]
62. Tian, S.; Ning, T.; Wang, Y.; Liu, Z.; Li, G.; Li, Z.; Lal, R. Crop yield and soil carbon responses to tillage method changes in North China. *Soil Tillage Res.* **2016**, *163*, 207–213. [CrossRef]
63. He, J.; Shi, Y.; Zhao, J.; Yu, Z. Strip rotary tillage with subsoiling increases winter wheat yield by alleviating leaf senescence and increasing grain filling. *Crop J.* **2019**, *8*, 327–340. [CrossRef]
64. Fen, W.; Zhai, L.; Xu, P.; Zhang, Z.; Baillo, E.; Tolosa, L.; Kimotho, R.; Jia, X.; Guo, H. Effects of deep vertical rotary tillage on the grain yield and resource use efficiency of winter wheat in the Huang-Huai-Hai Plain of China. *J. Integr. Agric.* **2021**, *20*, 593–605. [CrossRef]
65. FAO. *The Future of Food and Agriculture–Trends and Challenges*; Food and Agriculture Organisation: Rome, Italy, 2017; ISBN 978-92-5-109551-5.
66. Hallett, P.; Balana, B.; Towers, W.; Moxey, A.; Chamen, T. *Studies to Inform Policy Development with Regard to Soil Degradation: Subproject A: Cost Curve for Mitigation of Soil Compaction*; The James Hutton Institute: Dundee, Scotland, 2012.
67. Mariotti, B.; Hoshika, Y.; Cambi, M.; Marra, E.; Feng, Z.; Paoletti, E.; Marchi, E. Vehicle-induced compaction of forest soil affects plant morphological and physiological attributes: A meta-analysis. *For. Ecol. Manag.* **2020**, *462*, 118004. [CrossRef]

68. Hoogmoed, W.; Berkhout, J.; Stroosnijder, L. Soil tillage options for water management under erratic-rainfall conditions. In Proceedings of the Tillage in Arid and Semi-Arid Areas, an International Seminar, Rabat, Morocco, 22–23 April 1992.
69. Keller, T.; Colombi, T.; Ruiz, S.; Manalili, M.P.; Rek, J.; Stadelmann, V.; Wunderli, H.; Breitenstein, D.; Reiser, R.; Oberholzer, H.; et al. Long-Term Soil Structure Observatory for Monitoring Post-Compaction Evolution of Soil Structure. *Vadose Zone J.* **2017**, *16*, 1–16. [CrossRef]
70. Ball, B.; Bingham, I.; Rees, R.M.; Watson, C.A.; Litterick, A. The role of crop rotations in determining soil structure and crop growth conditions. *Can. J. Soil Sci.* **2005**, *85*, 557–577. [CrossRef]

land

MDPI

Review

A Review of Crop Husbandry and Soil Management Practices Using Meta-Analysis Studies: Towards Soil-Improving Cropping Systems

René Rietra *, Marius Heinen and Oene Oenema

Wageningen Environmental Research, Wageningen University & Research, Droevendaalsesteeg 3, 6708 PB Wageningen, The Netherlands; marius.heinen@wur.nl (M.H.); oene.oenema@wur.nl (O.O.)
* Correspondence: rene.rietra@wur.nl

Abstract: Coherent improvements in crop varieties and crop husbandry and soil management practices are needed to increase global crop production in a sustainable manner. However, these practices are often discussed separately, and as a result there is little overview. Here, we present a database and synthesis of 154 meta-analysis studies related to ten main crop husbandry and soil management practices, including crop type and rotations, tillage, drainage, nutrient management, irrigation and fertigation, weed management, pest management, crop residue management, mechanization and technology, and landscape management. Most meta-analysis studies were related to tillage (55), followed by crop type and rotations (32), nutrient management (25), crop residue management (19), and irrigation and fertigation (18). Few studies were related to landscape management (6) and mechanization and technology (2). In terms of outcome, studies focused on crop yield and quality (81), soil quality (73), and environmental impacts (56), and little on economic effects (7) or resource use efficiency (24). Reported effects of alternative practices, relative to conventional practice, were positive in general. Effect sizes were relatively large for environmental effects (nutrient leaching, greenhouse gas emissions), and small for soil quality (except for soil life) and crop yield. Together, meta-analysis studies indicate that there is large scope for increasing cropland productivity and minimizing environmental impacts. A roadmap is provided for integration and optimization of all ten practices, and recommendations are formulated to address the gaps in meta-analysis studies.

Keywords: crop residue; crop rotation; crop yield; environmental effects; irrigation; nutrient management; resource use; soil-improving cropping systems; soil quality; tillage

Citation: Rietra, R.; Heinen, M.; Oenema, O. A Review of Crop Husbandry and Soil Management Practices Using Meta-Analysis Studies: Towards Soil-Improving Cropping Systems. *Land* **2022**, *11*, 255. https://doi.org/10.3390/land11020255

Academic Editors: Guido Wyseure, Julián Cuevas González and Jean Poesen

Received: 13 January 2022
Accepted: 30 January 2022
Published: 8 February 2022

1. Introduction

Global yields of main crops (wheat, rice, maize, and soybean) have increased by an average 1 to 2% per year during the last decades [1,2], in response to the increasing global food and feed demands, and facilitated through technological improvements. Forecasts suggest that mean crop yields per ha of cropland have to increase by as much as 2.4% per year to be able to meet the food and feed demands by the human population in 2050, also because further expansion of global cropland area and/or increased frequency of harvesting are not feasible [3,4]. The slow-moving mean increase in global crop yields during recent decades are in part related to areas where crop yields have been stagnating and to areas where crop yields have not increased at all or have fallen. Recent analyses suggest that crop yields are not increasing on 25 to 40% of the harvested global cropland area [1]. Yield increases of wheat, maize, and rice tend to be lowest in low-income countries because of lack of resources and poor crop husbandry practices. In high-income countries, yield increases may be less than average when actual yields approach attainable crop yields, suggesting that yields reach biophysical limits [5,6]. Crop yields may also stagnate in some countries because of climate change and environmental regulations [7–9] and soil degradation [10–12].

The yield increases per unit of surface area during recent decades have mainly been the result of improved germplasm and improved crop husbandry and soil management practices, including inputs of fertilizers, irrigation, and pesticides [13]. Availability of high-yielding cultivars, fertilizers, irrigation water, and pesticides are commonly considered to be the dominant yield-controlling factors, next to climate and soil quality. However, the importance of precise timing and careful execution of the various crop husbandry practices in the proper order should not be neglected [13]. The crop husbandry and soil management practices together determine how far actual crop yields deviate from attainable crop yields and from potential crop yields [14]. Attainable crop yields, defined as the best yield achieved by the best farms through skillful use of the best available technology [14], are on average 70 to 80% of the potential yield. Potential crop yields are commonly defined as the yields obtained when cultivars adapted to the local environmental conditions are grown with minimal stress, achieved with best management practices [11,15,16]. Actual yields on farmers' fields range from 30 to 100% of attainable yields, depending on region [1].

Crop husbandry and soil management practices also influence the environmental sustainability of crop production systems, especially in cases where the pressures to increase crop yields are high. Concerns have arisen about intensive crop production systems with poor crop husbandry and soil management practices, as these pollute groundwater and surface waters with nitrogen (N), phosphorus (P) and pesticides, and emit greenhouse gases and ammonia (NH_3) into the atmosphere [15–17]. There are also concerns about soil degradation through processes such as erosion, salinization, compaction, and declines of soil organic matter content and soil biodiversity [10]. The United Nations (UN) Sustainable Development Goals (SDGs) address essentially all of these concerns and indirectly guide the actions of nations in the pursuit of a more sustainable world. Of the 17 SDGs, at least five have a direct relation with cropping systems and soils, while others have a more indirect relation [18]. SDG-2 aims to 'end hunger, achieve food and nutrition security, and promote sustainable agriculture' and is key to the success of the SDG agenda [19].

While there are several spatially explicit assessments of changes in crop yields over time (e.g., [2,20], there are no spatially explicit, integrated assessments of the sustainability of crop husbandry and soil management practices. The main reason for this lack of assessments is the diversity of crop husbandry and soil management practices, and the lack of methods and procedures for making such integrated assessments. Wezel et al. [21] analyzed 15 agroecological cropping practices qualitatively in terms of possible advantages and drawbacks, for temperate areas. Others have reviewed the impacts of one or a few specific crop husbandry practices (e.g., [22–24], often on the basis of a meta-analysis of published studies. There is as yet no coherent overview and comparison of the effects of all main crop husbandry and soil management practices.

The aim of this study was to provide a review of crop husbandry and soil management practices on the basis of meta-analysis studies. Meta-analysis papers commonly analyze and synthesize many experimental studies related to topical research questions and/or ambiguous research findings. The term 'meta-analysis' was first used in 1976 and referred to 'the statistical analysis of a large collection of analysis results from individual studies for the purpose of integrating the findings' [25,26]. Most meta-analysis studies related to crop husbandry and soil management practices date from the last 10 to 20 years, and the cumulative number has increased exponentially (Figure 1).

Thus, our main hypothesis is that meta-analysis studies summarize and synthesize vast amounts of research results, and unravel underlying mechanisms of variations, and thereby provide overview. By reviewing and synthesizing meta-analysis studies related to several crop husbandry and soil management practices, we aimed to (i) summarize the main impacts of these crop husbandry and soil management practices, (ii) identify the most topical research areas, and (iii) suggest guidelines for 'sustainable cropping systems'. The crop husbandry and soil management practices examined were assessed in terms of (a) crop yield and quality, (b) soil quality, (c) economic effects, (d) resource use efficiency,

(e) environmental effects, and (f) human health impacts. However, none of the reviewed studies addressed human health impacts; as a consequence, this aspect is not reported here.

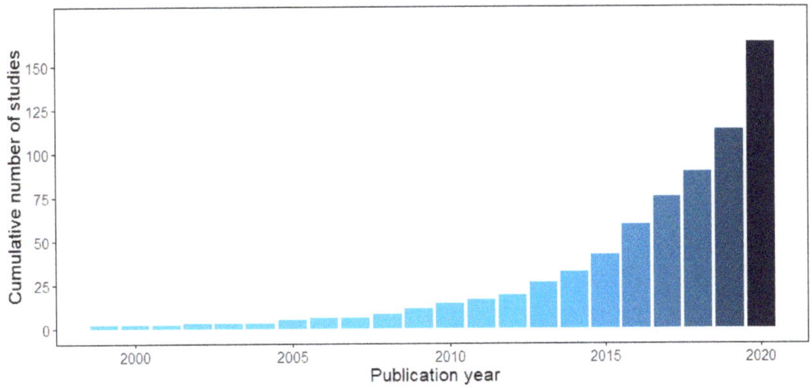

Figure 1. Exponential increase in time of published studies used in our overview (the bar left of 2000 refers to all studies before the year 2000). In total 163 unique studies (peer-reviewed publications) were used: 154 meta-analysis studies and 9 reviews.

2. Materials and Methods

2.1. Data Collection

We reviewed meta-analyses studies related to crop husbandry and soil management practices (henceforth 'practices'). A total of ten main categories of practices were examined: (1) crop type and crop rotations including intercropping, cover crops, (2) nutrient management, (3) irrigation and fertigation, (4) controlled drainage, (5) tillage practices, (6) pest management, (7) weed management, (8) crop residue management including mulching, (9) mechanization including precision technology, and (10) landscape management, including hedgerows, tree lines, buffer strips. These ten categories of practices relate to the main crop yield defining, limiting, and reducing factors [11,15,16], and with climate-related factors (not included here) have very dominant effects on crop yield and quality, soil quality, and the environmental impacts of crop production.

For each of these categories, quantitative effects of specific practices were distilled from the meta-analysis studies. In most studies, an improved or modified practice was compared with the conventional practice. We focused on the following five outcomes (impacts): (a) crop yield and quality, (b) soil quality, (c) farm income, (d) resource use efficiency, and (e) environmental effects. We attributed the indicators that were used in the meta-analysis to these five outcomes. For crop yield and quality, farm income, and resource use efficiency, a limited number of straightforward indicators were used commonly, but for soil quality and environmental effects a wide range of indicators have been reported. We made no selection in these indicators. We focused on effect sizes defined as the standardized mean difference between the effect of a specific treatment practice relative to that of the control treatment. It is often given as the response ratio (RR) which is the ratio of the effect of a specific treatment (Xt) and the control treatment (Xc), with or without natural log of the ratio.

We collected data from peer-reviewed meta-analysis publications only. The publications were identified using the online database Scopus (https://www.scopus.com/sources, accessed on 5 May 2021) within the period 1997–2020. Publications were searched using the keywords "meta-analysis" and the results were refined using additional words—"crop type", "crop rotation", "nutrient management", "fertilization", "irrigation", "fertigation", "drainage", "tillage", "pest management", "disease management", "weed management", "crop residue management", "mulching", "mechanization", "landscape management", "hedgerows", and "buffer strips". These keywords were searched for in the title, abstract, and keywords. Additionally, we used the forward and backward snowballing technique when applicable and a few review studies that presented quantitative data, such as a meta-analysis, were included as well. Further information about the selection and analysis of data is provided in the Supplementary Materials.

A list of abbreviations used in this publication is given in Table A1 (Appendix A).

2.2. Data Compilation and Analysis

Data from the meta-analysis studies were compiled in Windows Excel. The region of the study, the specific practice, the conventional (control) practice, the results, the units and the number of observations were recorded. No further data processing and analyses of the data were undertaken. The Windows Excel database, with all results extracted from the meta-analyses studies, is in the Supplementary Materials.

3. Results

3.1. Overview

Table 1 presents an overview of the meta-analysis studies across categories of practices. Most of the meta-analysis studies dealt with tillage practices (55). Crop type and crop rotations (32), nutrient management (25), irrigation and fertigation (18), and crop residue management (19) have also been analyzed frequently. In contrast, only two studies related to mechanization and (precision) technology.

Table 1. Summary of the number of reviewed meta-analysis studies across crop husbandry and soil management practices, and across aspects (outcome). Note that the sum of the studies for the different aspects can be larger than the number of meta-analysis studies since some studies reported on several aspects.

		Number of Meta-Analysis Studies per Aspect				
Crop Husbandry and Soil Management Practices	**Total**	**Crop Yield & Quality**	**Soil Quality**	**Resource Use Efficiency**	**Economic Aspects**	**Environmental Impacts**
1 Crop type and crop rotations	32	12	12	2	1	14
2 Nutrient management #	25	12	9	0	1	7
3 Irrigation and fertigation	18	12	2	11	0	4
4 Drainage	6	1	1	0	1	4
5 Tillage	55	19	36	5	2	14
6 Pest management	7	3	3	0	0	1
7 Weed management	4	2	2	0	0	0
8 Crop residue and mulching	19	14	5	6	1	8
9 Mechanization and technology	2	3	1	0	1	0
10 Landscape management	6	3	2	0	0	4
Total	174 &	81	73	24	7	56

#: one reference included the human health related aspect survival time of zoonotic pathogens; &: The total number of studies reported here consisted of 163 unique publications, some of which considered more than one crop husbandry or soil management practice.

Most meta-analysis studies examined the effects of specific practices on crop yield and quality (81). For soil quality (73 studies), soil organic matter content was the main focus. For environmental effects (56 studies), the focus was mainly on greenhouse gas emissions and nitrate leaching. Resource use efficiency was examined mainly for irrigation and fertigation, nutrient management and tillage. Only seven meta-analysis studies included economic aspects (Table 1).

The meta-analysis studies reviewed covered a large number of experimental studies and practices in different parts of the world. Each meta-analysis study was based on a large number of underlying studies (on average more than 100; range 8 to 678). In order to attain an impression of how often literature sources have been used in multiple meta-analysis studies, we collected and examined the literature sources of the 55 meta analyses on tillage. In most cases, references to the original studies were provided in the supporting information, but for seven out of the 55 meta-analyses studies no references were made available. For the remaining 48 studies we collected in total 5465 references to original studies. These were then manually checked on replicate use. Over two-third of these references were used only once, 26% of these were used in two meta-analysis studies, 4% were used three times, and 2% were used four times. Three references were used in seven meta-analyses studies. We conclude that essentially all meta-analyses related to tillage were based on unique studies, which replicated use of original studies is relatively small (given the large number of meta-analyses related to tillage), and that the results of these meta-analyses are largely independent on each other therefore. We did not check repeated use of original studies for other categories of practices.

3.2. Crop Type and Crop Rotation

Selecting the proper crop varieties and crop rotations is often farm and region-specific and key to successful crop farming. Crop rotation is the practice of planting different crops sequentially on the same field, mainly to combat pest and weed pressures and improve soil quality, and thereby to enhance crop yield sustainably. Crop rotations have been the subject of many meta-analysis (Table 1), whereby almost equal attention has been given to crop yield, soil quality, and environmental effects, but little attention to the economic aspects and to resource use efficiency. Specific crop varieties and cultivars have not been the subject of meta-analysis.

Effects of crop rotation, intercropping, and cover crops on crop yield, soil quality, and the environment were positive in almost all studies (Figure 2). Pre-crops before wheat [27] and especially legumes as pre-crops [28] had positive effects on wheat yield, soil quality, and pesticide use [29]. However, effects of pre-crops depend on the nitrogen fertilization rate: yield benefits are highest under low nitrogen fertilization [28]. Indeed, interactions with other crop husbandry and soil management hold for many crop rotation effects; nutrient management, irrigation, pest, disease, and weed management all have a large impact on the effect size of crop rotations [30–32].

The simultaneous cultivation of two or more crop species within one field for at least a part of the growing period (intercropping) also has positive effects on crop yield, but the effect size strongly depends on the crop types and intercropping patterns [30,33]. Growing cover or catch crops after the main crop reduces soil erosion and nitrate leaching and contributes to soil carbon sequestration [34,35], but requires labor and the suppressive effects on pest, diseases and weeds are not always positive. Growing mixtures of varieties of cereals [36] or mixtures of grasses [37] has positive effects on yield (stability) and nitrogen use efficiency.

Effects of crop rotations on GHG emission are variable [38]; this holds also for the effects on cover crops on GHG emissions [39].

3.3. Nutrient Management

The 25 meta-analysis studies related to nutrient management have paid more or less equal attention to crop yield and quality, soil quality, and environmental effects, but little or no attention to economic effects and resource use efficiency (Table 1). Almost all studies reported significant positive effect sizes of the studied nutrient management practices relative to conventional practices (Figure 3).

A main focus has been the characterization of differences between fertilizer types, especially between organic and mineral fertilizers [65,66,78,80] and between 'conventional' fertilizers and fertilizers with inhibitors [71,74], in relation to fertilizer effectiveness, soil quality, and environmental impacts. Deriving the optimal nutrient application rates have been the topic of many experimental studies in the past, and this has also been the subject of several meta-analysis studies [64,68,70]. Better timing of fertilization and placement of fertilizers gave positive effects on yields in most cases [75,84]. Soil liming increased crop yields, especially when pH was low [67]. Positive effects of organic soil amendments and mineral fertilizers on soil biological activity and microbial biomass were found, while the response of soil enzyme activity depended on enzyme type [68,70,80,81]. Nitrous oxide emissions from cropland increase with nitrogen fertilization, but the increase can be mitigated through better compliance with fertilizer recommendations, and the use of nitrification inhibitors and biochar [72]. Slurry acidification, deep placement, and urease inhibitors decreases ammonia emissions from slurries and urea fertilizers applied to soil [76,86]. No meta-analysis studies related to the effectiveness of manure products from different manure processing techniques [87]. Increasing grazing intensity of pastures increased C, N, and P losses from these pastures ([83] as well as the transfer of zoonotic pathogens to water courses [77]). Only few studies pointed at the effects of interactions between categories of practices, including interactions between intercropping, tillage, and fertilizers types in fruit yield [63], interactions between fertilization, and irrigation in fruit yields [70,88] and in maize yields [70].

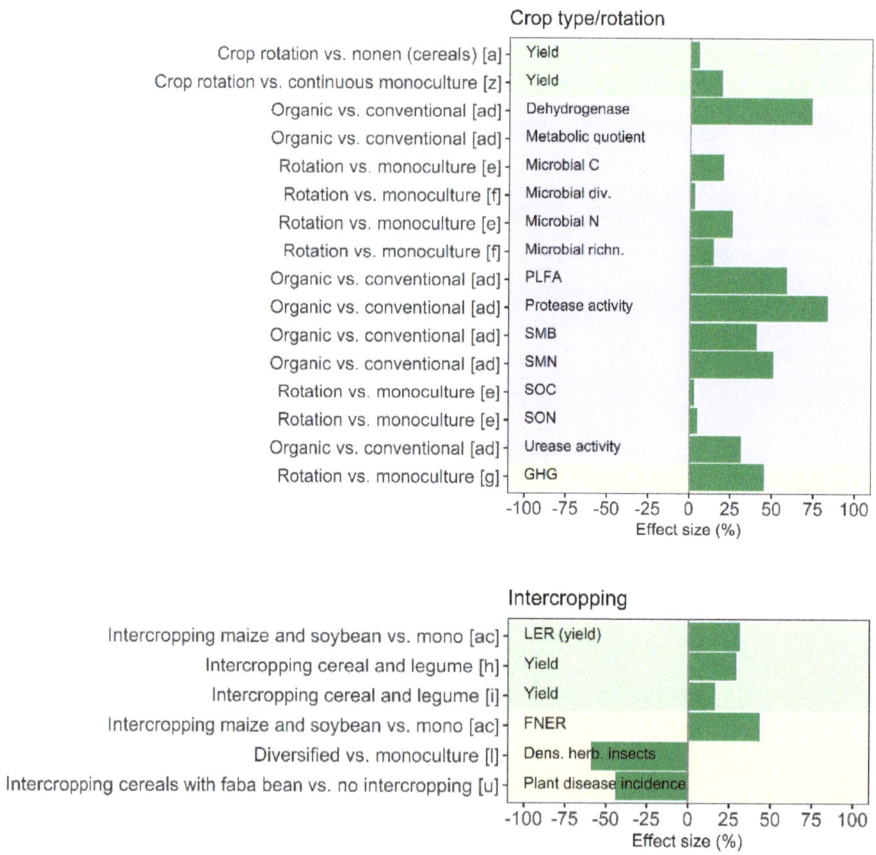

Figure 2. *Cont.*

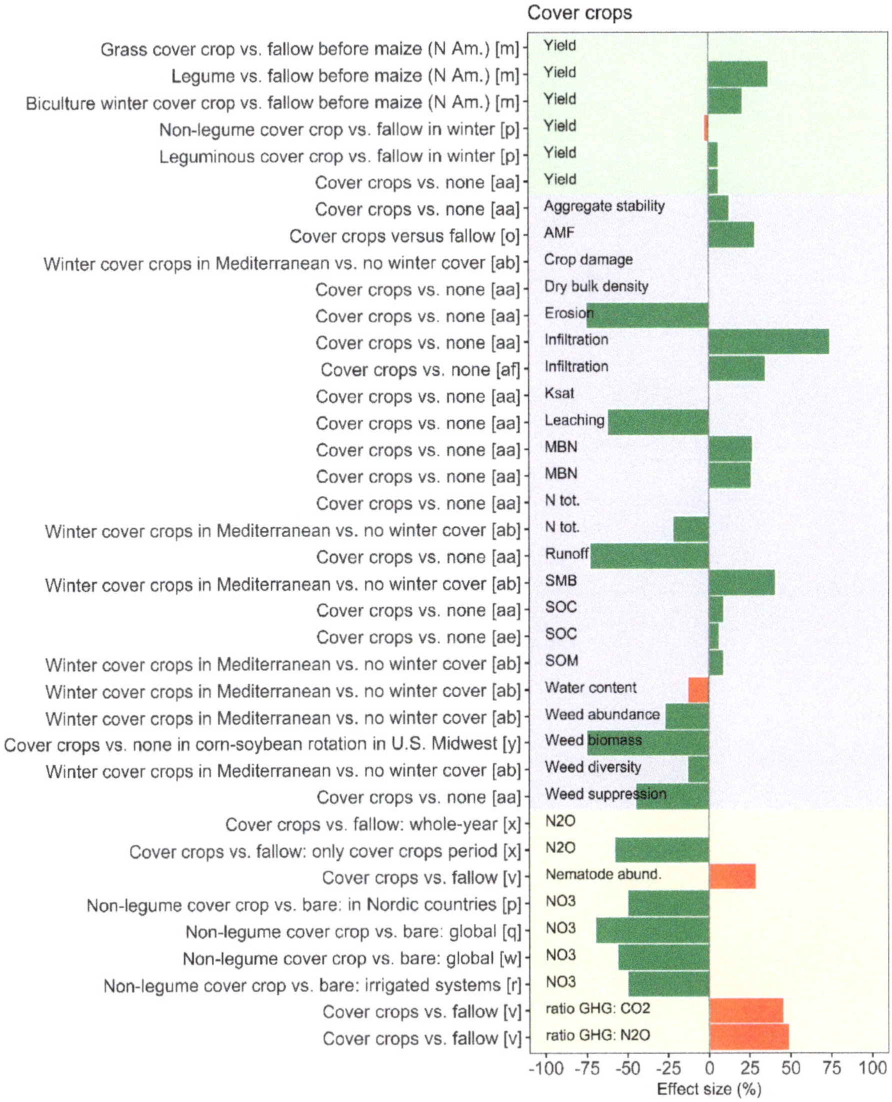

Figure 2. Overall effect sizes (response ratios) reported in meta-analysis studies on cropping (split in crop types and crop rotation, cover crops, intercropping, and perennial crops) grouped per area of interest (light green: agronomic; light blue: soil quality; light orange: resource use efficiency; light brown: economic (no data); light yellow: environmental impacts). The management or treatment comparison is indicated outside the y-axis and the variable to which the data refer are listed inside the y-axis (abbreviations can be found in Table A1). Green bars indicate improvement, red bars indicate worsening. See also Table S1. [a]: [24]; [b]: [28]; [c]: [27]; [d]: [40]; [e]: [41]; [f]: [42]; [g]: [38]; [h]: [43]; [i]: [44]; [j]: [30]; [k]: [45]; [l]: [46]; [m]: [47]; [n]: [48]; [o]: [49]; [p]: [32]; [q]: [31]; [r]: [50]; [s]: [51]; [t]: [52]; [u]: [33]; [v]: [53]; [w]: [54]; [x]: [39]; [y]: [55]; [z]: [56]; [aa]: [57]; [ab]: [58]; [ac]: [59]; [ad]: [60]; [ae]: [61]; [af]: [62].

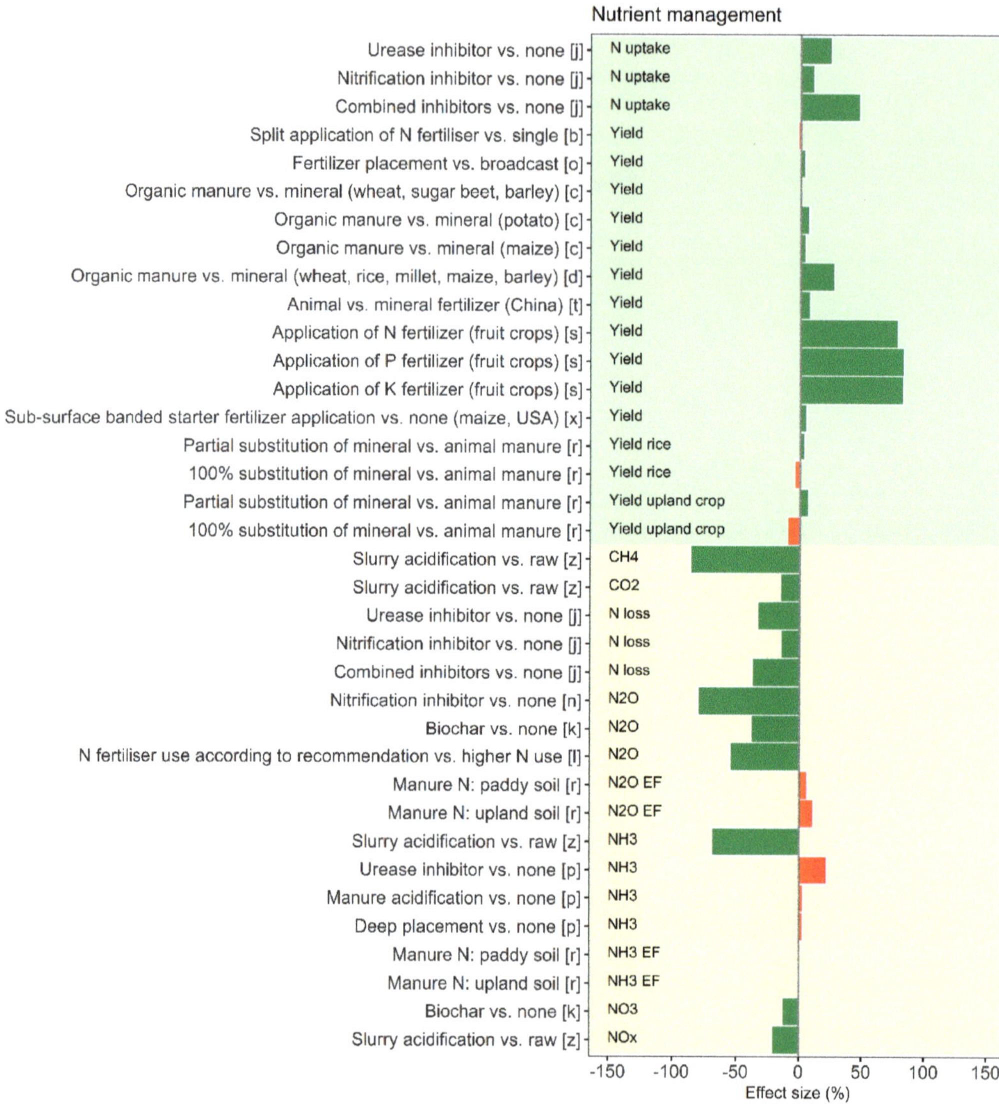

Figure 3. *Cont.*

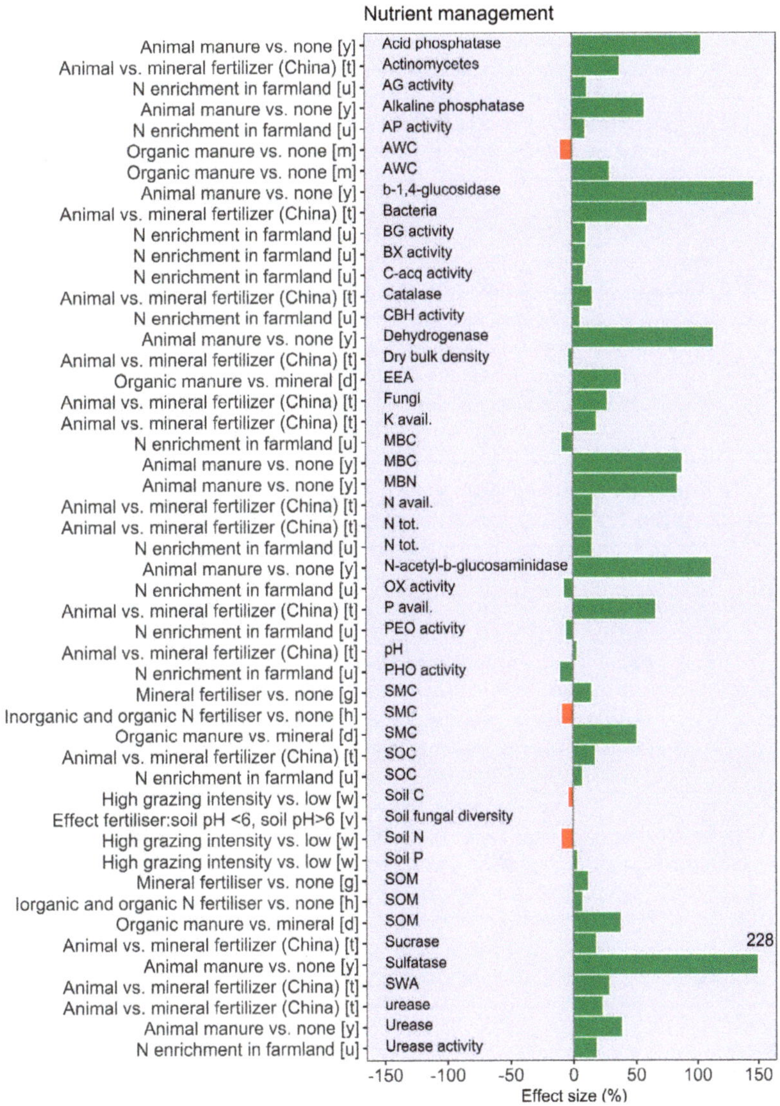

Figure 3. Overall response ratios reported in meta-analysis studies on nutrient management grouped per area of interest (light green: agronomic; light blue: soil quality; light orange: resource use efficiency (no data); light brown: economic (no data); light yellow: environmental impacts). The management or treatment comparison is indicated outside the y-axis and the variable to which the data refer are listed inside the y-axis (abbreviations can be found in Table A1). Green bars indicate improvement, red bars indicate worsening. See also Table S2. [a]: [63]; [b]: [64]; [c]: [65]; [d]: [66]; [e]: [67]; [f]: [68]; [g]: [69]; [i]: [70]; [j]: [71]; [k]: [72]; [l]: [39]; [m]: [73]; [n]: [74]; [o]: [75]; [p]: [76]; [q]: [77]; [r]: [78]; [s]: [79]; [t]: [80]; [u]: [81]; [v]: [82]; [w]: [83]; [x]: [84]; [y]: [85]; [z]: [86].

3.4. Irrigation and Fertigation

A total of 18 meta-analysis studies related to irrigation and/or fertigation, mainly examining the effects of irrigation methods and amounts on crop yield and water use efficiency for different cropping systems and regions (Table 1). The relative strong focus on water use efficiency reflects that irrigation water is a scarce resource. Most studies reported positive effect sizes of irrigation practices on crop yield and water use efficiency relative to conventional irrigation practices (Figure 4). Effect sizes of water productivity of optimal irrigation and deficit irrigation ranged from 20 to 80%. However, some studies also reported negative effects of deficit irrigation practices relative to conventional irrigation practices, possibly because irrigation was reduced too much in deficit irrigation treatments.

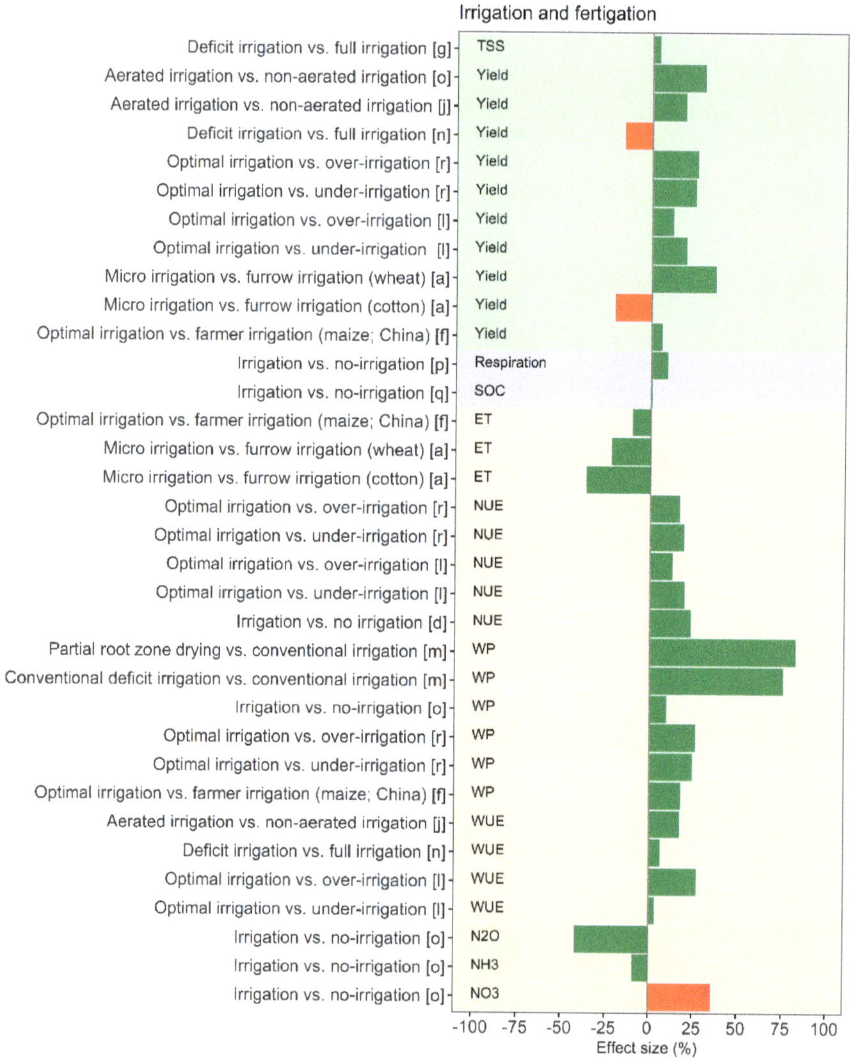

Figure 4. Overall response ratios reported in meta-analysis studies on irrigation and fertigation grouped per area of interest (light green: agronomic; light blue: soil quality; light orange: resource use efficiency; light brown: economic (no data); light yellow: environmental impacts). The management or treatment comparison is indicated outside the y-axis and the variable to which the data refer are listed inside the y-axis (abbreviations can be found in Table A1). Green bars indicate improvement, red bars indicate worsening. See also Table S3. [a]: [94]; [b]: [90]; [c]: [95]; [d]: [96]; [e]: [91]; [f]: [97]; [g]: [98]; [h]: [99]; [i]: [89]; [j]: [100]; [k]: [101]; [l]: [88]; [m]: [102]; [n]: [103]; [o]: [104]; [p]: [92]; [q]: [93]; [r]: [70].

Two meta-analysis studies examined the effects of irrigation method on emissions of N_2O [89,90], while one study examined flood irrigation practices on emissions of methane from paddy rice [91]. Two studies examined effect of irrigation on soil respiration and soil carbon contents [92,93]. Only two studies pointed at large effects of interactions between irrigation and fertilization in water use efficiency and nutrient use efficiency [70,88].

3.5. Controlled Drainage

Effects of controlled drainage on the loss of water, nutrients, and greenhouse gases were assessed through six meta-analysis studies (Table 1). Controlled drainage is defined as the use of adjustable head structures to prevent discharge when the water table is lower than the outlet elevation. In this way the loss of water and nutrients may be altered, depending on the target. The quantitative effects of controlled drainage on reducing drainage volumes, N losses, and methane emissions were relatively large (range 17 to 85%) (Table 2). Generally, controlled drainage resulted in reduced drainage volumes, depending on soil type [105]. Controlled drainage also reduced N-losses via drainage water to surface water [106–108] and methane (CH_4) emissions from peat lands [109]. No impact on yield was found by [108]. Alternating wetting and drying cycles in paddy rice greatly decreased CH_4 emissions, but increased N_2O emission; yet total greenhouse gas emissions decreased through improved water management [110].

Table 2. Controlled drainage: effect sizes as reported in meta-analysis studies. See also Table S4.

Parameter	Comparison of Treatments	Main Results
Yield	Drainage vs. none	not significant [e]
Economic benefit	Drainage vs. none	9 to 37 $ ha^{-1} yr^{-1} [c]
CH4 emission from paddy rice field	Wetting and drying vs. continuous flooding	−35% [f] −29% for CH4+N2O (net GWP) [f]
CH4 emission from peat	Drainage vs. none	−84% [a]
Drainage volume	Drainage vs. none	−47% [b]; −17% to -85% [d]; -19% [e]
N-load	Drainage vs. none	−41% [c]; −18% to −85% [d]; −32% [e]
P-load	Drainage vs. none	−19% [e]

[a]: [109]; [b]: [105]; [c]: [106]; [d]: [107]; [e]: [108]; [f]: [110].

3.6. Tillage

Tillage refers to the preparation of the soil for growing crops, with or without incorporation of crop residues in the soil and/or weed control. In conventional or traditional tillage (TT), the topsoil (usually the upper 15 to 25 cm) is turned and/or milled. Conservation tillage (including no-tillage (NT) or reduced tillage (RT)) is the practice of minimizing soil disturbance, whereby crop residues commonly remain on the soil surface to protect the soil, while herbicides or precision mechanical weeding tools are used to control weeds. Tillage practices are debated because of high fossil energy and labor costs, and their effects on soil erosion, crop yield, soil organic carbon, and soil biodiversity. This debate is reflected in the high number (55) of meta-analysis studies (Table 1). The focus of most meta-analysis studies has been on soil quality (36), followed by crop yield effects (19) and environmental effects (14). Two studies synthesized economic implications of different tillage practices (Table 1). Of the total number of studies, 20 were global studies, 14 studies related to (parts of) China, 5 to the Mediterranean, 3 to US, 1 to South Asia, 1 to Brazil, 1 to Europe and none to Africa.

Overall, conservation tillage decreased crop yields, increased soil organic carbon contents in the topsoil, increased soil biodiversity and the abundance of soil organisms, and increased N_2O emissions relative to conventional tillage, but the magnitude of the differences depended on climate and the particular study (Figure 5). Yield penalties of no-till depended on crop residue return and crop rotation and were larger in tropical than temperate regions, and tended to decrease with an increase in the duration of no-till [22,111]. The South-Asian study was probably the most integrated one, as it examined effect sizes of crop yield, water use, soil organic C sequestration, emissions of CO_2, CH_4, and N_2O, and economic costs [112]. The cost of production was significantly lower under no-till than under conventional tillage in all the selected crops, and the net economic returns increased by 5 to 32%. Manley et al. [113] examined the economic cost of soil carbon sequestration in the US through no-till. They found that the additional carbon sequestration of no-till compared to conventional till was small and variable, and as a result, the net economic benefit also varied widely.

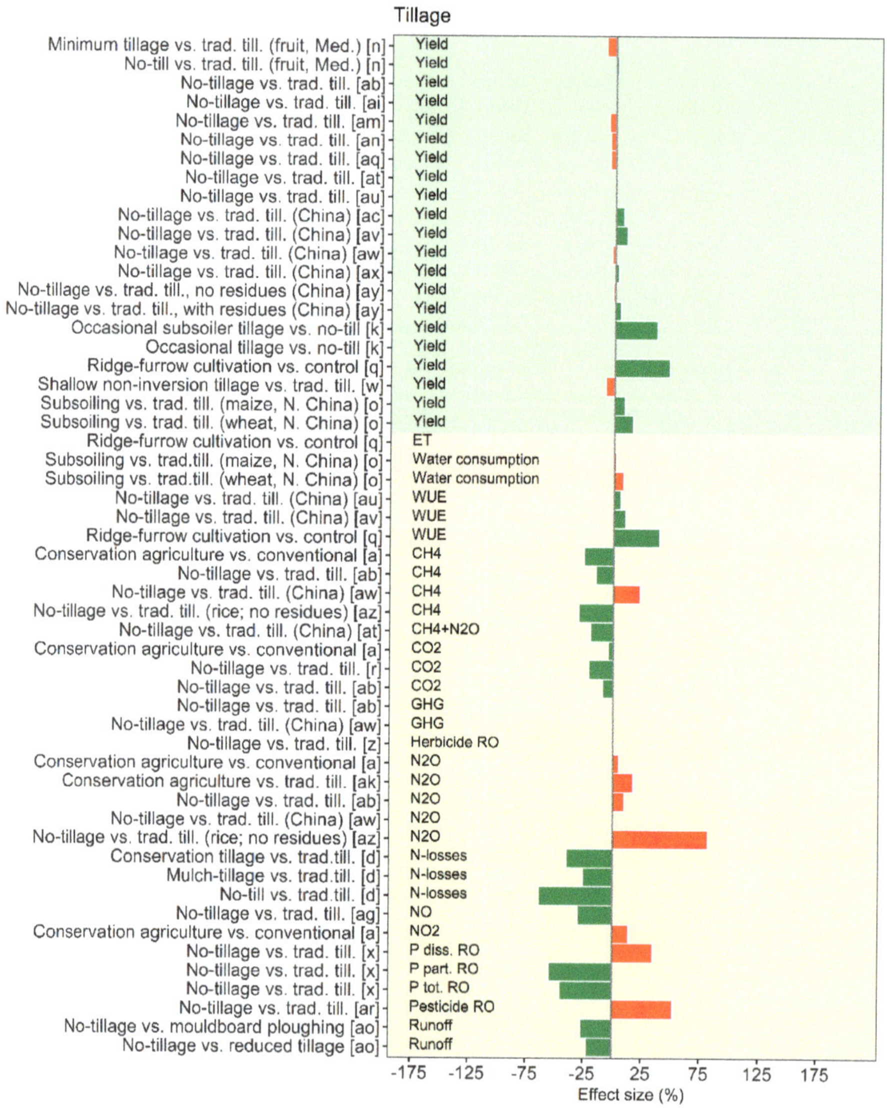

Figure 5. *Cont.*

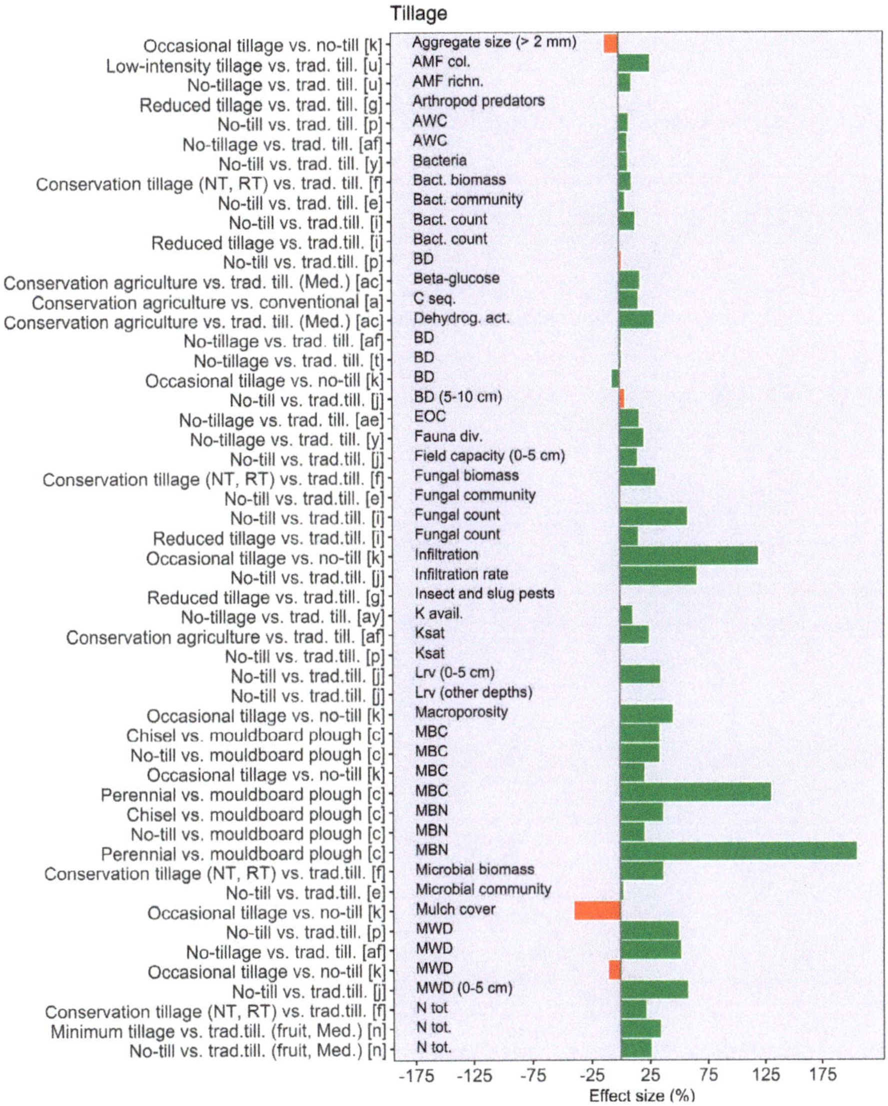

Figure 5. *Cont.*

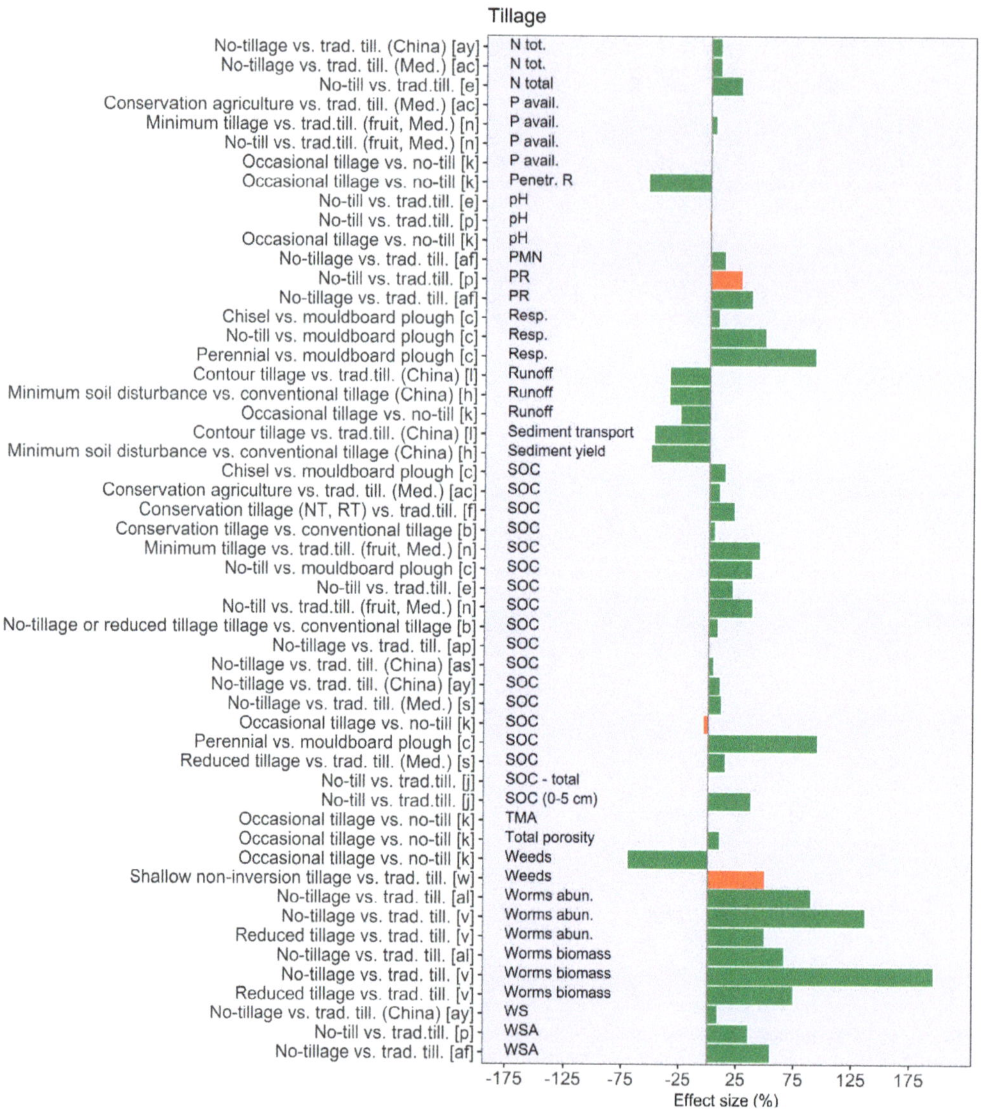

Figure 5. Overall response ratios reported in meta-analysis studies on tillage grouped per area of interest (light green: agronomic; light blue: soil quality; light orange: resource use efficiency; light brown: economic (no data); light yellow: environmental impacts). The management or treatment comparison is indicated outside the y-axis and the variable to which the data refer are listed inside the y-axis (abbreviations can be found in Table A1). Green bars indicate improvement, red bars indicate worsening. See also Table S5. [a]: [112]; [b]: [61]; [c]: [123]; [d]: [123]; [e]: [124]; [f]: [125]; [g]: [124]; [h]: [120]; [i]: [115]; [j]: [118]; [k]: [119]; [l]: [126]; [m]: [127]; [n]: [63]; [o]: [128]; [p]: [129]; [q]: [130]; [r]: [131]; [s]: [132]; [t]: [133]; [u]: [49]; [v]: [114]; [w]: [134]; [x]: [53]; [y]: [135]; [z]: [122]; [aa]: [136]; [ab]: [137]; [ac]: [138]; [ad]: [139]; [ae]: [140]; [af]: [117]; [ag]: [74]; [ah]: [141]; [ai]: [142]; [aj]: [113]; [ak]: [143]; [al]: [116]; [am]: [111]; [an]: [22]; [ao]: [121]; [ap]: [144]; [aq]: [24]; [ar]: [145]; [as]: [146]; [at]: [147]; [au]: [148]; [av]: [149]; [aw]: [150]; [ax]: [151]; [ay]: [152]; [az]: [153].

Several studies found positive effect sizes of no-till versus conventional till for number of earthworms and for the diversity of the (micro) biological community (e.g., [114–116]). No-till tended to increase the bulk density in the lower part of the topsoil (10–20 cm) and the water infiltration rate significantly [117,118]. However, no-till combined with occasional conventional tillage decreased soil bulk density compared to conventional tillage [119]. Effects of no-till on erosion are strongly affected by crop type and soil surface mulching; on average no-till and conservation agriculture reduced erosion [120,121], but pesticides in runoff tended to increase [122].

3.7. Pest Management

Pest management refers to the control of the number of undesirable organisms (pathogens, pest organisms) below an acceptable threshold, which is often based on economic principles. Methods of control can be crop rotation, chemical, biological, physical/mechanical, and/or genetic. There are often interactions with crop residue management, tillage, nutrient management, irrigation, and landscape management [154]. We found seven meta-analysis studies related to pest management (Table 1), of which three were in the context of comparing organic versus conventional agriculture (Table 3). Muneret et al. [155] found that organic farming experiences higher levels of pest infestation, but is able to match or outperform conventional pest control practices against plant pathogens and animal pests. Lesur-Dumoulin et al. [156] found that yields in organic horticulture were on average 10 to 32% lower than yields in conventional horticulture (Table 3). Garratt et al. [157] observed that organic farming practices can increase natural enemy numbers and also pest responses. Fertilization tends to increase insects and fungal plant pathogens [158,159]. Biofumigation through incorporating Brassicaceae plants and crop residues, which release glucosinolates and isothiocyanates, in soil reduced pest abundance and subsequently increased crop yield by 30% [160]. Anaerobic soil disinfestation, through temporal soil sealing following incorporation of labile organic carbon in the soil, is also effective against soil borne pathogens [161]. Furthermore, it has been indicated that addition of organic amendments and improving soil quality and biodiversity may result in fewer pests [162].

Table 3. Pest management: main effects as reported in meta-analysis studies. See also Table S6.

Parameter	Management Practices	Result
Yield	Biofumigation	Abs. diff.: 29% [a]
Yield	Anaerobic soil disinfestation	Abs. diff.: 30% [b]
Suppression of pathogens	Anaerobic soil disinfestation	Abs. diff.: 70% [b]
Yield	Organic/conventional	Ratio: 0.83 [c]
Disease severity response by fungal plant pathogens	Fertilized vs. unfertilized	Increase 0.3 ± 0.1 [d]
Change in insect population	Fertilization	Increase/decrease 175/78 [e]
Change in pest population	Organic/non-organic	Increase/decrease 42/26 [f]

[a]: [160]; [b]: [161]; [c] [156]; [d]: [159]; [e]: [158]; [f]: [157].

3.8. Weed Management

Weed management refers to the control of the number of weed plants (especially noxious weeds) to below an acceptable threshold, as weeds compete with the crop for light, water and nutrients. Weed management often includes a number of methods, including crop rotation/intercropping/cover crops, soil cultivation (weeding, hoeing), mulching (crop residues or plastic covers), herbicides spraying, and burning. We found four meta-analysis studies related to weed control (Table 2).

Verret et al. [163] found that intercropping with legume companion plants enhanced weed control, generally without reducing the yield of the main crop (Table 4). Cover crops can also decrease the incidence of weeds and may have other ecosystem services [164]. Crop rotation with different planting dates and crop diversification, combined with limited soil disturbance, can disrupt weed-crop associations in addition to reducing yield loss and rebuilding soil fertility [165–167]. Glyphosate is the most used chemical weed control agent [168], but is debated because of its effects on soil biodiversity and soil microbial respiration [169] and human health [168].

Table 4. Weed management: main effects as reported in meta-analysis studies. See also Table S7.

Parameter	Comparison of Treatments	Result
Weed biomass	Legume intercropping vs. conventional, both non-weeded and weeded	−56%, −42% [a]
Weed density, biomass Parasitic nematodes	Cover crops vs. traditional tillage	−10%, −5%, +29% [b]
Number of studies with increase soil organic matter	Reduced tillage vs. traditional tillage	+40 and −7 out of 78 studies [c]
Soil microbial respiration Soil microbial biomass	Glyphosate vs. no use, <10 mg kg Glyphosate vs. no use, >10 mg kg	logarithm of ratio: 0.064 ± 0.126, 0.04 ± 0.09 [d]

[a]: [163]; [b]: [53]; [c]: [170]; [d]: [169].

3.9. Crop Residue Management

Crop residues may be left on the soil surface, incorporated in the soil, burned or removed from the field for use as livestock feed or biofuel. Evidently, there are trade-offs in managing crop residues [171]. Conservation agriculture promotes the return of the crop residues to the soil to increase soil quality and reduce soil erosion, often in combination with zero-tillage or reduced tillage (Section 3.6). In this review, we distinguished crop residue management as a separate management practice, because of the relatively large number (19) of meta-analysis studies related to just crop residues (Table 1). Crop yield, water and nitrogen use efficiency, emissions of N_2O, and soil carbon sequestration were the main topics of these studies.

In most cases, crop residue management and mulching increased crop yields, and water and nitrogen use efficiencies by 0 to 50% (Figure 6). Mulching greatly reduced soil evaporation and thereby provided a greater fraction of soil water to the crop, which boosted crop yields. Crop residue return has a positive effect on soil carbon sequestration and soil microbial activity, but N_2O emissions increased as well. Nine out of the 19 meta-analysis studies dealt with soil mulching effects in China, as it is a common practice in dry-land farming in China (and India). One study examined the performance of biodegradable plastics to determine the optimal type of mulching for maize, wheat, potato, and cotton [172], and another [173] compared the performance of biodegradable films relative to polyethylene films.

3.10. Mechanization

Mechanization has greatly increased labor productivity in modern crop production systems, especially during the last century, and thereby has greatly contributed to farm-scale enlargement and withdrawal of labor from agriculture [189]. However, mechanization has also contributed to increased fossil fuel use and increased soil compaction [190]. During the last decades, research emphasis has shifted to precision technology, controlled traffic, and robotization. However, only two meta-analysis studies have touched mechanization, precision technology and robotization (Table 1). Ampoorter et al. [191] concluded on the basis of an analysis of 11 studies with 35 forest stands that mechanical harvesting of trees has led to the compaction of the top 30 cm of forest soils, with the largest effects on the top 10 cm. One study was of a different nature: It examined the change in the ratio of maize grain yield to labor input following the introduction of specific sustainable intensification practices technologies in sub-Saharan countries [192]. No firm conclusions could be derived because of lack of sufficient empirical studies.

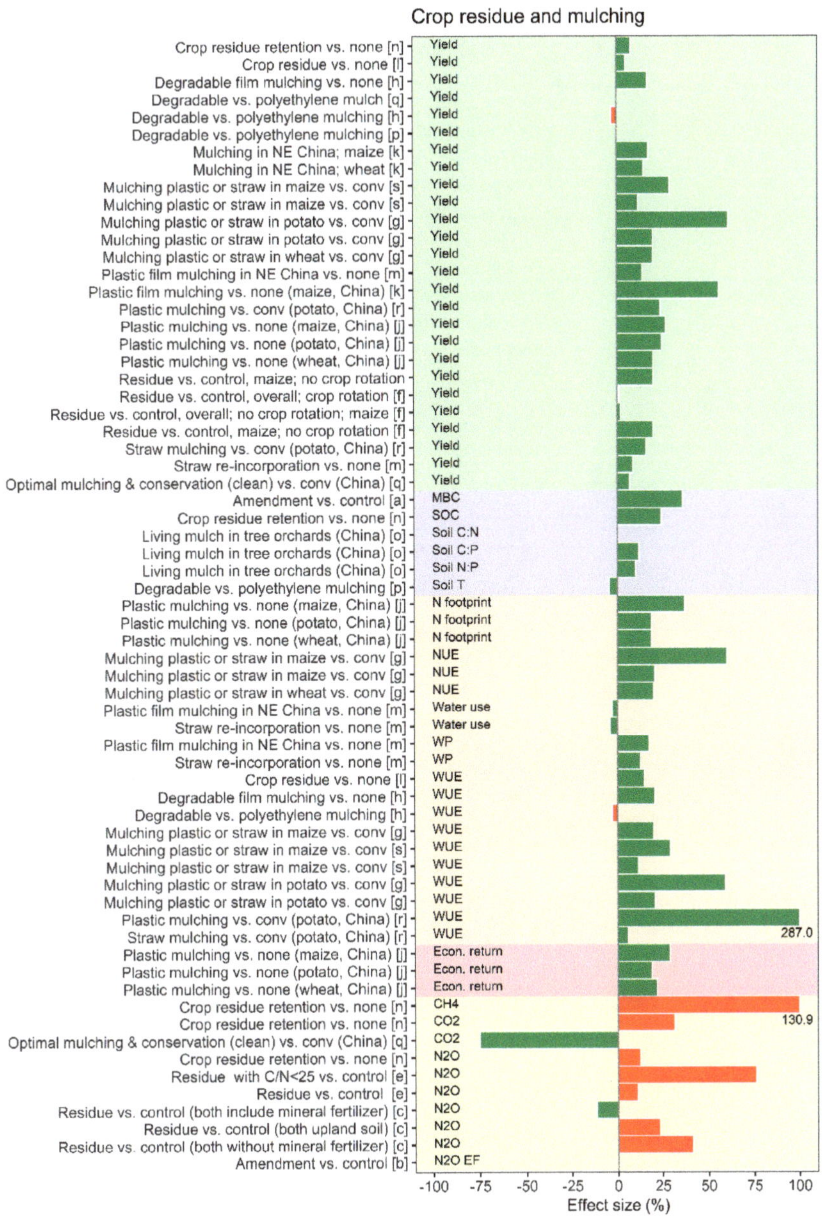

Figure 6. Overall response ratios reported in meta-analysis studies on crop residues and mulching grouped per area of interest (light green: agronomic; light blue: soil quality; light orange: resource use efficiency; light brown: economic; light yellow: environmental impacts). The management or treatment comparison is indicated outside the y-axis and the variable to which the data refer are listed inside the y-axis (abbreviations can be found in Table A1). Green bars indicate improvement, red bars indicate worsening. See also Table S8. [a]: [174]; [b] [175]; [c]: [176]; [d]: [177]; [e]: [178]; [f]: [22]; [g]: [179]; [h]: [172]; [i]: [180]; [j]: [181]; [k]: [182]; [l]: [183]; [m]: [97]; [n]: [184]; [o]: [185]; [p]: [173]; [q]: [186]; [r]: [187]; [s]: [188].

3.11. Landscape Management

Landscape management is a relatively new concept and has increased in importance following the approval of the UN Sustainable Development Goals and the recognition that the landscape is often the best scale for managing interactions, synergies, and trade-offs for natural resource management [193–195]. Landscape management in the context of sustainable food production may include hydrological measures, terracing, hedgerows, tree lines, wind breaks, flower strips, corridors, and agroforestry, depending on the landscape, environmental conditions, and stakeholders.

We identified six meta-analysis studies related to landscape management practices (Tables 1 and 5). Three of these quantified the benefits of windbreaks on crop yields [196–198]. Three studies analyzed the effects of hedgerows and flower strips on pollination, pest control, and crop yield [197–199], and two studies analyzed the effects of hedgerows on runoff and erosion [198,200]. Both large positive effects and negative effects have been reported (Table 5).

Table 5. Landscape management: summary of the results as reported by meta-analysis studies. See also Table S9.

Parameter	Management Practices	Result
Crop yield increase	Wind breaks	Spring wheat +8%, winter wheat +23%, barley +25%, oats +6%, rye +19%, millet +44%, corn +12%, alfalfa +99%, hay +20% [a]
Crop yield	Hedgerows vs. control; next to hedge until twice the height; beyond twice the height until 20 times the height	−29%, +6% [b]
Soil organic matter in crop field	Hedgerows vs. control	6% [b]
Interception of N, P, suspended solids from soil surface flow	Hedgerows Grass strips	69%, 67%, 91% [b] 67%, 73%, 90% [b]
Crop yield	Hedge rows, flower strips vs. none	ns [c]
Pest control	Hedge rows, flower strips vs. none	ns, −16% [c]
Pollination	Hedge rows, flower strips vs. none	ns [c]
Abundance, richness of pollinators in crop	Flower strips vs. none	ns, ns [d]
Pollinator species richness	Effect of agri-environment management, type landscape: Small, simple Small, complex Large, simple Large, complex	Hedge's d: sign. [f] ns [f] sign. [f] sign. [f]
Soil SOM, total N, total P, alkali N, available P, readily available K, total K	Hedge rows vs. none	Hedge's d sign. [e] Hedge's d ns [e]

[a]: [196]; [b]: [198]; [c]: [197]; [d]: [199]; [e]: [200]; [f]: [201].

4. Discussion

4.1. Main Findings

Most meta-analysis studies reported positive effects of alternative/improved practices relative to conventional practices. The 32 studies related to crop type and crop rotation clearly indicated the positive effects of crop rotations versus continuous cropping, legumes in crop rotations versus no legumes in crop rotations, intercropping versus monocultures, and cover cropping versus no cover cropping on crop yield and soil quality. Positive effects of especially cover crops and perennial crops on erosion control and minimizing nitrate leaching were also found, depending on, e.g., N fertilization.

The 25 studies related to nutrient management examined a diversity of nutrient sources, and application methods, timing, and strategies. Most studies reported positive effects of alternative/modified practices on crop yield and on minimizing environmental pollution, relative to conventional practices. Impacts of nutrient management strongly depended on environmental conditions.

The 18 studies related to irrigation and fertigation focused on the method, timing and volume of irrigation. Drip irrigation, deficit irrigation, and subsoil irrigation were all effective in increasing water use efficiency compared to sprinkling irrigation and especially flood irrigation. No economic assessments were made, and long-term impacts on soil quality and environmental pollution were also not reported.

Six studies related to drainage, with a focus on controlled drainage in response to variable rainfall patterns. Results indicate that controlled drainage increased farm income when compared to no human-induced drainage.

A total of 55 meta-analysis studies were devoted to tillage practices. Reduced tillage tended to reduce crop yields, but increased farm income (one study only), water use efficiency, soil carbon contents, and emissions of nitrous oxide (N_2O), which is a potent greenhouse gas. Reduced tillage in combination with crop residue return (mulching) and crop rotation had a slight positive effect on crop yield compared to conventional tillage.

Most of the seven studies related to pest management compared organic farming and conventional farming management practices. In general, organic farming management practices greatly decreased the use of pesticides, but lowered crop yields as well, depending on crop type and rotation, N application rate, soil quality, and (soil) biodiversity.

The four studies related to weed management did not provide a coherent view. Legume intercropping, cover cropping, and reduced tillage had positive effects on soil carbon contents but the effects on weed and crop yield were not clear.

The 19 studies related to crop residue management and mulching in dryland and/or irrigated conditions reported in general positive effects of mulching on crop yield and water use efficiency, but also increases in N_2O emissions, which are unwanted.

The six studies related to landscape management reported positive effects of windbreaks and hedgerows on crop yields and erosion control, but depending on site specific conditions, and provided that the surface area of windbreaks and hedgerows is in balance with the cropping area.

Evidently, most of the studies reported positive effects of the examined alternative/improved practices, relative to the common practice, on either crop yield, soil quality, resource use efficiency, and the environment (decreased emissions). While global assessment studies often paint rather pessimistic views on the state of food production, agriculture, and the environment [10,16,202–204], it is clear that the 174 studies reviewed here present a picture of optimism and hope. Indeed, there is large body of scientific/empirical evidence that some specific practices are more effective than others, i.e., have positive effect sizes relative to conventional practices (Figures 1–6; Tables 2–5), and that these positive effects may contribute to the sustainability of crop and food production. However, large steps still have to be made to integrate, optimize, and transfer the scientific findings of meta-analysis in current practice. We note that only few meta-analysis studies examined interactions between categories of practices, while essentially no meta-analysis study made in-depth comparisons at cropping system level in which all ten categories of crop husbandry and soil management practices had been optimized. Hence, there is need for further integration and optimization of all ten crop husbandry and soil management practices, and show the effectiveness of optimized practices through experimental studies and ultimately meta-analysis studies. There is also a need to transfer the positive messages of meta-analysis studies to practice through demonstration, extension services and possibly economic incentives. Cropping systems with all crop husbandry and soil management practices optimized may be termed 'soil-improving cropping systems', to emphasize the two-way interaction between soil and crop (see Section 4.3).

4.2. Uneven Coverage of Meta-Analysis Studies

Some crop husbandry and soil management practices have been studied extensively and repeatedly, while some other practices have received little research attention (Table 1).

Further, most studies have examined the effects of practices on crop yield, soil quality and environmental effects, while farm income (cost-benefit ratios) and resource use efficiency have received less attention (and human health aspects not at all). Evidently, the coverage of meta-analysis studies across practices and outcomes has been uneven; 75% of all studies addressed four practices, in the order: soil tillage > crop type and crop rotations > nutrient management > irrigation/fertigation (Table 1).

The large interest in soil tillage (55 meta-analyses studies) is certainly related to the importance of soil conservation, and the envisaged reduction in soil erosion, net greenhouse gas emissions, energy use, and labor through minimum or zero tillage. The effect-size of tillage practices were relatively small (0–10%) for crop yield, modest (0–50%) for greenhouse gas emissions and nutrient leaching, and relatively large and positive for soil quality, especially for soil life (0–150%).

The relatively large attention for nutrient management and irrigation/fertigation is related to the role of nutrients and water in boosting crop yields across the world (e.g., [205], to the depletion of fresh water resources [206] and rock phosphorus resources [207,208], and to the ecological impacts of excess nitrogen and phosphorus in the environment [16,209]). Nutrient and irrigation water inputs often form a relatively large economic cost to farmers, especially in developing countries, but this aspect has not been addressed.

We found only two meta-analyses related to mechanization and technology in agriculture (including forestry). However, several recent textbooks on precision technology for cropping systems do address the possible economic and environmental impacts of technological applications for sensing, field operations, and data handling, analysis, and control (e.g., [210–212]. Indeed, mechanization has revolutionized crop production systems during the past century but differently in different regions of the world. It has made large-scale crop production systems possible, has led to an exodus of laborers, has contributed to international trade of food and feed, and has indirectly affected essentially all crop husbandry and soil management practices. Robotization goes a step further and may revolutionize crop production systems again in the near future; it also offers the opportunity the reduce the impact of heavy machines on soil compaction. Keller et al. [190] estimated that the increase in weight of agricultural vehicles has caused an increase in soil bulk density, and thereby decreased root growth, crop yields, and soil hydraulic properties. They speculate that heavy machinery has contributed to yield stagnation and increased flooding in Europe [190].

We recommend that future meta-analysis studies related to crop husbandry and soil management practices should pay more attention to the socio-economic impacts of practices including possible barriers and constraints for their implementation in practice. Next, we recommend that more emphasis has to be given to interactions between multiple crop husbandry and soil management practices, and to comparisons of region-specific optimized packages of these practices. Further, Africa should not be neglected, as much of the increased food demand (and food production) during the next few decades will occur in Africa.

4.3. *Towards High-Yielding, Soil-Improving, and Environmentally Sound Cropping Systems*

The effect of specific crop yield defining, yield limiting, or yield reducing factors is largest when all other crop yield defining, limiting, or reducing factors are optimal, i.e., at a level where these do not affect crop yield [213]. This 'law of the optimum' may have also influenced the outcomes of meta-analyses studies; optimality of all factors will have enhanced the effect size of an alternative practice relative to the control practice, and vice versa. We have no insight in the degree of optimality of yield factors in the studies underlying the reviewed meta-analyses, but simply note here that there is often a gap between actual and attainable yields, and between actual and attainable environmental performances in practice. These gaps have to be narrowed to be able to produce adequate amounts of food in a sustainable and region-specific way [214].

The reviewed meta-analysis studies provide many suggestions for improved practices, but the optimization of all practices has to be done for specific regions, at farm level and/or regional levels. The possible steps in the optimization process have been summarized in Figure 7; it provides a roadmap for developing high-yielding, soil-improving and environmental-sound cropping systems. Steps 1 and 2 deal with the analyses and description of the current cropping systems, including its socio-economic and environmental environments. Steps 3 to 12 then deal with the selection and optimization of the 10 main specific crop husbandry and soil management practices, while taking the results of steps 1 and 2 into account. The actual process of optimization will be iterative, until the most optimal combination of practices has been identified. Variants of this road-map have been tested within the EU-funded project SoilCare, and results are presented in this special issue.

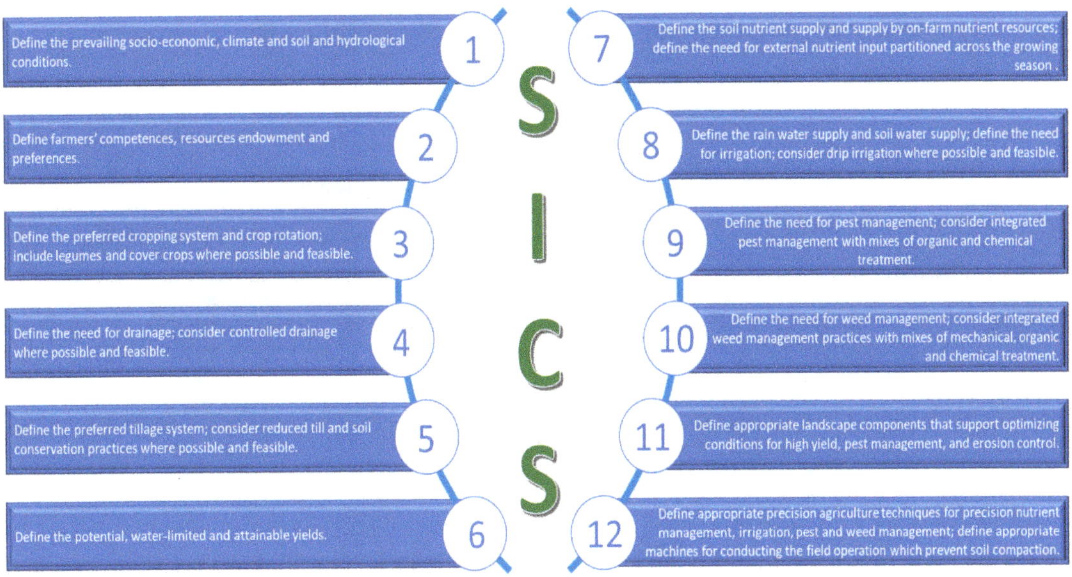

Figure 7. Towards sustainable cropping systems; a step-wise roadmap for developing high-yielding, soil-improving and environmentally sound cropping systems. The steps (1 to 12) have to be taken in a consecutive-iterative manner so as to find the optimal combination of practices.

4.4. Concluding Remarks

Crop husbandry and soil management practices are of critical importance for closing yield gaps, raising farm income and soil quality, and minimizing the environmental impacts of cropping systems in the world. We identified ten categories of crop husbandry and soil management practices, based on the concept of crop yield defining, limiting and reducing factors, and tried to quantify the effects of improved or modified practices relative to conventional practices, by using results of meta-analysis studies.

Our review was based on the premise that meta-analysis papers and reviews synthesize large numbers of experimental studies related to topical research questions and important research findings. For example, closing yield gaps and decreasing environmental impacts are topical, and thus we expected that in the course of the last 20 years when meta-analysis studies blossomed, a wealth of synthesized information would become accessible to help improve crop husbandry and soil management practices and thereby increase crop yield and soil quality, and decrease the environmental impact of crop production. The meta-analysis studies reviewed covered a huge number of experimental studies and practices in different parts of the world, albeit uneven. The number of studies per category of practices seem to reflect topics of hot societal debates and/or studies with controversial research findings. The number of meta-analysis studies per category of practices seem not to reflect those topics and practices that have largest impacts on crop yields, soil quality, and the environment.

Most meta-analysis studies reported positive effects of specific practices relative to conventional practices, on crop yield, soil quality and the environment. However, most meta-analysis studies examined single practices, with limited emphasis on interactions between categories of practices, and on the optimization across practices. Further, the coverage of studies was uneven, both in terms of practices, sustainability aspects and world regions. Notably, economic aspects were rarely addressed.

Based on this review, we derived a roadmap with twelve steps for integrating and optimization of all main crop husbandry and soil management practices, so as to develop high-yielding, soil-improving, and environmentally-sound cropping systems. We call these 'soil-improving cropping systems' to emphasize that cropping systems must maintain and improve soil quality to remain sustainable. This roadmap has been tested in practice and some results are presented in other papers of this special issue. We also made a number of recommendations.

Supplementary Materials: The following supporting information can be downloaded at: https://www.mdpi.com/article/10.3390/land11020255/s1, Table S1: Cropping: effects on (a) crop yield and quality, (b) soil quality, (c) economic effects, (d) resource use efficiency, (e) environmental effects, and (f) human health impacts as reported in meta analysis studies; aoi = area of interest. Table S2: Nutrient management: effects on (a) crop yield and quality, (b) soil quality, (c) economic effects, (d) resource use efficiency, (e) environmental effects, and (f) human health impacts as reported in meta analysis studies; aoi = area of interest. Table S3: Irrigation and fertigation: effects on (a) crop yield and quality, (b) soil quality, (c) economic effects, (d) resource use efficiency, (e) environmental effects, and (f) human health impacts as reported in meta analysis studies; aoi = area of interest. DI = deficit irrigation, PRD = partial rootzone drying, FI = full irrigation, AI = aerated irrigation, NAI non aerated irrigation, RDI = regulated deficit irrigation, CDI = conventional deficit irrigation, CI = conventional irrigation, OI = over irrigation, UI = under irrigation, OPTI = optimal irrigation. Table S4: Controlled drainage: effects on (a) crop yield and quality, (b) soil quality, (c) economic effects, (d) resource use efficiency, (e) environmental effects, and (f) human health impacts as reported in meta analysis studies; aoi = area of interest (see Table 1). Table S5: Soil tillage: effects on (a) crop yield and quality, (b) soil quality, (c) economic effects, (d) resource use efficiency, (e) environmental effects, and (f) human health impacts as reported in meta analysis studies; aoi = area of interest. NT = no tillage, TT = traditional tillage, CA = conservation agriculture, RT = reduced tillage, MT = minimum tillage. Table S6: Pest management: effects on (a) crop yield and quality, (b) soil quality, (c) economic effects, (d) resource use efficiency, (e) environmental effects, and (f) human health impacts as reported in meta analysis studies; aoi = area of interest. Table S7: Weed management: effects on (a) crop yield and quality, (b) soil quality, (c) economic effects, (d) resource use efficiency, (e) environmental effects, and (f) human health impacts as reported in meta analysis studies; aoi = area of interest. Table S8: Crop residue management & mulching: effects on (a) crop yield and quality, (b) soil quality, (c) economic effects, (d) resource use efficiency, (e) environmental effects, and (f) human health impacts as reported in meta analysis studies; aoi = area of interest. Table S9: Landscape management: effects on (a) crop yield and quality, (b) soil quality, (c) economic effects, (d) resource use efficiency, (e) environmental effects, and (f) human health impacts as reported in meta analysis studies; aoi = area of interest. Table S10: Explanation of the main columns in the accompanying Excel sheet.

Author Contributions: Conceptualization and writing by all authors; data collection and data analysis by R.R. + M.H. (contributed equally to this study). Review and editing were done by all authors. All authors have read and agreed to the published version of the manuscript.

Funding: The present study was carried out in the EU project SoilCare ("*Soil care for profitable and sustainable crop production in Europe*"), EU grant agreement 677407; https://www.soilcare-project.eu (accessed on 12 January 2022).

Institutional Review Board Statement: Not applicable.

Informed Consent Statement: Not applicable.

Data Availability Statement: All tabulated data are available in the Supplemental Information (Word document with tables; Excel file with extended data).

Acknowledgments: This study builds on SoilCare report 07 (https://www.soilcare-project.eu/resources/deliverables; accessed on 12 January 2022); we thank the co-authors of that report for their contributions.

Conflicts of Interest: The authors declare no conflict of interest.

Appendix A

Table A1. List of abbreviations.

Abbreviation	Meaning	Abbreviation	Meaning
abund.	Abundance	N_2O	Nitrous oxide (emission)
act.	Activity	NH_3	Ammonia (emission)
AG activity	Activity of α-1,4-glucosidase	NO_3	Nitrate (leaching)
AMF	Arbuscular mycorrhizal fungi	NOx	Nitrogen oxides
aoi	Area of interest	NUE	Nitrogen (nutrient) use efficiency
AWC	Available water content	OA	Organic agriculture
Bact.	Bacteria	OX activity	Oxidative decomposition
BD	Dry bulk density	P	Phosphorus
BG activity	Activity of β-1,4-glucosidase	part.	Particulate
BX activity	Activity of β-1,4-xylosidase	Penetr. R	Penetration resistance
C	Carbon	PEO activity	Peroxidase activity
C-acq. activity	Hydrolytic C acquisition enzymes	PHO activity	Phenol oxidase activity
CBH activity	Activity of β-D-cellobiosidase	PLFA	Phospholipid fatty-acids
CH_4	Methane	PMN	Potentially mineralizable N
CO_2	Carbon dioxide emission	PR	Penetration resistance
col.	Colonies	Resp.	Respiration
Dehydrog.	Dehydrogenase activity	richn.	Richness
Dens. herb.	Density herbivorous insects	RO	Runoff
diss.	Dissolved	RR	Response ratio: RR = (Xt − Xc)/Xc
div.	Diversity	SDG	Sustainable development goals
Econ. return	Economic return	seq.	Sequestration
EEA	Soil extracellular enzyme activity	SICS	Soil-improving cropping systems
EF	Emission factor	SMB	Soil microbial biomass
EF_{ad}	Additional N2O emission factor #	SMC	Soil microbial C
EOC	Extractable organic carbon	SOC	Soil organic C
ET	Evapotranspiration or water use	Soil T	Soil temperature
FNER	Fertilizer N equivalent ratio	SOM	Soil organic matter
GHG	Greenhouse gas	SON	Soil organic N
GWP	Global warming potential $	SWA	Soil water-stable aggregate
K	Potassium	TMA	Total microbial activity
Ksat	Hydraulic conductivity at saturation	tot.	Total
LER	Land equivalent ratio	TSS	Total soluble solids
Lrv	Root length density	WP	Water productivity
Max. econ. return	Maximum economic return	WS	Water storage
MBC	Microbial biomass C	WSA	Water stable aggregates
MBN	Microbial biomass N	WUE	Water use efficiency
MWD	Aggregate mean weight diameter	Xc	Effect (value) of control treatment
N	Nitrogen	Xt	Effect (value) of specific treatment

#: which is the conservation tillage-induced change in N_2O emission compared to conventional tillage when N fertilizer is applied; $: CH_4 and N_2O emissions per unit yield.

References

1. Iizumi, T.; Kotoku, M.; Kim, W.; West, P.C.; Gerber, J.S.; Brown, M.E. Uncertainties of potentials and recent changes in global yields of major crops resulting from census- and satellite-based yield datasets at multiple resolutions. *PLoS ONE* **2018**, *13*, e0203809. [CrossRef] [PubMed]
2. Ray, D.K.; Ramankutty, N.; Mueller, N.D.; West, P.C.; Foley, J.A. Recent patterns of crop yield growth and stagnation. *Nat Commun.* **2012**, *3*, 1293. [CrossRef] [PubMed]
3. Foley, J.A.; Ramankutty, N.; Brauman, K.A.; Cassidy, E.S.; Gerber, J.S.; Johnston, M.; Mueller, N.D.; O'Connell, C.; Ray, D.K.; West, P.C.; et al. Solutions for a cultivated planet. *Nature* **2011**, *478*, 337–342. [CrossRef] [PubMed]
4. Ray, D.K.; Mueller, N.D.; West, P.C.; Foley, J.A. Yield Trends Are Insufficient to Double Global Crop Production by 2050. *PLoS ONE* **2013**, *8*, e66428. [CrossRef] [PubMed]
5. Lobell, D.B.; Cassman, K.G.; Field, C.B. Crop Yield Gaps: Their Importance, Magnitudes, and Causes. *Annu. Rev. Environ. Resour.* **2009**, *34*, 179–204. [CrossRef]
6. Van Ittersum, M.K.; Cassman, K.G.; Grassini, P.; Wolf, J.; Tittonell, P.; Hochman, Z. Yield gap analysis with local to global relevance—A review. *Field Crop. Res.* **2013**, *143*, 4–17. [CrossRef]
7. Brisson, N.; Gate, P.; Gouache, D.; Charmet, G.; Oury, F.-X.; Huard, F. Why are wheat yields stagnating in Europe? A comprehensive data analysis for France. *Field Crop. Res.* **2010**, *119*, 201–212. [CrossRef]
8. Schauberger, B.; Ben-Ari, T.; Makowski, D.; Kato, T.; Kato, H.; Ciais, P. Yield trends, variability and stagnation analysis of major crops in France over more than a century. *Sci. Rep.* **2018**, *8*, 16865. [CrossRef]
9. Schils, R.; Olesen, J.E.; Kersebaum, K.-C.; Rijk, B.; Oberforster, M.; Kalyada, V.; Khitrykau, M.; Gobin, A.; Kirchev, H.; Manolova, V.; et al. Cereal yield gaps across Europe. *Eur. J. Agron.* **2018**, *101*, 109–120. [CrossRef]
10. Amundson, R.; Berhe, A.A.; Hopmans, J.W.; Olson, C.; Sztein, A.E.; Sparks, D.L. Soil science. Soil and human security in the 21st century. *Science* **2015**, *348*, 1261071. [CrossRef]
11. Cassman, K.G. Ecological intensification of cereal production systems: Yield potential, soil quality, and precision agriculture. *Proc. Natl. Acad. Sci. USA* **1999**, *96*, 5942–5959. [CrossRef]
12. Cassman, K.G.; Dobermann, A.; Walters, D.T.; Yang, H. Meeting cereal demand while protecting natural resources And improving environmental quality. *Annu. Rev. Environ. Resour.* **2003**, *28*, 315–358. [CrossRef]
13. Connor, D.J.; RLoomis, R.S.; Cassman, K.G. *Crop Ecology—Productivity and Management in Agricultural Systems*, 2nd ed.; Cambridge University Press: Cambridge, UK, 2011.
14. FAO; DWFI. *Yield Gap Analysis of Field Crops—Methods and Case Studies*; Sadras, V.O., Cassman, K.G.G., Grassini, P., Hall, A.J., Bastiaanssen, W.G.M., Laborte, A.G., Milne, A.E., Sileshi, G., Steduto, P., Eds.; FAO Water Reports No. 41; FAO: Rome, Italy, 2015.
15. Snyder, C.S.; Bruulsema, T.W.; Jensen, T.L.; Fixen, P.E. Review of greenhouse gas emissions from crop production systems and fertilizer management effects. *Agric. Ecosyst. Environ.* **2009**, *133*, 247–266. [CrossRef]
16. Steffen, W.; Richardson, K.; Rockstrom, J.; Cornell, S.E.; Fetzer, I.; Bennett, E.M.; Biggs, R.; Carpenter, S.R.; de Vries, W.; de Wit, C.A.; et al. Planetary boundaries: Guiding human development on a changing planet. *Science* **2015**, *347*, 1259855. [CrossRef] [PubMed]
17. Tilman, D.; Balzer, C.; Hill, J.; Befort, B.L. Global food demand and the sustainable intensification of agriculture. *Proc. Natl. Acad. Sci. USA* **2011**, *108*, 20260–20264. [CrossRef] [PubMed]
18. Bouma, J. Soil Security in Sustainable Development. *Soil Syst.* **2019**, *3*, 5. [CrossRef]
19. Gil, J.D.B.; Daioglou, V.; van Ittersum, M.; Reidsma, P.; Doelman, J.C.; van Middelaar, C.E.; van Vuuren, D.P. Reconciling global sustainability targets and local action for food production and climate change mitigation. *Glob. Environ. Change* **2019**, *59*, 101983. [CrossRef]
20. Iizumi, T.; Sakai, T. The global dataset of historical yields for major crops 1981–2016. *Sci. Data* **2020**, *7*, 97. [CrossRef]
21. Wezel, A.; Casagrande, M.; Celette, F.; Vian, J.-F.; Ferrer, A.; Peigné, J. Agroecological practices for sustainable agriculture. A review. *Agron. Sustain. Dev.* **2013**, *34*, 1–20. [CrossRef]
22. Pittelkow, C.M.; Linquist, B.A.; Lundy, M.E.; Liang, X.; van Groenigen, K.J.; Lee, J.; van Gestel, N.; Six, J.; Venterea, R.T.; van Kessel, C. When does no-till yield more? A global meta-analysis. *Field Crop. Res.* **2015**, *183*, 156–168. [CrossRef]
23. Reckling, M.; Bergkvist, G.; Watson, C.A.; Stoddard, F.L.; Zander, P.M.; Walker, R.L.; Pristeri, A.; Toncea, I.; Bachinger, J. Trade-Offs between Economic and Environmental Impacts of Introducing Legumes into Cropping Systems. *Front. Plant Sci.* **2016**, *7*, 669. [CrossRef] [PubMed]
24. Van den Putte, A.; Govers, G.; Diels, J.; Gillijns, K.; Demuzere, M. Assessing the effect of soil tillage on crop growth: A meta-regression analysis on European crop yields under conservation agriculture. *Eur. J. Agron.* **2010**, *33*, 231–241. [CrossRef]
25. Gurevitch, J.; Koricheva, J.; Nakagawa, S.; Stewart, G. Meta-analysis and the science of research synthesis. *Nature* **2018**, *555*, 175–182. [CrossRef] [PubMed]
26. O'Rourke, K. An historical perspective on meta-analysis: Dealing quantitatively with varying study results. *J. R. Soc. Med.* **2007**, *100*, 579–582. [CrossRef]
27. Angus, J.F.; Kirkegaard, J.A.; Hunt, J.R.; Ryan, M.H.; Ohlander, L.; Peoples, M.B. Break crops and rotations for wheat. *Crop. Pasture Sci.* **2015**, *66*, 523–552. [CrossRef]
28. Preissel, S.; Reckling, M.; Schläfke, N.; Zander, P. Magnitude and farm-economic value of grain legume pre-crop benefits in Europe: A review. *Field Crop. Res.* **2015**, *175*, 64–79. [CrossRef]

29. Andert, S.; Bürger, J.; Stein, S.; Gerowitt, B. The influence of crop sequence on fungicide and herbicide use intensities in North German arable farming. *Eur. J. Agron.* **2016**, *77*, 81–89. [CrossRef]
30. Li, C.; Hoffland, E.; Kuyper, T.W.; Yu, Y.; Zhang, C.; Li, H.; Zhang, F.; van der Werf, W. Syndromes of production in intercropping impact yield gains. *Nat. Plants* **2020**, *6*, 653–660. [CrossRef]
31. Tonitto, C.; David, M.B.; Drinkwater, L.E. Replacing bare fallows with cover crops in fertilizer-intensive cropping systems: A meta-analysis of crop yield and N dynamics. *Agric. Ecosyst. Environ.* **2006**, *112*, 58–72. [CrossRef]
32. Valkama, E.; Lemola, R.; Känkänen, H.; Turtola, E. Meta-analysis of the effects of undersown catch crops on nitrogen leaching loss and grain yields in the Nordic countries. *Agric. Ecosyst. Environ.* **2015**, *203*, 93–101. [CrossRef]
33. Zhang, C.; Dong, Y.; Tang, L.; Zheng, Y.; Makowski, D.; Yu, Y.; Zhang, F.; van der Werf, W. Intercropping cereals with faba bean reduces plant disease incidence regardless of fertilizer input; a meta-analysis. *Eur. J. Plant Pathol.* **2019**, *154*, 931–942. [CrossRef]
34. Vannoppen, W.; Vanmaercke, M.; De Baets, S.; Poesen, J. A review of the mechanical effects of plant roots on concentrated flow erosion rates. *Earth-Sci. Rev.* **2015**, *150*, 666–678. [CrossRef]
35. Munkholm, L.J.; Heck, R.J.; Deen, B. Long-term rotation and tillage effects on soil structure and crop yield. *Soil Tillage Res.* **2013**, *127*, 85–91. [CrossRef]
36. Kiær, L.P.; Skovgaard, I.M.; Østergård, H. Grain yield increase in cereal variety mixtures: A meta-analysis of field trials. *Field Crop. Res.* **2009**, *114*, 361–373. [CrossRef]
37. Finn, J.A.; Kirwan, L.; Connolly, J.; Sebastià, M.T.; Helgadottir, A.; Baadshaug, O.H.; Bélanger, G.; Black, A.; Brophy, C.; Collins, R.P.; et al. Ecosystem function enhanced by combining four functional types of plant species in intensively managed grassland mixtures: A 3-year continental-scale field experiment. *J. Appl. Ecol.* **2013**, *50*, 365–375. [CrossRef]
38. Sainju, U.M. A global meta-analysis on the impact of management practices on net global warming potential and greenhouse gas intensity from cropland soils. *PLoS ONE* **2016**, *11*, e0148527. [CrossRef]
39. Han, Z.; Walter, M.T.; Drinkwater, L.E. N_2O emissions from grain cropping systems: A meta-analysis of the impacts of fertilizer-based and ecologically-based nutrient management strategies. *Nutr. Cycl. Agroecosyst.* **2017**, *107*, 335–355. [CrossRef]
40. West, T.O.; Post, W.M. Soil organic carbon sequestration rates by tillage and crop rotation: A global data analysis. *Soil Sci. Soc. Am. J.* **2002**, *66*, 1930–1946. [CrossRef]
41. McDaniel, M.D.; Tiemann, L.K.; Grandy, A.S. Does agricultural crop diversity enhance soil microbial biomass and organic matter dynamics? A meta-analysis. *Ecol. Appl.* **2014**, *24*, 560–570. [CrossRef]
42. Venter, Z.S.; Jacobs, K.; Hawkins, H.-J. The impact of crop rotation on soil microbial diversity: A meta-analysis. *Pedobiologia* **2016**, *59*, 215–223. [CrossRef]
43. Bedoussac, L.; Journet, E.-P.; Hauggaard-Nielsen, H.; Naudin, C.; Corre-Hellou, G.; Jensen, E.S.; Prieur, L.; Justes, E. Ecological principles underlying the increase of productivity achieved by cereal-grain legume intercrops in organic farming. A review. *Agron. Sustain. Dev.* **2015**, *35*, 911–935. [CrossRef]
44. Yu, Y.; Stomph, T.-J.; Makowski, D.; Zhang, L.; van der Werf, W. A meta-analysis of relative crop yields in cereal/legume mixtures suggests options for management. *Field Crop. Res.* **2016**, *198*, 269–279. [CrossRef]
45. Dassou, A.G.; Tixier, P. Response of pest control by generalist predators to local-scale plant diversity: A meta-analysis. *Ecol. Evol.* **2016**, *6*, 1143–1153. [CrossRef]
46. Tonhasca, A., Jr.; Byrne, D.N. The effects of crop diversification on herbivorous insects: A meta-analysis approach. *Ecol. Entomol.* **1994**, *19*, 239–244. [CrossRef]
47. Miguez, F.E.; Bollero, G.A. Review of Corn Yield Response under Winter Cover Cropping Systems Using Meta-Analytic Methods. *Crop Sci.* **2005**, *45*, 2318–2329. [CrossRef]
48. Poeplau, C.; Don, A. Carbon sequestration in agricultural soils via cultivation of cover crops—A meta-analysis. *Agric. Ecosyst. Environ.* **2015**, *200*, 33–41. [CrossRef]
49. Bowles, T.M.; Jackson, L.E.; Loeher, M.; Cavagnaro, T.R.; Nuñez, M. Ecological intensification and arbuscular mycorrhizas: A meta-analysis of tillage and cover crop effects. *J. Appl. Ecol.* **2017**, *54*, 1785–1793. [CrossRef]
50. Quemada, M.; Baranski, M.; Nobel-de Lange, M.N.J.; Vallejo, A.; Cooper, J.M. Meta-analysis of strategies to control nitrate leaching in irrigated agricultural systems and their effects on crop yield. *Agric. Ecosyst. Environ.* **2013**, *174*, 203–206. [CrossRef]
51. Vicente-Vicente, J.L.; García-Ruiz, R.; Francaviglia, R.; Aguilera, E.; Smith, P. Soil carbon sequestration rates under Mediterranean woody crops using recommended management practices: A meta-analysis. *Agric. Ecosyst. Environ.* **2016**, *235*, 204–214. [CrossRef]
52. Poeplau, C.; Don, A. Soil carbon changes under Miscanthus driven by C4accumulation and C3decompostion—Toward a default sequestration function. *GCB Bioenergy* **2014**, *6*, 327–338. [CrossRef]
53. Daryanto, S.; Wang, L.; Jacinthe, P.A. Meta-Analysis of Phosphorus Loss from No-Till Soils. *J. Environ. Qual.* **2017**, *46*, 1028–1037. [CrossRef] [PubMed]
54. Thapa, R.; Mirsky, S.B.; Tully, K.L. Cover Crops Reduce Nitrate Leaching in Agroecosystems: A Global Meta-Analysis. *J. Environ. Qual.* **2018**, *47*, 1400–1411. [CrossRef] [PubMed]
55. Nichols, V.; Martinez-Feria, R.; Weisberger, D.; Carlson, S.; Basso, B.; Basche, A. Cover crops and weed suppression in the U.S. Midwest: A meta-analysis and modeling study. *Agric. Environ. Lett.* **2020**, *5*, e20022. [CrossRef]
56. Zhao, J.; Yang, Y.; Zhang, K.; Jeong, J.; Zeng, Z.; Zang, H. Does crop rotation yield more in China? A meta-analysis. *Field Crop. Res.* **2020**, *245*, 107659. [CrossRef]

57. Jian, J.; Lester, B.J.; Du, X.; Reiter, M.S.; Stewart, R.D. A calculator to quantify cover crop effects on soil health and productivity. *Soil Tillage Res.* **2020**, *199*, 104575. [CrossRef]
58. Shackelford, G.E.; Kelsey, R.; Dicks, L.V. Effects of cover crops on multiple ecosystem services: Ten meta-analyses of data from arable farmland in California and the Mediterranean. *Land Use Policy* **2019**, *88*, 104204. [CrossRef]
59. Xu, Z.; Li, C.; Zhang, C.; Yu, Y.; van der Werf, W.; Zhang, F. Intercropping maize and soybean increases efficiency of land and fertilizer nitrogen use; A meta-analysis. *Field Crop. Res.* **2020**, *246*, 107661. [CrossRef]
60. Lori, M.; Symnaczik, S.; Mäder, P.; De Deyn, G.; Gattinger, A. Organic farming enhances soil microbial abundance and activity—A meta-analysis and meta-Regression. *PLoS ONE* **2017**, *12*, e0180442. [CrossRef]
61. Bai, X.; Huang, Y.; Ren, W.; Coyne, M.; Jacinthe, P.A.; Tao, B.; Hui, D.; Yang, J.; Matocha, C. Responses of soil carbon sequestration to climate-smart agriculture practices: A meta-analysis. *Glob. Change Biol.* **2019**, *25*, 2591–2606. [CrossRef]
62. Basche, A.D.; DeLonge, M.S. Comparing infiltration rates in soils managed with conventional and alternative farming methods: A meta-analysis. *PLoS ONE* **2019**, *14*, e0215702. [CrossRef]
63. Morugán-Coronado, A.; Linares, C.; Gómez-López, M.D.; Faz, Á.; Zornoza, R. The impact of intercropping, tillage and fertilizer type on soil and crop yield in fruit orchards under Mediterranean conditions: A meta-analysis of field studies. *Agric. Syst.* **2020**, *178*, 102736. [CrossRef]
64. Fernandez, J.A.; DeBruin, J.; Messina, C.D.; Ciampitti, I.A. Late-season nitrogen fertilization on maize yield: A meta-analysis. *Field Crop. Res.* **2020**, *247*, 107586. [CrossRef]
65. Hijbeek, R.; van Ittersum, M.K.; ten Berge, H.F.M.; Gort, G.; Spiegel, H.; Whitmore, A.P. Do organic inputs matter—A meta-analysis of additional yield effects for arable crops in Europe. *Plant Soil* **2016**, *411*, 293–303. [CrossRef]
66. Luo, G.; Li, L.; Friman, V.-P.; Guo, J.; Guo, S.; Shen, Q.; Ling, N. Organic amendments increase crop yields by improving microbe-mediated soil functioning of agroecosystems: A meta-analysis. *Soil Biol. Biochem.* **2018**, *124*, 105–115. [CrossRef]
67. Li, Y.; Cui, S.; Chang, S.X.; Zhang, Q. Liming effects on soil pH and crop yield depend on lime material type, application method and rate, and crop species: A global meta-analysis. *J. Soils Sediments* **2019**, *19*, 1393–1406. [CrossRef]
68. Jaggard, K.W.; Qi, A.; Armstrong, M.J. A meta-analysis of sugarbeet yield responses to nitrogen fertilizer measured in England since 1980. *J. Agric. Sci.* **2009**, *147*, 287–301. [CrossRef]
69. Geisseler, D.; Scow, K.M. Long-term effects of mineral fertilizers on soil microorganisms—A review. *Soil Biol. Biochem.* **2014**, *75*, 54–63. [CrossRef]
70. Li, Y.; Li, Z.; Cui, S.; Chang, S.X.; Jia, C.; Zhang, Q. A global synthesis of the effect of water and nitrogen input on maize (Zea mays) yield, water productivity and nitrogen use efficiency. *Agric. For. Meteorol.* **2019**, *268*, 136–145. [CrossRef]
71. Sha, Z.; Ma, X.; Wang, J.; Lv, T.; Li, Q.; Misselbrook, T.; Liu, X. Effect of N stabilizers on fertilizer-N fate in the soil-crop system: A meta-analysis. *Agric. Ecosyst. Environ.* **2020**, *290*, 106763. [CrossRef]
72. Borchard, N.; Schirrmann, M.; Cayuela, M.L.; Kammann, C.; Wrage-Monnig, N.; Estavillo, J.M.; Fuertes-Mendizabal, T.; Sigua, G.; Spokas, K.; Ippolito, J.A.; et al. Biochar, soil and land-use interactions that reduce nitrate leaching and N_2O emissions: A meta-analysis. *Sci. Total Environ.* **2019**, *651*, 2354–2364. [CrossRef]
73. Eden, M.; Gerke, H.H.; Houot, S. Organic waste recycling in agriculture and related effects on soil water retention and plant available water: A review. *Agron. Sustain. Dev.* **2017**, *37*, 11. [CrossRef]
74. Liu, S.; Lin, F.; Wu, S.; Ji, C.; Sun, Y.; Jin, Y.; Li, S.; Li, Z.; Zou, J. A meta-analysis of fertilizer-induced soil NO and combined NO+N_2O emissions. *Glob. Change Biol.* **2017**, *23*, 2520–2532. [CrossRef] [PubMed]
75. Nkebiwe, P.M.; Weinmann, M.; Bar-Tal, A.; Müller, T. Fertilizer placement to improve crop nutrient acquisition and yield: A review and meta-analysis. *Field Crop. Res.* **2016**, *196*, 389–401. [CrossRef]
76. Ti, C.; Xia, L.; Chang, S.X.; Yan, X. Potential for mitigating global agricultural ammonia emission: A meta-analysis. *Environ. Pollut.* **2019**, *245*, 141–148. [CrossRef]
77. Tran, D.T.Q.; Bradbury, M.I.; van Ogtrop, F.F.; Bozkurt, H.; Jones, B.J.; McConchie, R. Environmental drivers for persistence of *Escherichia coli* and salmonella in manure-amended soils: A meta-analysis. *J. Food Prot.* **2020**, *83*, 1268–1277. [CrossRef]
78. Zhang, X.; Fang, Q.; Zhang, T.; Ma, W.; Velthof, G.L.; Hou, Y.; Oenema, O.; Zhang, F. Benefits and trade-offs of replacing synthetic fertilizers by animal manures in crop production in China: A meta-analysis. *Glob. Change Biol.* **2020**, *26*, 888–900. [CrossRef]
79. Li, W.; Yang, M.; Wang, J.; Wang, Z.; Fan, Z.; Kang, F.; Wang, Y.; Luo, Y.; Kuang, D.; Chen, Z.; et al. Agronomic responses of major fruit crops to fertilization in China: A meta-analysis. *Agronomy* **2020**, *10*, 15. [CrossRef]
80. Du, Y.; Cui, B.; Zhang, Q.; Wang, Z.; Sun, J.; Niu, W. Effects of manure fertilizer on crop yield and soil properties in China: A meta-analysis. *Catena* **2020**, *193*, 104617. [CrossRef]
81. Jian, S.; Li, J.; Chen, J.; Wang, G.; Mayes, M.A.; Dzantor, K.E.; Hui, D.; Luo, Y. Soil extracellular enzyme activities, soil carbon and nitrogen storage under nitrogen fertilization: A meta-analysis. *Soil Biol. Biochem.* **2016**, *101*, 32–43. [CrossRef]
82. Ye, G.; Lin, Y.; Luo, J.; Di, H.J.; Lindsey, S.; Liu, D.; Fan, J.; Ding, W. Responses of soil fungal diversity and community composition to long-term fertilization: Field experiment in an acidic Ultisol and literature synthesis. *Appl. Soil Ecol.* **2020**, *145*, 103305. [CrossRef]
83. He, M.; Zhou, G.; Yuan, T.; van Groenigen, K.J.; Shao, J.; Zhou, X. Grazing intensity significantly changes the C: N: P stoichiometry in grassland ecosystems. *Glob. Ecol. Biogeogr.* **2020**, *29*, 355–369. [CrossRef]
84. Quinn, D.J.; Lee, C.D.; Poffenbarger, H.J. Corn yield response to sub-surface banded starter fertilizer in the U.S.: A meta-analysis. *Field Crop. Res.* **2020**, *254*, 107834. [CrossRef]

85. Liu, S.; Wang, J.; Pu, S.; Blagodatskaya, E.; Kuzyakov, Y.; Razavi, B.S. Impact of manure on soil biochemical properties: A global synthesis. *Sci. Total Environ.* **2020**, *745*, 141003. [CrossRef] [PubMed]
86. Emmerling, C.; Krein, A.; Junk, J. Meta-analysis of strategies to reduce NH_3 emissions from slurries in European agriculture and consequences for greenhouse gas emissions. *Agronomy* **2020**, *10*, 1633. [CrossRef]
87. Foged, H.; Flotats Ripoll, X.; Bonmatí Blasi, A.; Palatsi Civit, J.; Magrí Aloy, A.; Schelde, K.M. *Inventory of Manure Processing Activities in Europe*; Technical Report No. I Concerning "Manure Processing Activities in Europe" to the European Commission; Directorate-General Environment: Brussels, Belgium, 2012; 138p.
88. Qin, W.; Assinck, F.B.T.; Heinen, M.; Oenema, O. Water and nitrogen use efficiencies in citrus production: A meta-analysis. *Agric. Ecosyst. Environ.* **2016**, *222*, 103–111. [CrossRef]
89. Cayuela, M.L.; Aguilera, E.; Sanz-Cobena, A.; Adams, D.C.; Abalos, D.; Barton, L.; Ryals, R.; Silver, W.L.; Alfaro, M.A.; Pappa, V.A.; et al. Direct nitrous oxide emissions in Mediterranean climate cropping systems: Emission factors based on a meta-analysis of available measurement data. *Agric. Ecosyst. Environ.* **2017**, *238*, 25–35. [CrossRef]
90. He, G.; Wang, Z.; Cui, Z. Managing irrigation water for sustainable rice production in China. *J. Clean. Prod.* **2020**, *245*, 118928. [CrossRef]
91. Jiang, Y.; Carrijo, D.; Huang, S.; Chen, J.; Balaine, N.; Zhang, W.; van Groenigen, K.J.; Linquist, B. Water management to mitigate the global warming potential of rice systems: A global meta-analysis. *Field Crop. Res.* **2019**, *234*, 47–54. [CrossRef]
92. Zhou, L.; Zhou, X.; Shao, J.; Nie, Y.; He, Y.; Jiang, L.; Wu, Z.; Hosseini Bai, S. Interactive effects of global change factors on soil respiration and its components: A meta-analysis. *Glob. Change Biol.* **2016**, *22*, 3157–3169. [CrossRef]
93. Zhou, X.; Zhou, L.; Nie, Y.; Fu, Y.; Du, Z.; Shao, J.; Zheng, Z.; Wang, X. Similar responses of soil carbon storage to drought and irrigation in terrestrial ecosystems but with contrasting mechanisms: A meta-analysis. *Agric. Ecosyst. Environ.* **2016**, *228*, 70–81. [CrossRef]
94. Fan, Y.; Wang, C.; Nan, Z. Determining water use efficiency of wheat and cotton: A meta-regression analysis. *Agric. Water Manag.* **2018**, *199*, 48–60. [CrossRef]
95. Mitchell-McCallister, D.; Cano, A.; West, C. Meta-analysis of crop water use efficiency by irrigation system in the Texas High Plains. *Irrig. Sci.* **2020**, *38*, 535–546. [CrossRef]
96. Liu, B.-Y.; Zhao, X.; Li, S.-S.; Zhang, X.-Z.; Virk, A.L.; Qi, J.-Y.; Kan, Z.-R.; Wang, X.; Ma, S.-T.; Zhang, H.-L. Meta-analysis of management-induced changes in nitrogen use efficiency of winter wheat in the North China Plain. *J. Clean. Prod.* **2020**, *251*, 119632. [CrossRef]
97. Zheng, H.; Shao, R.; Xue, Y.; Ying, H.; Yin, Y.; Cui, Z.; Yang, Q. Water productivity of irrigated maize production systems in Northern China: A meta-analysis. *Agric. Water Manag.* **2020**, *234*, 106119. [CrossRef]
98. Adu, M.O.; Yawson, D.O.; Abano, E.E.; Asare, P.A.; Armah, F.A.; Opoku, E.K. Does water-saving irrigation improve the quality of fruits and vegetables? Evidence from meta-analysis. *Irrig. Sci.* **2019**, *37*, 669–690. [CrossRef]
99. Adu, M.O.; Yawson, D.O.; Armah, F.A.; Asare, P.A.; Frimpong, K.A. Meta-analysis of crop yields of full, deficit, and partial root-zone drying irrigation. *Agric. Water Manag.* **2018**, *197*, 79–90. [CrossRef]
100. Du, Y.-D.; Niu, W.-Q.; Gu, X.-B.; Zhang, Q.; Cui, B.-J.; Zhao, Y. Crop yield and water use efficiency under aerated irrigation: A meta-analysis. *Agric. Water Manag.* **2018**, *210*, 158–164. [CrossRef]
101. Lu, J.; Shao, G.; Cui, J.; Wang, X.; Keabetswe, L. Yield, fruit quality and water use efficiency of tomato for processing under regulated deficit irrigation: A meta-analysis. *Agric. Water Manag.* **2019**, *222*, 301–312. [CrossRef]
102. Sadras, V.O. Does partial root-zone drying improve irrigation water productivity in the field? A meta-analysis. *Irrig. Sci.* **2008**, *27*, 183–190. [CrossRef]
103. Yu, L.; Zhao, X.; Gao, X.; Siddique, K.H.M. Improving/maintaining water-use efficiency and yield of wheat by deficit irrigation: A global meta-analysis. *Agric. Water Manag.* **2020**, *228*, 105906. [CrossRef]
104. Zheng, H.; Ying, H.; Yin, Y.; Wang, Y.; He, G.; Bian, Q.; Cui, Z.; Yang, Q. Irrigation leads to greater maize yield at higher water productivity and lower environmental costs: A global meta-analysis. *Agric. Ecosyst. Environ.* **2019**, *273*, 62–69. [CrossRef]
105. Amenumey, S.; Sands, G.; Wilson, B.; Mulla, D.; Nieber, J.; Swenson, J. Meta-analysis as a statistical tool for evaluating the hydrologic effects of water table management. In Proceedings of the American Society of Agricultural and Biological Engineers Annual International Meeting 2009, ASABE 2009, Reno, Nevada, 21–24 June 2009; pp. 2787–2804.
106. Christianson, L.; Knoot, T.; Larsen, D.; Tyndall, J.; Helmers, M. Adoption potential of nitrate mitigation practices: An ecosystem services approach. *Int. J. Agric. Sustain.* **2013**, *12*, 407–424. [CrossRef]
107. Skaggs, R.W.; Youssef, M.A.; Gilliam, J.W.; Evans, R.O. Effect of controlled drainage on water and nitrogen balances in drained lands. *Trans. ASABE* **2010**, *53*, 1843–1850. [CrossRef]
108. Wang, Z.; Shao, G.; Lu, J.; Zhang, K.; Gao, Y.; Ding, J. Effects of controlled drainage on crop yield, drainage water quantity and quality: A meta-analysis. *Agric. Water Manag.* **2020**, *239*, 106253. [CrossRef]
109. Abdalla, M.; Hastings, A.; Truu, J.; Espenberg, M.; Mander, U.; Smith, P. Emissions of methane from northern peatlands: A review of management impacts and implications for future management options. *Ecol. Evol.* **2016**, *6*, 7080–7102. [CrossRef]
110. Yagi, K.; Sriphirom, P.; Cha-un, N.; Fusuwankaya, K.; Chidthaisong, A.; Damen, B.; Towprayoon, S. Potential and promisingness of technical options for mitigating greenhouse gas emissions from rice cultivation in Southeast Asian countries. *Soil Sci. Plant Nutr.* **2019**, *66*, 37–49. [CrossRef]

111. Pittelkow, C.M.; Liang, X.; Linquist, B.A.; van Groenigen, K.J.; Lee, J.; Lundy, M.E.; van Gestel, N.; Six, J.; Venterea, R.T.; van Kessel, C. Productivity limits and potentials of the principles of conservation agriculture. *Nature* **2015**, *517*, 365–368. [CrossRef]
112. Kiran Kumara, T.M.; Kandpal, A.; Pal, S. A meta-analysis of economic and environmental benefits of conservation agriculture in South Asia. *J. Environ. Manag.* **2020**, *269*, 110773. [CrossRef]
113. Manley, J.; Van Kooten, G.C.; Moeltner, K.; Johnson, D.W. Creating carbon offsets in agriculture through no-till cultivation: A meta-analysis of costs and carbon benefits. *Clim. Change* **2005**, *68*, 41–65. [CrossRef]
114. Briones, M.J.I.; Schmidt, O. Conventional tillage decreases the abundance and biomass of earthworms and alters their community structure in a global meta-analysis. *Glob. Change Biol.* **2017**, *23*, 4396–4419. [CrossRef]
115. Li, Y.; Zhang, Q.; Cai, Y.; Yang, Q.; Chang, S.X. Minimum tillage and residue retention increase soil microbial population size and diversity: Implications for conservation tillage. *Sci. Total Environ.* **2020**, *716*, 137164. [CrossRef] [PubMed]
116. Moos, J.H.; Schrader, S.; Paulsen, H.M. Reduced tillage enhances earthworm abundance and biomass in organic farming: A meta-analysis. *Landbauforschung* **2017**, *67*, 123–128. [CrossRef]
117. Li, Y.; Li, Z.; Cui, S.; Jagadamma, S.; Zhang, Q. Residue retention and minimum tillage improve physical environment of the soil in croplands: A global meta-analysis. *Soil Tillage Res.* **2019**, *194*, 104292. [CrossRef]
118. Mondal, S.; Chakraborty, D.; Bandyopadhyay, K.; Aggarwal, P.; Rana, D.S. A global analysis of the impact of zero-tillage on soil physical condition, organic carbon content, and plant root response. *Land Degrad. Dev.* **2020**, *31*, 557–567. [CrossRef]
119. Peixoto, D.S.; Silva, L.; Melo, L.B.B.; Azevedo, R.P.; Araujo, B.C.L.; Carvalho, T.S.; Moreira, S.G.; Curi, N.; Silva, B.M. Occasional tillage in no-tillage systems: A global meta-analysis. *Sci. Total Environ.* **2020**, *745*, 140887. [CrossRef] [PubMed]
120. Jia, L.; Zhao, W.; Fu, B.; Daryanto, S.; Wang, S.; Liu, Y.; Zhai, R. Effects of minimum soil disturbance practices on controlling water erosion in China's slope farmland: A meta-analysis. *Land Degrad. Dev.* **2019**, *30*, 706–716. [CrossRef]
121. Sun, Y.; Zeng, Y.; Shi, Q.; Pan, X.; Huang, S. No-tillage controls on runoff: A meta-analysis. *Soil Tillage Res.* **2015**, *153*, 1–6. [CrossRef]
122. Elias, D.; Wang, L.; Jacinthe, P.A. A meta-analysis of pesticide loss in runoff under conventional tillage and no-till management. *Environ. Monit. Assess.* **2018**, *190*, 79. [CrossRef]
123. Nunes, M.R.; Karlen, D.L.; Veum, K.S.; Moorman, T.B.; Cambardella, C.A. Biological soil health indicators respond to tillage intensity: A US meta-analysis. *Geoderma* **2020**, *369*, 114335. [CrossRef]
124. Li, Y.; Song, D.; Liang, S.; Dang, P.; Qin, X.; Liao, Y.; Siddique, K.H.M. Effect of no-tillage on soil bacterial and fungal community diversity: A meta-analysis. *Soil Tillage Res.* **2020**, *204*, 104721. [CrossRef]
125. Chen, H.; Dai, Z.; Veach, A.M.; Zheng, J.; Xu, J.; Schadt, C.W. Global meta-analyses show that conservation tillage practices promote soil fungal and bacterial biomass. *Agric. Ecosyst. Environ.* **2020**, *293*, 106841. [CrossRef]
126. Jia, L.; Zhao, W.; Zhai, R.; An, Y.; Pereira, P. Quantifying the effects of contour tillage in controlling water erosion in China: A meta-analysis. *Catena* **2020**, *195*, 104829. [CrossRef]
127. Sun, W.; Canadell, J.G.; Yu, L.; Yu, L.; Zhang, W.; Smith, P.; Fischer, T.; Huang, Y. Climate drives global soil carbon sequestration and crop yield changes under conservation agriculture. *Glob. Change Biol.* **2020**, *26*, 3325–3335. [CrossRef] [PubMed]
128. Wang, J.; Pan, Z.; Pan, F.; He, D.; Pan, Y.; Han, G.; Huang, N.; Zhang, Z.; Yin, W.; Zhang, J.; et al. The regional water-conserving and yield-increasing characteristics and suitability of soil tillage practices in Northern China. *Agric. Water Manag.* **2020**, *228*, 105883. [CrossRef]
129. Li, Y.; Li, Z.; Cui, S.; Zhang, Q. Trade-off between soil pH, bulk density and other soil physical properties under global no-tillage agriculture. *Geoderma* **2020**, *361*, 114099. [CrossRef]
130. Wang, Y.; Guo, T.; Qi, L.; Zeng, H.; Liang, Y.; Wei, S.; Gao, F.; Wang, L.; Zhang, R.; Jia, Z. Meta-analysis of ridge-furrow cultivation effects on maize production and water use efficiency. *Agric. Water Manag.* **2020**, *234*, 106144. [CrossRef]
131. Abdalla, K.; Chivenge, P.; Ciais, P.; Chaplot, V. No-tillage lessens soil CO_2 emissions the most under arid and sandy soil conditions: Results from a meta-analysis. *Biogeosciences* **2016**, *13*, 3619–3633. [CrossRef]
132. Aguilera, E.; Lassaletta, L.; Gattinger, A.; Gimeno, B.S. Managing soil carbon for climate change mitigation and adaptation in Mediterranean cropping systems: A meta-analysis. *Agric. Ecosyst. Environ.* **2013**, *168*, 25–36. [CrossRef]
133. Angers, D.A.; Eriksen-Hamel, N.S. Full-Inversion Tillage and Organic Carbon Distribution in Soil Profiles: A Meta-Analysis. *Soil Sci. Soc. Am. J.* **2008**, *72*, 1370–1374. [CrossRef]
134. Cooper, J.; Baranski, M.; Stewart, G.; Nobel-de Lange, M.; Bàrberi, P.; Fließbach, A.; Peigné, J.; Berner, A.; Brock, C.; Casagrande, M.; et al. Shallow non-inversion tillage in organic farming maintains crop yields and increases soil C stocks: A meta-analysis. *Agron. Sustain. Dev.* **2016**, *36*, 22. [CrossRef]
135. de Graaff, M.-A.; Hornslein, N.; Throop, H.L.; Kardol, P.; van Diepen, L.T.A. Effects of agricultural intensification on soil biodiversity and implications for ecosystem functioning: A meta-analysis. *Adv. Agron.* **2019**, *155*, 1–44. [CrossRef]
136. González-Sánchez, E.J.; Ordóñez-Fernández, R.; Carbonell-Bojollo, R.; Veroz-González, O.; Gil-Ribes, J.A. Meta-analysis on atmospheric carbon capture in Spain through the use of conservation agriculture. *Soil Tillage Res.* **2012**, *122*, 52–60. [CrossRef]
137. Huang, Y.; Ren, W.; Wang, L.; Hui, D.; Grove, J.H.; Yang, X.; Tao, B.; Goff, B. Greenhouse gas emissions and crop yield in no-tillage systems: A meta-analysis. *Agric. Ecosyst. Environ.* **2018**, *268*, 144–153. [CrossRef]
138. Lee, H.; Lautenbach, S.; Nieto, A.P.G.; Bondeau, A.; Cramer, W.; Geijzendorffer, I.R. The impact of conservation farming practices on Mediterranean agro-ecosystem services provisioning—A meta-analysis. *Reg. Environ. Change* **2019**, *19*, 2187–2202. [CrossRef]

139. Li, M.; Wang, J.; Guo, D.; Yang, R.; Fu, H. Effect of land management practices on the concentration of dissolved organic matter in soil: A meta-analysis. *Geoderma* **2019**, *344*, 74–81. [CrossRef]
140. Li, S.; Zheng, X.; Liu, C.; Yao, Z.; Zhang, W.; Han, S. Influences of observation method, season, soil depth, land use and management practice on soil dissolvable organic carbon concentrations: A meta-analysis. *Sci. Total Environ.* **2018**, *631–632*, 105–114. [CrossRef] [PubMed]
141. Luo, Z.; Wang, E.; Sun, O.J. Can no-tillage stimulate carbon sequestration in agricultural soils? A meta-analysis of paired experiments. *Agric. Ecosyst. Environ.* **2010**, *139*, 224–231. [CrossRef]
142. Mahal, N.K.; Castellano, M.J.; Miguez, F.E. Conservation Agriculture Practices Increase Potentially Mineralizable Nitrogen: A Meta-Analysis. *Soil Sci. Soc. Am. J.* **2018**, *82*, 1270–1278. [CrossRef]
143. Mei, K.; Wang, Z.; Huang, H.; Zhang, C.; Shang, X.; Dahlgren, R.A.; Zhang, M.; Xia, F. Stimulation of N_2O emission by conservation tillage management in agricultural lands: A meta-analysis. *Soil Tillage Res.* **2018**, *182*, 86–93. [CrossRef]
144. Ugarte, C.M.; Kwon, H.; Andrews, S.S.; Wander, M.M. A meta-analysis of soil organic matter response to soil management practices: An approach to evaluate conservation indicators. *J. Soil Water Conserv.* **2014**, *69*, 422–430. [CrossRef]
145. Velthof, G.; Commelin, M.; Ros, M.; Oenema, O.; Klages, S.; Tendler, L.; Rowbottom, J.; Wright, I.; Doody, D.; Farrow, L.; et al. *Identification of Most Promising Measures and Practices*; Report D4.3, Fairway Report Series; Wageningen University and Research: Wageningen, The Netherlands, 2020.
146. Du, Z.; Angers, D.A.; Ren, T.; Zhang, Q.; Li, G. The effect of no-till on organic C storage in Chinese soils should not be overemphasized: A meta-analysis. *Agric. Ecosyst. Environ.* **2017**, *236*, 1–11. [CrossRef]
147. Feng, J.; Chen, C.; Zhang, Y.; Song, Z.; Deng, A.; Zheng, C.; Zhang, W. Impacts of cropping practices on yield-scaled greenhouse gas emissions from rice fields in China: A meta-analysis. *Agric. Ecosyst. Environ.* **2013**, *164*, 220–228. [CrossRef]
148. Wang, Y.; Zhang, Y.; Zhou, S.; Wang, Z. Meta-analysis of no-tillage effect on wheat and maize water use efficiency in China. *Sci. Total Environ.* **2018**, *635*, 1372–1382. [CrossRef]
149. Wei, H.; Wang, S.; Yang, W.; Sun, H.; Yin, L.; Deng, X. Meta analysis on impact of no-tillage and subsoiling tillage on spring maize and winter wheat yield and water use efficiency on the loess plateau. *Sci. Agric. Sin.* **2017**, *50*, 461–473. [CrossRef]
150. Xu, C.; Han, X.; Bol, R.; Smith, P.; Wu, W.; Meng, F. Impacts of natural factors and farming practices on greenhouse gas emissions in the North China Plain: A meta-analysis. *Ecol. Evol.* **2017**, *7*, 6702–6715. [CrossRef]
151. Yin, M.; Li, Y.; Chen, P.; Xu, L.; Shen, S.; Wang, X. Effect of no-tillage on maize yield in northern region of China-a meta-analysis. *Sci. Agric. Sin.* **2018**, *51*, 843–854. [CrossRef]
152. Zhao, X.; Liu, S.-L.; Pu, C.; Zhang, X.-Q.; Xue, J.-F.; Ren, Y.-X.; Zhao, X.-L.; Chen, F.; Lal, R.; Zhang, H.-L. Crop yields under no-till farming in China: A meta-analysis. *Eur. J. Agron.* **2017**, *84*, 67–75. [CrossRef]
153. Zhao, X.; Liu, S.L.; Pu, C.; Zhang, X.Q.; Xue, J.F.; Zhang, R.; Wang, Y.Q.; Lal, R.; Zhang, H.L.; Chen, F. Methane and nitrous oxide emissions under no-till farming in China: A meta-analysis. *Glob. Change Biol.* **2016**, *22*, 1372–1384. [CrossRef]
154. Karp, D.S.; Chaplin-Kramer, R.; Meehan, T.D.; Martin, E.A.; DeClerck, F.; Grab, H.; Gratton, C.; Hunt, L.; Larsen, A.E.; Martinez-Salinas, A.; et al. Crop pests and predators exhibit inconsistent responses to surrounding landscape composition. *Proc. Natl. Acad. Sci. USA* **2018**, *115*, E7863–E7870. [CrossRef]
155. Muneret, L.; Mitchell, M.; Seufert, V.; Aviron, S.; Djoudi, E.A.; Pétillon, J.; Plantegenest, M.; Thiéry, D.; Rusch, A. Evidence that organic farming promotes pest control. *Nat. Sustain.* **2018**, *1*, 361–368. [CrossRef]
156. Lesur-Dumoulin, C.; Malézieux, E.; Ben-Ari, T.; Langlais, C.; Makowski, D. Lower average yields but similar yield variability in organic versus conventional horticulture. A meta-analysis. *Agron. Sustain. Dev.* **2017**, *37*, 45. [CrossRef]
157. Garratt, M.P.D.; Wright, D.J.; Leather, S.R. The effects of farming system and fertilisers on pests and natural enemies: A synthesis of current research. *Agric. Ecosyst. Environ.* **2011**, *141*, 261–270. [CrossRef]
158. Butler, J.; Garratt, M.P.D.; Leather, S.R. Fertilisers and insect herbivores: A meta-analysis. *Ann. Appl. Biol.* **2012**, *161*, 223–233. [CrossRef]
159. Veresoglou, S.D.; Barto, E.K.; Menexes, G.; Rillig, M.C. Fertilization affects severity of disease caused by fungal plant pathogens. *Plant Pathol.* **2013**, *62*, 961–969. [CrossRef]
160. Morris, E.K.; Fletcher, R.; Veresoglou, S.D. Effective methods of biofumigation: A meta-analysis. *Plant Soil* **2019**, *446*, 379–392. [CrossRef]
161. Shrestha, U.; Auge, R.M.; Butler, D.M. A Meta-Analysis of the Impact of Anaerobic Soil Disinfestation on Pest Suppression and Yield of Horticultural Crops. *Front. Plant Sci.* **2016**, *7*, 1254. [CrossRef]
162. Bonanomi, G.; Antignani, V.; Capodilupo, M.; Scala, F. Identifying the characteristics of organic soil amendments that suppress soilborne plant diseases. *Soil Biol. Biochem.* **2010**, *42*, 136–144. [CrossRef]
163. Verret, V.; Gardarin, A.; Pelzer, E.; Médiène, S.; Makowski, D.; Valantin-Morison, M. Can legume companion plants control weeds without decreasing crop yield? A meta-analysis. *Field Crop. Res.* **2017**, *204*, 158–168. [CrossRef]
164. Daryanto, S.; Fu, B.; Wang, L.; Jacinthe, P.-A.; Zhao, W. Quantitative synthesis on the ecosystem services of cover crops. *Earth-Sci. Rev.* **2018**, *185*, 357–373. [CrossRef]
165. Anderson, R.L. Integrating a complex rotation with no-till improves weed management in organic farming. A review. *Agron. Sustain. Dev.* **2015**, *35*, 967–974. [CrossRef]
166. Nichols, V.; Verhulst, N.; Cox, R.; Govaerts, B. Weed dynamics and conservation agriculture principles: A review. *Field Crop. Res.* **2015**, *183*, 56–68. [CrossRef]

167. Peigné, J.; Ball, B.; Roger-Estrade, J.; David, C. Is conservation tillage suitable for organic farming? A review. *Soil Use Manag.* **2007**, *23*, 129–144. [CrossRef]
168. Zhang, L.; Rana, I.; Shaffer, R.M.; Taioli, E.; Sheppard, L. Exposure to glyphosate-based herbicides and risk for non-Hodgkin lymphoma: A meta-analysis and supporting evidence. *Mutat. Res. Rev. Mutat. Res.* **2019**, *781*, 186–206. [CrossRef] [PubMed]
169. Nguyen, D.B.; Rose, M.T.; Rose, T.J.; Morris, S.G.; Van Zwieten, L. Impact of glyphosate on soil microbial biomass and respiration: A meta-analysis. *Soil Biol. Biochem.* **2016**, *92*, 50–57. [CrossRef]
170. Govaerts, B.; Verhulst, N.; Castellanos-Navarrete, A.; Sayre, K.D.; Dixon, J.; Dendooven, L. Conservation Agriculture and Soil Carbon Sequestration: Between Myth and Farmer Reality. *Crit. Rev. Plant Sci.* **2009**, *28*, 97–122. [CrossRef]
171. Turmel, M.-S.; Speratti, A.; Baudron, F.; Verhulst, N.; Govaerts, B. Crop residue management and soil health: A systems analysis. *Agric. Syst.* **2015**, *134*, 6–16. [CrossRef]
172. Gu, X.; Cai, H.; Fang, H.; Li, Y.; Chen, P.; Li, Y. Effects of degradable film mulching on crop yield and water use efficiency in China: A meta-analysis. *Soil Tillage Res.* **2020**, *202*, 104676. [CrossRef]
173. Tofanelli, M.B.D.; Wortman, S.E. Benchmarking the agronomic performance of biodegradable mulches against polyethylene mulch film: A meta-analysis. *Agronomy* **2020**, *10*, 1618. [CrossRef]
174. Kallenbach, C.; Grandy, A.S. Controls over soil microbial biomass responses to carbon amendments in agricultural systems: A meta-analysis. *Agric. Ecosyst. Environ.* **2011**, *144*, 241–252. [CrossRef]
175. Charles, A.; Rochette, P.; Whalen, J.K.; Angers, D.A.; Chantigny, M.H.; Bertrand, N. Global nitrous oxide emission factors from agricultural soils after addition of organic amendments: A meta-analysis. *Agric. Ecosyst. Environ.* **2017**, *236*, 88–98. [CrossRef]
176. Shan, J.; Yan, X. Effects of crop residue returning on nitrous oxide emissions in agricultural soils. *Atmos. Environ.* **2013**, *71*, 170–175. [CrossRef]
177. Chen, H.; Li, X.; Hu, F.; Shi, W. Soil nitrous oxide emissions following crop residue addition: A meta-analysis. *Glob. Change Biol.* **2013**, *19*, 2956–2964. [CrossRef] [PubMed]
178. Essich, L.; Nkebiwe, P.M.; Schneider, M.; Ruser, R. Is Crop Residue Removal to Reduce N_2O Emissions Driven by Quality or Quantity? A Field Study and Meta-Analysis. *Agriculture* **2020**, *10*, 546. [CrossRef]
179. Qin, W.; Hu, C.; Oenema, O. Soil mulching significantly enhances yields and water and nitrogen use efficiencies of maize and wheat: A meta-analysis. *Sci. Rep.* **2015**, *5*, 16210. [CrossRef]
180. Mo, F.; Yu, K.L.; Crowther, T.W.; Wang, J.Y.; Zhao, H.; Xiong, Y.C.; Liao, Y.C. How plastic mulching affects net primary productivity, soil C fluxes and organic carbon balance in dry agroecosystems in China. *J. Clean. Prod.* **2020**, *263*, 121470. [CrossRef]
181. Wang, L.; Coulter, J.A.; Li, L.; Luo, Z.; Chen, Y.; Deng, X.; Xie, J. Plastic mulching reduces nitrogen footprint of food crops in China: A meta-analysis. *Sci. Total Environ.* **2020**, *748*, 141479. [CrossRef]
182. Wang, N.; Ding, D.; Malone, R.W.; Chen, H.; Wei, Y.; Zhang, T.; Luo, X.; Li, C.; Chu, X.; Feng, H. When does plastic-film mulching yield more for dryland maize in the Loess Plateau of China? A meta-analysis. *Agric. Water Manag.* **2020**, *240*, 106290. [CrossRef]
183. Lu, X. A meta-analysis of the effects of crop residue return on crop yields and water use efficiency. *PLoS ONE* **2020**, *15*, e0231740. [CrossRef]
184. Zhao, X.; Liu, B.Y.; Liu, S.L.; Qi, J.Y.; Wang, X.; Pu, C.; Li, S.S.; Zhang, X.Z.; Yang, X.G.; Lal, R.; et al. Sustaining crop production in China's cropland by crop residue retention: A meta-analysis. *Land Degrad. Dev.* **2020**, *31*, 694–709. [CrossRef]
185. Chen, G.; Liu, S.; Xiang, Y.; Tang, X.; Liu, H.; Yao, B.; Luo, X. Impact of living mulch on soil C:N:P stoichiometry in orchards across China: A meta-analysis examining climatic, edaphic, and biotic dependency. *Pedosphere* **2020**, *30*, 181–189. [CrossRef]
186. Xiao, L.; Zhao, R.; Zhang, X. Crop cleaner production improvement potential under conservation agriculture in China: A meta-analysis. *J. Clean. Prod.* **2020**, *269*, 122262. [CrossRef]
187. Li, Q.; Li, H.; Zhang, L.; Zhang, S.; Chen, Y. Mulching improves yield and water-use efficiency of potato cropping in China: A meta-analysis. *Field Crop. Res.* **2018**, *221*, 50–60. [CrossRef]
188. Gao, H.; Yan, C.; Liu, Q.; Li, Z.; Yang, X.; Qi, R. Exploring optimal soil mulching to enhance yield and water use efficiency in maize cropping in China: A meta-analysis. *Agric. Water Manag.* **2019**, *225*, 105741. [CrossRef]
189. Mazoyer, M.; Roudart, L. *A History of World Agriculture—From the Neolithic Age to the Current Crisis*; Earthscan: London, UK, 2006.
190. Keller, T.; Sandin, M.; Colombi, T.; Horn, R.; Or, D. Historical increase in agricultural machinery weights enhanced soil stress levels and adversely affected soil functioning. *Soil Tillage Res.* **2019**, *194*, 104293. [CrossRef]
191. Ampoorter, E.; de Schrijver, A.; van Nevel, L.; Hermy, M.; Verheyen, K. Impact of mechanized harvesting on compaction of sandy and clayey forest soils: Results of a meta-analysis. *Ann. For. Sci.* **2012**, *69*, 533–542. [CrossRef]
192. Dahlin, A.S.; Rusinamhodzi, L. Yield and labor relations of sustainable intensification options for smallholder farmers in sub-Saharan Africa. A meta-analysis. *Agron. Sustain. Dev.* **2019**, *39*, 32. [CrossRef]
193. Arts, B.; Buizer, M.; Horlings, L.; Ingram, V.; van Oosten, C.; Opdam, P. Landscape Approaches: A State-of-the-Art Review. *Annu. Rev. Environ. Resour.* **2017**, *42*, 439–463. [CrossRef]
194. FAO. *Landscapes for Life: Integrated Management of Landscapes and Seascapes for Sustainable Food and Agriculture*; FAO: Rome, Italy, 2017.
195. Freeman, O.E.; Duguma, L.A.; Minang, P.A. Operationalizing the integrated landscape approach in practice. *Ecol. Soc.* **2015**, *20*, 24. [CrossRef]
196. Kort, J. Benefits of windbreaks to field and forage crops. *Agric. Ecosyst. Environ.* **1988**, *22–23*, 165–190. [CrossRef]

197. Albrecht, M.; Kleijn, D.; Williams, N.M.; Tschumi, M.; Blaauw, B.R.; Bommarco, R.; Campbell, A.J.; Dainese, M.; Drummond, F.A.; Entling, M.H. The effectiveness of flower strips and hedgerows on pest control, pollination services and crop yield: A quantitative synthesis. *Ecol. Lett.* **2020**, *23*, 1488–1498. [CrossRef]
198. Van Vooren, L.; Reubens, B.; Broekx, S.; De Frenne, P.; Nelissen, V.; Pardon, P.; Verheyen, K. Ecosystem service delivery of agri-environment measures: A synthesis for hedgerows and grass strips on arable land. *Agric. Ecosyst. Environ.* **2017**, *244*, 32–51. [CrossRef]
199. Zamorano, J.; Bartomeus, I.; Grez, A.A.; Garibaldi, L.A. Field margin floral enhancements increase pollinator diversity at the field edge but show no consistent spillover into the crop field: A meta-analysis. *Insect Conserv. Divers.* **2020**, *13*, 519–531. [CrossRef]
200. Zheng, Y.; Wang, H.; Qin, Q.; Wang, Y. Effect of plant hedgerows on agricultural non-point source pollution: A meta-analysis. *Environ. Sci. Pollut. Res.* **2020**, *27*, 24831–24847. [CrossRef]
201. Marja, R.; Kleijn, D.; Tscharntke, T.; Klein, A.M.; Frank, T.; Batáry, P. Effectiveness of agri-environmental management on pollinators is moderated more by ecological contrast than by landscape structure or land-use intensity. *Ecol. Lett.* **2019**, *22*, 1493–1500. [CrossRef] [PubMed]
202. FAO. *The State of Food and Agriculture 2020: Overcoming Water Challenges in Agriculture*; FAO: Rome, Italy, 2020.
203. IPBES. *Global Assessment Report on Biodiversity and Ecosystem Services of the Intergovernmental Science-Policy Platform on Biodiversity and Ecosystem Services*; IPBES Secretariat: Bonn, Germany, 2019.
204. UN. *Transforming Our World: The 2030 Afenda for Sustainable Development*; United Nations: New York, NY, USA, 2015.
205. Mueller, N.D.; Gerber, J.S.; Johnston, M.; Ray, D.K.; Ramankutty, N.; Foley, J.A. Closing yield gaps through nutrient and water management. *Nature* **2012**, *490*, 254–257. [CrossRef] [PubMed]
206. Mekonnen, M.M.; Hoekstra, A.Y. Four billion people facing severe water scarcity. *Sci. Adv.* **2016**, *2*, e1500323. [CrossRef]
207. Cordell, D.; Drangert, J.-O.; White, S. The story of phosphorus: Global food security and food for thought. *Glob. Environ. Change* **2009**, *19*, 292–305. [CrossRef]
208. van Kauwenbergh, S.J. *World Phosphate Rock Reserves and Resources*; International Fertilizer Development Center (IFDC): Muscle Shoals, AL, USA, 2010.
209. Sutton, M.A.; Bleeker, A.; Howard, C.M.; Bekunda, M.; Grizzetti, B.; de Vries, W.; van Grinsven, H.J.M.; Abrol, Y.P.; Adhya, T.K.; Billen, G.; et al. *Our Nutrient World: The Challenge to Produce More Food and Energy with Less Pollution*; Centre for Ecology and Hydrology (CEH): Edinburgh, UK, 2013.
210. Shannon, D.K.; Clay, D.E.; Kitchen, N.R. *Precision Agriculture Basics*; American Society of Agronomy: Madison, WI, USA, 2018.
211. Stafford, J.V. *Precisoin Agriculture'19*; Wageningen Academic Publisher: Wageningen, The Netherlands, 2019.
212. Zhang, Q. *Precision Agriculture Technology for Crop Farming*; CRC Press: Boca Raton, FL, USA, 2016.
213. De Wit, C.T. Resource Use Efficiency in Agriculture. *Agric. Syst.* **1992**, *40*, 125–151. [CrossRef]
214. Willett, W.; Rockström, J.; Loken, B.; Springmann, M.; Lang, T.; Vermeulen, S.; Garnett, T.; Tilman, D.; DeClerck, F.; Wood, A.; et al. Food in the Anthropocene: The EAT–Lancet Commission on healthy diets from sustainable food systems. *Lancet* **2019**, *393*, 447–492. [CrossRef]

MDPI
St. Alban-Anlage 66
4052 Basel
Switzerland
Tel. +41 61 683 77 34
Fax +41 61 302 89 18
www.mdpi.com

Land Editorial Office
E-mail: land@mdpi.com
www.mdpi.com/journal/land

www.ingramcontent.com/pod-product-compliance
Lightning Source LLC
Chambersburg PA
CBHW061608220926
51850CB00005B/450
* 9 7 8 3 0 3 6 5 6 2 5 3 7 *